The Immersed Interface Method

siam FRONTIERS
IN APPLIED MATHEMATICS

The SIAM series on Frontiers in Applied Mathematics publishes monographs dealing with creative work in a substantive field involving applied mathematics or scientific computation. All works focus on emerging or rapidly developing research areas that report on new techniques to solve mainstream problems in science or engineering.

The goal of the series is to promote, through short, inexpensive, expertly written monographs, cutting edge research poised to have a substantial impact on the solutions of problems that advance science and technology. The volumes encompass a broad spectrum of topics important to the applied mathematical areas of education, government, and industry.

BOOKS PUBLISHED IN FRONTIERS IN APPLIED MATHEMATICS

Batzel, Jerry J.; Kappel, Franz; Schneditz, Daniel; and Tran, Hien T., *Cardiovascular and Respiratory Systems: Modeling, Analysis, and Control*

Li, Zhilin and Ito, Kazufumi, *The Immersed Interface Method: Numerical Solutions of PDEs Involving Interfaces and Irregular Domains*

Smith, Ralph C., *Smart Material Systems: Model Development*

Iannelli, M.; Martcheva, M.; and Milner, F. A., *Gender-Structured Population Modeling: Mathematical Methods, Numerics, and Simulations*

Pironneau, O. and Achdou, Y., *Computational Methods in Option Pricing*

Day, William H. E. and McMorris, F. R., *Axiomatic Consensus Theory in Group Choice and Biomathematics*

Banks, H. T. and Castillo-Chavez, Carlos, editors, *Bioterrorism: Mathematical Modeling Applications in Homeland Security*

Smith, Ralph C. and Demetriou, Michael, editors, *Research Directions in Distributed Parameter Systems*

Höllig, Klaus, *Finite Element Methods with B-Splines*

Stanley, Lisa G. and Stewart, Dawn L., *Design Sensitivity Analysis: Computational Issues of Sensitivity Equation Methods*

Vogel, Curtis R., *Computational Methods for Inverse Problems*

Lewis, F. L.; Campos, J.; and Selmic, R., *Neuro-Fuzzy Control of Industrial Systems with Actuator Nonlinearities*

Bao, Gang; Cowsar, Lawrence; and Masters, Wen, editors, *Mathematical Modeling in Optical Science*

Banks, H. T.; Buksas, M. W.; and Lin, T., *Electromagnetic Material Interrogation Using Conductive Interfaces and Acoustic Wavefronts*

Oostveen, Job, *Strongly Stabilizable Distributed Parameter Systems*

Griewank, Andreas, *Evaluating Derivatives: Principles and Techniques of Algorithmic Differentiation*

Kelley, C. T., *Iterative Methods for Optimization*

Greenbaum, Anne, *Iterative Methods for Solving Linear Systems*

Kelley, C. T., *Iterative Methods for Linear and Nonlinear Equations*

Bank, Randolph E., *PLTMG: A Software Package for Solving Elliptic Partial Differential Equations. Users' Guide 7.0*

Moré, Jorge J. and Wright, Stephen J., *Optimization Software Guide*

Rüde, Ulrich, *Mathematical and Computational Techniques for Multilevel Adaptive Methods*

Cook, L. Pamela, *Transonic Aerodynamics: Problems in Asymptotic Theory*

Banks, H. T. , *Control and Estimation in Distributed Parameter Systems*

Van Loan, Charles, *Computational Frameworks for the Fast Fourier Transform*

Van Huffel, Sabine and Vandewalle, Joos, *The Total Least Squares Problem: Computational Aspects and Analysis*

Castillo, José E., *Mathematical Aspects of Numerical Grid Generation*

Bank, R. E., *PLTMG: A Software Package for Solving Elliptic Partial Differential Equations. Users' Guide 6.0*

McCormick, Stephen F., *Multilevel Adaptive Methods for Partial Differential Equations*

Grossman, Robert, *Symbolic Computation: Applications to Scientific Computing*

Coleman, Thomas F. and Van Loan, Charles, *Handbook for Matrix Computations*

McCormick, Stephen F., *Multigrid Methods*

Buckmaster, John D., *The Mathematics of Combustion*

Ewing, Richard E., *The Mathematics of Reservoir Simulation*

The Immersed Interface Method
Numerical Solutions of PDEs Involving Interfaces and Irregular Domains

Zhilin Li
Kazufumi Ito

North Carolina State University
Raleigh, North Carolina

siam
Society for Industrial and Applied Mathematics
Philadelphia

10 9 8 7 6 5 4 3 2

Library of Congress Cataloging-in-Publication Data:

Li, Zhilin, 1956-
The immersed interface method : numerical solutions of PDEs involving interfaces and irregular domains / Zhilin Li, Kazufumi Ito.
p. cm. — (Frontiers in applied mathematics)
Includes bibliographical references and index.
ISBN 978-0-898716-09-2 (pbk.)
1. Differential equations, Partial—Numerical solutions. 2. Numerical analysis.
3. Interfaces (Physical sciences)—Mathematics. I. Ito, Kazufumi. II. Title. III. Series.

QA374.L42 2006
518'.64—dc22

2006044254

To our families:

Xiaoyun, Mike, and Matthew

Junko, Yuka, and Satoru

Contents

Preface

Interface problems arise in many applications. For example, when there are two different materials, such as water and oil, or the same material but at different states, such as water and ice, we are dealing with an interface problem. If partial or ordinary differential equations are used to model these applications, the parameters in the governing differential equations are typically discontinuous across the interface separating two materials or two states, and the source terms are often singular to reflect source/sink distributions along codimensional interfaces. Because of these irregularities, the solutions to the differential equations are typically nonsmooth, or even discontinuous as in the example of the pressure inside and outside an inflated balloon. As a result, many standard numerical methods based on the assumption of smoothness of solutions do not work or work poorly for interface problems.

Another type of problem involves differential equations defined on irregular domains. Examples include underground water flow passing through different objects such as stones, sponges, etc. In a free boundary problem, not only is the domain arbitrary but it also changes with time.

For interface problems and problems defined on irregular domains, analytic solutions are rarely available. The rapid development of modern computers has made it possible to find numerical solutions of these problems. Standard finite difference methods based on simple grids will likely lead to loss of accuracy in a neighborhood of interfaces or near irregular boundaries. While there are some sophisticated methods and software packages for interface and irregular domain problems, the complexity and/or the extra effort needed for learning these methods and software packages are obstacles for nonexperts. The cost and limitations of possible mesh generation processes for complicated geometries at every or every other time step are also major concerns for moving interface or free boundary problems.

In this monograph, we introduce the *immersed interface method* (IIM) developed for interface problems and problems defined on irregular domains. This method is based on uniform or adaptive grids or triangulation in Cartesian, polar, or spherical coordinates. Standard finite difference or finite element methods are used away from interfaces or boundaries. The finite difference or finite element schemes are modified locally near or on the interfaces or boundaries according to the interface relations so that high-order accuracy can be obtained in the entire domain. Since interfaces or irregular boundaries are one dimension lower than solution domains, the extra costs in dealing with interfaces or irregular boundaries are generally insignificant. Furthermore, many available software packages based on uniform Cartesian, polar, or spherical grids, such as the fast Fourier transform (FFT) and fast Poisson solvers, can be applied easily with the immersed interface method. The immersed interface method is designed to be simple enough so that it can be implemented by researchers and graduate students who have reasonable backgrounds in finite difference or finite element methods, but it is powerful enough to solve complicated problems with good accuracy.

The immersed interface method has been used in conjunction with evolution schemes, such as the level set method and the front-tracking method, to solve a number of moving interface and free boundary problems. Particularly, we will discuss in this monograph its applications to Stefan problems and unstable crystal growth, incompressible Stokes and Navier–Stokes flows with moving interfaces, an inverse problem of identifying unknown shapes in a region, a nonlinear interface problem of magnetorheological fluids containing iron particles, and other problems.

This monograph is based on the results of the authors' research in this area, and of course materials the authors have used in teaching advanced graduate numerical analysis. It also contains some recent research results such as fourth-order compact schemes for interface problems and problems defined on irregular domains, and a fast iterative method for Stokes equations with a discontinuous viscosity.

A Web site, http://www4.ncsu.edu/~zhilin/IIM/index.html, has been set up to post or link the recent computer codes/packages of the immersed interface method. This site can also be accessed from http:/www.siam.org/books/fr33, which will redirect you to our site.

We would like to thank the United States National Science Foundation (NSF), United States Army Research Office (ARO), North Carolina State University (NCSU), University of California Los Angeles (UCLA), and other universities and institutions for their support. We also thank Drs. Randall J. LeVeque, Loyce Adams, Ralph Smith, Xiaobiao Lin, Stanley Osher, Hongkai Zhao, Sharon Lubkin, Tao Lin, and many others for research collaborations and their support. We are also thankful to Xiaohai Wan and Sheng Xu for their partial proofreading of the monograph.

Zhilin Li
Kazufumi Ito

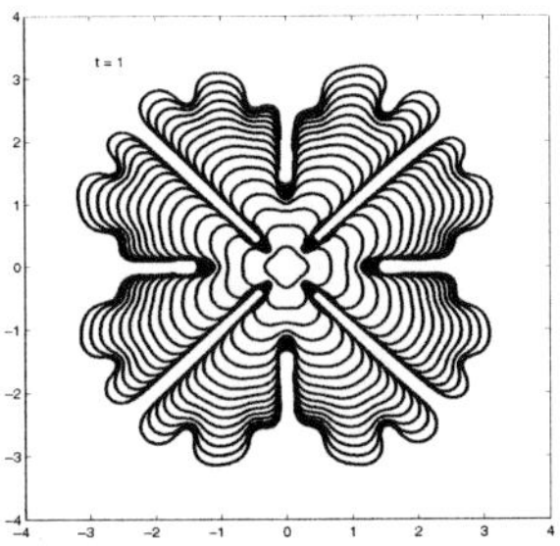

Chapter 1

Introduction

Fixed or moving interface problems, free boundary problems, and problems defined on irregular domains have many applications but are challenging. They have attracted much attention from theorists and numerical analysts over the years. Mathematically, interface problems usually lead to differential equations whose input data and solutions have discontinuities or nonsmoothness across interfaces. The study of the regularity of the solutions for these problems is complicated by the presence of interfaces, discontinuities in the coefficients, and singular source terms. Computationally, many numerical methods designed for smooth solutions do not work, or work poorly, for these problems due to their irregularities.

In this monograph, we introduce the immersed interface method (IIM) based on uniform or adaptive grids or triangulation in Cartesian, polar, or spherical coordinates for solving various interface problems and problems defined on irregular domains. Beginning with fixed interfaces and irregular boundaries, we will discuss how to solve the governing differential equations accurately and efficiently. Away from the interfaces, the IIM takes advantage of standard finite difference or finite element methods that use a uniform grid or triangulation. The IIM modifies the numerical schemes near or on the interfaces to treat the irregularities. Since the dimension of the interfaces is one dimension lower than that of the solution domain, such modifications generally do not increase computational costs significantly.

We will also discuss how the IIM is used for moving interface and free boundary problems. The strategy is based on the common approach called the splitting method in which the governing differential equations are solved first with the interface or boundary fixed. The velocity field is then computed from the solution of the governing equations and used to evolve the interface or boundary with an evolution scheme. Such processes can be combined with Runge–Kutta methods or implicit time-stepping schemes to increase the accuracy of the solution and the motion of the interface in time. The IIM has been applied, in conjunction with the level set and the front-tracking methods, to various problems including the simulation of electromigration of voids, nonlinear interface problems, Stefan problems and crystal growth, and incompressible flows with moving interfaces modeled by Stokes and Navier–Stokes equations. We will describe a few applications of the IIM in the last chapter.

In this chapter, we present some model problems to show the importance and characteristics of the problems discussed in this monograph. We give a minireview of other finite difference methods for interface problems and problems defined on irregular domains. We also introduce notation and other information used in this monograph.

1.1 A one-dimensional model problem

Consider an elastic string with two ends fixed and an external force; see Figure 1.1 for an illustration. It is well known that the displacement of the string can be modeled as the solution of the following two-point boundary value problem:

$$(\tau u_x)_x = f(x), \qquad 0 < x < 1, \qquad u(0) = 0, \qquad u(1) = 0, \tag{1.1}$$

where τ, assumed to be a constant, is the surface tension coefficient of the string. If $f(x)$ is a unit point force at some point α, $0 < \alpha < 1$, then

$$\int_0^1 f(x)\phi(x)dx = \lim_{\epsilon\to 0}\int_0^1 \delta_\epsilon(x-\alpha)\phi(x)dx = \int_0^1 \delta(x-\alpha)\phi(x)dx = \phi(\alpha) \tag{1.2}$$

for all $\phi(x) \in C^1[0,1]$ vanishing at $x = 0$ and $x = 1$, where $\delta_\epsilon(x)$ is a continuous nonnegative function with a compact support such that $\int_{-\infty}^{\infty} \delta_\epsilon(x)dx = 1$; see Figure 1.5 for two examples of such $\delta_\epsilon(x)$. Such a function $\delta(x)$ is called the Dirac delta function, which is not a standard function and is defined in the sense of the distribution. Note that the differential equation in (1.1) is simply $u_{xx} = 0$ in the subdomains $(0,\alpha)$ and $(\alpha, 1)$. While the solution (the displacement of the string) is continuous, its first-order derivative is not. In fact, if we integrate (1.1) from the left to the right of α, we have

$$\int_{\alpha^-}^{\alpha^+} (\tau u_x)_x \, dx = \int_{\alpha^-}^{\alpha^+} \delta(x-\alpha)\, dx = 1.$$

This leads to $\tau u_x(\alpha^+) = \tau u_x(\alpha^-) + 1$ and the exact solution

$$u(x) = \begin{cases} -\dfrac{(1-\alpha)x}{\tau} & \text{if } 0 \le x \le \alpha, \\[2ex] -\dfrac{\alpha(1-x)}{\tau} & \text{if } \alpha < x \le 1. \end{cases} \tag{1.3}$$

In this example, due to the singular delta function source, the solution is not smooth at $x = \alpha$. However, the solution is piecewise smooth in each subdomain $(0,\alpha)$ and $(\alpha, 1)$. There are no irregularities in the differential equation and the solution in each subdomain. The solution in one subdomain is coupled with the solution from the other side of the interface α by the following relations:

$$u(\alpha^+) = u(\alpha^-), \qquad u_x(\alpha^+) = u_x(\alpha^-) + \frac{1}{\tau}. \tag{1.4}$$

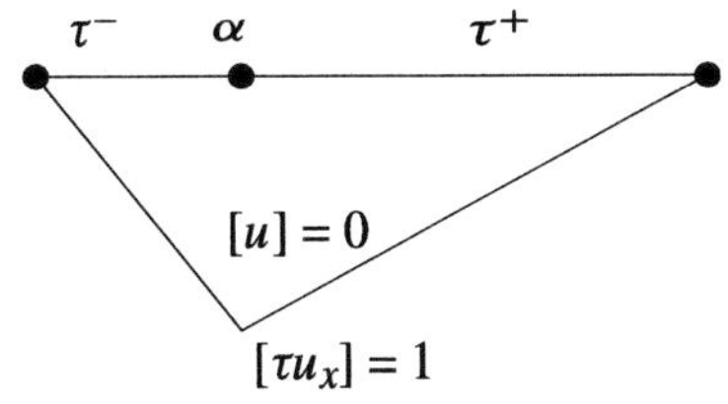

Figure 1.1. *A diagram of the solution of the one-dimensional model problem* (1.1). *The solution is not smooth at the interface* $x = \alpha$ *due to the singular delta function source or the discontinuity in the coefficient* τ.

We will call these two relations the *jump conditions* across the interface α. The jump conditions are defined by

$$[u]_{x=\alpha} \stackrel{\text{def}}{=} u(\alpha^+) - u(\alpha^-) \stackrel{\text{def}}{=} \lim_{x\to\alpha^+} u(x) - \lim_{x\to\alpha^-} u(x), \tag{1.5}$$

$$[u_x]_{x=\alpha} \stackrel{\text{def}}{=} u_x(\alpha^+) - u_x(\alpha^-) \stackrel{\text{def}}{=} \lim_{x\to\alpha^+} u_x(x) - \lim_{x\to\alpha^-} u_x(x). \tag{1.6}$$

In some of the literature, they are also called the internal boundary conditions. Often we omit the subscript ($x = \alpha$) in the jump conditions for simplicity of notation.

If the string is made of two different materials at the point $x = \alpha$, then τ is also discontinuous at $x = \alpha$ and the jump conditions become

$$[u]_{x=\alpha} = 0, \qquad [\tau u_x]_{x=\alpha} = 1. \tag{1.7}$$

In other words, we cannot move τ out of the bracket; see Figure 1.1.

1.2 A two-dimensional example of heat propagation in a heterogeneous material

Consider two materials with different heat conductivities and that come in contact with each other along an interface, for example, in a circle, as shown in Figure 1.2(a). The temperature distributions along the four sides of the far-field boundary are fixed. Initially we assume the temperature is zero everywhere. The mathematical description of the problem is the following:

$$\begin{aligned}
& u_t = \nabla\cdot(\beta\nabla u), \qquad -1 \le x,\, y \le 1,\\
& \beta = \begin{cases} 1 & \text{if } x^2+y^2 \le \dfrac{1}{4},\\ 100 & \text{otherwise,}\end{cases}\\
& \text{BC:}\quad u(-1,y,t) = u(x,-1,t) = 0,\\
& \qquad u(x,1,t) = \sin\left((x+1)\frac{\pi}{4}\right), \qquad u(1,y,t) = \sin\left((y+1)\frac{\pi}{4}\right),\\
& \text{IC:}\ u(x,y,0) = 0.
\end{aligned} \tag{1.8}$$

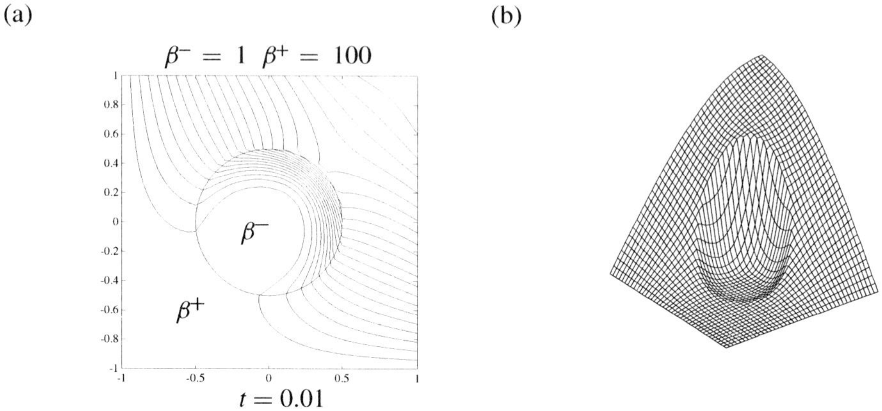

Figure 1.2. *Heat propagation in two different materials as modeled by* (1.8). (a) *A contour plot of the temperature* $u(x, y, t)$ *at* $t = 0.01$. (b) *A mesh plot of the solution at* $t = 0.01$.

The heat propagates with time and travels faster in the material with larger heat conductivity than in the material with smaller heat conductivity. Figure 1.2(a) shows a contour plot of the temperature distribution after a short time, while Figure 1.2(b) is a mesh plot of the solution.

In this example, the solution $u(x, y, t)$ is the temperature which is continuous across the interface. Since there are no external heat sources or sinks across the interface, the heat flux is continuous. Therefore we have

$$[u]_\Gamma = 0, \qquad \left[\beta \frac{\partial u}{\partial \mathbf{n}}\right]_\Gamma = 0, \tag{1.9}$$

where Γ is the interface, the circle $x^2 + y^2 = 0.5^2$, $\mathbf{n}$ is the unit normal direction of the interface pointing outward, and $\frac{\partial u}{\partial \mathbf{n}}$ is the normal derivative of the solution $u(x, y, t)$. The jumps are defined as the difference of the limiting values from each side of the interface. For example, the jump in the flux $[\beta \frac{\partial u}{\partial \mathbf{n}}]_\Gamma$ at a point $\mathbf{X}$ on the interface is defined as

$$\left[\beta \frac{\partial u}{\partial \mathbf{n}}\right]_{\mathbf{X}\in\Gamma} \stackrel{\text{def}}{=} \left(\beta^+ \lim_{\mathbf{x}\to\mathbf{X},\mathbf{x}\in\Omega^+} \nabla u(\mathbf{x}) - \beta^- \lim_{\mathbf{x}\to\mathbf{X},\mathbf{x}\in\Omega^-} \nabla u(\mathbf{x})\right) \cdot \mathbf{n}, \tag{1.10}$$

where $\Omega^\pm$ is the domain outside/inside the interface, which is the circle in this example. The jump conditions in (1.9) are called the natural jump conditions, or natural internal boundary conditions in some of the literature. For simplicity, we will omit the subscripts Γ and $\mathbf{X}$ if no confusion occurs. We will use the notation $u_\mathbf{n} = \frac{\partial u}{\partial \mathbf{n}} = \nabla u \cdot \mathbf{n}$ for the directional derivative of u in the normal direction. Since the heat conductivity is discontinuous, from the flux condition

$$[\beta u_\mathbf{n}]_\Gamma = \beta^+ u_\mathbf{n}^+ - \beta^- u_\mathbf{n}^- = \beta^+ [u_\mathbf{n}]_\Gamma + [\beta]_\Gamma u_\mathbf{n}^- = 0,$$

we can conclude that $[u_\mathbf{n}] = -[\beta] u_\mathbf{n}^- / \beta^+$, which is nonzero in this example because both $[\beta]$ and $u_\mathbf{n}^-$ are nonzero. We will use similar notation in this monograph and will not repeat the definitions.

1.3 Examples of irregular domains and free boundary problems

Consider a conduct line in an integrated circuit. Due to manufacturing processes and other factors, some voids (nonconductive regions) can develop within the conduct line. These voids, while evolving very slowly, can move, grow, merge, and may eventually cause failure of the conduct line. The motion of voids depends on the surface Laplacian of the electrical and chemical potentials; see [180]. The electrical potential is the solution of the Laplacian equation exterior to all voids; see Figure 1.3 for an illustration. The IIM for such a Poisson equation on an irregular domain is explained in §6.2. In §11.4, we will show another application of the fast Poisson solver on irregular domains using the IIM for an inverse problem of shape identification.

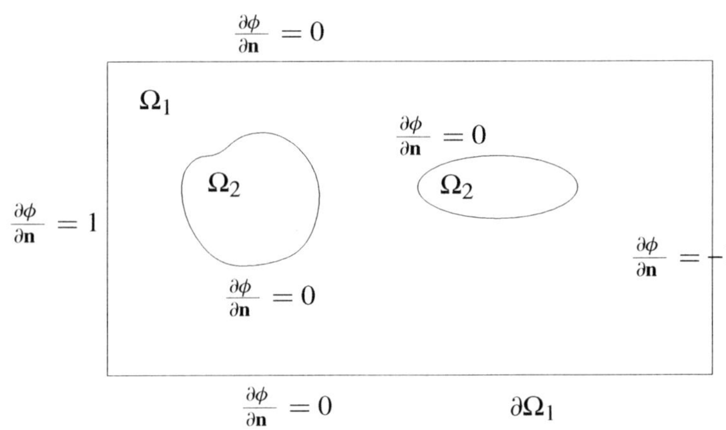

Figure 1.3. *A sketch of a potential problem defined on an irregular domain. The regions of Ω_2 are voids, which are insulators; see* [180].

Stefan problems and unstable crystal growth are examples of free boundary problems. Consider an undercooled seed with initial temperature lower than the melting temperature. The solidification process will be initiated around the seed; it is intrinsically unstable. The moving front develops unstable dendrites. In Figure 1.4, we show a solidification process at different times. More details will be explained in §11.3; see also [175].

1.4 The scope of the monograph and the methodology

The biggest chunk of this monograph will be devoted to interface problems. There are many different kinds of interface problems. In this monograph, we will discuss interface problems that have one or several of the following features:

- The coefficients of differential equations, such as conductivity, viscosity, permeability, etc., may be discontinuous across some arbitrary interfaces.

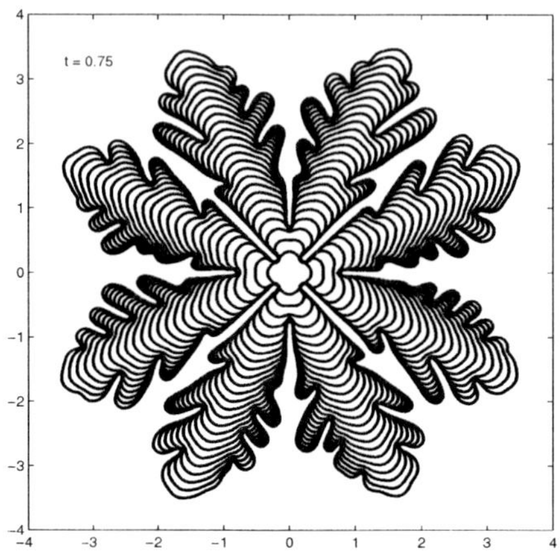

Figure 1.4. *An expanding crystal at different times. The simulation is taken from* [175]; *see also* §11.3.

- The source terms may have a finite jump or a delta function singularity along some arbitrary interfaces.

- The solution to an interface problem may be nonsmooth across the interface (i.e., the gradient or the first partial derivatives are discontinuous) or even discontinuous. But we will assume that the solution is bounded and has certain regularities (i.e., the solution has continuous partial derivatives up to some order) away from interfaces or boundaries.

- We have some prior knowledge of the jump conditions of the solution and the flux across interfaces. The jump conditions usually can be obtained from the underlying physics, as in the example of heat propagation, or from the governing differential equations, as in the examples in §1.1 and §1.2.

- Interfaces or boundaries may be fixed or continuously moving with time.

- There can be one or several interfaces in the solution domain.

For a problem defined on an irregular domain, we often use an embedding technique. The problem then can be treated as a special interface problem. The technique will be explained in detail in §6.2. Thus, we will simply use the terminology of *interface problems* to include problems defined on irregular domains in this monograph.

The IIM is designed to solve interface problems including moving interface and free boundary problems, and problems on irregular domains using uniform or adaptive grids or triangulation in Cartesian, polar, or spherical coordinates.

1.4.1 Jump conditions

Generally, the domain for an interface problem with a bounded solution can be divided into several regions. The solutions in different regions are continuously differentiable to a certain degree and they are coupled by some interface relations, which are called the jump conditions across the interfaces. It is crucial for the IIM to have a prior knowledge of these jump conditions either from physical reasoning or from the governing differential equations. In the example of the differential equation $(\beta u_x)_x = v\,\delta(x-\alpha)$, the jump relations $[u]_{x=\alpha} = 0$ and $[\beta\, u_x]_{x=\alpha} = v$ can be derived easily from the differential equation itself. With a little effort we can prove that the jump conditions for the differential equation

$$(\beta u_x)_x + \left(\beta u_y\right)_y = \int_\Gamma v(s)\,\delta\Big((x,y) - (X(s), Y(s))\Big)\, ds \tag{1.11}$$

are $[u]_\Gamma = 0$ and $[\beta u_n]_\Gamma = v(s)$ at each point $(X(s), Y(s))$ on the interface Γ, where δ now is the Dirac delta function in two dimensions, Γ is an arbitrary interface, and s is the arc-length parameter of Γ. However, it is not always easy to derive jump conditions. The derivation of the jump conditions for Stokes or Navier–Stokes equations involving an interface in [144] is not trivial.

From another point of view, the jump conditions can be regarded as internal boundary conditions that make a problem well-posed. Consider the partial differential equation (PDE) (1.11) in reference to the diagram in Figure 1.6 with a Dirichlet boundary condition on the outside boundary $\partial\Omega$. In the interior of Ω excluding Γ, the PDE is simply $\Delta u = 0$. However, the PDE (1.11) is not well-posed unless we specify two conditions along Γ. Different jump conditions often correspond to different applications. For many applications, the solution is continuous and the flux is the source strength, which gives $[u]_\Gamma = 0$ and $[\beta u_n]_\Gamma = v(s)$. The problem is then well-posed and has a unique solution.

For many applications we have enough information to determine the jump conditions. For instance, in the example of the heat propagation, we know that both the temperature and the heat flux are continuous across the interface, so we have the jump conditions $[u]_\Gamma = 0$ and $[\beta u_n]_\Gamma = 0$ at every point of the interface. In the ice melting problem, for example, the value of the temperature on the interface is known to be the melting temperature.

1.4.2 The choice of grids

To solve an interface problem numerically, it is necessary to have a computational grid or mesh. While there are a few choices, such as a body-fitted grid or a meshless method, in this monograph, we will use fixed and uniform grids or triangulation in Cartesian, polar, or spherical coordinates.

One obvious advantage of using a fixed and uniform grid is that there is almost no cost in the grid generation process. Furthermore, conventional numerical schemes can be used at most grid points (called regular grid points) that are away from interfaces, since there are no irregularities at those grid points. Only those grid points near or on the interfaces, which are usually fewer than those regular grid points, need special attention. The simple data structure of a fixed and uniform grid makes it easy to use the method to solve complicated interface problems with reasonable cost and given accuracy.

Another advantage of using a fixed and uniform grid is that we can take advantage of many software packages and methods developed for uniform grids or triangulation in Cartesian, polar, or spherical coordinates, for example, the fast Poisson solver [252], Clawpack [153], Amrclaw [18], the level set method [206, 207, 238], the structured multigrid solver MGD9V [62, 5], and many others.

As a particular example, for the elliptic interface problem (1.11), if β is constant but $v(s) \neq 0$, the solution is nonsmooth, that is, the gradient has a nonzero jump at the interface. We will see in Chapter 2 that the IIM uses the standard 5-point central finite difference scheme at all grid points and only adds a nonzero correction term to the right-hand side of the finite difference equations at grid points near or on the interface Γ. This means that a fast Poisson solver based on a uniform Cartesian grid can still be used to solve the linear system of equations—an advantage that would be lost if a different grid were used.

Even if β is discontinuous so that the coefficients in the linear system must be modified, the system obtained using the IIM described in Chapter 2, §4 maintains the same block structure as in the case in which β is a constant. One can use available software packages designed for uniform rectangular grids; for example, the multigrid methods [5, 6, 62].

More important, for moving interface and free boundary problems, although it is possible to develop moving mesh methods that conform to the interfaces in each time step or every other time step, this is generally more complicated than simply allowing the interface to move relative to a fixed underlying uniform grid.

1.5 A minireview of some popular finite difference methods for interface problems

There is a vast collection of research papers in the literature that address interface problems. In the discussion below, we discuss a few commonly used finite difference methods for interface problems.

1.5.1 The smoothing method for discontinuous coefficients

In one space dimension, let $\beta(x)$ be a function having a finite jump at $x = \alpha$, i.e., $[\beta]_\alpha = \lim_{x\to\alpha^+}\beta(x) - \lim_{x\to\alpha^-}\beta(x) \neq 0$. Define

$$\beta^-(x) = \begin{cases} \beta(x), & \text{if } x \le \alpha, \\ \beta(\alpha-) + \beta'(\alpha-)(x-\alpha), & \text{if } x > \alpha, \end{cases} \tag{1.12a}$$

$$\beta^+(x) = \begin{cases} \beta(\alpha+) + \beta'(\alpha+)(x-\alpha), & \text{if } x < \alpha, \\ \beta(x), & \text{if } x \ge \alpha. \end{cases} \tag{1.12b}$$

(If $\beta'(\alpha\mp)$ does not exist, then we can use $\beta^{\mp}(x) = \beta(\alpha\mp)$ for the extension.) We can smooth $\beta(x)$ using

$$\beta_\epsilon(x) = \beta^-(x) + (\beta^+(x) - \beta^-(x))\, H_\epsilon(x-\alpha), \tag{1.13}$$

where $H_\epsilon(x)$ is the smoothed Heaviside function,

$$H_\epsilon(x) = \begin{cases} 0 & \text{if } x < -\epsilon, \\ \dfrac{1}{2}\left(1 + \dfrac{x}{\epsilon} + \dfrac{1}{\pi}\sin\dfrac{\pi x}{\epsilon}\right) & \text{if } |x| \le \epsilon, \\ 1 & \text{if } x > \epsilon, \end{cases} \tag{1.14}$$

and $\epsilon > 0$ is a small number depending on the mesh size of a numerical scheme; see, for example, [251]. The coefficient in the front of the sine function is chosen so that $H_\epsilon(x)$ is both continuous and smooth at $x = \pm\epsilon$. Notice that the smoothing function $H_\epsilon(x)$ is an antiderivative of the discrete cosine delta function (1.20) if we choose $\epsilon := 2\epsilon$. Another smoothing function corresponding to the discrete hat delta function (1.19) is

$$H_\epsilon(x) = \begin{cases} 0 & \text{if } x < -\epsilon, \\ \dfrac{(x+\epsilon)^2}{2\epsilon^2} & \text{if } -\epsilon \le x \le 0, \\ 1 - \dfrac{(\epsilon - x)^2}{2\epsilon^2} & \text{if } 0 \le x \le \epsilon, \\ 1 & \text{if } x > \epsilon. \end{cases} \tag{1.15}$$

The smoothing method is easy to implement in one space dimension but may not be very accurate; see, for example, Figure 2.2 in Chapter 2, where the error is visible for a simple interface problem. The smoothing method generally will smear the solution as well.

For two- and three-dimensional problems, the smoothing method may not be so easy to implement unless the interface is expressed as the zero level set of a Lipschitz continuous function $\varphi(\mathbf{x})$. For example, let the zero level set $\{\mathbf{x},\ \varphi(\mathbf{x}) = 0\}$ be the interface, and let the entire domain be divided into two disjoint parts $\Omega^- = \{\mathbf{x},\ \varphi(\mathbf{x}) < 0\}$ and $\Omega^+ = \{\mathbf{x},\ \varphi(\mathbf{x}) > 0\}$. Then the smoothing function of a discontinuous function $\beta(\mathbf{x})$ is

$$\beta_\epsilon(\mathbf{x}) = \beta^-(\mathbf{x}) + (\beta^+(\mathbf{x}) - \beta^-(\mathbf{x}))\, H_\epsilon(\varphi(\mathbf{x})), \tag{1.16}$$

where $\beta^-(\mathbf{x})$ and $\beta^+(\mathbf{x})$ are zero outside their domains of definition.

1.5.2 The harmonic averaging for discontinuous coefficients

For elliptic interface problems, another method that is more accurate than the smoothing method for discontinuous coefficients is *harmonic averaging*; see [17, 240, 255]. Take the one-dimensional problem $(\beta u_x)_x = f(x)$ as an example. The discrete form of $(\beta u_x)_x = f$ can be written as

$$\frac{1}{h^2}\left(\beta_{i+\frac{1}{2}}(u_{i+1} - u_i) - \beta_{i-\frac{1}{2}}(u_i - u_{i-1})\right) = f(x_i), \tag{1.17}$$

where $h = x_i - x_{i-1}$ is the mesh size of a uniform grid in the x-direction. If β is smooth, then we can take $\beta_{i+\frac{1}{2}} = \beta(x_{i+\frac{1}{2}})$, where $x_{i+\frac{1}{2}} = x_i + \frac{h}{2}$, and the discretization is second-order accurate. If β is discontinuous in (x_{i-1}, x_{i+1}), then the harmonic average of $\beta(x)$ is

$$\beta_{i+\frac{1}{2}} = \left(\frac{1}{h}\int_{x_i}^{x_{i+1}} \beta^{-1}(x)\, dx\right)^{-1}. \tag{1.18}$$

This can be justified by the homogenization theory for problems in which $\beta(x)$ varies rapidly on the scale of the grid cells. The finite difference scheme (1.17) using the harmonic averaging (1.18) is second-order accurate in the maximum norm for the one-dimensional elliptic interface problems with $[u]_{x=\alpha} = 0$, $[\beta u_x]_{x=\alpha} = 0$, and $[f]_{x=\alpha} = 0$, due primarily to the result of fortuitous cancellation; see [160]. However, we need to calculate the integral (1.18) accurately enough to guarantee second-order accuracy.

In two space dimensions, the harmonic averaging is also commonly used to deal with discontinuous coefficients [17, 240], now integrating over squares to obtain the harmonic average of $\beta(x, y)$. In this case, however, the method does not give second-order accurate results in general because the cancellations are unlikely to take place for arbitrary interfaces; see, for example, [228]. It is not a trivial task to compute the integrals accurately near the interface in two space dimensions when β is discontinuous.

1.5.3 Peskin's immersed boundary (IB) method

This method was originally developed by Peskin to model blood flow in a human heart (see, for example, [209, 210, 213, 214]) and has been applied to many other problems, particularly in biophysics. We refer readers to the recent review article [211] for the model/method and its applications. One of the important ideas in the immersed boundary (IB) method is the use of a discrete delta function to distribute a singular source to nearby grid points. There are several discrete delta functions in the literature. The commonly used ones include the hat function

$$\delta_\epsilon(x) = \begin{cases} (\epsilon - |x|)/\epsilon^2 & \text{if } |x| < \epsilon, \\ 0 & \text{if } |x| \ge \epsilon, \end{cases} \tag{1.19}$$

and Peskin's original discrete cosine delta function

$$\delta_\epsilon(x) = \begin{cases} \dfrac{1}{4\epsilon}\left(1 + \cos\left(\dfrac{\pi x}{2\epsilon}\right)\right) & \text{if } |x| < 2\epsilon, \\ 0 & \text{if } |x| \ge 2\epsilon. \end{cases} \tag{1.20}$$

Figure 1.5 shows diagrams of these two discrete delta functions; they are both continuous. The second one, first introduced by Peskin and used most often in the literature, is also smooth. The discrete delta function approach is robust and simple to implement. In high dimensions, the discrete delta function often used in the literature is the product of one-dimensional discrete delta functions, for example, $\delta_\epsilon(x, y) = \delta_\epsilon(x)\delta_\epsilon(y)$. With Peskin's discrete delta function approach, one can discretize the right-hand side of (1.11) at a grid point (x_i, y_j) as

$$f_{ij} = \sum_{k=1}^{N_b} v(s_k)\, \delta_h(x_i - X_k)\, \delta_h(y_j - Y_k) \Delta s, \tag{1.21}$$

where N_b is the number of discrete points $\{(X_k, Y_k)\}$ on the interface, and h is the mesh spacing. In this way, the singular source is distributed to nearby grids points in a neighborhood of the interface Γ.

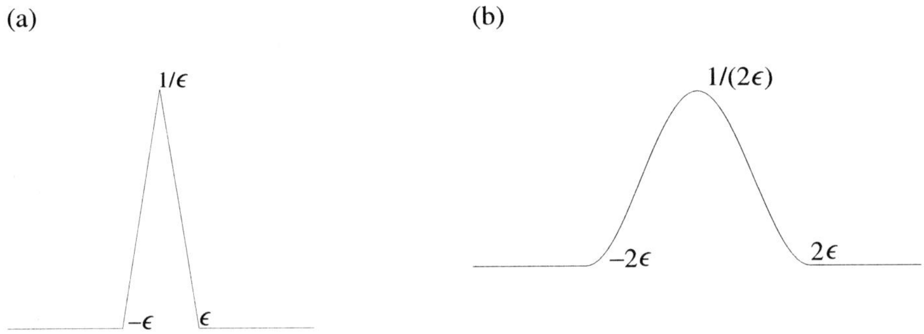

Figure 1.5. *Two typical discrete delta functions.* (a) *Discrete hat delta function.* (b) *Peskin's discrete delta function.*

In one space dimension, using the discrete delta function to solve the model interface problem,

$$u_{xx} = v\,\delta(x-\alpha), \qquad x=\alpha \in (0,1), \qquad u(0)=u(1)=0,$$

is easy to analyze. In this case the interface reduces to a single point. The finite difference method

$$(U_{j+1} - 2U_j + U_{j-1})/h^2 = v\,\delta_h(x_j - \alpha), \tag{1.22}$$

with δ_h given by the discrete hat function (1.19), turns out to be very accurate; in fact it produces the exact solution $U_i = u(x_j)$ at all grid points in spite of the nonsmoothness of $u(x)$; see [23]. However, if (1.20) is used in (1.22), then the computed solution is only first-order accurate and is smeared in the neighborhood of the interface.

Beyer and LeVeque [23] have also analyzed time-dependent versions of the problems in one space dimension and have shown that second-order accuracy can still be obtained with the choice of the hat discrete function. The moments relations for high-order discrete delta functions are first derived for the interpolation

$$u(X) = \int_I u(x)\delta(x-X) \approx \sum_i u(x_i)\delta_h(x_i - X), \tag{1.23}$$

where I is the entire domain.

Various studies have been conducted to find the best way to discretize a Dirac delta function with different representations of interfaces (where the source is distributed) and discover how it would affect the accuracy of numerical methods for PDEs that involve delta function source distribution. Some recent research [215, 257, 258] has focused on the accuracy of the distribution, that is, on getting an accurate approximation to the quadrature of

$$\begin{aligned} &\iint_\Omega \phi(x,y)\left\{\int_\Gamma C(s)\,\delta\Big((x,y)-(X(s),Y(s))\Big)\,ds\right\}dx\,dy \\ &\quad = \int_\Gamma C(s)\,\phi\,(X(s),Y(s))\,ds \end{aligned} \tag{1.24}$$

for any smooth test function ϕ, or to the interpolation (1.23) in one dimension, or

$$\begin{aligned} u(X,Y) &= \iint_{\Omega} u(x,y)\,\delta\Big((x,y)-(X,Y)\Big)\,dxdy \\ &\approx \sum_{ij} u(x_i,y_j)\delta_h(x_i-X)\delta_h(y_j-Y)h_x h_y \end{aligned} \tag{1.25}$$

in two dimensions where h_x and h_y are mesh spacing in the x- and y-direction respectively.

When a high-order discrete delta function is used to solve PDEs with delta function singularities, it may maintain the order of accuracy of the computed solution in an average norm such as the L^1 or L^2 norm. However, it seems unlikely that any discrete delta function with a closed form can yield a second-order accurate solution in the maximum norm for PDEs with delta function singularities except in a few special situations, e.g., when the interface is aligned with grid lines. This has been confirmed in many examples in the literature. An intuitive explanation is that the expression (1.21), as a discrete form of the right-hand side of (1.11), is independent of the derivative of $v(s)$ and the curvature of the interface, which seem to be crucial factors in obtaining second-order accuracy in the IIM. The problem is further complicated by other factors in the governing equations such as the coefficients of the PDEs and the type of PDEs (elliptic, parabolic, hyperbolic, etc.). The technique discussed in §§ 5.2 and 5.3 gives a method of discrete delta functions, not in a closed form, which can provide pointwise second-order accurate solutions for certain interface problems.

The IB method is simple and robust. It has been combined with vortex methods [54, 58, 197] and with adaptive mesh refinement [224, 225]. It has been applied to many problems in mathematical biology and computational fluid mechanics; see, for example, [10, 22, 27, 55, 66, 77, 79, 80, 82, 246, 265, 291]. Various work has been developed to improve the accuracy of the IB method [56, 57, 142, 216], and it has also been parallelized [199]. An IB package is also available in [76]. As pointed out in [211], there is not yet any complete convergence proof for the IB method; however, stability analysis of the IB method is given in [242, 243] for a membrane problem.

1.5.4 Numerical methods based on integral equations

Greenbaum, Mayo, and their collaborators [188, 189, 190] are among a few who first combined integral equations based on the single and double layer theory with finite difference methods to solve a Poisson equation on an irregular domain. The irregular domain is embedded into a larger rectangle, and then the problem is recast as an elliptic interface problem such that the solution is harmonic in the rectangle, excluding the boundary. Taylor expansions at irregular grid points and the integral representation of the particular solution near the irregular boundary are used. The source strength is determined from an integral equation. The jump conditions are derived from the integral equation and are used to derive the finite difference schemes at all grid points in the rectangular domain so that a fast Poisson solver can be used. The hybrid method of the integral equation approach and the finite difference method has been generalized to different problems such as biharmonic equations [95, 192, 194] and Stokes and Navier–Stokes equations [24, 97, 191, 193], and has been parallelized [195]. This approach can be accelerated if combined with the fast multipole method [95, 96, 98, 198].

Mayo [190] and Mayo and Greenbaum [194] have also derived an integral equation for elliptic interface problems with piecewise coefficients. By solving the integral equation, they solved interface problems to second-order accuracy in the L^∞ norm using the techniques developed by Mayo in [188, 190] for solving Poisson and biharmonic equations on irregular regions. The total cost includes solving the integral equation and a regular Poisson equation. The possibility of extension to variable coefficients is mentioned in [190]. The methods based on integral equations are most effective for homogeneous source terms and certain boundary conditions. They probably still can be applied with some extra effort for nonhomogeneous source terms and different boundary conditions. The implementations of these methods, especially when they are coupled with the fast multipole method, however, are not trivial.

1.5.5 The ghost fluid method

After the original IIM [154, 156, 160] was developed, various numerical methods emerged for interface problems. A notable one is the ghost fluid method (GFM). The GFM was first used to properly treat boundary conditions and remove spurious oscillations for hyperbolic systems. Liu, Fedkiw, and Kang [183] and Liu and Sideris [184] developed the GFM for elliptic interface problems.

One of the motivations of the GFM is to simplify and symmetrize the IIM. The GFM is a sharp interface method because it builds on the jump conditions in the finite difference scheme as the IIM does. However, the GFM is generally first-order accurate in the maximum norm for interface problems. Essentially, the GFM decomposes the flux jump in each axis direction so that the problem can be treated dimension by dimension. The main errors of the GFM come from the decomposition of the flux jump conditions, because it is hard to do it accurately, and from ignoring second-order derivatives terms if a local truncation error analysis is conducted.

The main advantage of the GFM is that it is simple and relatively easier to implement. The system of finite difference equations from the GFM is symmetric for self-adjoint elliptic problems. The GFM has been used in multiphase incompressible flows [136], two phase incompressible flame simulations [203], and other applications [71]. The GFM has also been applied to irregular domain problems with Dirichlet boundary conditions [46, 86, 183]. Second-order accuracy can be achieved using a one-sided interpolation. The approach is essentially in the same spirit as a direct discretization (see, for example [202]) with a symmetric system of equations for the self-adjoint linear operator.

One of the disadvantages of the GFM is that a fast Poisson solver cannot be used because the finite difference coefficients have been changed at grid points near the boundary. Usually a preconditioned conjugate gradient (CG) method or an algebraic multigrid (AMG) method is applied to solve the entire system of equations. These iterative methods are usually slower than fast Poisson solvers. The approach may be only first-order accurate for a mixed or nonhomogeneous Neumann boundary condition. If a high-order scheme is used, especially a one-sided interpolation scheme, then the stability of the linear system may be deteriorated.

1.5.6 Finite difference and finite volume methods

There are a few other approaches developed for interface problems. Finite difference methods derived from finite volume formulation have been developed for interface problems and problems defined on irregular domains; see, for example, [7, 8, 74, 126]. One advantage of this approach is that it is possible to get a second-order flux as well. In a finite volume approach, some efforts are needed to evaluate the curvature information of the interface, to reconstruct the interface to high-order accuracy, and to overcome the overflow/underflow of the finite volumes. Those considerations and quite a few possibilities of flow directions and geometries may complicate the implementation of the algorithms. In §11.6, we also briefly explain other finite difference methods related to the IIM.

1.6 Conventions and notation

Here we summarize some commonly used notation and computational frames used in this monograph so that we do not have to repeat them later. We will introduce other notation that is less frequently used whenever it appears.

1.6.1 Cartesian grids

For interface problems, we usually assume that the domain Ω is a rectangular region, that is, $\Omega = [a, b]$ in one dimension, $\Omega = [a, b] \times [c, d]$ in two dimensions, and $\Omega = [a, b] \times [c, d] \times [r, s]$ in three dimensions. For problems defined on an irregular domain, we will embed the irregular domain into a rectangular one $R \supseteq \Omega$. In either case, a uniform Cartesian grid can be generated as

$$x_i = a + ih_x, \qquad h_x = \frac{b-a}{M}, \qquad i = 0, 1, \ldots, M; \tag{1.26}$$

$$y_j = c + jh_y, \qquad h_y = \frac{d-c}{N}, \qquad j = 0, 1, \ldots, N; \tag{1.27}$$

$$z_l = r + lh_z, \qquad h_z = \frac{s-r}{L}, \qquad l = 0, 1, \ldots, L. \tag{1.28}$$

It is often convenient to set $h_x = h_y = h_z = h$ for simplicity.

We will use the notation $\Gamma \subset \Omega$ to denote the interface which typically divides the domain Ω into two parts Ω^+ and Ω^-; see Figure 1.6 for an illustration.

In general, we use uppercase letters such as U_i, U_{ij}, etc. for the discrete approximations to the PDEs at grid points and use lowercase letters such as u, v, etc. for the exact solutions. In other words, we have $U_i \approx u(x_i)$, $U_{ij} \approx u(x_i, y_j)$, and so on. We use boldface letters for vectors in space; for example, $\mathbf{x} = (x, y)$ and a velocity field $\mathbf{u} = (u, v)$ in two space dimensions. Uppercase letters are also used to represent the matrices.

1.6.2 Limiting values and jump conditions

Referring to the diagram in Figure 1.6, an interface Γ divides the domain into two parts that we will denote as the "+" and "−" sides if there is no confusion. Given a piecewise smooth

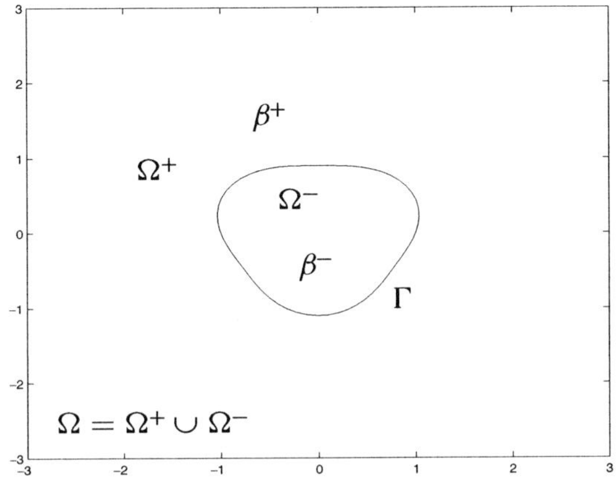

Figure 1.6. *A diagram of a rectangular domain $\Omega = \Omega^+ \cup \Omega^-$ with an interface Γ. The coefficients $\beta(\mathbf{x})$ may have a finite jump across the interface Γ.*

function $u(\mathbf{x})$ that can have a finite jump discontinuity across the interface Γ, the limiting values in $u(\mathbf{x})$ and its normal derivative at a point $\mathbf{X}$ on the interface Γ are defined as

$$u^{\pm}(\mathbf{X}) \stackrel{\text{def}}{=} \lim_{\mathbf{x}\to\mathbf{X},\mathbf{x}\in\Omega^{\pm}} u(\mathbf{x}), \tag{1.29}$$

$$u_{\mathbf{n}}^{\pm}(\mathbf{X}) = \frac{\partial u^{\pm}}{\partial \mathbf{n}}(\mathbf{X}) = \nabla u^{\pm}(\mathbf{X}) \cdot \mathbf{n}(\mathbf{X}) \stackrel{\text{def}}{=} \left(\lim_{\mathbf{x}\to\mathbf{X},\mathbf{x}\in\Omega^{\pm}} \nabla u(\mathbf{x}) \right) \cdot \mathbf{n}(\mathbf{X}), \tag{1.30}$$

where $\mathbf{n}$ is the unit normal direction pointing to a preselected side. The jumps in $u(\mathbf{x})$ and in its normal derivative are then defined as

$$[u]_{\mathbf{x}=\mathbf{X}} \stackrel{\text{def}}{=} u^{+}(\mathbf{X}) - u^{-}(\mathbf{X}), \tag{1.31}$$

$$[u_{\mathbf{n}}]_{\mathbf{x}=\mathbf{X}} \stackrel{\text{def}}{=} u_{\mathbf{n}}^{+}(\mathbf{X}) - u_{\mathbf{n}}^{-}(\mathbf{X}). \tag{1.32}$$

For simplicity, we often omit the subscript $\mathbf{X}$ or use the notation $[u]_\Gamma$ to express the jump of u across the interface. Note that the "+" and "−" sides are used to distinguish the two different regions for convenience. We should be able to switch between the two sides without confusion.

Note also that the jump conditions are codimension one functions defined only along the interface. In two dimensions, an interface is a curve. It is often convenient to express the interface using the arc-length parameterization as

$$\Gamma = \Gamma(s) = \left\{ (x(s),\ y(s)),\ \ 0 \le s \le L_\Gamma \right\}, \tag{1.33}$$

where L_Γ is the length of Γ. The jumps $[u]$, $[u_{\mathbf{n}}]$, $[\beta u_{\mathbf{n}}]$, ... are functions of the arc length s and can be written as $[u](s)$, $[u_{\mathbf{n}}](s)$, $[\beta u_{\mathbf{n}}](s)$, and so forth.

1.6.3 The local coordinates

Since the flux jump condition is often given in the normal direction, it is more convenient to use the local coordinates in the normal and tangential directions. Given a point (X, Y) on the interface, the local coordinate system in the normal and the tangential directions is defined as (see Figure 1.7 for an illustration)

$$\begin{cases} \xi = (x - X)\cos\theta + (y - Y)\sin\theta, \\ \eta = -(x - X)\sin\theta + (y - Y)\cos\theta, \end{cases} \tag{1.34}$$

where θ is the angle between the x-axis and the normal direction, pointing to the direction of a specified side, say the "+" side in Figure 1.6. At the point (X, Y), the interface can be written as

$$\xi = \chi(\eta) \quad \text{with} \quad \chi(0) = 0, \;\; \chi'(0) = 0. \tag{1.35}$$

The curvature of the interface at (X, Y) is χ''.

The three-dimensional local coordinates are defined in (4.3)–(4.5) in Chapter 4.

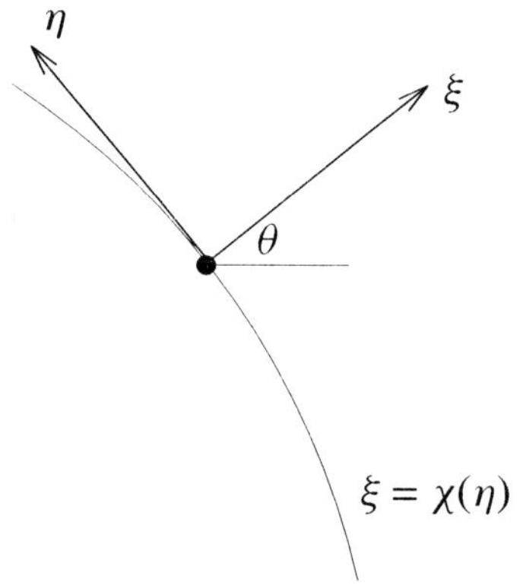

Figure 1.7. *A diagram of the local coordinates in the normal and tangential directions, where θ is the angle between the x-axis and the normal direction.*

1.6.4 Interface representations

To solve interface problems numerically, we need some information about the interface such as position, tangential and normal directions, and sometimes its curvature. Some common approaches used to express the interface are the following.

Analytic expressions

If the interface is fixed, we may have an analytic expression for it. An analytic representation is useful, especially for testing purpose. We have analytic expressions for circles, ellipses, quadratic curves, and some other curves. If an analytic expression is too complicated, then a discrete method to compute interface quantities, such as normal and tangential directions, curvature, etc. may be preferred. For moving interface problems, analytic expressions are rarely available even if we have an analytic expression initially.

Lagrangian frames using control points

In this approach, a set of control points on the interface, say (X_k, Y_k), $k = 1, 2, \ldots, N_b$, in two space dimensions is given. The interface then can be regarded as the function of the arc length, which can be approximated by

$$s_0 = 0, \qquad \Delta s_k = |\mathbf{X}_{k+1} - \mathbf{X}_k|, \tag{1.36}$$

$$s_{k+1} = s_k + \Delta s_k, \quad k = 1, 2, \ldots. \tag{1.37}$$

There are two ways to get derivatives information of the interface which is needed to compute the normal and tangential directions, and the curvature if needed. The first approach is to use a direct discretization, such as a central finite difference formula, to get the required derivatives from $\mathbf{X}_k$ and Δs_k. This approach has been widely used in implementing the IB method for many applications. However, we must balance the needs of accuracy and stability in this approach. Usually higher-order accurate finite difference formulas, or too many control points, will destabilize the algorithm and worsen the condition of the resulting linear system of equations; see, for example, [108].

A better approach is to use piecewise polynomial or trigonometric interpolations, for example, a cubic spline interpolation, to get an approximate expression of the interface $(s, \mathbf{X}(s))$. Then we can obtain the information about the interface from the analytic expression of the interpolated interface. For example, the tangential direction and the curvature can be determined according to the formulas, respectively,

$$\boldsymbol{\tau} = \left(\dot{X}, \dot{Y}\right)/\sqrt{\dot{X}^2 + \dot{Y}^2}, \tag{1.38}$$

$$\kappa = \left(\ddot{X}\dot{Y} - \ddot{Y}\dot{X}\right) / \left(\dot{X}^2 + \dot{Y}^2\right)^{3/2}. \tag{1.39}$$

Using the definition above, a circle will have a negative curvature. This approach works well for many test problems including Stokes flows with a moving interface. One advantage is that we can take relatively few control points on the interface if the interface is smooth. A cubic spline package [160, 165] has been developed and intensively used for a number of applications in two dimensions. However, this approach is difficult to apply for problems with multiconnected domains and three-dimensional problems. In these cases, the level set function approach is a better choice.

The level set function approach

In this approach, an interface is represented by the zero level set of a Lipschitz continuous function $\varphi(\mathbf{x})$ defined on the entire domain or a computational tube satisfying

$$\varphi(\mathbf{x}) > 0 \qquad \text{for } (\mathbf{x}) \in \Omega^-, \tag{1.40a}$$

$$\varphi(\mathbf{x}) = 0 \qquad \text{for } (\mathbf{x}) \in \Gamma, \tag{1.40b}$$

$$\varphi(\mathbf{x}) < 0 \qquad \text{for } (\mathbf{x}) \in \Omega^+. \tag{1.40c}$$

The signed distance function is an example of such a level set function. Given an interface $\Gamma \in C^0$ in a domain Ω and a positive direction of Γ, toward which the normal vector $\mathbf{n}$ is pointing, the signed distance function is defined as $\varphi(\mathbf{x}) = dist(\mathbf{x}, \Gamma)$ if $\mathbf{x}$ is on the positive side of Γ, otherwise $\varphi(\mathbf{x}) = -dist(\mathbf{x}, \Gamma)$. Thus a signed distance function is well defined for any interface $\Gamma \in C^0$. Note that if an interface is smooth, that is, the unit normal direction is well defined at every point on Γ, then the signed distance function is differentiable at least in a neighborhood of Γ (the set of $\varphi(\mathbf{x}) = 0$). If Γ has a finite number of corners, then the signed distance function is still continuous, but not differentiable along some curves and surfaces, which are called characteristics. For example, the characteristics of a square are the two diagonals, where the signed distance function is nondifferentiable.

Often, the interface is given by a set of ordered points. Various approaches can be used to construct a level set function using either a PDE method or a geometric approach; see, for example, [47, 238, 261, 284, 285].

Numerically, the underlying grid and the interface are both defined on the same Eulerian coordinates system at grid points. In other words, the interface is implicitly defined by $\varphi_{ij} = \varphi(x_i, y_j)$ in two dimensions and $\varphi_{ijk} = \varphi(x_i, y_j, z_k)$ in three dimensions. Below is a sample MATLAB® code to generate a level set function representation for two ellipses in the domain $[-1, 1] \times [-1, 1]$.

```
N=40; h=2/N;                          % Domain: [-1, 1] x [-1, 1]
[x,y] = meshgrid(-1:h:1, -1:h:1);
phi1 = sqrt( (x-0.35).^2/0.35^2 + (y-0.35).^2/0.25^2)-1;
phi2 = sqrt( (x+0.4).^2/0.2^2 + (y+0.45).^2/0.25^2)-1;
phi  = min(phi1,phi2);
figure(1); z=zeros(N+1,N+1); contour(x,y,phi',z);
axis('square')
figure(2); mesh(x,y,phi')
```

The level set function is defined on a 41×41 uniform grid. The code also generates two plots: one is the zero level set that represents the two ellipses in Figure 1.8(a), and the other is a mesh plot of the level set function in the domain in Figure 1.8(b). The level set function is differentiable almost everywhere except for one characteristics curve.

The interface information, such as the normal and tangential directions, the curvature, etc., can be easily computed from

$$\mathbf{n} = \frac{\nabla\varphi}{|\nabla\varphi|}, \tag{1.41a}$$

$$\kappa = \frac{\varphi_{xx}\varphi_y^2 - 2\varphi_x\varphi_y\varphi_{xy} + \varphi_{yy}\varphi_x^2}{\left(\varphi_x^2 + \varphi_y^2\right)^{3/2}} \tag{1.41b}$$

at grid points where $\varphi(\mathbf{x})$ has appropriate partial derivatives. Those partial derivatives can be computed using central finite difference schemes at grid points. Below is a MATLAB code to compute the unit normal direction (n_x, n_y) of a level function $\varphi(x_i, y_j)$ at a grid point (x_i, y_j) with a uniform mesh size h.

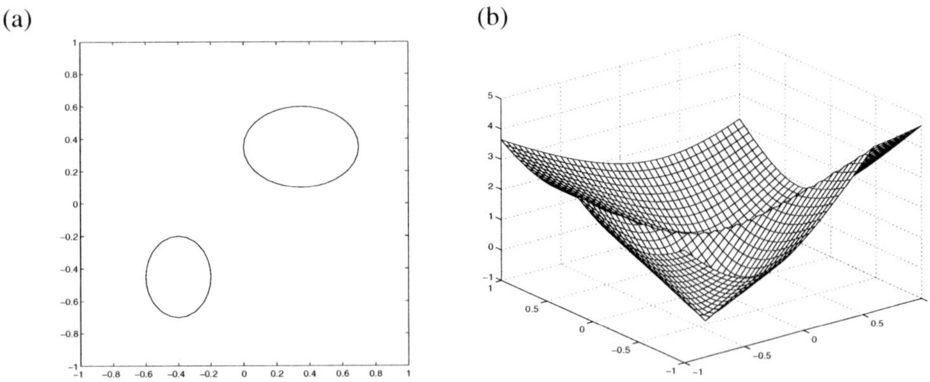

Figure 1.8. (a) *A contour plot of the zero level set of* φ_{ij}. (b) *A mesh plot of the level set function.*

```
phx = (phi(i+1,j)-phi(i-1,j))/(2*h);
phy = (phi(i,j+1)-phi(i,j-1))/(2*h);
phn = (phx^2 + phy^2 + 1e-15); nx = phx/phn; ny = phy/phn;
```

We add a tolerance $\epsilon = 1e - 15$ in the denominator to prevent possible breakdowns at centers of the level set function where $\nabla\varphi = 0$.

If (X, Y) is on the interface ($\varphi(X, Y) = 0$) but is not a grid point, the normal and tangential derivatives of the interface can be approximated using those points from the four corners of the rectangle that contains (X, Y) via the bilinear interpolation; see §11.1.3.

The orthogonal projections of a grid point on the interface

In the level set function representation of an interface, the interface is implicitly defined by the values of the level set function at the grid points. In the IIM, it is sometimes necessary to find the interface information on the interface. The bridge is made up of the orthogonal projections of certain grid points ($|\varphi| \le Ch$) on the interface. Assume that $\mathbf{x}$ is a grid point near the interface. Let $\mathbf{X}^*$ be the orthogonal projection of $\mathbf{x}$ on the interface. We can write

$$\mathbf{X}^* = \mathbf{x} + \alpha\mathbf{p}, \tag{1.42}$$

where

$$\mathbf{p} = \begin{bmatrix} \varphi_x(\mathbf{x}) \\ \varphi_y(\mathbf{x}) \end{bmatrix} \tag{1.43}$$

in Cartesian coordinates. Since $\mathbf{x}$ is close to the interface, α is small. Using the Taylor expansion of $\varphi(\mathbf{X}^*) = \varphi(\mathbf{x} + \alpha\mathbf{p}) = 0$ at $\mathbf{x}$, we get the following quadratic equation for the unknown scalar α:

$$\varphi(\mathbf{x}) + (\nabla\varphi(\mathbf{x}) \cdot \mathbf{p})\,\alpha + \frac{1}{2}\left(\mathbf{p}^T \mathrm{He}(\varphi(\mathbf{x}))\,\mathbf{p}\right)\alpha^2 = 0, \tag{1.44}$$

where

$$\mathbf{p}^T \mathrm{He}(\varphi)\,\mathbf{p} = \varphi_x^2\,\varphi_{xx} + 2\,\varphi_x\varphi_y\varphi_{xy} + \varphi_y^2\,\varphi_{yy} \tag{1.45}$$

in Cartesian coordinates. The partial derivatives $\nabla\varphi(\mathbf{x})$ and the Hessian matrix $\mathrm{He}(\varphi)$ are computed at the grid point $\mathbf{x}$. The computed projections have third-order accuracy if $\nabla\varphi(\mathbf{x})$ and $\mathrm{He}(\varphi)$ are computed using the standard centered 5-point finite difference formulas; see Table 5.1 in Chapter 5.

The choice of $\varphi(\mathbf{x})$ obviously is not unique. A unit circle in two dimensions, for example, is the zero level set of the smooth functions

$$\varphi(x, y) = \left(x^2 + y^2\right)^m - 1$$

for any $m > 0$. It has been known in the literature that the best choice is the signed distance function ($m = 1/2$ in the unit circle case) in which $|\nabla\varphi| = 1$. This avoids having the level set function be too steep or too flat. For a level set function which is far from the distance function, a reinitialization process should be applied to get an approximation to the distance function while keeping the interface unchanged; see, for example, [47, 123, 250]. For free boundary and moving interface problems, a reinitialization process has been found to be crucial in obtaining accurate simulations.

Note that for a given level set function $\varphi(\mathbf{x})$, any level set $\varphi(\mathbf{x}) = C$ is a curve in two dimensions, and a surface in three dimensions. In other words, the geometric information such as the normal and tangential directions and the curvature are continuously defined everywhere. In other words, the geometric information of an interface can be continuously extended along the normal lines of the interface at least in a neighborhood of the interface where the normal lines do not intersect. Similarly, a function defined along an interface, particularly, the velocity of the interface in a moving interface problem, can be extended along normal lines. Usually, we are interested only in information in a neighborhood of the zero level set $\varphi(\mathbf{x}) = 0$. In the literature, it is called a computational domain if $|\varphi(\mathbf{x})| \le \delta \sim h$ for a given number δ. Within the computational domain, one can easily get the geometric information of the level set function at grid points using standard central finite difference schemes from (1.41a) and (1.41b). For a point that is not a grid point, a bilinear interpolation (see (11.7) in two dimensions in §11.1.3, and (4.36) in three dimensions in §4.3.4) can be used to interpolate the geometric information at the neighboring grid points.

We will review some evolution schemes for free boundary and moving interface problems in Chapter 11.

1.7 What is the IIM?

The IIM has been developed as a sharp interface method which can accurately capture discontinuities in the solution and the flux. The first IIM paper was published in 1994 [154]. We call a numerical method the IIM if the method has the following properties.

- A uniform or adaptive grid or triangulation in Cartesian, polar, or spherical coordinates is used instead of a body-fitted grid.

- Prior knowledge of jump conditions (or internal boundary conditions) exists either from physical reasoning or from the governing differential equations. More interface relations often need to be derived from the given jump conditions and governing PDEs; see, for example, §3.1.

- Away from the interface, standard finite difference or finite element methods are used in the discretization. The numerical methods are modified according to the jump conditions only at grid points or elements near or on the interface.
- We emphasize pointwise convergence, that is, we are more concerned with errors in the infinity norm L^∞ instead of an average norm such as L^1 and L^2. This is because an average norm cannot exactly reflect errors near the interface. The accuracy of solutions near or at the interfaces is often the main interest for moving interface and free boundary problems. The IIM generally has the same global order accuracy, often second order, as that when the method is applied to a regular problem without an interface. Note that, while the global errors have the same order of accuracy at all grid points for the IIM, the local truncation errors may be one order lower at grid points or elements that are near the interface than that at regular grid points or elements that are away from the interface.
- The method becomes the standard one if the discontinuities in the coefficients, in the solution, and in the flux disappear.

The term "immersed interface" is used because the method is motivated by Peskin's "immersed boundary method" and the method is first developed for interface problems. For a problem defined on an irregular domain, or for a free boundary problem, the domain is embedded into a larger rectangular box and the problem is treated as an interface problem. There are remarkable differences between the IB method and the IIM. The IB method is a smoothing method with a transition region that smears discontinuities, while the IIM is a sharp interface method in which the discontinuities or the jump conditions are enforced either exactly or approximately. For some application problems, a smoothing method may be more appropriate, while in other applications, a sharp interface method is preferred.

For interface problems, the regularity of the solutions is difficult; we refer readers to [11, 30, 44, 73, 112] for some discussions. In the discussion of finite difference methods for interface problems, it is reasonable to assume that the solutions are *piecewise* smooth. The discontinuities occur only at interfaces. In the discussion of finite element methods, we assume that the solutions are in proper Sobolev spaces.

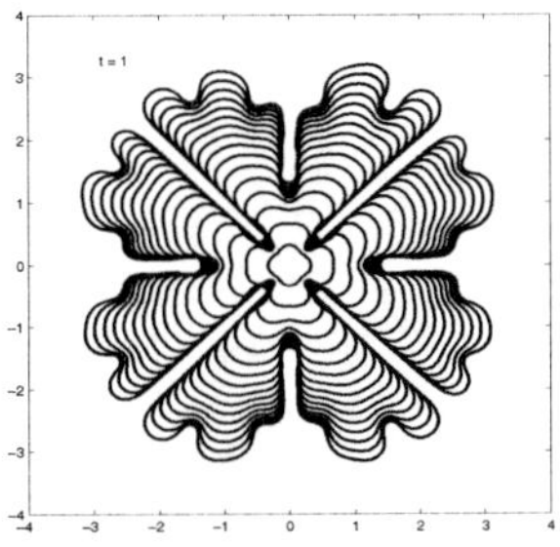

Chapter 2

The IIM for One-Dimensional Elliptic Interface Problems

We begin with the one-dimensional model problem

$$(\beta u_x)_x - \sigma u = f + v\,\delta(x - \alpha), \quad 0 < x < 1, \quad 0 < \alpha < 1, \tag{2.1}$$

with specified boundary conditions of $u(x)$ at $x = 0$ and $x = 1$. The function $\beta(x)$ is allowed to be discontinuous at $x = \alpha$. While the method and analysis are simple for the one-dimensional problem, they still illustrate the main ideas of the IIM.

2.1 Reformulating the problem using the jump conditions

If $f(x) \in L^2(0, 1)$, $\sigma(x) \in C(0, 1)$, and $\beta(x) \in C^1(0, \alpha) \cup C^1(\alpha, 1)$, then $u(x) \in C(0, 1)$, that is, $u(x)$ is continuous. By integrating (2.1) from $x = \alpha^-$ to $x = \alpha^+$, we get

$$[\beta u_x]_{x=\alpha} = \beta^+ u_x^+ - \beta^- u_x^- = v \quad \text{in addition to} \quad [u]_{x=\alpha} = u^+ - u^- = 0. \tag{2.2}$$

An alternative way to state problem (2.1) is to require that $u(x)$ satisfy the equation

$$(\beta u_x)_x - \sigma u = f, \quad x \in (0, \alpha) \cup (\alpha, 1), \tag{2.3}$$

excluding the interface α, together with the two internal boundary conditions (2.2) at $x = \alpha$. When $f(x)$ is continuous, we also have

$$\beta_x^+ u_x^+ + \beta^+ u_{xx}^+ - \sigma^+ u^+ = \beta_x^- u_x^- + \beta^- u_{xx}^- - \sigma^- u^-.$$

For simplicity we start with the assumption that $\sigma(x)$ and $f(x)$ are smooth functions and β is a piecewise constant with a finite jump at α. Thus we have $\sigma^+ = \sigma^-$, $\beta_x^- = \beta_x^+ = 0$, and $u^+ = u^-$. We can express the limiting values from the "+" side, where $x > \alpha$, in terms of those from the "−" side, where $x < \alpha$, to get

$$u^+ = u^-, \quad u_x^+ = \frac{\beta^-}{\beta^+} u_x^- + \frac{v}{\beta^+}, \quad u_{xx}^+ = \frac{\beta^- u_{xx}^-}{\beta^+}. \tag{2.4}$$

2.2 The IIM for the simple one-dimensional model equation

A finite difference method for a linear differential equation usually involves the following procedures: (1) generating a grid; (2) substituting the derivatives with finite difference approximations at all grid points, where the solution is unknown, to get a linear system of equations; (3) solving the linear system of equations to get an approximation to the original differential equation; (4) carring out the error analysis. The IIM follows the same procedure.

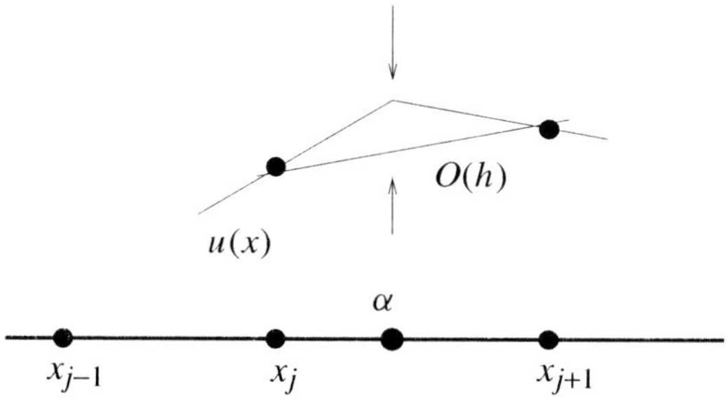

Figure 2.1. *A diagram of a one-dimensional grid and the interface α between x_j and x_{j+1}.*

The algorithm of the IIM for (2.3) and (2.2) is outlined below.

Step 1: Generate a Cartesian grid, say

$$x_i = ih, \quad i = 0, 1, \ldots, N,$$

where $h = 1/N$. The point α will typically fall between the grid points, say $x_j \le \alpha < x_{j+1}$; see Figure 2.1 for an illustration. The grid points x_j and x_{j+1} are called *irregular grid points* if a standard 3-point central finite difference stencil is used at grid points away from the interface α. The other grid points are called regular grid points.

Step 2: Determine the finite difference scheme at regular grid points. At a grid point x_i, $i \ne j, j+1$, the standard 3-point central finite difference approximation

$$\frac{1}{h^2}\Big(\beta_{i+\frac{1}{2}}(U_{i+1}-U_i) - \beta_{i-\frac{1}{2}}(U_i - U_{i-1})\Big) - \sigma_i U_i = f_i \tag{2.5}$$

is used, where $\beta_{i+\frac{1}{2}} = \beta(x_{i+\frac{1}{2}})$, $\sigma_i = \sigma(x_i)$, $f_i = f(x_i)$.

Step 3: Determine the finite difference equations at irregular grid points x_j and x_{j+1}. The finite difference equations are determined from the method of undetermined coefficients,

$$\begin{aligned}\gamma_{j,1}U_{j-1} + \gamma_{j,2}U_j + \gamma_{j,3}U_{j+1} - \sigma_j U_j &= f_j + C_j,\\ \gamma_{j+1,1}U_j + \gamma_{j+1,2}U_{j+1} + \gamma_{j+1,3}U_{j+2} - \sigma_{j+1}U_{j+1} &= f_{j+1} + C_{j+1}.\end{aligned} \tag{2.6}$$

For the simple model problem in which $\sigma \equiv 0$, $[f] = 0$, and β is a piecewise constant, the coefficients of the finite difference equations at x_j and x_{j+1} have the following closed form:

$$\begin{cases} \gamma_{j,1} = (\beta^- - [\beta](x_j - \alpha)/h)/D_j, \\ \gamma_{j,2} = (-2\beta^- + [\beta](x_{j-1} - \alpha)/h)/D_j, \\ \gamma_{j,3} = \beta^+/D_j, \end{cases}$$
$$\begin{cases} \gamma_{j+1,1} = \beta^-/D_{j+1}, \\ \gamma_{j+1,2} = (-2\beta^+ + [\beta](x_{j+2} - \alpha)/h)/D_{j+1}, \\ \gamma_{j+1,3} = (\beta^+ - [\beta](x_{j+1} - \alpha)/h)/D_{j+1}, \end{cases} \tag{2.7}$$

where

$$\begin{aligned} D_j &= h^2 + [\beta](x_{j-1} - \alpha)(x_j - \alpha)/2\beta^-, \\ D_{j+1} &= h^2 - [\beta](x_{j+2} - \alpha)(x_{j+1} - \alpha)/2\beta^+. \end{aligned} \tag{2.8}$$

The derivation is given in the next section. It will be shown later that $D_j \neq 0$ and $D_{j+1} \neq 0$ if $\beta^-\beta^+ > 0$. The correction terms are

$$C_j = \gamma_{j,3}\,(x_{j+1} - \alpha)\,\frac{v}{\beta^+}, \qquad C_{j+1} = \gamma_{j+1,1}\,(\alpha - x_j)\,\frac{v}{\beta^-}. \tag{2.9}$$

For more general one-dimensional interface problems, the coefficients $\{\gamma_{j,k}\}$ and $\{\gamma_{j+1,k}\}$ are determined from a system of equations (see (2.14) and (2.16) in §2.3) since the closed forms of the finite difference coefficients are complicated and unnecessary.

Step 4: Solve the system of equations (2.5)–(2.6), whose coefficient matrix is tridiagonal, to get an approximate solution of $u(x)$ at all grid points.

2.2.1 The derivation of the finite difference scheme at an irregular grid point

We illustrate the idea of the IIM in determining the finite difference coefficients $\gamma_{j,1}$, $\gamma_{j,2}$, and $\gamma_{j,3}$ in (2.6) for the simple case where $\sigma = 0$, β is a piecewise constant, and $f(x)$ is continuous. The criterion in determining the finite difference coefficients is to minimize the magnitude of the local truncation error

$$T_j = \gamma_{j,1}u(x_{j-1}) + \gamma_{j,2}u(x_j) + \gamma_{j,3}u(x_{j+1}) - f(x_j) - C_j. \tag{2.10}$$

The main idea is to expand the solution $u(x_{j-1})$, $u(x_j)$, $u(x_{j+1})$, $f(x_j)$ at the interface α from each side of the interface and then use the interface relations (2.2) or (2.4) to express $u^\pm(\alpha)$, $u_x^\pm(\alpha)$, and $u_{xx}^\pm(\alpha)$ in terms of the quantities from one particular side; finally, we match the expansion against the differential equation to the leading terms to get a system of equations for the finite difference coefficients.

Using the Taylor expansion for $u(x_{j+1})$ at α, we have

$$u(x_{j+1}) = u^+(\alpha) + (x_{j+1} - \alpha)\,u_x^+(\alpha) + \frac{1}{2}\,(x_{j+1} - \alpha)^2\,u_{xx}^+(\alpha) + O(h^3).$$

Using the jump relations (2.4), the expression above can be written as

$$u(x_{j+1}) = u^-(\alpha) + (x_{j+1} - \alpha)\left(\frac{\beta^-}{\beta^+}u_x^-(\alpha) + \frac{v}{\beta^+}\right)$$
$$+ \frac{1}{2}(x_{j+1} - \alpha)^2 \frac{\beta^-}{\beta^+}u_{xx}^-(\alpha) + O(h^3).$$

The Taylor expansions of $u(x_{j-1})$ and $u(x_j)$ at α have the following expression:

$$u(x_l) = u^-(\alpha) + (x_l - \alpha)u_x^-(\alpha) + \frac{1}{2}(x_l - \alpha)^2 u_{xx}^-(\alpha) + O(h^3), \quad l = j-1,\ j.$$

Thus the local truncation error of the finite difference scheme (2.6) at $x = x_j$ has the following form:

$$\begin{aligned}
T_j &= \gamma_{j,1}u(x_{j-1}) + \gamma_{j,2}u(x_j) + \gamma_{j,3}u(x_{j+1}) - f(x_j) - C_j \\
&= \left(\gamma_{j,1} + \gamma_{j,2} + \gamma_{j,3}\right)u^-(\alpha) + \gamma_{j,3}(x_{j+1} - \alpha)\frac{v}{\beta^+} \\
&\quad + \left((x_{j-1} - \alpha)\gamma_{j,1} + (x_j - \alpha)\gamma_{j,2} + \frac{\beta^-}{\beta^+}(x_{j+1} - \alpha)\gamma_{j,3}\right)u_x^-(\alpha) \\
&\quad + \frac{1}{2}\left((x_{j-1} - \alpha)^2\gamma_{j,1} + (x_j - \alpha)^2\gamma_{j,2} + \frac{\beta^-}{\beta^+}(x_{j+1} - \alpha)^2\gamma_{j,3}\right)u_{xx}^-(\alpha) \\
&\quad - f(\alpha) - O(h) - C_j + O\left(\max_{1\le l\le 3}|\gamma_{j,l}|\,h^3\right),
\end{aligned}$$

after collecting terms for $u^-(\alpha)$, $u_x^-(\alpha)$, and $u_{xx}^-(\alpha)$.

By minimizing the magnitude of T_j and using the differential equation at α from the "−" side,[1] we get the following system of equations for the coefficients $\{\gamma_{j,k}\}$:

$$\left\{\begin{aligned}
&\gamma_{j,1} + \gamma_{j,2} + \gamma_{j,3} = 0, \\
&(x_{j-1} - \alpha)\,\gamma_{j,1} + (x_j - \alpha)\,\gamma_{j,2} + \frac{\beta^-}{\beta^+}(x_{j+1} - \alpha)\gamma_{j,3} = 0, \\
&\frac{1}{2}(\alpha - x_{j-1})^2\,\gamma_{j,1} + \frac{1}{2}(\alpha - x_j)^2\,\gamma_{j,2} + \frac{\beta^-}{2\beta^+}(x_{j+1} - \alpha)^2\,\gamma_{j,3} = \beta^-.
\end{aligned}\right. \tag{2.11}$$

It is easy to verify that the $\{\gamma_{j,k}\}$'s in the first three equations in (2.7) satisfy the system above when $\sigma = 0$ and β is a piecewise constant. Once those $\{\gamma_{j,k}\}$'s have been computed, it is easy to set the correction term

$$C_j = \gamma_{j,3}(x_{j+1} - \alpha)\frac{v}{\beta^+}, \tag{2.12}$$

which matches the remaining leading terms in the local truncation error T_j above.

---

[1] It is also possible to further expand at $x = x_j$ to match the differential equation at $x = x_j$. The finite difference equations will be changed slightly at the two irregular grid points, but the order of convergence will be the same.

It is worth pointing out the following important properties and special cases of the finite difference coefficients $\{\gamma_{j,k}\}$ derived from the IIM.

- If β is a continuous constant, then by solving system (2.11), we recover the standard 3-point central finite difference scheme with $\gamma_{j1} = \gamma_{j3} = \beta/h^2$ and $\gamma_{j2} = -2\beta/h^2$.

- In the case when β is a piecewise constant, the harmonic averaging coefficients satisfy the first and second equations of (2.11) but not the third, indicating that the truncation error of this method at x_j and x_{j+1} is $O(1)$. But the method has global second-order accuracy in the infinity norm, due to the cancellation of the errors. It has been shown that $T_j = -T_{j+1} + O(h)$ for the harmonic averaging method [160, 228].

- If $v = 0$, then $C_j = C_{j+1} = 0$. In this case a discontinuity in β affects only the coefficients, but not the right-hand side.

- If β is a constant and $\sigma = 0$, then

$$C_j = \frac{v}{h^2}(x_{j+1} - \alpha) = v\,\delta_h(x_j - \alpha); \qquad C_{j+1} = v\,\delta_h(x_{j+1} - \alpha), \tag{2.13}$$

  where δ_h is the discrete hat delta function (1.19). In this case we can view the finite difference scheme as a direct discretization of the equation

$$\beta u''(x) = f(x) + v\delta(x - \alpha).$$

- The coefficients $\{\gamma_{j,k}\}$ depend only on the function $\beta(x)$ and the position of α relative to the grid, but not on v.

2.3 The IIM for general one-dimensional elliptic interface problems

For a general interface problem in which all the coefficients, $\beta(x)$, $\sigma(x)$, and $f(x)$, may have a finite jump at $x = \alpha$, and the solution itself may have a jump $[u] = w$, in addition to the flux jump condition $[\beta u_x] = v$, the IIM has been developed in [160]. The finite difference coefficients at $x = x_j$ are the solution of the following linear system:

$$\left\{\begin{array}{ll} \gamma_{j,1} + \gamma_{j,2} + \left(1 + \dfrac{(x_{j+1}-\alpha)^2}{2\beta^+}[\sigma]\right)\gamma_{j,3} & = 0, \\[2ex] (x_{j-1}-\alpha)\,\gamma_{j,1} + (x_j - \alpha)\,\gamma_{j,2} & \\ \quad + \left\{\dfrac{\beta^-}{\beta^+}(x_{j+1}-\alpha) + \left(\dfrac{\beta_x^-}{\beta^+} - \dfrac{\beta^-\beta_x^+}{(\beta^+)^2}\right)\dfrac{(x_{j+1}-\alpha)^2}{2}\right\}\gamma_{j,3} = \beta_x^-, & \\[2ex] \dfrac{(x_{j-1}-\alpha)^2}{2}\gamma_{j,1} + \dfrac{(x_j-\alpha)^2}{2}\gamma_{j,2} + \dfrac{(x_{j+1}-\alpha)^2\,\beta^-}{2\,\beta^+}\gamma_{j,3} & = \beta^-. \end{array}\right. \tag{2.14}$$

Once these $\{\gamma_{j,k}\}$'s have been computed, we can determine the correction term at $x = x_j$ as

$$C_j = \gamma_{j,3}\left\{ w + (x_{j+1}-\alpha)\,\frac{v}{\beta^+} - \frac{(x_{j+1}-\alpha)^2}{2}\left(\frac{\beta_x^+ v}{(\beta^+)^2} - \sigma^+\frac{w}{\beta^+} - \frac{[f]}{\beta^+}\right)\right\}. \tag{2.15}$$

The linear system of equations for the coefficients of the finite difference equation at $x = x_{j+1}$ is

$$\begin{cases} \left(1 - \dfrac{(x_j-\alpha)^2}{2\beta^-}[\sigma]\right)\gamma_{j+1,1} + \gamma_{j+1,2} + \gamma_{j+1,3} = 0, \\ \left\{\dfrac{\beta^+}{\beta^-}(x_j-\alpha) + \left(\dfrac{\beta_x^+}{\beta^-} - \dfrac{\beta_x^-\beta^+}{(\beta^-)^2}\right)\dfrac{(x_j-\alpha)^2}{2}\right\}\gamma_{j+1,1} \\ \qquad + (x_{j+1}-\alpha)\gamma_{j+1,2} + (x_{j+2}-\alpha)\gamma_{j+1,3} = \beta_x^+, \\ \dfrac{(x_j-\alpha)^2}{2}\dfrac{\beta^+}{\beta^-}\gamma_{j+1,1} + \dfrac{(x_{j+1}-\alpha)^2}{2}\gamma_{j+1,2} + \dfrac{(x_{j+2}-\alpha)^2}{2}\gamma_{j+1,3} = \beta^+, \end{cases} \tag{2.16}$$

and the correction term C_{j+1} is

$$C_{j+1} = \gamma_{j+1,1}\left\{ -w - (x_j-\alpha)\frac{v}{\beta^-} - \frac{1}{2}(x_j-\alpha)^2\left(-\frac{\beta_x^- v}{(\beta^-)^2} + \sigma^-\frac{w}{\beta^-} + \frac{[f]}{\beta^-}\right)\right\}. \tag{2.17}$$

Note that the system of equations for the finite difference equations and the correction terms are the same at x_j and x_{j+1} if we switch "+" and "−" symbols.

2.4 The error analysis of the IIM for one-dimensional interface problems

The systematic error analysis of the IIM for one-dimensional problems is given in [110, 160] in which the authors show that the finite difference scheme obtained from the IIM satisfies the discrete maximum principle and the computed solution has second-order convergence in the infinity norm. We list some main results in this section with a sketch of the proofs but we omit the details.

Theorem 2.1. *Assume that* $\beta^-\beta^+ > 0$ *and* $\sigma = 0$. *If* $\beta_x^- = 0$ *and* $\beta_x^+ = 0$, *then the system of equations* (2.11) *has a unique solution. For more general problems, the system* (2.14) *is guaranteed to have a unique solution if* h *is sufficiently small. The theorem also applies to the other irregular grid point* x_{j+1}.

Proof: Let det(A) be the determinant of the coefficient matrix of the system (2.11). With the help of the symbolic software package *Mathematica*® *or* Maple®, we obtain that

$$\det(A) = \begin{cases} C_1\, D_j & \text{if } \beta_x^-(\alpha) = 0 \text{ and } \beta_x^+(\alpha) = 0, \\ C_1\left(D_j + O(h^3)\right) & \text{otherwise,} \end{cases}$$

where C_1 is a nonzero constant. Without loss of generality, we assume $\beta^- > 0$ and $\beta^+ > 0$. Notice that $(x_{j-1} - \alpha)(x_j - \alpha) \geq 0$, and h is small, so if $[\beta] \geq 0$, then $|\det(A)| > |C_1 D_j|/2 \geq |C_1| h^2/2 > 0$, and hence the theorem is true. If $[\beta] < 0$, i.e., $\beta^- > \beta^+$, then, since $\beta^-\beta^+ > 0$, we have

$$\left| \frac{\beta^- - \beta^+}{\beta^-} \right| < 1.$$

Hence

$$\begin{aligned} h^2 + \frac{[\beta]}{2\beta^-}(x_{j-1} - \alpha)(x_j - \alpha) &> h^2 - \frac{1}{2}(x_{j-1} - \alpha)(x_j - \alpha) \\ &> h^2 - \frac{1}{2}(2h)h = 0, \end{aligned}$$

where we have used the fact that the interface α is between x_j and x_{j+1}, that is, $x_j \leq \alpha < x_{j+1}$. The rest of the proof is trivial. □

If $\beta^+\beta^- < 0$, then the systems may be singular even though this can rarely happen. In this case, we cannot even guarantee that the differential equation has a unique solution.

The following theorem shows that the coefficients of the finite difference equations obtained from the IIM satisfy the discrete maximum principle and are bounded.

Theorem 2.2. *Assume that the conditions in Theorem* 2.1 *are satisfied and* $\beta(x) \geq \beta_{min} > 0$. *If h is sufficiently small, then the coefficients of the difference scheme obtained from the IIM satisfy*

1. $\dfrac{\bar{C}}{h^2} \leq |\gamma_{i,k}| \leq \dfrac{C}{h^2}$, *where* $C = \dfrac{2\,\beta_{max}^2}{\beta_{min}}$ *and* $\bar{C} = \dfrac{\beta_{min}^2}{\beta_{max}}$.
2. $\gamma_{i,1} > 0,\ \gamma_{i,3} > 0$, *and* $\gamma_{i,2} < 0,\ \gamma_{i,1} + \gamma_{i,3} \leq |\gamma_{i,2}|$,

for $i = 1, 2, \ldots, M$, *where*

$$\beta_{max} = \max_{0 \leq x \leq 1} \{\beta(x)\}, \qquad \beta_{min} = \min_{0 \leq x \leq 1} \{\beta(x)\}.$$

We omit the proof here because it is quite long and technical. Interested readers can find the proof in [110].

The local truncation error of the finite difference method is $O(h^2)$ at regular grid points, and is $O(h)$ at the two irregular grid points x_j and x_{j+1}. We define the following notation:

$$\begin{aligned} \gamma_{min} &= \min_{1 \leq j \leq M-1}\ \min_{k=1,2,3} \left\{ \left|\gamma_{j,k}\, h^2\right| \right\}, \\ \gamma_{max} &= \max_{1 \leq j \leq M-1}\ \max_{k=1,2,3} \left\{ \left|\gamma_{j,k}\, h^2\right| \right\}, \end{aligned} \tag{2.18}$$

$$M_{xxx} = \max\left\{ \max_{x<\alpha} |u'''(x)|, \quad \max_{x>\alpha} |u'''(x)| \right\},$$
$$M_{xxxx} = \max\left\{ \max_{x<\alpha} |u''''(x)|, \quad \max_{x>\alpha} |u''''(x)| \right\}. \tag{2.19}$$

Define also a nonnegative comparison function,

$$\phi(x) = \begin{cases} \dfrac{(x-\alpha)^2}{2\beta^-} + \dfrac{E_1(1-\alpha)(\alpha-x)}{\beta^-} + \dfrac{(\alpha-x_j)(x_j-x)}{\gamma_{j,1}h^3} & \text{if } x \le \alpha, \\[2ex] \dfrac{(x-\alpha)^2}{2\beta^+} + \dfrac{E_1\alpha(x-\alpha)}{\beta^+} + \dfrac{(x_{j+1}-\alpha)(x-x_{j+1})}{\gamma_{j+1,3}h^3} & \text{if } x > \alpha, \end{cases} \tag{2.20}$$

where $E_1 = \max\{\beta^-, \beta^+\}/\gamma_{min}$. Then we have the following error estimates.

Theorem 2.3. *Assume that the conditions in Theorem* 2.1 *are satisfied and* $\beta(x)$ *is a piecewise constant; then*

$$|T_i| \le \frac{\beta_{max}\, M_{xxxx}}{12} h^2, \quad i = 1, 2, \ldots, j-1, \quad j+2, \ldots, M-1,$$
$$|T_j| \le \gamma_{max} M_{xxx} \left(\frac{(\alpha - x_{j-1})^3 + (\alpha - x_j)^3 + (x_{j+1}-\alpha)^3}{6h^2} \right), \tag{2.21}$$
$$|T_{j+1}| \le \gamma_{max} M_{xxx} \left(\frac{(\alpha - x_j)^3 + (x_{j+1}-\alpha)^3 + (x_{j+2}-\alpha)^3}{6h^2} \right),$$

and

$$||u(x_i) - U_i||_\infty \le \frac{5\,\gamma_{max}\,\phi_{max}\,C}{3} h^2, \tag{2.22}$$

where U_i *is the approximated solution from the finite difference equations,* $u(x_i)$ *is the exact solution at* $x = x_i$, *and*

$$\phi_{max} = \max\left(\phi(0), \phi(1)\right), \qquad C = \max\left(M_{xxx}, M_{xxxx}\right), \tag{2.23}$$

where ϕ *is given by* (2.20).

Again we refer interested readers to [110] for the proof.

2.5 One-dimensional numerical examples and a comparison with other methods

In this section we provide two examples to numerically verify second-order convergence of the IIM and compare them with other commonly used numerical methods for interface problems.

Example 2.1. In Figure 2.2, we show a comparison of the numerical results obtained from the IIM and the smoothing method with the discrete cosine delta function (1.20) for one-dimensional interface problem (2.1) with $\sigma = 0$. The source term is $f(x) = \delta(x - \alpha)$. The

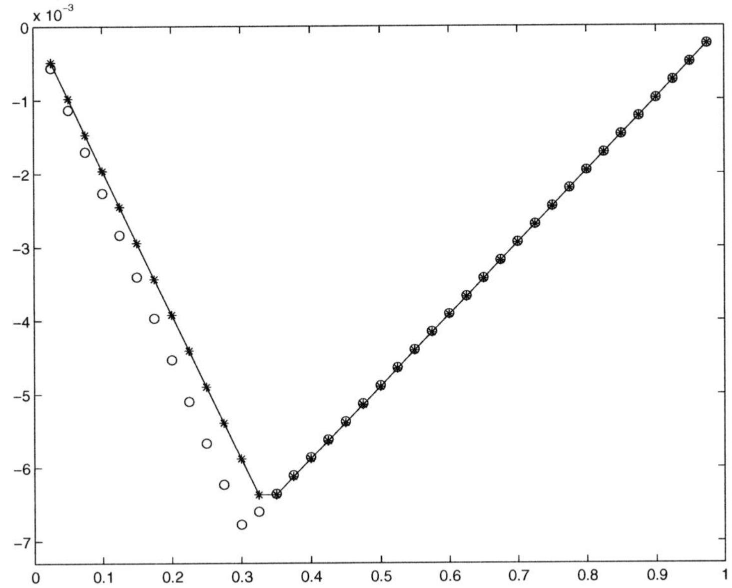

Figure 2.2. *A comparison of the computed solutions. The solid line is the exact solution. The " $*$ " is the result obtained from the IIM which is exact. The "o" is the result obtained from the smoothing method* (1.13) *with* $\epsilon = 2h$ *combined with the discrete cosine delta function. The mesh size is* $h = 1/40$.

boundary condition is $u(0) = u(1) = 0$. It is easy to check that the exact solution is

$$u(x) = \begin{cases} Bx(1-\alpha) & \text{if } 0 \le x \le \alpha, \\ B\alpha(1-x) & \text{if } \alpha < x \le 1, \end{cases}$$

where $B = -1/\left(\beta^+\alpha + \beta^-(1-\alpha)\right)$. The parameters are $\alpha = 1/3$, $\beta^- = 1$, $\beta^+ = 100$. In this case, the IIM gives the exact solution, while the result computed from the smoothing method (1.13) with $\epsilon = 2h$ combined with the cosine discrete delta function (1.20) is first-order accurate, and the location of the discontinuity is also shifted.

Example 2.2. In this example, we use the IIM to solve the following problem:

$$(\beta u_x)_x = 12x^2, \quad 0 < x < 1, \quad \beta = \begin{cases} \beta^- & \text{if } x < \alpha, \\ \beta^+ & \text{if } x > \alpha, \end{cases}$$

$$u(0) = 0, \quad u(1) = \frac{1}{\beta^+} + \left(\frac{1}{\beta^-} - \frac{1}{\beta^+}\right)\alpha^4.$$

In this example, $f(x) = 12x^2$ is continuous and the natural jump conditions $[u] = 0$ and $[\beta u_x] = 0$ are satisfied across the interface α. The exact solution is

Table 2.1. *A grid refinement analysis of the IIM and the harmonic averaging (HV) method for Example 2.2 with $\alpha = 1/2$, $\beta^- = 1$, $\beta^+ = 2$. Second-order convergence in the infinity norm is verified.*

M	$\|E_M\|_\infty(IIM)$	Ratio1 (order1)	Ratio2	$\|E_M\|_\infty$(HV)	Ratio3
20	2.6285^{-4}			5.0683^{-4}	
40	5.3523^{-5}	4.9110 (2.2960)		1.2787^{-4}	3.9634
80	1.5980^{-5}	3.3493 (1.7439)	16.4485	3.1842^{-5}	4.0161
160	3.3802^{-6}	4.7276 (2.2411)	15.8342	7.9779^{-6}	3.9912
320	9.9130^{-7}	3.4099 (1.7697)	16.1206	1.9923^{-6}	4.0043
640	2.1176^{-7}	4.6811 (2.2268)	15.9622	4.9835^{-7}	3.9978

$$u(x) = \begin{cases} \dfrac{x^4}{\beta^-} & \text{if } x < \alpha, \\ \dfrac{x^4}{\beta^+} + \left(\dfrac{1}{\beta^-} - \dfrac{1}{\beta^+}\right)\alpha^4 & \text{if } x > \alpha. \end{cases}$$

Table 2.1 shows the results of a grid refinement analysis of the IIM and the harmonic averaging method, with $\alpha = 1/2$, $\beta^- = 1$, $\beta^+ = 2$. The error is measured in the infinity norm,

$$\|E_M\|_\infty = \max_i \{\, |u(x_i) - U_i| \,\}, \tag{2.24}$$

where $u(x_i)$ is the exact solution at $x = x_i$. The ratios are defined as

$$\text{Ratio1 (Ratio 3)} = \|E_{2M}\|_\infty / \|E_M\|_\infty, \qquad \text{Ratio2} = \|E_{4M}\|_\infty / \|E_M\|_\infty. \tag{2.25}$$

It is well known that the ratio of two consecutive errors approaches 4 for a second-order method. However, for an interface problem, the errors of the computed results using the IIM usually do not decrease monotonically, but rather oscillate. In the fourth column, we list the results of the ratio with mesh size decreased by a factor of 4. The ratio for a second-order method will approach 16. A more accurate approach is to use the linear regression analysis to find an approximate order of convergence, which will be seen later in this monograph (see, for example, section 6.1.6). In the parentheses, we list the results of the corresponding convergence order defined as $\log(\text{ratio})/\log(2)$.

The results from the harmonic averaging approach also give second-order convergence; see the fifth and sixth columns. But the errors are about $2 \sim 4$ times larger than those obtained from the IIM. The local truncation errors $T(x_j)$ and $T(x_{j+1})$ of the harmonic averaging method at irregular grid points x_j and x_{j+1} ($x_j \le \alpha < x_{j+1}$) are $O(1)$ but they satisfy $T(x_j) = -T(x_{j+1}) + O(h)$. In other words, the errors cancel each other out, which leads to second-order global accuracy for the computed solution. For nonhomogeneous jump conditions, or discontinuous coefficients, or in high dimensions, the errors may not fully cancel each other out, which usually leads to deterioration of global errors; see also [160, 228].

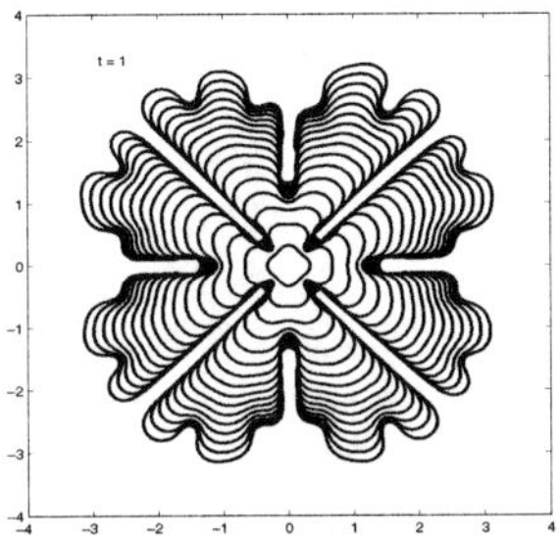

Chapter 3

The IIM for Two-Dimensional Elliptic Interface Problems

In this chapter we discuss the IIM for solving two-dimensional elliptic interface problems,

$$(\beta u_x)_x + (\beta u_y)_y - \sigma(x, y)\, u = f(x, y), \quad (x, y) \in \Omega = \Omega^+ \cap \Omega^-, \tag{3.1}$$

with a prescribed boundary condition on $\partial\Omega$, where $\beta \geq \beta_{min} > 0$ and σ and f are *piecewise* continuous but may have a finite jump discontinuity across some interface (a curve in two dimensions) $\Gamma \in C^2$ within the domain Ω; see Figure 1.6 for an illustration. Two interface conditions, or internal boundary conditions, are needed in advance to make the problem well-posed. We assume locally that they are defined by

$$[u] = u^+ - u^- = w, \tag{3.2a}$$

$$[\beta u_n] = \beta^+ \frac{\partial u^+}{\partial n} - \beta^- \frac{\partial u^-}{\partial n} = v, \tag{3.2b}$$

where w and v are two functions defined only along the interface Γ. When $w \equiv 0$ and $v \equiv 0$, such jump conditions are called natural interface conditions and are often implied naturally instead of specified explicitly in (3.2a)–(3.2b). Note that if $w \equiv 0$ and σ is continuous, then the solution to the interface problem is equivalent to the solution of the single equation in the entire domain,

$$\nabla \cdot (\beta \nabla u(\mathbf{x})) - \sigma\, u(\mathbf{x}) = f(\mathbf{x}) + \int_\Gamma v\, \delta(\mathbf{x} - \mathbf{X}(s)) ds, \tag{3.3}$$

where δ is the Dirac delta function in two-dimensional space. The second term on the right-hand side is a distribution that satisfies

$$\iint_\Omega \int_\Gamma v\, \delta(\mathbf{x} - \mathbf{X}(s))\, \Psi(\mathbf{x}) ds\, d\mathbf{x} = \int_\Gamma v\, \Psi(\mathbf{X}(s)) ds \tag{3.4}$$

for any smooth function $\Psi(\mathbf{x})$.

The discussion of the existence and the regularity of the solution can be found, for example, in [11, 44]. In general, if β, σ, and f are piecewise smooth in Ω, $w = 0$, and v is differentiable along Γ, then the solution to the interface problem exists and is in $H^1(\Omega)$.

The problem of how to solve the interface problem (3.1)–(3.2b) efficiently has attracted much attention from numerical analysts for a long time due to its enormous applications. The problem itself describes many important applications in multiphase flows, potential theory, ideal flows, and many others. The most expensive step of several well-known efficient methods for Navier–Stokes equations (for example, the projection method [16, 34, 48, 49, 16, 134, 222] and the method using the vorticity stream-function formulation [36, 70, 109]) is solving one or several elliptic interface problems.

Before we explain the IIM, we first provide some theoretical preparations.

3.1 Interface relations for two-dimensional elliptic interface problems

From the jump conditions (3.2a)–(3.2b) and the PDE (3.1), we can derive the following interface relations that represent the limiting values from one side in terms of the other using the local coordinates (1.34); see Figures 1.6 and 1.7 for illustrations.

Theorem 3.1. *Let (X, Y) be a point on the interface Γ. Assume that $\Gamma \in C^2$ in a neighborhood of (X, Y) corresponding to the local coordinates* (1.34) *at $(0, 0)$. Then from the jump conditions* (3.2a)–(3.2b) *and the PDE* (3.1), *we have the following interface relations:*

$$
\begin{aligned}
u^+ &= u^- + w, \\
u_\xi^+ &= \rho\, u_\xi^- + \frac{v}{\beta^+}, \\
u_\eta^+ &= u_\eta^- + w', \\
u_{\xi\xi}^+ &= \left(\frac{\beta_\xi^-}{\beta^+} - \chi'' \right) u_\xi^- + \left(\chi'' - \frac{\beta_\xi^+}{\beta^+} \right) u_\xi^+ + \frac{\beta_\eta^-}{\beta^+} u_\eta^- - \frac{\beta_\eta^+}{\beta^+} u_\eta^+ \\
&\quad + (\rho - 1)\, u_{\eta\eta}^- + \rho u_{\xi\xi}^- - w'' + \frac{[f]}{\beta^+} + \frac{[\sigma] u^- + \sigma^+ [u]}{\beta^+}, \\
u_{\eta\eta}^+ &= u_{\eta\eta}^- + (u_\xi^- - u_\xi^+)\, \chi'' + w'', \\
u_{\xi\eta}^+ &= \frac{\beta_\eta^-}{\beta^+} u_\xi^- - \frac{\beta_\eta^+}{\beta^+} u_\xi^+ + \left(u_\eta^+ - \rho u_\eta^- \right) \chi'' + \rho\, u_{\xi\eta}^- + \frac{v'}{\beta^+},
\end{aligned}
\tag{3.5}
$$

where $\rho = \beta^-/\beta^+$ and w', v' and w'', v'' are the first- and second-order surface derivatives of w and v at (X, Y) on the interface.

Proof: In a neighborhood of (X, Y), the interface can be expressed as $\xi = \chi(\eta)$ with $\chi(0) = 0$ and $\chi'(0) = 0$. The jump conditions w and v are then functions of η. For simplicity, we still use the notation $[u] = w(\eta)$ and $[\beta u_n] = v(\eta)$ in the local coordinate system. Differentiating (3.2a) with respect to η along the interface, we get

$$
[u_\xi]\, \chi' + [u_\eta] = w'(\eta). \tag{3.6}
$$

Setting $\eta = 0$, we get the second equality in (3.5). Differentiating the equation above again with respect to η, we obtain

$$[u_\xi]\,\chi'' + \chi'\frac{d}{d\eta}[u_\xi] + [u_{\xi\eta}]\,\chi' + [u_{\eta\eta}] = w''(\eta). \tag{3.7}$$

Setting $\eta = 0$, we get the fifth equality in (3.5). Notice that in the local coordinates, (3.2b) can be written as

$$\beta^+(u_\xi^+ - u_\eta^+\chi') = \beta^-(u_\xi^- - u_\eta^-\chi') + v\sqrt{1+(\chi')^2}. \tag{3.8}$$

Differentiating this with respect to η along the interface, we have

$$\begin{aligned}
(\beta_\xi{}^+\chi' + \beta_\eta{}^+)\,(u_\xi^+ - u_\eta^+\chi') &+ \beta^+\left(u_{\xi\xi}^+\chi' + u_{\xi\eta}^+ - \frac{d}{d\eta}(u_\eta^+)\,\chi' - u_\eta^+\chi''\right)\\
&= (\beta_\xi{}^-\chi' + \beta_\eta{}^-)\,(u_\xi^- - u_\eta^-\chi')\\
&\quad + \beta^-\left(u_{\xi\xi}^-x' + u_{\xi\eta}^- - \frac{d}{d\eta}(u_\eta^-)\,\chi' - u_\eta^-\chi''\right)\\
&\quad + v'(\eta)\left(\sqrt{1+(\chi')^2} + \frac{v(\eta)\chi'\chi''}{\sqrt{1+(\chi')^2}}\right).
\end{aligned} \tag{3.9}$$

Setting $\eta = 0$, we get the last equality in (3.5). Since the PDE (3.1) is invariant under the transformation (1.34), from (3.1) we have

$$\left[\,\beta(u_{\xi\xi} + u_{\eta\eta}) + \beta_\xi u_\xi + \beta_\eta u_\eta - \sigma u\,\right] = [f]\,.$$

Expanding the jumps using $[(\,)] = (\,)^+ - (\,)^-$ and solving $u_{\xi\xi}^+$ from the expression above, we get

$$\begin{aligned}
u_{\xi\xi}^+ &= \frac{\beta^-}{\beta^+}u_{\xi\xi}^- + \frac{\beta^-}{\beta^+}u_{\eta\eta}^- - u_{\eta\eta}^+ + \frac{\beta_\xi{}^-}{\beta^+}u_\xi^- - \frac{\beta_\xi{}^+}{\beta^+}u_\xi^+\\
&\quad + \frac{\beta_\eta{}^-}{\beta^+}u_\eta^- - \frac{\beta_\eta{}^+}{\beta^+}u_\eta^+ + \frac{[f]}{\beta^+} - \frac{\sigma^-u^- - \sigma^+u^+}{\beta^+}.
\end{aligned} \tag{3.10}$$

The numerator of the last term can be rewritten as

$$\sigma^-u^- - \sigma^+u^+ = -[\sigma]\,u^- - [u]\,\sigma^+. \tag{3.11}$$

This gives the third equality in (3.5). □

These interface relations are used in deriving the finite difference method in the next section.

3.2 The finite difference scheme of the IIM in two dimensions

Given a Cartesian grid (x_i, y_j), $i = 0, 1, \ldots, M$, $j = 0, 1, \ldots, N$, the finite difference scheme for (3.1) has the following generic form:

$$\sum_k^{n_s} \gamma_k\, U_{i+i_k, j+j_k} - \sigma_{ij}\,U_{ij} = f_{ij} + C_{ij} \tag{3.12}$$

at any grid point (x_i, y_j), where $u(x_i, y_j)$ is unknown. In the finite difference scheme above, n_s is the number of grid points involved in the finite difference stencil and U_{ij} is an approximation to the solution $u(x, y)$ of (3.1) at (x_i, y_j). The sum over k involves a finite number of grid points neighboring (x_i, y_j). So each i_k and j_k will take values in the set $\{0, \pm 1, \pm 2, \ldots\}$. The coefficients $\{\gamma_k\}$ and the indexes i_k and j_k depend on (i, j), so they should really be labeled as γ_{ijk}, etc. But for simplicity of notation, we will concentrate on a single grid point (x_i, y_j) and drop the dependency.

The local truncation error at a grid point (x_i, y_j) is defined as

$$T_{ij} = \sum_{k}^{n_s} \gamma_k\, u\left(x_{i+i_k}, y_{j+j_k}\right) - \sigma(x_i, y_j)\, u(x_i, y_j) - f(x_i, y_j) - C_{ij}. \tag{3.13}$$

A grid point (x_i, y_j) is called a *regular grid point* in reference to the standard 5-point finite difference stencil centered at (i, j) if all five grid points are on the same side of the interface. At regular grid points, the local truncation errors are $O(h^2)$ if the standard centered 5-point $(n_s = 5)$ finite difference formula,

$$\begin{aligned} &\frac{1}{h}\left\{\left(\beta_{i+1/2,j}\frac{(U_{i+1,j}-U_{ij})}{h} - \beta_{i-1/2,j}\frac{(U_{i,j}-U_{i-1,j})}{h}\right)\right. \\ &\left.+\left(\beta_{i,j+1/2}\frac{(U_{i,j+1}-U_{i,j})}{h} - \beta_{i,j-1/2}\frac{(U_{i,j}-U_{i,j-1})}{h}\right)\right\} - \sigma_{ij}U_{ij} = f_{ij}, \end{aligned} \tag{3.14}$$

is used, where

$$\beta_{ij} = \beta(x_i, y_j), \quad \beta_{i+\frac{1}{2},j} = \beta\left(x_i + \frac{h}{2}, y_j\right),$$

and so on. At an irregular grid point, the correction term is simply zero, i.e., $C_{ij} = 0$.

If (x_i, y_j) is an *irregular grid point*, that is, the grid points in the centered 5-point stencil are from both sides of the interface, then an undetermined coefficients method is used to set up a linear system of equations for the finite difference coefficients $\{\gamma_k\}$ in (3.12). The correction term C_{ij} can be determined after the $\{\gamma_k\}$'s are obtained. With the assumption that the solution is piecewise smooth, a point (x_i^*, y_j^*) on the interface Γ near the grid point (x_i, y_j) is chosen so that the Taylor expansion can be carried out from each side of the interface. Usually, (x_i^*, y_j^*) is chosen either as the orthogonal projection of (x_i, y_j) on the interface or as the intersection of the interface and one of the axes.

Let the local coordinates of (x_{i+i_k}, y_{j+j_k}) be (ξ_k, η_k). The idea is to minimize the magnitude of the local truncation error T_{ij} in (3.13) by matching the finite difference equation to the differential equation up to all second-order partial derivatives. Thus, the local truncation error would be zero if the exact solution is a piecewise quadratic function, which implies second-order convergence if the stability condition is also satisfied.

The Taylor expansion of $u(x_{i+i_k}, y_{j+j_k})$ about (x_i^*, y_j^*) under the local coordinates is

$$\begin{aligned} u(x_{i+i_k}, y_{j+j_k}) &= u(\xi_k, \eta_k) = u^{\pm} + \xi_k u_\xi^{\pm} + \eta_k u_\eta^{\pm} \\ &+ \frac{1}{2}\xi_k^2 u_{\xi\xi}^{\pm} + \xi_k\eta_k u_{\xi\eta}^{\pm} + \frac{1}{2}\eta_k^2 u_{\eta\eta}^{\pm} + O(h^3), \end{aligned} \tag{3.15}$$

where the "+" or "−" sign is chosen depending on whether (ξ_k, η_k) lies on the "+" or "−" side of Γ. After the expansions of all terms, $u(x_{i+i_k}, y_{j+j_k})$, used in the finite difference equation (3.12), the local truncation error T_{ij} can be expressed as a linear combination of the values $u^{\pm}$, $u^{\pm}_{\xi}$, $u^{\pm}_{\eta}$, $u^{\pm}_{\xi\xi}$, $u^{\pm}_{\xi\eta}$, $u^{\pm}_{\eta\eta}$ as follows:

$$\begin{aligned} T_{ij} &= a_1\, u^- + a_2\, u^+ + a_3\, u^-_{\xi} + a_4\, u^+_{\xi} + a_5\, u^-_{\eta} + a_6 u^+_{\eta} \\ &\quad + a_7\, u^-_{\xi\,\xi} + a_8\, u^+_{\xi\,\xi} + a_9\, u^-_{\eta\,\eta} + a_{10}\, u^+_{\eta\,\eta} + a_{11}\, u^-_{\xi\,\eta} \\ &\quad + a_{12}\, u^+_{\xi\,\eta} - \sigma^- u^- - f^- - C_{ij} + \max\{|\gamma_k|\}\, O(h^3). \end{aligned} \tag{3.16}$$

The quantities $f^{\pm}$, $\sigma^{\pm}$, and $\beta^{\pm}$ are the limiting values of the functions at (x_i^*, y_j^*) from the "+" or "−" side of the interface. The coefficients $\{a_j\}$ depend only on the position of the stencil relative to the interface. They are independent of the PDE, β, u, σ, f, and the jump conditions w and v. If we define the index sets K^+ and K^- by

$$K^{\pm} = \{k : \ (\xi_k, \eta_k) \text{ is on the "}\pm\text{" side of } \Gamma\},$$

then the $\{a_j\}$'s are given by

$$\begin{aligned} a_1 &= \sum_{k\in K^-} \gamma_k, & a_2 &= \sum_{k\in K^+} \gamma_k, \\ a_3 &= \sum_{k\in K^-} \xi_k\gamma_k, & a_4 &= \sum_{k\in K^+} \xi_k\gamma_k, \\ a_5 &= \sum_{k\in K^-} \eta_k\gamma_k, & a_6 &= \sum_{k\in K^+} \eta_k\gamma_k, \\ a_7 &= \frac{1}{2}\sum_{k\in K^-} \xi_k^2\gamma_k, & a_8 &= \frac{1}{2}\sum_{k\in K^+} \xi_k^2\gamma_k, \\ a_9 &= \frac{1}{2}\sum_{k\in K^-} \eta_k^2\gamma_k, & a_{10} &= \frac{1}{2}\sum_{k\in K^+} \eta_k^2\gamma_k, \\ a_{11} &= \sum_{k\in K^-} \xi_k\eta_k\gamma_k, & a_{12} &= \sum_{k\in K^+} \xi_k\eta_k\gamma_k. \end{aligned} \tag{3.17}$$

Using the interface relations (3.5), we eliminate the quantities from one side, say the "+" side, using the quantities from the other side, say the "−" side, and collect terms to get an expression of the form[2]

[2]A more subtle approach is to expand all $u(x_{i+i_k}, y_{j+j_k})$ at the grid point (x_i, y_j). If (x_{i+i_k}, y_{j+j_k}) is a grid point on the side opposite (x_i, y_j), we can first expand $u(x_{i+i_k}, y_{j+j_k})$ at (x_i^*, y_j^*) on the interface; then express all the quantities up to second-order derivatives in terms of those on the other side using the jump relations (3.5); and then expand those quantities again at (x_i, y_j). This approach gives slightly better results (with a smaller error constant).

$$
\begin{aligned}
T_{ij} = {} & \left(a_1 + \frac{a_8\,[\sigma]}{\beta^+} + a_2\right) u^- \\
& + \left\{a_3 + a_8\left(\frac{\beta_\xi^{\,-}}{\beta^+} - \chi''\right) + a_{10}\chi'' + a_{12}\frac{\beta_\eta^{\,-}}{\beta^+}\right. \\
& \qquad \left. + \rho\left(a_4 + a_8\left(\chi'' - \frac{\beta_\xi^{\,+}}{\beta^+}\right) - a_{10}\chi'' - a_{12}\frac{\beta_\eta^{\,+}}{\beta^+}\right) - \beta_\xi^{\,-}\right\} u_\xi^- \\
& + \left\{a_5 + a_6 + a_8\left(\frac{\beta_\eta^{\,-}}{\beta^+} - \frac{\beta_\eta^{\,+}}{\beta^+}\right) + a_{12}(1-\rho)\,\chi'' - \beta_\eta^{\,-}\right\} u_\eta^- \\
& + \left\{a_7 + a_8\rho - \beta^-\right\} u_{\xi\xi}^- + \left\{a_9 + a_{10} + a_8\,(\rho - 1) - \beta^-\right\} u_{\eta\eta}^- \\
& + \left\{a_{11} + a_{12}\rho\right\} u_{\xi\eta}^- - \sigma^- u^- - f^- \\
& + \left(\hat{T}_{ij} - C_{ij}\right) + \max\{|\gamma_k|\}\, O(h^3),
\end{aligned}
\tag{3.18}
$$

where

$$
\begin{aligned}
\hat{T}_{ij} = {} & a_2\, w + a_{12}\,\frac{v'}{\beta^+} + \left(a_6 - \frac{a_8\,\beta_\xi^{\,+}}{\beta^+} + a_{12}\chi''\right) w' \\
& + a_{10}\, w'' + \frac{1}{\beta^+}\left(a_4 + a_8\left(\chi'' - \frac{\beta_\xi^{\,+}}{\beta^+}\right) - a_{10}\chi'' - a_{12}\frac{\beta_\eta^{\,+}}{\beta^+}\right) v \\
& + a_8\left\{\frac{[f]}{\beta^+} + \frac{\sigma^+\, w}{\beta^+} - w''\right\}.
\end{aligned}
\tag{3.19}
$$

Assuming the finite difference scheme is stable, we can guarantee second-order accuracy of the approximate solution by requiring the coefficients of u^-, u_ξ^-, u_η^-, $\ldots$, $u_{\xi\eta}^-$ to vanish,

$$
\begin{aligned}
a_1 + a_2 + a_8\,\frac{[\sigma]}{\beta^+} &= 0, \\
a_3 + \rho a_4 + a_8\frac{\beta_\xi^{\,-} - \rho\beta_\xi^{\,+} - [\beta]\chi''}{\beta^+} + a_{10}\frac{[\beta]\chi''}{\beta^+} & \\
+ a_{12}\frac{\beta_\eta^{\,-} - \rho\beta_\eta^{\,+}}{\beta^+} &= \beta_\xi^{\,-}, \\
a_5 + a_6 - a_8\frac{[\beta_\eta]}{\beta^+} + a_{12}(1-\rho)\,\chi'' &= \beta_\eta^{\,-}, \\
a_7 + a_8\,\rho &= \beta^-, \\
a_9 + a_{10} + a_8\,(\rho - 1) &= \beta^-, \\
a_{11} + a_{12}\,\rho &= 0,
\end{aligned}
\tag{3.20}
$$

where $\rho = \beta^-/\beta^+$. Once the $\{\gamma_k\}$'s are computed, we can easily obtain C_{ij} as

$$
C_{ij} = \hat{T}_{ij}, \tag{3.21}
$$

where $\hat{T}_{ij}$ is given by (3.19) since

$$
\beta^-\left(u_{\xi\xi}^- + u_{\eta\eta}^-\right) + \beta_\xi^- u_\xi^- + \beta_\eta^- u_\eta^- - \sigma^- u^- - f^- = 0
$$

at the point (x_i^*, y_j^*). The local truncation error of the derived finite difference equation is generally $O(h)$ at an irregular grid point.

3.3 The 6-point finite difference stencil at irregular grid points

In order to get second-order methods, we require all six equations in (3.20) to be satisfied. Generally speaking, we need six or more grid points in the finite difference stencil. In the original IIM [154, 160], a grid point, in addition to the standard central 5-point stencil, but within the compact 9-point stencil, is used to get a second-order finite difference scheme. An optimization technique is used to chose the sixth grid point in [81].

The exact nature of the finite difference coefficients at irregular grid points depends on how large the jump in β is. For mild discontinuities, we observed the following.

- The contributions to the finite difference schemes at irregular points are mainly from the standard 5-point stencil. These coefficients are $O(1/h^2)$, while the contributions from the "additional points" are typically much smaller. The magnitude depends on the jump in β and on the geometry.
- All the coefficients, except occasionally γ_6, have the same sign, "$-$" sign for the diagonal and "$+$" sign for the off-diagonal, as in the classic 5-point finite difference formula. Since the contribution from the sixth point is much smaller than that from the standard five grid points, we expect the classical theoretical analysis to be applicable to the resulting linear system with slight modifications. In particular, the system is nearly diagonally dominant, and strictly so if all γ_6's are positive.

Note that the resulting linear system of equations is still block tridiagonal. Most sparse solvers or iterative methods can be applied. However, the coefficient matrix may no longer be an M-matrix.

3.4 The fast Poisson solver for problems with only singular sources

If the coefficients β and σ are constants in (3.1), then the finite difference coefficients obtained from (3.20) are simply that of the standard 5-point central finite difference scheme; see the proof in [160]. That is, the coefficients are $\gamma_k = \beta/h^2$ for the four neighbors of (x_i, y_j), and $-4\beta/h^2$ for the master grid point (x_i, y_j). There is no need to solve (3.20).

For this type of interface problem, the irregularities are from the singular source distribution in (3.3). Typically, a jump in the flux $[u_{\mathbf{n}}] = v/\beta$ is due to a source distribution along the interface, while a jump in the solution $[u] = w$ is due to a dipole distribution along the interface; see, for example, [138]. For this type of interface problem, while the standard 5-point central finite difference coefficients can be used, we still need to add correction terms C_{ij} according to (3.19), which is also much simplified now. The finite difference scheme is simply

$$\beta \frac{U_{i-1,j} + U_{i+1,j} + U_{i,j-1} + U_{i,j+1} - 4U_{i,j}}{h^2} - \sigma U_{ij} = f_{ij} + C_{ij}. \tag{3.22}$$

The correction term at an irregular grid point (x_i, y_j) now is

$$\begin{aligned} C_{ij} &= a_2\, w + a_{12}\, \frac{v'}{\beta} + \left(a_6 + a_{12}\chi''\right)\, w' + a_{10}\, w'' \\ &+ \frac{1}{\beta}\left\{ a_4 + (a_8 - a_{10})\chi'' \right\}\, v + a_8 \left\{ \frac{[f]}{\beta} + \frac{\sigma\, w}{\beta} - w'' \right\}. \end{aligned} \tag{3.23}$$

Note that if $w \equiv 0$, then C_{ij} can be regarded as a second-order discretization of the source distribution in (3.3) that contains a two-dimensional Dirac delta function. Note that C_{ij} depends on the curvature ($\kappa = \chi''$) of the interface, which means it is difficult to get an analytic expression (or closed form) for the correction terms. In Chapter 5, we explain another method that can transform the interface problem into a new one with a smooth solution using a level set function. The approach there provides a simpler way to compute the correction terms without evaluating surface derivatives of the jump conditions and the curvature.

Since the standard discrete Laplacian can be used and only the right-hand sides of the finite difference equations need to be modified, a fast Poisson solver such as Fishpack [2] can be applied to solve the system of the finite difference equations. This makes the IIM an efficient method since the computational cost spent on irregular grid points is only a fraction of that for a fast Poisson solver or a multigrid method; see §3.8 at the end of this chapter for a distribution of the CPU times.

This special case is the foundation of the augmented approach discussed in Chapter 6 for the fast IIM for elliptic interface problems (3.1)–(3.2b) with a piecewise constant coefficient β, and fast solvers for generalized Helmholtz equations defined on irregular domains.

3.5 Enforcing the discrete maximum principle

In this section, we discuss the general case when $\beta(x, y)$ is a function of x and y and has a finite jump across the interface. Since the PDE (3.1) satisfies the maximum principle, it is desirable that the finite difference scheme satisfy the discrete maximum principle; see §6.5 of Morton and Mayers [202] for the definition. This will guarantee that the coefficient matrix of the finite difference equations is an M-matrix which is diagonally dominant and invertible. Most of the iterative methods are guaranteed to converge for M-matrices. For this purpose, we impose the sign restrictions on the coefficients $\{\gamma_k\}$ in (3.12),

$$\begin{aligned} \gamma_k &\geq 0 \quad \text{if} \quad (i_k, j_k) \neq (0, 0), \\ \gamma_k &< 0 \quad \text{if} \quad (i_k, j_k) = (0, 0), \end{aligned} \tag{3.24}$$

along the equality constraints in (3.20). Such a finite difference method is called a *maximum principle preserving scheme*. The coefficient matrix of the system of the finite difference equations of such a scheme is diagonally dominant and invertible. At regular grid points, the standard central finite difference scheme satisfies the sign restriction and the equations in (3.20). So the discussion is needed only for irregular grid points. At an irregular grid point, (x_i, y_j), to enforce the equality and inequality constraints, we form the following quadratic constrained optimization problem to determine the coefficients of the finite difference scheme:

$$\min_{\boldsymbol{\gamma}} \left\{ \frac{1}{2} \boldsymbol{\gamma}^T H \boldsymbol{\gamma} - \boldsymbol{\gamma}^T \mathbf{g} \right\}, \tag{3.25}$$

$$\text{s.t.} \quad \begin{cases} A\boldsymbol{\gamma} = \mathbf{b}, \\ \gamma_k \geq 0 \quad \text{if} \quad (i_k, j_k) \neq (0, 0), \\ \gamma_k < 0 \quad \text{if} \quad (i_k, j_k) = (0, 0), \end{cases} \tag{3.26}$$

where $\boldsymbol{\gamma} = [\,\gamma_1, \gamma_2, \ldots, \,]^T$ is the vector composed of the coefficients of the finite difference equation, H is a symmetric positive definite matrix, and $\mathbf{g} \in R^{n_s}$. $A\boldsymbol{\gamma} = \mathbf{b}$ is the system of linear equations (3.20). Naturally, we want to choose $\{\gamma_k\}$ in such a way that the finite difference equation becomes the standard 5-point central finite difference scheme if there is no interface. This can be done by minimizing

$$\min_{\gamma_k} \frac{1}{2} \sum_k (\gamma_k - g_k)^2, \tag{3.27}$$

where

$$\begin{aligned} g_k &= \frac{2\beta_{i+i_k, j+j_k}}{h_x^2 + h_y^2}, && (i_k, j_k) \in \{\,(-1,0), (1,0), (0,-1), (0,1)\,\}, \\ g_k &= -\frac{8\beta_{i,j}}{h_x^2 + h_y^2}, && (i_k, j_k) = (0,0), \\ g_k &= 0, && \text{otherwise}, \end{aligned} \tag{3.28}$$

where h_x and h_y are the mesh spacing in the x- and y-directions. The matrix H in (3.25) is often chosen as the identity matrix. Another choice of $\mathbf{g}$ is $\mathbf{g} = A^+\mathbf{b}$, where A^+ is the pseudoinverse of A and $\mathbf{g} = A^+\mathbf{b}$ is the least squares solution to the system of equations $A\mathbf{x} = \mathbf{b}$. We can also choose some combination of (3.28) and $A^+\mathbf{b}$.

In the optimization algorithm (3.25)–(3.26), we need to select a set of grid points (x_{i_k}, y_{j_k}). The solution $\{\gamma_k\}$ to the constrained optimization problem is composed of the coefficients of the finite difference scheme at the particular irregular grid point.

3.5.1 Choosing the finite difference stencil

If we use the standard 5-point stencil centered at (x_i, y_j), first-order accuracy can be guaranteed by enforcing the first three equations in (3.20) plus the sign constraint. The convergence proof, along with some numerical examples, is given in [166].

In order to get second-order methods, we require all six equations in (3.20) to be satisfied, plus the sign restrictions (3.24) to be satisfied, for the optimization problem. Thus we should choose $n_s \geq 6$. Since the symmetry in the linear system of the finite difference equations is not required (and may be difficult to enforce), we expect that the optimization problem has solutions if the standard 9-point compact stencil ($n_s = 9$) is chosen. This has been numerically verified in [166] and will be discussed later in the next section. Moreover, the standard 9-point compact stencil is preferred because the resulting linear system of equations is block tridiagonal and the multigrid solver DMGD9V (developed for the standard 9-point compact stencil) [62] can be used.

3.5.2 Solving the optimization problem

There are several commercial and educational software packages that are designed to solve constrained quadratic optimization problems. For example, the *QP* function in MATLAB; the *QL* subroutine using the FORTRAN computer language developed by K. Schittkowski [234]; and the *IQP* FORTRAN code from PORT managed by Lucent Technologies. Information about these software packages can be found on the Web.[3]

Most quadratic optimization solvers require users to provide an initial guess, lower and upper bounds, and other information. A good choice of an initial guess is $\mathbf{g}$ in (3.28). Reasonable lower and upper bounds of the solution are the following:

$$\begin{aligned} 0 < \gamma_k < \frac{2\,\beta_{\max}}{h_x^2 + h_y^2} \quad &\text{if} \quad (i_k, j_k) \neq (0,0); \\ -\frac{8\,\beta_{\max}}{h_x^2 + h_y^2} < \gamma_k < 0 \quad &\text{if} \quad (i_k, j_k) = (0,0), \end{aligned} \tag{3.29}$$

where $\beta_{\max}$ is an estimation of the upper bound of the coefficient $\beta(x, y)$. Since the size of the optimization problem is small, the total cost in finding the coefficients is only a small portion compared with that needed for solving the linear system of equations. The numerical tests using a multigrid linear solver show that the extra time needed in dealing with interfaces including solving the optimization problem is only about 5–8% in the entire solution process; see, for example, Table 3.4 in §3.8.

In the case when the optimization solver fails to give a solution or provides a wrong solution, we can either add a few more grid points that are closer to the interface or switch to a first-order scheme at the particular grid point. The breakdown happens only at a few grid points when the jump ratio is large or the grid is rather coarse. Turning to a first-order scheme at a few grid points usually does not affect global second-order accuracy due to the nature of the ellipticity.

3.6 The error analysis of the maximum principle preserving scheme

If enough grid points are enclosed such that the six equations in (3.20) and the sign property are satisfied, the maximum principle preserving scheme is second-order accurate, which

[3]www.mathworks.com
http://www.uni-bayreuth.de/departments/math/~Kschittkowski/gl.htm
http://www.bell-labs.com/project/PORT

will be proved in this section. What the minimum number of grid points is and which grid points should be included are still open questions. To be cautious and to reduce the grid orientation effects, we recommend taking a standard compact 9-point stencil. The solution to the corresponding optimization problem has been shown to exist and also to be bounded, by numerical verification.

3.6.1 Existence of the solution to the optimization problem

Without loss of generality, we assume that $\sigma = 0$ in (3.1), and h is small enough that the interface behaves like a straight line relative to the underlying grid. Under these conditions, the terms that contain χ'' in (3.20) are high-order terms of h compared to those in a_3 and ρa_4 and therefore can be neglected. For simplicity, we also assume that β is a piecewise constant. With the standard compact 9-point finite difference stencil, the following conjecture has been numerically verified.

Conjecture 3.1. *Let (x_i, y_j) be an irregular grid point, and let (x_i^*, y_j^*) be its orthogonal projection on the interface. Then the optimization problem defined in (3.25)–(3.26), with six equalities and the sign constraints, has solutions. The solution of the coefficients $\{\gamma_k\}$ also satisfies*

$$\left|\frac{\gamma_k}{\beta^-}\right| \le \frac{C}{h^2}, \qquad k = 1, \ldots, 9. \tag{3.30}$$

Furthermore, there is at least one γ_k from each side of the interface such that

$$\frac{\gamma_k}{\beta^-} \ge \frac{C_1}{h^2}, \tag{3.31}$$

and thus

$$\sum_{\xi_k \ge 0} \gamma_k \xi_k \ge \frac{C_2}{h} \quad \text{if} \quad \max_{\xi_k \ge 0} \{\xi_k\} > C_3 h \tag{3.32}$$

for some $C_3 > 0$. The constants are $O(1)$ which depend on the coefficient β.

Numerical verification of Conjecture 3.1

To numerically verify the conjecture, we first shift and scale the problem in the following way:

$$\bar{x} = \frac{x - x_i}{h}, \qquad \bar{y} = \frac{y - y_j}{h}, \qquad \bar{\gamma}_k = \frac{\gamma_k}{h^2}.$$

For simplicity, we use the same notation without bars. The compact 9-point stencil then is in the square $-1 \le x, y \le 1$. With the local coordinate system, it is enough to consider the case where the projection is in the first quadrant.

Given any point (x^*, y^*) and an angle θ,

$$0 \le x^* \le 1, \qquad 0 \le y^* \le 1 - x^*, \qquad 0 \le \theta \le \frac{\pi}{2}, \tag{3.33}$$

the straight line

$$\cos\theta\,(x - x^*) + \sin\theta\,(y - y^*) = 0 \tag{3.34}$$

is a good approximation to the interface if $h > 0$ is sufficiently small so that $\chi'' h^2$ is negligible.

The interface cuts the unit square $0 \le x, y \le 1$ into two parts. We denote the side which contains the origin as the "$-$" side, and the other side as the "$+$" side. We also scale the coefficient β in such a way that either $\beta^- = 1$, $\beta^+ = 1/\rho$ or $\beta^+ = 1$, $\beta^- = \rho$.

The optimization problem then is

$$\min_{\gamma_k} \frac{1}{2}\sum_k (\gamma_k - g_k)^2,$$

$$\text{s.t.} \quad \begin{cases} A\boldsymbol{\gamma} = \mathbf{b}, \\ \gamma_k \ge 0 \quad \text{if} \quad (i_k, j_k) \ne (0,0), \\ \gamma_k < 0 \quad \text{if} \quad (i_k, j_k) = (0,0), \end{cases}$$

where

$$\begin{aligned} g_k &= \beta^{\pm}, && (i_k, j_k) \in \{(-1,0), (1,0), (0,-1), (0,1)\}, \\ g_k &= -4, && (i_k, j_k) = (0,0), \\ g_k &= 0, && \text{otherwise}, \end{aligned}$$

and $A\boldsymbol{\gamma} = \mathbf{b}$ is the following system of equations from (3.20):

$$\begin{aligned} &a_1 + a_2 = 0, && a_3 + \rho a_4 = 0, && a_5 + a_6 = 0, \\ &a_7 + a_8\,\rho = \beta^-, && a_9 + a_{10} + a_8\,(\rho - 1) = \beta^-, && a_{11} + a_{12}\,\rho = 0. \end{aligned}$$

To solve the above constrained optimization problem numerically, we use a uniform grid on the unit square $0 \le r, \theta \le 1$,

$$\begin{aligned} r_i &= i\,\Delta r, && \Delta r = \frac{1}{M}, && i = 0, 1, \dots, M, \\ \theta_j &= j\,\Delta\theta, && \Delta\theta = \frac{\pi}{2N}, && j = 0, 1, \dots, N. \end{aligned}$$

We also choose a discrete set of the jump ratio $\beta^+/\beta^- = \rho$,

$$\rho_k = 10^{-N_1 + k\,\Delta\rho}, \qquad \Delta\rho = \frac{N_2 + N_1}{L}, \qquad k = 0, 1, \dots, L,$$

to verify the conjecture. The orthogonal projection of the origin on the interface then is $x_i^* = r_i\,\cos\theta_j$, $y_j^* = r_i\,\sin\theta_j$ excluding those $y_j^* > 1 - x_i^*$ that are outside of the 5-point stencil. We also define

$$\gamma_{max}(\rho) = \max_{r_i,\theta_j}\max_{1\le k\le 9}\frac{|\gamma_k|}{\beta^-}, \qquad \gamma_{min}(\rho) = \min_{r_i,\theta_j}\min_{1\le k\le 9}\frac{|\gamma_k|}{\beta^-},$$

$$S_{min}(\rho) = \min_{\substack{r_i,\theta_j,\max\{\xi_k\}\ge C_3 h \\ \xi_k \ge 0}} \sum_{\xi_k \ge 0} \xi_k \gamma_k.$$

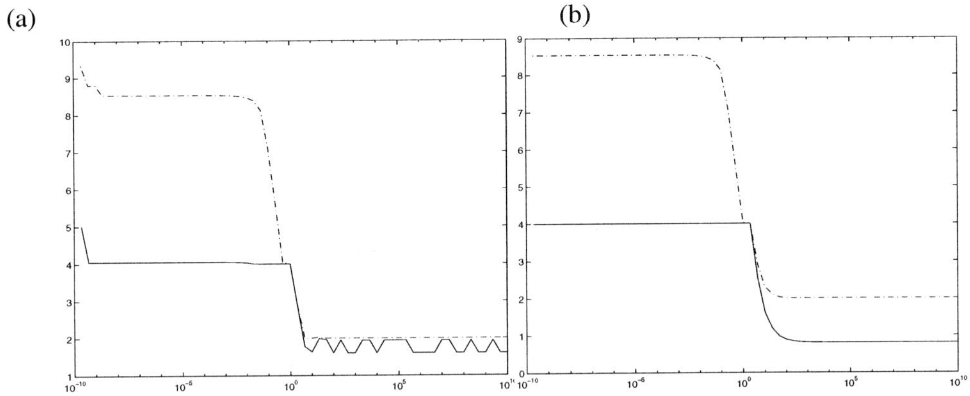

Figure 3.1. *The computed $\gamma_{max}(\rho)$ and $\gamma_{min}(\rho)$ with $M = N = L = 60$, $N_1 = 9$, and $N_2 = 10$. (a) $\beta^- = 1$, $\beta^+ = 1/\rho$. (b) $\beta^+ = 1$, $\beta^- = \rho$. The x-axis is between 10^{-10} and 10^{+10}; the y-axis is about between 0 and 10.*

The numerical tests show that the solution to the optimization problem always exists. Figures 3.1 and 3.2 summarize the numerical verification results for Conjecture 3.1. In Figures 3.1(a) and (b), the dashed line is $\gamma_{max}(\rho)$ and it is bounded by $|\gamma_k|\,h^2 \le 10$; the solid line is $\gamma_{min}(\rho)$ and it is bounded by $|\gamma_5|\,h^2 \ge 1$. If $\beta^- = \beta^+$, we have $\gamma_5 h^2 = 4$ exactly as we can see from Figure 3.1. Figures 3.1(a) and (b) confirm the inequalities (3.30) and (3.31).

In Figure 3.2, we plot $h S_{min}(\rho)/C_2$ for the case $C_3 = 1$ and

$$C_2 = \frac{\max\left\{\beta^-/\beta^+,\, \beta^+/\beta^-\right\}}{\max\{1,\ \beta^-\}\ \max\{1,\ \beta^+\}}. \tag{3.35}$$

This constant was found by numerical experiments as well. The minimum of $\sum_{\xi_k \ge 0} \gamma_k \xi_k$ is taken from all the cases except for the point $(1, 0)$ where the interface is actually $x = 1$. In this case, the grid point touches the interface and the finite difference scheme is the standard centered finite scheme with 5-point stencil plus a possible nonzero correction C_{ij} for the jump in the solution and the flux. In Figures 3.2(a) and (b), we have

$$\sum_{\xi_k \ge 0} \gamma_k \xi_k \ge \frac{0.01\, C_2}{h} \quad \text{if} \quad \sqrt{\sum_{\xi_k \ge 0} \xi_k^2} \ge Ch.$$

Thus the numerical verification confirms inequalities (3.30)–(3.32). We have tried different grid sizes and all the results showed the same conclusions. The ratio ρ of the jump in β ranges from 10^{-9} to 10^{10}, which should cover most applications.

The complete theoretical proof of the conjecture is difficult although we are able to prove the conjecture for special values of ρ, for example, $\rho \ge 1$.

3.6.2 The proof of the convergence of the finite difference scheme

The following lemma, which is a generalization of Theorems 6.1 and 6.2 of Morton and Mayer [202] for multiple subregions J_i, is used to prove the convergence of the maximum preserving IIM.

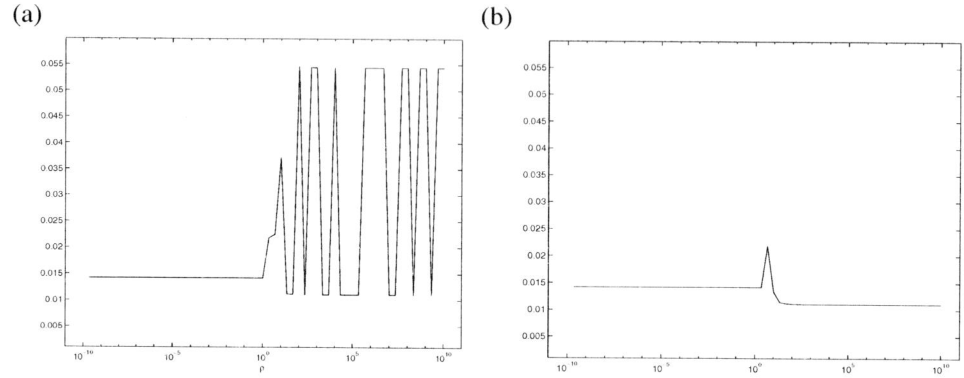

Figure 3.2. *The computed $S_{min}(\rho)$ with $M = N = 60 = L = 60$, $N_1 = 9$, and $N_2 = 10$. (a) $\beta^- = 1$, $\beta^+ = 1/\rho$. (b) $\beta^+ = 1$, $\beta^- = \rho$. The x-axis is between 10^{-10} and 10^{+10}; the y-axis is about between 0 and 0.06.*

Lemma 3.2. *Given a finite difference scheme L_h defined on a discrete set of interior points J_Ω for an elliptic PDE with a Dirichlet boundary condition, assume that the following conditions hold.*

1. *J_Ω can be partitioned into a number of disjoint regions,*

$$J_\Omega = J_1 \cup J_2 \cup J_3 \cdots \cup J_{N_\Omega}, \qquad J_i \cap J_k = \emptyset \quad \text{if } i \neq k. \tag{3.36}$$

2. *The truncation error of the finite difference scheme at a grid point p satisfies*

$$|T_p| \leq T_i \quad \forall p \in J_i, \qquad i = 1, 2, \ldots, N_\Omega. \tag{3.37}$$

3. *There exists a nonnegative mesh function ϕ defined on $\cup_{i=1}^{s} J_i$ satisfying*

$$L_h \phi_p \geq C_i > 0 \quad \forall p \in J_i, \qquad i = 1, 2, \ldots, N_\Omega. \tag{3.38}$$

Then the global error of the approximate solution $\{U_{ij}\}$ from the finite difference scheme at the mesh points is bounded by

$$||E_h||_\infty \leq \left(\max_{A \in J_{\partial\Omega}} \phi_A \right) \max_{1 \leq i \leq N_\Omega} \left\{ \frac{T_i}{C_i} \right\}, \tag{3.39}$$

where E_h is the difference between the exact solution of the differential equation and the approximate solution of the finite difference equations at the mesh points, and $J_{\partial\Omega}$ is the set that contains the boundary points.

The proof of this lemma is trivial and is omitted. Using the lemma above, we can prove the following error estimate for the maximum principle preserving scheme.

Theorem 3.3. *Let $u(x, y)$ be the exact solution to* (3.1) *and* (3.2a)–(3.2b) *with $\sigma \geq 0$ and a Dirichlet boundary condition. Assume the following:*

(1) *the optimization problem* (3.25)–(3.26), *with the constraints* (3.20) *using the standard compact* 9-*point stencil, has a solution* $\{\gamma_k\}$ *at every irregular grid point;*

(2) *the solution* $u(x, y)$ *has up to third-order piecewise continuous partial derivatives;*

(3) *the mesh spacing* h *is sufficiently small;*

(4) *the following inequalities are true:*

$$|\gamma_k| \le \frac{C_1}{h^2} \quad \textit{and} \quad \sum_{\xi_k \ge 0} \gamma_k \xi_k \ge \frac{C_2}{h}. \tag{3.40}$$

Then we have the following error estimate for $\{U_{ij}\}$*, the solution of the finite difference scheme obtained from the maximum principle IIM*

$$\|u(x_i, y_j) - U_{ij}\|_\infty \le Ch^2, \tag{3.41}$$

where the constant C *depends on the underlying grid and interface, as well as on* u, f, *and* β.

Proof: Consider the solution to the following interface problem:

$$\begin{aligned} \nabla \cdot \beta \nabla \phi = 1, \\ [\phi] = 0, \quad [\beta \phi_{\boldsymbol{n}}] = 1, \quad \phi_{\partial\Omega} = 1. \end{aligned} \tag{3.42}$$

From the results in [11, 44], we know that the solution ϕ exists, and it is unique and piecewise continuous. Therefore the solution is also bounded. Let

$$\bar{\phi}(x, y) = \phi(x, y) + \left| \min_{(x,y)\in\Omega} \phi(x, y) \right|. \tag{3.43}$$

Note that the second term in the right-hand side is a constant. If (3.40) is true, then we know that

$$L_h\bar{\phi}(x_i, y_j) \ge \begin{cases} 1 + O(h^2) & \text{if } (x_i, y_j) \text{ is a regular grid point,} \\ \displaystyle\sum_{\xi_k \ge 0} \gamma_k \xi_k \ge \frac{C_2}{h} + O(1) & \text{if } (x_i, y_j) \text{ is an irregular grid point.} \end{cases}$$

Note that the second inequality above is due to the jump in the flux in $\bar{\phi}$ and, at some irregular grid points, $L_h\bar{\phi}(x_i, y_j)$ can be very large, but it is nonnegative. Thus, the first inequality above still holds. At regular grid points, we have

$$\frac{|T_{ij}|}{L_h\bar{\phi}(x_i, y_j)} \le \frac{C_3 h^2}{1}.$$

At irregular grid points, where (3.40) is satisfied, we have

$$\frac{|T_{ij}|}{L_h\bar{\phi}(x_i, y_j)} \le \frac{C_4 h}{C_2/h} = \frac{C_4}{C_2} h^2$$

since the local truncation errors at irregular grid points are bounded by

$$|T_{ij}| \le C_4 h$$

for some constant C_4 that depends on the second derivatives of the solution on each side of the interface. Thus, from Lemma 3.2, we have proved second-order convergence. □

Remark 3.1. *The key to the convergence theorem is that the solution to the optimization problem exists and the inequalities in* (3.40) *hold, which have been numerically verified. The second condition in* (3.40) *may be violated when the interface is very close to a grid point, other than* (x_i, y_j), *involved in the finite difference stencil. In this case, the finite difference scheme is actually very close to the standard finite difference scheme using the 5-point stencil with a correction term on the right-hand side. Therefore second-order convergence is still true. This fact can be stated in the following theorem (the proof of the theorem is given in* [166]).

Theorem 3.4. *If conditions* (1)–(3) *in Theorem* 3.3 *are satisfied and either* (3.40) *or*

$$\max_{\xi_k \ge 0}\{\xi_k\} \le C_5 h^{1+\varepsilon}, \quad \text{with } C_5 > 0 \text{ and } \varepsilon > 0, \tag{3.44}$$

is true, then

$$\|u(x_i, y_j) - U_{ij}\|_\infty \le C h^2, \tag{3.45}$$

where the constants depend on the underlying grid, the interface, u, f, *and* β.

Note that we can choose the constants C_2 and C_5 so that one of the conditions in (3.40) and (3.44) is true. Even if they are both violated at a few grid points, the errors from these points are $O(h^2 \log h)$ and the global accuracy is still almost second order.

3.7 Some numerical examples for two-dimensional elliptic interface problems

We show some numerical experiments using the maximum principle preserving IIM for two-dimensional elliptic interface problems. The results agree with the analysis in §3.6.2. The linear system of equations is solved using the multigrid method DMGD9V developed by De Zeeuw [62]. The interface is a closed curve in the solution domain. More examples can be found in [154, 160, 166].

Example 3.1. In this example, the interface is the circle $x^2 + y^2 = 1/4$ within the domain $-1 \le x, y \le 1$. The equations are

$$(\beta u_x)_x + (\beta u_y)_y = f(\mathbf{x}) + C \int_\Gamma \delta(\mathbf{x} - \mathbf{X}(s))\, ds \tag{3.46}$$

with $\quad f(\mathbf{x}) = 8\,(x^2 + y^2) + 4 \quad$ and

$$\beta(\mathbf{x}) = \begin{cases} x^2 + y^2 + 1 & \text{if } x^2 + y^2 \le \dfrac{1}{4}, \\[2mm] b & \text{if } x^2 + y^2 > \dfrac{1}{4}. \end{cases}$$

Table 3.1. *A grid refinement analysis of the maximum principle preserving scheme for Example* 3.1 *with* $b = 10$, $C = 0.1$, *and* $N_{coarse} = 6$. *Average second-order convergence is confirmed.*

N_{finest}	N_b	n_l	$\| E_N \|_\infty$	Order
42	40	4	$4.8638e\ 10^{-4}$	
82	80	5	$1.4476e\ 10^{-4}$	1.7484
162	160	6	$3.0120\ 10^{-5}$	2.2649
322	320	7	$8.2255\ 10^{-6}$	1.8726
642	640	8	$2.0599\ 10^{-6}$	1.9975

Table 3.2. *A grid refinement analysis of the maximum principle preserving scheme for Example* 3.1 *with* $N_{coarse} = 9$. *Second-order convergence is confirmed.*

			$b = 1000,\ C = 0.1$		$b = 0.001,\ C = 0.1$	
N_{finest}	N_b	n_l	$\| E_N \|_\infty$	Order	$\| E_N \|_\infty$	Order
34	40	3	$5.1361\ 10^{-4}$		9.3464	
66	80	4	$8.2345\ 10^{-5}$	2.7598	2.0055	2.3204
130	160	5	$1.8687\ 10^{-5}$	2.1878	$5.8084\ 10^{-1}$	1.8280
258	320	6	$4.0264\ 10^{-6}$	2.2394	$1.3741\ 10^{-1}$	2.1031
514	640	7	$9.430\ 10^{-7}$	2.1059	$3.5800\ 10^{-2}$	1.9514

The Dirichlet boundary condition is determined from the exact solution

$$u(\mathbf{x}) = \begin{cases} r^2 & \text{if } r \le \dfrac{1}{2}, \\ \left(1 - \frac{1}{8b} - \frac{1}{b}\right)/4 + \left(\frac{r^4}{2} + r^2\right)/b + C\,\log(2r)/b & \text{if } r > \dfrac{1}{2}, \end{cases} \tag{3.47}$$

where $r = \sqrt{x^2 + y^2}$. In this example, we have a *variable* and discontinuous coefficient $\beta(\mathbf{x})$. Table 3.1 (with modest jump ratio in β) and Table 3.2 (with large jump ratio in β) show the results of a grid refinement analysis for different choices of b and C. The maximum error over all grid points,

$$\| E_N \|_\infty = \max_{i,j} \mid u(x_i, y_j) - U_{ij} \mid, \tag{3.48}$$

is presented. The order of convergence is computed from

$$\text{order} = \left| \frac{\log\left(\| E_{N_1} \|_\infty / \| E_{N_2} \|_\infty\right)}{\log(N_1/N_2)} \right|, \tag{3.49}$$

Table 3.3. *A grid refinement analysis of maximum principle preserving scheme for Example 3.2 with $N_{coarse} = 9$. Average second-order convergence is confirmed.*

			$\beta^+ = 1000,\ \beta^- = 1$		$\beta^+ = 1,\ \beta^- = 1000$	
N_{finest}	N_b	n_l	$\Vert E_N \Vert_\infty$	Order	$\Vert E_N \Vert_\infty$	Order
34	40	3	$1.8322\ 10^{-1}$		$8.0733\ 10^{-3}$	
66	80	4	$3.5224\ 10^{-3}$	5.9574	$3.0371\ 10^{-3}$	1.4739
130	160	5	$4.5814\ 10^{-5}$	3.0090	$7.1981\ 10^{-4}$	2.1238
258	320	6	$1.4240\ 10^{-5}$	1.7049	$1.6876\ 10^{-4}$	2.1162
514	640	7	$3.1501\ 10^{-6}$	2.1887	$2.7407\ 10^{-5}$	2.6371

which is the solution of the equation

$$\Vert E_N \Vert_\infty = C\, h^{\text{order}} \tag{3.50}$$

with two different N's (h's).

In the tables of this section, N_b is the number of roughly equally spaced control points used to represent the interface Γ; N_{coarse} and N_{finest} are the number of the coarsest and finest grid lines, respectively, when the multigrid solver DMGD9V is used; and n_l is the number of levels used for the multigrid method.

As explained in §6.1.6 (see also [163]), for interface problems, the errors usually do not decline monotonously. Instead the error depends on the relative location of the underlying grid and the interface. Nevertheless, the average of the convergence order approaches 2 in Tables 3.1 and 3.2. Compared with the results in [154] using a 6-point stencil, the maximum principle preserving scheme gives a slightly better result. Notice that as the parameter b becomes smaller, both the solution and its gradient in the outside of the interface become larger in magnitude and the problem becomes harder to solve. But the maximum principle preserving scheme still converges quadratically.

Example 3.2. In this example, the coefficient β is a piecewise constant and $\sigma = 0$. The PDE is $\nabla \cdot (\beta \nabla u) = f$. The jumps $[u]$ in the solution, $[\beta u_n]$ in the flux, and $[f]$ in the source term are determined from the exact solution,

$$u(x, y) = \begin{cases} x^2 - y^2 & \text{if } x^2 + 4y^2 \le 1, \\ \sin(x)\cos(y) & \text{if } x^2 + 4y^2 > 1. \end{cases} \tag{3.51}$$

Unlike in Example 3.1, the solution in this example is *discontinuous*. Table 3.3 shows the results of a grid refinement analysis using the maximum principle preserving scheme. Again we see clearly second-order convergence. Figure 3.3(a) is a plot of the solution which is composed of two pieces.

The finite difference scheme using a 6-point stencil is straightforward and easier to implement. However, we have neither an estimate of the eigenvalues of the coefficient matrix nor the condition number. With the maximum principle preserving scheme, the coefficient matrix is an M-matrix and diagonally dominant. As a result, standard iterative

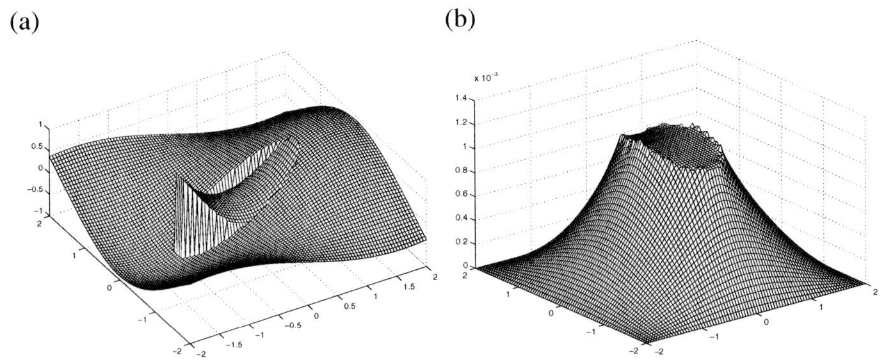

Figure 3.3. (a) *The solution of Example* 3.2 *with jumps in the solution as well as in the normal derivative. The parameters are* $\beta^+ = 1$, $\beta^- = 100$, *and* $N_{finest} = 82$. (b) *The error plot with the same parameters. The error distribution is better than that obtained from the 6-point IMM.*

methods such as an SOR or the multigrid DMGD9V method are guaranteed to converge. Furthermore, the errors of the solution obtained from the maximum principle preserving scheme are usually more evenly distributed; see, for example, Figure 3.3(b).

3.8 Algorithm efficiency analysis

A natural concern about the maximum principle preserving scheme is how much extra cost is needed in solving the quadratic optimization problem at each irregular grid point. In Figure 3.4(a), we plot the percentage of the CPU time used in the interface treatment versus the ratio of the jump in the coefficients $\log(\beta^+/\beta^-)$. The interface in polar coordinates is

$$r = r(\theta) = \frac{1}{2} + 0.1\,\sin(5\theta), \quad 0 \le \theta \le 2\pi. \tag{3.52}$$

For this interface, the curvature is quite large; see Figure 3.4(b). The cost for dealing with irregular grid points includes solving the quadratic optimization problem, indexing grid points, and finding the orthogonal projections (x_i^*, y_j^*).

For regular problems, the multigrid solver DMGD9V is comparable to a fast Poisson solver using an FFT.[4] The CPU time for the multigrid method, however, does depend on the jump in β. The dependence on β can be reduced and even eliminated by using better multigrid methods, as described in the next section. In almost all the numerical tests, the cost of the IIM in dealing with irregular grid points is less than 10% of the total CPU time. The percentage decreases as the mesh become finer. When $\beta^- = \beta^+$, the finite difference coefficients become the standard 5-point stencil scheme and the cost for the interface treatment reaches its minimum.

[4]Generally the fast Poisson solver using FFT can be used only for constant coefficients.

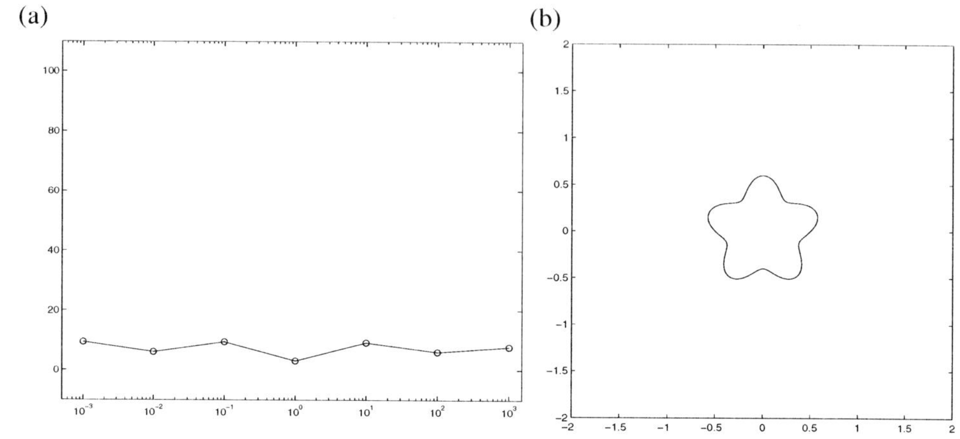

Figure 3.4. (a) *A plot of the percentage of the CPU time used for dealing with interfaces versus* $\log(\beta^+/\beta^-)$. *The axes are about* $[\,10^{-3}, 10^3\,] \times [\,0, 100\,]$. (b) *The domain of the test example on the square is* $[\,-2, 2\,] \times [\,-2, 2\,]$.

Table 3.4. *The CPU time for Example* 3.2 *with different parameters using an IBM SP2 machine. The outputs vary with machines.*

N_{finest}	N_b	n_l	N_{coarse}	β^-	β^+	CPU time (s)
130×130	160	5	9	10	1	0.03
258×258	320	6	9	1	1	0.03
258×258	320	6	9	1	100	0.05
258×258	320	6	9	1	10000	0.06
258×258	320	6	9	100	1	0.06
258×258	320	6	9	10000	1	3.29
514×514	640	7	9	1	1000	0.15
514×514	640	7	9	1000	1	0.35

The CPU time used in the entire solution process depends on the geometry and the jump in the coefficient β. Table 3.4 lists some statistics for Example 3.2 on an IBM SP2 machine. In this example, when $\beta^- < \beta^+$, the CPU time is just a little more than that needed for one fast Poisson solver. When $\rho = \beta^-/\beta^+ > 1$ gets bigger, we see the CPU time grows slowly. Note that the DMGD9V may fail if $\max\{\beta^-/\beta^+,\ \beta^+/\beta^-\}$ is very large, say 10^6 for Example 3.1.

Remark 3.2. *The linear system of equations using the maximum principle preserving scheme is irreducible and diagonally dominant. The multigrid solver DMGD9V is designed for a system of equations with a standard centered compact 9-point stencil. The method requires the system of equations to have positive/negative symmetric parts and it works well for problems with large variation in the coefficients. So it is natural to use DMGD9V. However, we do observe occasionally that the multigrid stops before it returns a convergent result. While fine tuning of the parameters of the multigrid method may make it work* *better, the multigrid methods described in the next section provide better alternatives for the linear system of finite difference equations obtained from the maximum principle preserving scheme.*

3.9 Multigrid solvers for large jump ratios

The problems encountered with DMGD9V with small values of $\beta^+ = b$ for Example 3.1, that is, $\rho_{cond} = \max\{\beta^-/\beta^+, \beta^+/\beta^-\}$ is large, can be overcome by using either of the two multigrid methods described in Adams and Chartier [3, 4].

The first method is the standard algebraic multigrid (AMG) method described in [227] and implemented as AMG1R6, version date 1997, by Ruge, Stüben, and Hempel. AMG is a black-box solver for the linear system $A^h u^h = f^h$. AMG uses the finest-grid matrix A^h to automatically deduce which equations are considered coarse grid equations at the next level, and to determine the interpolation operator P and the restriction operator R. The coarse grid system $A^{2h} e^{2h} = r^{2h}$ is determined by the Galerkin choice $A^{2h} = RA^h P$, and r^{2h} is computed as $r^{2h} = Rr^h$, where r^h is the residual based on an approximation of u^h on the fine grid. Once e^{2h} is found, the approximation of u^h is updated by the error approximation Pe^{2h}. The process is applied recursively. Since AMG automatically determines the coarse grid equations, the stencil can become more dense than the original 9-point stencil of the finest grid. The results in [4] show that for this test problem, AMG does not coarsen uniformly near the interface—more points near the interface are kept in the coarser grids. This leads to more computational expense than that used in the second method described below. Since the maximum principle preserving IIM described in this chapter was used to produce the original matrix A^h, it is an M-matrix. The AMG does not necessarily preserve this property on the coarser grids. AMG does, however, take note of positive off-diagonal elements when they occur and does an additional pass to re-examine its choices for coarse grid equations. But even for the smallest values of b, it was reported in [4] that less than .05% of the nonzero elements in the coarse grid matrices became positive, and that convergence was not adversely affected.

The second method, the geometric multigrid method, described in Adams and Chartier [3, 4], gave notable improvements over the version published by Adams and Li in [6]. The purpose is to develop a multigrid method explicitly for interface problems by including knowledge, such as the jump conditions in §3.1, at the interface. Furthermore, the hope was that with such knowledge, one could use simple standard coarsening (the coarse grid is taken to be every other grid point in the coordinate directions) and simple Gauss–Seidel relaxation, and could maintain the simplicity of the 9-point stencil for the coarser grid matrices. Like AMG, the coarser grid matrices would be determined in the Galerkin fashion with $A^{2h} = RA^h P$. The choices for R and P that Adams and Chartier proposed in [3, 4] performed very well for Example 3.1, and are briefly described below.

Let the matrix A^h be partitioned as

$$\begin{pmatrix} A_{cc} & A_{cf} \\ A_{fc} & A_{ff} \end{pmatrix}, \tag{3.53}$$

where A_{cc} contains the connections of the next coarser grid unknowns to each other, and A_{cf} contains the connections of these unknowns to the fine grid unknowns (that are not also coarse unknowns). Likewise, A_{ff} contains the connections of the fine grid unknowns (that are not also coarse unknowns) to each other and A_{fc} contains the connections of these unknowns to the coarser grid unknowns. The restriction operator that restricts the residual r^h on grid h to grid $2h$ is taken to be $R = -A_{cf} D_{ff}^{-1}$ where D_{ff} is the diagonal of A_{ff}.

The interpolation operator P is chosen differently for grid points on the fine grid that do not have any connections to the immersed interface (regular points) and for those that have connections to both sides of the immersed interface (irregular points). Since on every grid, the coefficient matrix is described by a 9-point stencil, the error residual equation at the center point (point 2) can be written as

$$\gamma_1 e_1 + \gamma_2 e_2 + \gamma_3 e_3 + \gamma_4 e_4 + \gamma_5 e_5 + \gamma_6 e_6 + \gamma_7 e_7 + \gamma_8 e_8 + \gamma_9 e_9 = r_2 \approx 0, \tag{3.54}$$

where e_1 and e_3 are the errors to the west and east, e_4 and e_5 are those to the north and south, and e_6, e_7, e_8, and e_9 are those to the southwest, southeast, northwest, and northeast, respectively. At regular points, the interpolated value for the error, e_2, at the midpoint of a vertical edge is expressed in terms of the coarse grid values e_4 and e_5 as

$$e_2 = c_4 e_4 + c_5 e_5, \tag{3.55}$$

where

$$c_4 = -(\gamma_6 + \gamma_4 + \gamma_7)/(\gamma_1 + \gamma_2 + \gamma_3) \tag{3.56}$$

and

$$c_5 = -(\gamma_8 + \gamma_5 + \gamma_9)/(\gamma_1 + \gamma_2 + \gamma_3). \tag{3.57}$$

Likewise, the interpolation formula for the error at the center of a horizontal edge in terms of the coarse grid values on either side is given by

$$e_2 = c_1 e_1 + c_3 e_3, \tag{3.58}$$

where

$$c_1 = -(\gamma_6 + \gamma_1 + \gamma_8)/(\gamma_4 + \gamma_2 + \gamma_5) \tag{3.59}$$

and

$$c_3 = -(\gamma_7 + \gamma_3 + \gamma_9)/(\gamma_4 + \gamma_2 + \gamma_5). \tag{3.60}$$

The values at the coarse grid points are simply copied from the coarse grid. Once the values at the centers of vertical and horizontal edges are computed, (3.54) is used to interpolate the center points of the coarse grid cells.

This is the same operator-induced interpolation used by many authors (see, for example [121, 32]). One can think of deriving them by using the Taylor approximation to e_8 and e_9 in terms of e_5, to e_6 and e_7 in terms of e_4, and to e_1 and e_3 in terms of e_2 for the vertical edge interpolation. Similar Taylor expansions are used to derive the formulas for centers of horizontal edges. However, these formulas are not accurate, when there is an interface cutting through the stencil, because the derivatives are not continuous and may be highly varying. For irregular grid points, we interpolate as described below.

First, we note that for this test problem, w and v are all zero. We make some assumptions about the error e after the prerelaxation step. We assume that the jump in the error at the interface $[e] = 0$, the jump in the flux of the error at the interface $[\beta e_{\mathbf{n}}] = 0$, and the jump in the error in the tangential direction at the interface $[e_\eta] = 0$. We also assume that the error at the interface varies more in the normal direction ξ than in the tangential direction η. These assumptions allow for large jumps in the error in the normal direction. Using similar ideas from the derivation of the IIM in the first part of this chapter, we can develop an interpolation formula for the center of a vertical edge e_2 in terms of the coarse grid errors e_4 and e_5 even when an interface cuts through this edge. We simply expand all three errors in terms of the error e^- where the interface cuts the vertical edge on the "−" side of the interface using the jump conditions. This gives

$$e^- + \xi_2\rho_2 e_\xi^- + \eta_2 e_\eta^- = c_4(e^- + \xi_4\rho_4 e_\xi^- + \eta_4 e_\eta^-) + c_5(e^- + \xi_5\rho_5 e_\xi^- + \eta_5 e_\eta^-) \tag{3.61}$$

to $O(h^2)$, where $\rho_i = 1$ if the grid point is inside or on the interface, and $\rho_i = \frac{\beta^-}{\beta^+}$ if the grid point is outside the interface. A similar equation holds for horizontal midpoints with the subscripts 4 and 5 replaced by 1 and 3.

In (3.61), we would like to be able to set the coefficients of e^-, e_ξ^-, and e_η^- to zero. This would give three equations, but we only have the two unknowns (c_4 and c_5 for vertical midpoints, or c_1 and c_3 for horizontal midpoints). Before going further, we write these three equations as

$$\begin{aligned} e^- &: c_4 + c_5 = 1, \\ e_\xi^- &: c_4\xi_4\rho_4 + c_5\xi_5\rho_5 = \xi_2\rho_2, \\ e_\eta^- &: c_4\eta_4 + c_5\eta_5 = \eta_2. \end{aligned}$$

For midpoints of both vertical and horizontal edges, we use the first and second equations to force the coefficients of e^- and e_ξ^- to vanish. The rationale is that we are assuming that the error will vary the most in the normal direction. Hence, since we cannot enforce all three equations, we hope that the change in the error in the tangential direction will be small compared with that in the normal direction. This gives the following values for c_4 and c_5:

$$c_5 = \frac{\rho_4(y^* - y_4) + \rho_2(y_2 - y^*)}{\rho_4(y^* - y_4) + \rho_5(y_5 - y^*)}, \qquad c_4 = 1 - c_5. \tag{3.62}$$

The values obtained for c_1 and c_3 are

$$c_1 = \frac{\rho_3(x^* - x_3) + \rho_2(x_2 - x^*)}{\rho_3(x^* - x_3) + \rho_1(x_1 - x^*)}, \qquad c_3 = 1 - c_1. \tag{3.63}$$

The only remaining issue is how to interpolate centers of the cells. Since we now have a formula for the cell corners (copy the coarse value) and the vertical and horizontal midpoints, we can solve (3.54) to find the value of e_2 for cell centers. This is the same strategy adopted for cell centers for regular points. Hence, the overall scheme takes advantage of the interface information as well as that of the PDE operator.

As mentioned earlier, A^h on the finest grid is an M-matrix. It is interesting to note that unlike AMG, the New method has M-matrices on both the finest and all coarser grids. This was the main reason reported in [3] for its superior performance compared to the Adams–Li method in [6].

Table 3.5 shows the results of a grid refinement analysis for these two methods applied to Example 3.1 when $b = 0.005$, 0.0005, and 0.00005 (that is, $\rho_{cond} = \max\{\beta^-/\beta^+,$

Table 3.5. *AMG and new comparisons for Example* 3.1.

$b = 0.005$		AMG				New			
n_{finest}	N_b	n_l	V's	*Rate*	f	n_l	V's	Rate	f
128	160	6	11	0.128	0.240	5	11	0.129	0.314
256	320	7	12	0.145	0.275	6	10	0.108	0.320
512	640	9	12	0.142	0.282	7	10	0.100	0.330
$b = 0.0005$		AMG				New			
128	160	6	14	0.170	0.285	5	13	0.152	0.330
256	320	8	14	0.168	0.288	6	13	0.147	0.350
512	640	9	14	0.167	0.304	7	13	0.150	0.366
$b = 0.00005$		AMG				New			
256	320	8	16	0.187	0.301	6	15	0.165	0.350
512	640	9	15	0.164	0.289	7	15	0.160	0.367

$\beta^+/\beta^-\} = 250$, 2500, and 25000, respectively). Here, n_l is the number of points taken to describe the interface, V is the number of V-cycles, and Rate is the average residual reduction factor across all V-cycles, and f is the reduction factor of the residual in the last iteration. Both methods were stopped when the residual scaled by the diagonal of A^h was less than 10^{-6}. The immersed interface multigrid method is referred to as New in the table.

Table 3.5 shows that both the New and AMG methods require a constant number of V-cycles as the problem size increases for a fixed value of b. The New method has a better average convergence rate, whereas AMG has a better reduction factor, f, in the last iteration. In all cases the New method was found to be slightly more efficient in terms of the accuracy produced per unit of work. For both methods, the number of V-cycles required only a slight increase (logarithmically) as b decreased.

More descriptions and comparisons of these methods can be found in [3, 4]. Also, preliminary results by Adams and Wiest (private communication) show that for both these methods, this dependence on b disappears if the methods are used as preconditioners for GMRES. In this context, AMG appears to have an advantage over the immersed interface multigrid method.

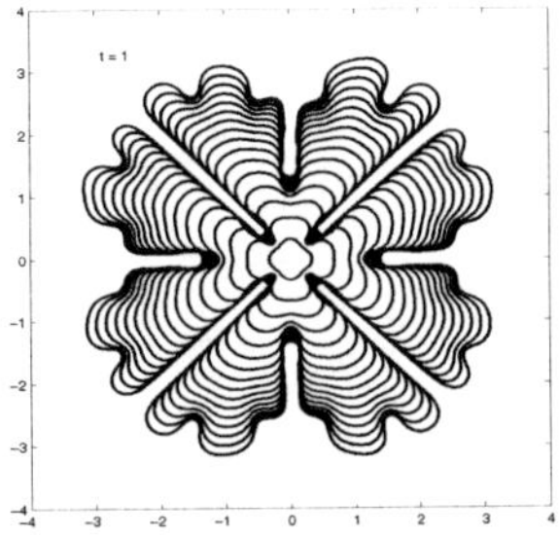

Chapter 4

The IIM for Three-Dimensional Elliptic Interface Problems

In this chapter, we discuss the IIM for the elliptic interface problem of the form

$$\nabla \cdot (\beta \nabla u) - \sigma u = f, \qquad (x, y, z) \in \Omega \setminus \Gamma \tag{4.1}$$

in three dimensions in a region Ω with a boundary condition on $\partial\Omega$, where all the coefficients β, σ, and f may be discontinuous across the interface Γ, which is a surface S: $x = x(s_1, s_2)$, $y = y(s_1, s_2)$, $z = z(s_1, s_2)$. To make the problem well-posed, we assume that we have the knowledge of the jump conditions in the solution and the flux,

$$[u] = w(s_1, s_2), \qquad [\beta u_n] = v(s_1, s_2), \tag{4.2}$$

where w and v are two known functions defined only on the interface Γ.

The IIM using a 10-point finite difference stencil was developed in [160, 161]. The maximum principle preserving scheme for three-dimensional elliptic interface problems was developed in [64, 65] and will be explained in this chapter. The idea and methodology are similar to that for two-dimensional problems. But there are some substantial differences and difficulties for three-dimensional problems. The finite difference schemes require computing the surface derivatives of the jump conditions. In Chapter 5, we will explain a strategy that can transform the interface problem into a new one with homogeneous jump conditions using a level set function.

Once again, we first discuss some theoretical issues for the elliptic interface problems in three dimensions.

4.1 A local coordinate system in three dimensions

Given a point (X, Y, Z) on the interface Γ, let $\boldsymbol{\xi}$ (with $\|\boldsymbol{\xi}\| = 1$) be the normal direction of Γ pointing to a specific side, say the "+" side; let $\boldsymbol{\eta}$ and $\boldsymbol{\tau}$ be two orthogonal unit vectors tangential to Γ; then a local coordinate transformation is defined as

$$\begin{cases} \xi = (x - X)\,\alpha_{x\xi} + (y - Y)\,\alpha_{y\xi} + (z - Z)\,\alpha_{z\xi}, \\ \eta = (x - X)\,\alpha_{x\eta} + (y - Y)\,\alpha_{y\eta} + (z - Z)\,\alpha_{z\eta}, \\ \tau = (x - X)\,\alpha_{x\tau} + (y - Y)\,\alpha_{y\tau} + (z - Z)\,\alpha_{z\tau}, \end{cases} \tag{4.3}$$

where $\alpha_{x\xi}$ represents the directional cosine between the x-axis and $\boldsymbol{\xi}$, and so forth. Note that the choice of the two orthogonal tangential vectors is not unique.

The three-dimensional coordinates transformation above can also be written in a matrix-vector form. Define the local transformation matrix as

$$A = \begin{pmatrix} \alpha_{x\xi} & \alpha_{y\xi} & \alpha_{z\xi} \\ \alpha_{x\eta} & \alpha_{y\eta} & \alpha_{z\eta} \\ \alpha_{x\tau} & \alpha_{y\tau} & \alpha_{z\tau} \end{pmatrix}; \tag{4.4}$$

then we have

$$\begin{pmatrix} \xi \\ \eta \\ \tau \end{pmatrix} = A \begin{pmatrix} x - X \\ y - Y \\ z - Z \end{pmatrix}. \tag{4.5}$$

Also, for any differentiable function $p(x, y, z)$, we have

$$\begin{pmatrix} \bar{p}_\xi \\ \bar{p}_\eta \\ \bar{p}_\tau \end{pmatrix} = A \begin{pmatrix} p_x \\ p_y \\ p_z \end{pmatrix}, \tag{4.6}$$

$$\begin{pmatrix} \bar{p}_{\xi\xi} & \bar{p}_{\xi\eta} & \bar{p}_{\xi\tau} \\ \bar{p}_{\eta\xi} & \bar{p}_{\eta\eta} & \bar{p}_{\eta\tau} \\ \bar{p}_{\tau\xi} & \bar{p}_{\tau\eta} & \bar{p}_{\tau\tau} \end{pmatrix} = A \begin{pmatrix} p_{xx} & p_{xy} & p_{xz} \\ p_{yx} & p_{yy} & p_{yz} \\ p_{zx} & p_{zy} & p_{zz} \end{pmatrix} A^T, \tag{4.7}$$

where $\bar{p}(\xi, \eta, \tau) = p(x, y, z)$ and A^T is the transpose of A. It is easy to verify that $A^T A = I$, the identity matrix. Note that under the local coordinates transformation (4.3), the PDE (4.1) is invariant. Therefore we will drop the bars for simplicity.

4.2 Interface relations for three-dimensional elliptic interface problems

Using the superscript "+" or "−" to denote the limiting values of a function from the Ω^+ side or the Ω^- side of the interface, respectively, we can write the limiting differential equation from the "−" side as

$$\beta^-(u^-_{\xi\xi} + u^-_{\eta\eta} + u^-_{\tau\tau}) + \beta^-_\xi u^-_\xi + \beta^-_\eta u^-_\eta + \beta^-_\tau u^-_\tau - \sigma^- u^- - f^- = 0 \tag{4.8}$$

under the local coordinate system. The interface under the local ξ-η-τ coordinate system can be expressed as

$$\xi = \chi(\eta, \tau) \quad \text{with} \quad \chi(0,0) = 0, \;\; \chi_\eta(0,0) = 0, \;\; \chi_\tau(0,0) = 0. \tag{4.9}$$

From the jump condition (4.2) and the differential equation (4.1), we can derive more interface relations, which are summarized in the following theorem.

Theorem 4.1. *Assume that the differential equation* (4.1) *has a solution* $u(\mathbf{x})$ *in a neighborhood of* Γ. *Assume also that* $u(\mathbf{x})$ *is a piecewise* C^2 *function in the neighborhood of* Γ *excluding the interface* Γ. *Then we have the interface relations*

$$u^+ = u^- + w, \tag{4.10a}$$

$$u_\xi^+ = \frac{\beta^-}{\beta^+} u_\xi^- + \frac{v}{\beta^+}, \tag{4.10b}$$

$$u_\eta^+ = u_\eta^- + w_\eta, \tag{4.10c}$$

$$u_\tau^+ = u_\tau^- + w_\tau, \tag{4.10d}$$

$$u_{\eta\tau}^+ = u_{\eta\tau}^- + (u_\xi^- - u_\xi^+)\chi_{\eta\tau} + w_{\eta\tau}, \tag{4.10e}$$

$$u_{\eta\eta}^+ = u_{\eta\eta}^- + (u_\xi^- - u_\xi^+)\chi_{\eta\eta} + w_{\eta\eta}, \tag{4.10f}$$

$$u_{\tau\tau}^+ = u_{\tau\tau}^- + (u_\xi^- - u_\xi^+)\chi_{\tau\tau} + w_{\tau\tau}, \tag{4.10g}$$

$$\begin{aligned} u_{\xi\eta}^+ = {} & \frac{\beta^-}{\beta^+} u_{\xi\eta}^- + \left(u_\eta^+ - \frac{\beta^-}{\beta^+} u_\eta^-\right) \chi_{\eta\eta} + \left(u_\tau^+ - \frac{\beta^-}{\beta^+} u_\tau^-\right) \chi_{\eta\tau} \\ & + \frac{\beta_\eta^-}{\beta^+} u_\xi^- - \frac{\beta_\eta^+}{\beta^+} u_\xi^+ + \frac{v_\eta}{\beta^+}, \end{aligned} \tag{4.10h}$$

$$\begin{aligned} u_{\xi\tau}^+ = {} & \frac{\beta^-}{\beta^+} u_{\xi\tau}^- + \left(u_\eta^+ - \frac{\beta^-}{\beta^+} u_\eta^-\right) \chi_{\eta\tau} + \left(u_\tau^+ - \frac{\beta^-}{\beta^+} u_\tau^-\right) \chi_{\tau\tau} \\ & + \frac{\beta_\tau^-}{\beta^+} u_\xi^- - \frac{\beta_\tau^+}{\beta^+} u_\xi^+ + \frac{v_\tau}{\beta^+}, \end{aligned} \tag{4.10i}$$

$$\begin{aligned} u_{\xi\xi}^+ = {} & \frac{\beta^-}{\beta^+} u_{\xi\xi}^- + \left(\frac{\beta^-}{\beta^+} - 1\right) u_{\eta\eta}^- + \left(\frac{\beta^-}{\beta^+} - 1\right) u_{\tau\tau}^- \\ & + u_\xi^+ \left(\chi_{\eta\eta} + \chi_{\tau\tau} - \frac{\beta_\xi^+}{\beta^+}\right) - u_\xi^- \left(\chi_{\eta\eta} + \chi_{\tau\tau} - \frac{\beta_\xi^-}{\beta^+}\right) \\ & + \frac{1}{\beta^+}\left(\beta_\eta^- u_\eta^- - \beta_\eta^+ u_\eta^+\right) + \frac{1}{\beta^+}\left(\beta_\tau^- u_\tau^- - \beta_\tau^+ u_\tau^+\right) \\ & + \frac{1}{\beta^+}\left([\sigma]u^- - \sigma^+[u]\right) + \frac{[f]}{\beta^+} - w_{\eta\eta} - w_{\tau\tau}. \end{aligned} \tag{4.10j}$$

Sketch of the proof: The first two interface conditions are the original jump conditions (4.2). By differentiating the first jump condition $[u] = w$ in (4.2) with respect to η and τ, respectively, we get

$$[u_\xi]\chi_\eta + [u_\eta] = w_\eta, \tag{4.11}$$

$$[u_\xi]\chi_\tau + [u_\tau] = w_\tau, \tag{4.12}$$

which give (4.10c) and (4.10d) if we evaluate the equations above at $(\xi, \eta, \tau) = (0, 0, 0)$ in the new coordinate system and use the fact $\chi_\eta(0, 0) = \chi_\tau(0, 0) = 0$. Differentiating (4.11) with respect to τ yields

$$\chi_\eta \frac{\partial}{\partial \tau}[u_\xi] + \chi_{\eta\tau}[u_\xi] + [u_{\eta\xi}]\chi_\tau + [u_{\eta\tau}] = w_{\eta\tau}, \tag{4.13}$$

from which we get (4.10e). Differentiating (4.11) with respect to η and differentiating (4.12) with respect to τ, respectively, we obtain

$$\chi_\eta \frac{\partial}{\partial \eta}[u_\xi] + \chi_{\eta\eta}[u_\xi] + \chi_\eta [u_{\eta\xi}] + [u_{\eta\eta}] = w_{\eta\eta}, \tag{4.14}$$

$$\chi_\tau \frac{\partial}{\partial \tau}[u_\xi] + \chi_{\tau\tau}[u_\xi] + \chi_\tau [u_{\tau\xi}] + [u_{\tau\tau}] = w_{\tau\tau}, \tag{4.15}$$

from which we get (4.10f) and (4.10g). Before differentiating the jump condition of the normal derivative $[\beta u_n] = v$ in (4.2), we first express the unit normal vector of the interface Γ as

$$\mathbf{n} = \frac{(1, \; -\chi_\eta, \; -\chi_\tau)}{\sqrt{1 + \chi_\eta^2 + \chi_\tau^2}}. \tag{4.16}$$

Thus, the second interface condition $[\beta u_{\mathbf{n}}] = v$ in (4.2) can be written as

$$\left[\beta\left(u_\xi - u_\eta \chi_\eta - u_\tau \chi_\tau\right)\right] = v(\eta, \tau)\sqrt{1 + \chi_\eta^2 + \chi_\tau^2}. \tag{4.17}$$

Differentiating this with respect to η, we get

$$\begin{aligned} &\left[\left(\beta_\xi \chi_\eta + \beta_\eta\right)\left(u_\xi - u_\eta \chi_\eta - u_\tau \chi_\tau\right)\right] \\ &\quad + \left[\beta\left(u_{\xi\xi}\chi_\eta + u_{\xi\eta} - \chi_\eta \frac{\partial}{\partial \eta}u_\eta - \chi_\tau \frac{\partial}{\partial \eta}u_\tau - u_\eta \chi_{\eta\eta} - u_\tau \chi_{\eta\tau}\right)\right] \\ &\quad = v_\eta \sqrt{1 + \chi_\eta^2 + \chi_\tau^2} + v\,\frac{\chi_\eta \chi_{\eta\eta}}{\sqrt{1 + \chi_\eta^2 + \chi_\tau^2}}, \end{aligned} \tag{4.18}$$

which gives (4.10h) at $(\xi, \eta, \tau) = (0, 0, 0)$. Similarly, differentiating (4.17) with respect to τ, we get the last interface relation (4.10j) by

$$\begin{aligned} &\left[\left(\beta_\xi \chi_\tau + \beta_\tau\right)\left(u_\xi - u_\eta \chi_\eta - u_\tau \chi_\tau\right)\right] \\ &\quad + \left[\beta\left(u_{\xi\xi}\chi_\tau + u_{\xi\tau} - \chi_\eta \frac{\partial}{\partial \tau}u_\eta - \chi_\tau \frac{\partial}{\partial \tau}u_\tau - u_\eta \chi_{\eta\tau} - u_\tau \chi_{\tau\tau}\right)\right] \\ &\quad = v_\tau \sqrt{1 + \chi_\eta^2 + \chi_\tau^2} + v\,\frac{\chi_\tau \chi_{\tau\tau}}{\sqrt{1 + \chi_\eta^2 + \chi_\tau^2}}, \end{aligned} \tag{4.19}$$

which gives (4.10i) at $(\xi, \eta, \tau) = (0, 0, 0)$.

To get the relation for $u^+_{\xi\xi}$ we need to use the differential equation (4.8) itself, from which we can write

$$\left[\beta\left(u_{\xi\xi} + u_{\eta\eta} + u_{\tau\tau}\right) + \beta_\xi u_\xi + \beta_\eta u_\eta + \beta_\tau u_\tau - \sigma u\right] = [f]. \tag{4.20}$$

Notice that

$$-\sigma^- u^- + \sigma^+ u^+ = -\sigma^- u^- + \sigma^+ u^- - \sigma^+ u^- + \sigma^+ u^+ = [\sigma] u^- + \sigma^+ [u]. \quad (4.21)$$

Rearranging (4.20) and using (4.21) above, we get

$$\begin{aligned} &\beta^+ \left(u^+_{\xi\xi} + u^+_{\eta\eta} + u^+_{\tau\tau}\right) + \beta^+_\xi u^+_\xi + \beta^+_\eta u^+_\eta + \beta^+_\tau u^+_\tau \\ &= \beta^- \left(u^-_{\xi\xi} + u^-_{\eta\eta} + u^-_{\tau\tau}\right) + \beta^-_\xi u^-_\xi \\ &\quad + \beta^-_\eta u^-_\eta + \beta^-_\tau u^-_\tau + [f] - \sigma^- u^- + \sigma^+ u^+. \end{aligned} \quad (4.22)$$

By solving $u^+_{\xi\xi}$ from the equation above, we get the last interface relation, (4.10j).

4.3 The finite difference scheme of the IIM in three dimensions

It is more convenient and also easier to use the zero level surface of a three-dimensional function $\varphi(x, y, z)$ to represent the interface Γ compared with other approaches. Let $\varphi(x, y, z)$ be a Lipschitz continuous function satisfying

$$\begin{array}{llll} \varphi(x, y, z) & < 0 & \text{if} & (x, y, z) \in \Omega^-, \\ \varphi(x, y, z) & = 0 & \text{if} & (x, y, z) \in \Gamma, \\ \varphi(x, y, z) & > 0 & \text{if} & (x, y, z) \in \Omega^+. \end{array} \quad (4.23)$$

Such a level set function is not unique but should be chosen as an approximation of the signed distance function to the interface Γ. We also assume that the level set function $\varphi(x, y, z)$ has up to second-order continuous partial derivatives in a neighborhood of the zero level set $\varphi = 0$. We refer the reader to [205, 238] for more information about the level set method.

Let the domain Ω be a rectangular domain and let $\{x_i, y_j, z_k\}$ be a uniform Cartesian grid. The level set function is defined at grid points $\varphi_{ijk} = \varphi(x_i, y_j, z_k)$. At a grid point $\mathbf{x}_{ijk}$, we define

$$\varphi^{min}_{ijk} = \min\left\{\varphi_{i\pm1,j,k}, \varphi_{i,j\pm1,k}, \varphi_{i,j,k\pm1}\right\}, \quad (4.24)$$

$$\varphi^{max}_{ijk} = \max\left\{\varphi_{i\pm1,j,k}, \varphi_{i,j\pm1,k}, \varphi_{i,j,k\pm1}\right\}. \quad (4.25)$$

A grid point $\mathbf{x}_{ijk}$ is called an *irregular grid point* if

$$\varphi^{max}_{ijk} \varphi^{min}_{ijk} \le 0 \quad (4.26)$$

in reference to the standard central 7-point finite difference stencil. Otherwise, $\mathbf{x}_{ijk}$ is called a *regular* grid point.

4.3.1 Finite difference equations at regular grid points

Similar to the IIM in two dimensions, the finite difference equations have the generic form

$$\sum_{m}^{n_s} \gamma_m U_{i+i_m,j+j_m,k+k_m} - \sigma_{ijk} U_{ijk} = f_{ijk} + C_{ijk} \tag{4.27}$$

at every grid point (x_i, y_j, z_k) where the solution is unknown, where i_m, j_m, k_m take values from $0, \pm 1, \pm 2, \dots$, that is, the summation is taken over a number of grid points neighboring (x_i, y_j, z_k). Note that we have omitted the dependency of m on i, j, and k for simplicity. At a regular grid point, the coefficients of the standard 7-point central finite difference scheme are

$$\begin{aligned} \gamma_{i\pm 1,j,k} &= \frac{\beta_{i\pm 1/2,j,k}}{h^2}, \qquad \gamma_{i,j\pm 1,k} = \frac{\beta_{i,j\pm 1/2,k}}{h^2}, \\ \gamma_{i,j,k\pm 1} &= \frac{\beta_{i,j,k\pm 1/2}}{h^2}, \\ \gamma_{i,j,k} &= -\left(\sum \gamma_{i\pm 1,j,k} + \sum \gamma_{i,j\pm 1,k} + \sum \gamma_{i,j,k\pm 1}\right), \end{aligned} \tag{4.28}$$

where $\beta_{i-1/2,j,k} = \beta(x_i - h_x/2, y_j, z_k)$, and so forth. Here, we have assumed that $h_x = h_y = h_z = h$. The correction term is simply $C_{ijk} = 0$ and the local truncation errors are $O(h^2)$.

4.3.2 Computing the orthogonal projection in a three-dimensional Cartesian grid

To derive the finite difference equation using the IIM at an irregular grid point $\mathbf{x}_{ijk} = (x_i, y_j, z_k)$, we need to choose a point $\mathbf{X}^*_{ijk} = (X^*_i, Y^*_j, Z^*_k)$ on the interface that is close to $\mathbf{x}_{ijk}$. At this point, we can use the Taylor expansion from each side of the interface so that the jump conditions (4.2) and the interface relations (4.10) can be utilized. While we can choose any point (on the interface) that is close to the grid point $\mathbf{x}_{ijk}$, it is natural to choose the orthogonal projection of $\mathbf{x}_{ijk}$ on the interface. Since $\pm\nabla\varphi(\mathbf{x}_{ijk})$ is the direction where the level set function has the largest rate increase/decline, the orthogonal projection can be determined from

$$\mathbf{X}^*_{ijk} = \mathbf{x}_{ijk} + \alpha\, \mathbf{p}, \tag{4.29}$$

where $\mathbf{p} = \nabla\varphi/|\nabla\varphi|$ and $|\alpha| \sim h$ is an approximation to the distance between the grid point $\mathbf{x}_{ijk}$ and $\mathbf{X}^*_{ijk}$. Neglecting $O(\alpha^3)$ and higher-order terms in the Taylor expansion of $\varphi(\mathbf{X}^*_{ijk}) = \varphi(\mathbf{x}_{ijk} + \alpha\,\mathbf{p}) = 0$ at $\mathbf{x}_{ijk}$, we get a quadratic equation for α,

$$\varphi(\mathbf{x}_{ijk}) + |\nabla\varphi(\mathbf{x}_{ijk})|\,\alpha + \frac{1}{2}\left(\mathbf{p}^T \operatorname{He}(\varphi(\mathbf{x}_{ijk}))\,\mathbf{p}\right)\alpha^2 = 0, \tag{4.30}$$

where $\operatorname{He}(\varphi(\mathbf{x}_{ijk}))$ is the Hessian matrix of φ

$$\operatorname{He}\left(\varphi(\mathbf{x}_{ijk})\right) = \begin{pmatrix} \varphi_{xx} & \varphi_{xy} & \varphi_{xz} \\ \varphi_{yx} & \varphi_{yy} & \varphi_{yz} \\ \varphi_{zx} & \varphi_{zy} & \varphi_{zz} \end{pmatrix} \tag{4.31}$$

evaluated at $\mathbf{x}_{ijk}$. The values of φ, φ_x, φ_y, φ_z, φ_{xx}, φ_{xy}, φ_{xz}, φ_{yy}, φ_{yz}, and φ_{zz} are evaluated at the grid point $\mathbf{x}_{ijk} = (x_i, y_j, z_k)$ and can be approximated using the standard central finite difference formulas, for example,

$$\varphi_x \approx \frac{\varphi_{i+1,j,k} - \varphi_{i-1,j,k}}{2h_x},$$

$$\varphi_{yy} \approx \frac{\varphi_{i,j-1,k} - 2\varphi_{i,j,k} + \varphi_{i,j+1,k}}{h_y^2},$$

$$\varphi_{xy} \approx \frac{\varphi_{i-1,j-1,k} + \varphi_{i+1,j+1,k} - \varphi_{i-1,j+1,k} - \varphi_{i+1,j-1,k}}{4h_x h_y}.$$

Note that the truncation error of the quadratic approximation is $O(\alpha^3) \sim O(h^3)$. The computed projections are typically third-order accurate.

4.3.3 Setting up a local coordinate system using a level set function

Given a point $\mathbf{X}^* = (X^*, Y^*, Z^*)$ on the interface, the unit normal direction of the interface at $\mathbf{X}^*$ is given by

$$\boldsymbol{\xi} = \frac{\nabla\varphi}{|\nabla\varphi|} = (\varphi_x,\ \varphi_y,\ \varphi_z)^T \Big/ \sqrt{\varphi_x^2 + \varphi_y^2 + \varphi_z^2},$$

where the partial derivatives are evaluated at (X^*, Y^*, Z^*). Note that at the interface Γ, the level set function should be chosen such that $|\nabla\varphi| \neq 0$. We prefer to choose the level set function φ as the signed distance function, which satisfies $|\nabla\varphi| = 1$ at least in a neighborhood of the interface Γ. The choice of the tangential axis vectors is not unique. It is recommended that if $\varphi_x^2 + \varphi_y^2 \geq \varphi_x^2 + \varphi_z^2$, the two tangential directions should be chosen as

$$\boldsymbol{\eta} = (\varphi_y,\ -\varphi_x,\ 0)^T \Big/ \sqrt{\varphi_x^2 + \varphi_y^2}, \tag{4.32}$$

$$\boldsymbol{\tau} = \frac{\mathbf{s}}{|\mathbf{s}|}, \qquad \text{where } \mathbf{s} = (\varphi_x\varphi_z,\ \varphi_y\varphi_z,\ -\varphi_x^2 - \varphi_y^2)^T. \tag{4.33}$$

Otherwise, the two tangential directions are chosen as

$$\boldsymbol{\eta} = (\varphi_z,\ 0,\ -\varphi_x)^T \Big/ \sqrt{\varphi_x^2 + \varphi_z^2}, \tag{4.34}$$

$$\boldsymbol{\tau} = \frac{\mathbf{t}}{|\mathbf{t}|}, \qquad \text{where } \mathbf{t} = (-\varphi_x\varphi_y,\ \varphi_x^2 + \varphi_z^2,\ -\varphi_y\varphi_z)^T. \tag{4.35}$$

The three unit orthogonal vectors $\boldsymbol{\xi}$, $\boldsymbol{\eta}$, and $\boldsymbol{\tau}$ form local coordinates in the normal and tangential directions.

4.3.4 The bilinear interpolation in three dimensions

The level set function is defined at grid points as φ_{ijk}. It is easy to approximate its partial derivatives and the principal curvatures of the level set function (second-order surface derivatives along a level surface $\varphi(x, y, z) = C$) at grid points using the central finite difference formulas. However, in order to determine the finite difference coefficients from

(4.43)–(4.52), we need the values of the interface information at the orthogonal projection $\mathbf{X}^*$. This can be done using the bilinear interpolation since we assume that the level set function has up to second-order continuous partial derivatives in a neighborhood of the interface Γ.

The bilinear interpolation in three dimensions uses eight grid points. Given any point (X^*, Y^*, Z^*) on the interface, we can find a cube containing the point with the eight vertices (x_0, y_0, z_0), (x_0, y_0, z_1), (x_0, y_1, z_0), (x_0, y_1, z_1), (x_1, y_0, z_0), (x_1, y_0, z_1), (x_1, y_1, z_0), and (x_1, y_1, z_1). Let Q_{ijk} be the function values at the eight vertices of a function $q(x, y, z)$. The 8-point bilinear interpolation for $q(x, y, z)$ is defined as

$$Q(x^*, y^*, z^*) = \frac{1}{8} \sum_{i,j,k=0}^{1} Q_{ijk}\, \bar{x}_i\, \bar{y}_j\, \bar{z}_k, \tag{4.36}$$

where

$$\begin{aligned} \bar{x}_i &= 1 + (2i-1)\left(\frac{2(x-x_0)}{h_x} - 1\right), \\ \bar{y}_j &= 1 + (2j-1)\left(\frac{2(y-y_0)}{h_y} - 1\right), \\ \bar{z}_k &= 1 + (2k-1)\left(\frac{2(z-z_0)}{h_z} - 1\right). \end{aligned} \tag{4.37}$$

Such an interpolation scheme has better than second-order accuracy if $\varphi(x, y, z)$ has up to second-order continuous partial derivatives.

4.4 Deriving the finite difference equation at an irregular grid point

At an irregular grid point, the method of undetermined coefficients is used to find the coefficients of the finite difference equation (4.27). Let $\mathbf{X}^*_{ijk}$ be the orthogonal projection of an irregular grid $(x_i, y_j, z_k) \in \Omega^-$ on the interface. We set a local coordinate system at $\mathbf{X}^*_{ijk}$ as we explained earlier and denote the local coordinates of a grid point $(x_{i+i_m}, y_{j+j_m}, z_{k+k_m})$ involved in the finite difference stencil as (ξ_m, η_m, τ_m). Using the Taylor expansion from both sides of the interface at $\mathbf{X}^*_{ijk}$, the local truncation error of the finite difference equation at (x_i, y_j, z_k) is

$$\begin{aligned} T_{ijk} &= \sum_m^{n_s} \gamma_m\, u(x_{i+i_m}, y_{j+j_m}, z_{k+k_m}) - \sigma_{ijk}\, u(x_i, y_j, z_k) - f_{ijk} - C_{ijk} \\ &= \sum_m^{n_s} \gamma_m \left(u^\pm + u^\pm_\xi \xi_m + u^\pm_\eta \eta_m + u^\pm_\tau \tau_m + \frac{\xi_m^2}{2} u^\pm_{\xi\xi} + \frac{\eta_m^2}{2} u^\pm_{\eta\eta} \right. \\ &\qquad \left. + \frac{\tau_m^2}{2} u^\pm_{\tau\tau} + \xi_m \eta_m\, u^\pm_{\xi\eta} + \xi_m \tau_m\, u^\pm_{\xi\tau} + \eta_m \tau_m\, u^\pm_{\eta\tau} \right) \\ &\qquad - \sigma_{ijk}\, (u_{ijk} + O(h)) - f_{ijk} - C_{ijk} + O\left(\max_m\{|\gamma_m|\}h^3\right), \end{aligned} \tag{4.38}$$

where the limiting values of the function and the derivatives are evaluated at $\mathbf{X}^*_{ijk}$ from the same side, where the grid point $(i + i_m, j + j_m, k + k_m)$ exists. If a grid point is on the interface, that is, $\varphi_{i+i_m, j+j_m, k+k_m} = 0$, then the values of $u, u_\xi, \ldots, u_{\eta\eta}$ are defined as the limiting values from a particular side. For simplicity, we use the limiting values from the "−" side. Using the interface relations (4.10a)–(4.10j), the expression (4.38) can be written as

$$
\begin{aligned}
T_{ijk} &= a_1\, u^- + a_2\, u^+ + a_3\, u^-_\xi + a_4\, u^+_\xi + a_5\, u^-_\eta + a_6\, u^+_\eta + a_7\, u^-_\tau \\
&+ a_8\, u^+_\tau + a_9\, u^-_{\xi\xi} + a_{10}\, u^+_{\xi\xi} + a_{11}\, u^-_{\eta\eta} + a_{12}\, u^+_{\eta\eta} + a_{13}\, u^-_{\tau\tau} \\
&+ a_{14}\, u^+_{\tau\tau} + a_{15}\, u^-_{\xi\eta} + a_{16}\, u^+_{\xi\eta} + a_{17}\, u^-_{\xi\tau} + a_{18}\, u^+_{\xi\tau} + a_{19}\, u^-_{\eta\tau} \\
&+ a_{20}\, u^+_{\eta\tau} - \sigma^- u^- - f^- - C_{ijk} + O\left(h + \max_m\{|\gamma_m\}h^3\right),
\end{aligned}
\tag{4.39}
$$

where the $\{a_i\}$ coefficients depend only on the position of the stencil relative to the interface. They are independent of the functions u, β, λ, and f. Similar to (3.17), if we define the index sets K^+ and K^- by

$$
K^\pm = \{m : (\xi_m, \eta_m, \tau_m) \text{ is on the " } \pm \text{ " side of } \Gamma\}, \tag{4.40}
$$

then the $\{a_i\}$'s with odd subscripts are given by

$$
\begin{aligned}
&a_1 = \sum_{m\in K^-} \gamma_m, \quad a_3 = \sum_{m\in K^-} \gamma_m\, \xi_m, \quad a_5 = \sum_{m\in K^-} \gamma_m\, \eta_m, \quad a_7 = \sum_{m\in K^-} \gamma_m\, \tau_m, \\
&a_9 = \frac{1}{2} \sum_{m\in K^-} \gamma_m\, \xi_m^2, \quad a_{11} = \frac{1}{2} \sum_{m\in K^-} \gamma_m\, \eta_m^2, \quad a_{13} = \frac{1}{2} \sum_{m\in K^-} \gamma_m\, \tau_m^2, \\
&a_{15} = \sum_{m\in K^-} \gamma_m\, \xi_m \eta_m, \quad a_{17} = \sum_{m\in K^-} \gamma_m\, \xi_m \tau_m, \quad a_{19} = \sum_{m\in K^-} \gamma_m\, \eta_m \tau_m.
\end{aligned}
\tag{4.41}
$$

The a_i's with even subscripts are exactly the same as above except the summation is from the subset K^+. Plugging the interface relations (4.10a)–(4.10j) into (4.39), we express all the quantities from the "+" side in terms of the quantities from the "−" side. Thus the right-hand side of (4.39) is represented by the linear combination of the quantities from the "−" side. After some manipulations, (4.39) can be written as

$$
\begin{aligned}
T_{ijk} &= (\quad) u^- + (\quad) u^-_\xi + (\quad) u^-_\eta + (\quad) u^-_\tau + (\quad) u^-_{\xi\xi} + (\quad) u^-_{\eta\eta} \\
&+ (\quad) u^-_{\tau\tau} + (\quad) u^-_{\xi\eta} + (\quad) u^-_{\xi\tau} + (\quad) u^-_{\eta\tau} - \sigma^- u^- - f^- \\
&+ (\hat{T}_{ijk} - C_{ijk}) + O\left(h + \max_m\{|\gamma_m|\}h^3\right).
\end{aligned}
\tag{4.42}
$$

The contents in the parentheses are the corresponding terms in the left-hand sides of (4.43)–(4.52). The term $\hat{T}_{ijk}$ given in (4.53) is a linear combination of the jump conditions and their surface derivatives in the solution and the flux from the interface relations (4.10a)–(4.10j).

By minimizing the magnitude of the local truncation error (4.39) or (4.42), that is, matching all the terms of u^-, u_ξ^-, ..., $u_{\eta\eta}^-$, up to second-order partial derivatives, with the PDE (4.8), we obtain the linear system of equations for the following finite difference coefficients $\{\gamma_m\}$:

$$a_1 + a_{10}\,\frac{[\sigma]}{\beta^+} + a_2 = 0, \tag{4.43}$$

$$\begin{aligned} a_3 &- a_{10}\left(\chi_{\eta\eta} + \chi_{\tau\tau} - \frac{\beta_\xi^-}{\beta^+}\right) + a_{12}\,\chi_{\eta\eta} + a_{14}\,\chi_{\tau\tau} + a_{16}\,\frac{\beta_\eta^-}{\beta^+} \\ &+ a_{18}\,\frac{\beta_\tau^-}{\beta^+} + a_{20}\,\chi_{\eta\tau} + \frac{\beta^-}{\beta^+}\left\{a_4 + a_{10}\left(\chi_{\eta\eta} + \chi_{\tau\tau} - \frac{\beta_\xi^+}{\beta^+}\right)\right. \\ &\left. - a_{12}\,\chi_{\eta\eta} - a_{14}\,\chi_{\tau\tau} - a_{16}\,\frac{\beta_\eta^+}{\beta^+} - a_{18}\,\frac{\beta_\tau^+}{\beta^+} - a_{20}\,\chi_{\eta\tau}\right\} = \beta_\xi^-, \end{aligned} \tag{4.44}$$

$$\begin{aligned} a_5 &+ a_{10}\frac{\beta_\eta^-}{\beta^+} - a_{16}\,\frac{\beta^-}{\beta^+}\,\chi_{\eta\eta} - a_{18}\,\frac{\beta^-}{\beta^+}\,\chi_{\eta\tau} \\ &+ a_6 - a_{10}\,\frac{\beta_\eta^+}{\beta^+} + a_{16}\,\chi_{\eta\eta} + a_{18}\,\chi_{\eta\tau} = \beta_\eta^-, \end{aligned} \tag{4.45}$$

$$\begin{aligned} a_7 &+ a_{10}\,\frac{\beta_\tau^-}{\beta^+} - a_{16}\,\frac{\beta^-}{\beta^+}\,\chi_{\eta\tau} - a_{18}\,\frac{\beta^-}{\beta^+}\,\chi_{\tau\tau} \\ &+ a_8 - a_{10}\,\frac{\beta_\tau^+}{\beta^+} + a_{16}\,\chi_{\eta\tau} + a_{18}\,\chi_{\tau\tau} = \beta_\tau^-, \end{aligned} \tag{4.46}$$

$$a_9 + a_{10}\,\frac{\beta^-}{\beta^+} = \beta^-, \tag{4.47}$$

$$a_{11} + a_{12} + a_{10}\left(\frac{\beta^-}{\beta^+} - 1\right) = \beta^-, \tag{4.48}$$

$$a_{13} + a_{14} + a_{10}\left(\frac{\beta^-}{\beta^+} - 1\right) = \beta^-, \tag{4.49}$$

$$a_{15} + a_{16}\,\frac{\beta^-}{\beta^+} = 0, \tag{4.50}$$

$$a_{17} + a_{18}\,\frac{\beta^-}{\beta^+} = 0, \tag{4.51}$$

$$a_{19} + a_{20} = 0. \tag{4.52}$$

If we can solve this linear system of equations to get $\{\gamma_m\}$'s, then C_{ijk}, which is a function of the coefficients of $\{\gamma_m\}$'s, should be chosen to offset all the jumps collected in $\hat{T}_{ijk}$ and is given by

$$
\begin{aligned}
C_{ijk} = {} & a_{10}\left(\frac{[f]}{\beta^+} + \frac{\sigma^+ w}{\beta^+} - w_{\eta\eta} - w_{\tau\tau}\right) + a_{12}\, w_{\eta\eta} + a_{14}\, w_{\tau\tau} \\
& + a_{16}\,\frac{v_\eta}{\beta^+} + a_{18}\,\frac{v_\tau}{\beta^+} + a_{20}\, w_{\eta\tau} + a_2\, w \\
& + \frac{1}{\beta^+}\left\{ a_4 + a_{10}\left(\chi_{\eta\eta} + \chi_{\tau\tau} - \frac{\beta_\xi^+}{\beta^+}\right) - a_{12}\,\chi_{\eta\eta} \right. \\
& \qquad \left. - a_{14}\,\chi_{\tau\tau} - a_{16}\,\frac{\beta_\eta^+}{\beta^+} - a_{18}\frac{\beta_\tau^+}{\beta^+} - a_{20}\,\chi_{\eta\tau} \right\} v \\
& + \left(a_6 - a_{10}\,\frac{\beta_\eta^+}{\beta^+} + a_{16}\,\chi_{\eta\eta} + a_{18}\,\chi_{\eta\tau}\right) w_\eta \\
& + \left(a_8 - a_{10}\,\frac{\beta_\tau^+}{\beta^+} + a_{16}\,\chi_{\eta\tau} + a_{18}\,\chi_{\tau\tau}\right) w_\tau .
\end{aligned} \tag{4.53}
$$

Computing the *principal curvatures* using the level set function

In order to determine the matrix entries of the linear system of equations (4.43)–(4.52) for the coefficients $\{\gamma_m\}$'s, we need to compute the second-order tangential derivatives $\chi_{\eta\eta}$, $\chi_{\tau\tau}$, $\chi_{\eta\tau}$ of the interface χ at $\mathbf{X}_{ijk}^*$. We call these quantities *principal curvatures.* These quantities are computed from the level set function. Since $\varphi(\chi(\eta,\tau),\eta,\tau) = 0$, it follows from the implicit function theory that

$$
\varphi_\eta + \varphi_\xi\,\chi_\eta = 0, \tag{4.54}
$$

$$
\varphi_\tau + \varphi_\xi\,\chi_\tau = 0, \tag{4.55}
$$

and

$$
\begin{aligned}
&\varphi_{\eta\eta} + \varphi_{\eta\xi}\,\chi_\eta + (\varphi_{\xi\eta} + \varphi_{\xi\xi}\,\chi_\eta)\chi_\eta + \varphi_\xi\,\chi_{\eta\eta} = 0, \\
&\varphi_{\eta\tau} + \varphi_{\eta\xi}\,\chi_\tau + (\varphi_{\xi\tau} + \varphi_{\xi\xi}\,\chi_\tau)\chi_\eta + \varphi_\xi\,\chi_{\eta\tau} = 0, \\
&\varphi_{\tau\tau} + \varphi_{\tau\xi}\,\chi_\tau + (\varphi_{\xi\tau} + \varphi_{\xi\xi}\,\chi_\tau)\chi_\tau + \varphi_\xi\,\chi_{\tau\tau} = 0.
\end{aligned}
$$

Thus from $\chi(0,0) = 0$, $\chi_\eta(0,0) = 0$, and $\chi_\tau(0,0) = 0$, we have

$$
\left\{
\begin{aligned}
\chi_{\eta\eta} &= -\varphi_{\eta\eta}/\varphi_\xi, \\
\chi_{\tau\tau} &= -\varphi_{\tau\tau}/\varphi_\xi, \\
\chi_{\eta\tau} &= -\varphi_{\eta\tau}/\varphi_\xi,
\end{aligned}
\right. \tag{4.56}
$$

where $(\varphi_\xi, \varphi_\eta, \varphi_\tau)$ and $(\varphi_{\xi\xi}, \varphi_{\eta\eta}, \varphi_{\tau\tau})$ can be computed from (4.6) and (4.7) (dropping the bar notation) in terms of the original coordinate system. The procedure is to first evaluate these quantities using the second-order central finite difference schemes at grid points, and then to evaluate them at the orthogonal projection using the bilinear interpolation.

4.4.1 Computing surface derivatives of interface quantities in three dimensions

If the jump conditions $[u] = w$ and $[\beta u_n] = v$ are not zero, then the first-order surface derivatives w_η, w_τ, v_η, v_τ and the second-order surface derivatives $w_{\eta\eta}$, $w_{\tau\tau}$, $w_{\eta\tau}$ are needed to evaluate the correction terms C_{ijk} in (4.53). These quantities can be obtained by the *least squares interpolation scheme* proposed in [163].

Letting $g(\mathbf{X}) = g(X(s_1, s_2), Y(s_1, s_2), Z(s_1, s_2))$ be a function defined on the interface, we therefore know its values at all projections $\{\mathbf{X}_p^*\}$. We explain below how to approximate the surface derivatives at a particular projection $\mathbf{X}^*$ using the values of g at the projections in a neighborhood of $\mathbf{X}^*$.

In a neighborhood of $\mathbf{X}^*$, the interface quantity $g(\mathbf{X})$ can be written as $g(\eta, \tau)$ using the local coordinates (4.3) centered at $\mathbf{X}^*$. The least squares interpolation for g_η, g_τ, and $g_{\eta\eta}$ at $\mathbf{X}^*$, for example, can be written as

$$\begin{aligned} g_\eta(\mathbf{X}^*) &= \sum_{|\mathbf{X}^* - \mathbf{X}_p^*| \le R_\epsilon,\, p \le N_p} \alpha_p\, g_p, \\ g_\tau(\mathbf{X}^*) &= \sum_{|\mathbf{X}^* - \mathbf{X}_p^*| \le R_\epsilon,\, p \le N_p} \lambda_p\, g_p, \\ g_{\eta\eta}(\mathbf{X}^*) &= \sum_{|\mathbf{X}^* - \mathbf{X}_p^*| \le R_\epsilon,\, p \le N_p} \mu_p\, g_p, \end{aligned} \tag{4.57}$$

where $g_p = g(\mathbf{X}_p^*)$ is the function value at the projection $\{\mathbf{X}_p^*\}$, R_ϵ is a prechosen parameter between $2.1h \sim 6.1h$, and N_p is the number of the orthogonal projections (surface points) involved in the interpolation. R_ϵ should be chosen such that at least N_p such surface points are involved. A good choice is $N_p = 9 \sim 12$. Usually, the closest N_p orthogonal projections to $\mathbf{X}^*$ in R_ϵ are chosen. We explain how to determine the coefficients α_p for $g_\eta(\mathbf{X}^*)$ as an illustration. It is based on the Taylor expansion and the singular value decomposition (SVD) to solve an underdetermined system of equations.

Under the local coordinates (4.3) centered at $\mathbf{X}^*$, we denote the (η, τ) coordinates of an orthogonal projection point $\mathbf{X}_p^*$, involved in the interpolation, as (η_p, τ_p). Using the Taylor expansion at $\mathbf{X}^*$, we have

$$\begin{aligned} &\sum_{|\mathbf{X}_p^* - \mathbf{X}^*| \le R_\epsilon} \alpha_p\, g_p \\ &= \sum_{|\mathbf{X}_p^* - \mathbf{X}^*| \le R_\epsilon} \alpha_p \left(g^* + \nabla g^* \cdot (\mathbf{X}_p^* - \mathbf{X}^*) + \frac{1}{2} (\mathbf{X}_p^* - \mathbf{X}^*)^T \mathrm{He}(g^*) (\mathbf{X}_p^* - \mathbf{X}^*) + \cdots \right) \\ &= (\quad) g^* + (\quad) g_\eta^* + (\quad) g_\tau^* + (\quad) g_{\eta\eta}^* + (\quad) g_{\eta\tau}^* + (\quad) g_{\tau\tau}^* + \cdots, \end{aligned}$$

where g^*, g_η^*, $\ldots$ are the function value and its surface derivatives g at $\mathbf{X}^*$, $\mathrm{He}(g^*)$ is the Hessian matrix of g^* in the local coordinates ξ-η-τ centered at the projection $\mathbf{X}^*$, and the contents in the parentheses are the corresponding right-hand sides in the system of equations below. Using the method of the undetermined coefficients, we set

$$\begin{aligned} &\sum_p \alpha_p = 0, \qquad \sum_p \alpha_p \eta_p = 1, \qquad \sum_p \alpha_p \tau_p = 0, \\ &\sum_p \alpha_p \eta_p^2 = 0, \qquad \sum_p \alpha_p \tau_p^2 = 0, \qquad \sum_p \alpha_p \eta_p \tau_p = 0, \end{aligned} \tag{4.58}$$

where the (η_p, τ_p)'s are the coordinates of the projections $\mathbf{X}_p^*$ on the interface in the parametric form $\xi = \chi(\eta, \tau)$ centered at $\mathbf{X}^*$. The underdetermined system should be solved by the SVD[5] which gives not only the least squares solution, but also the solution with the least 2-norm among all solutions. The other tangential derivatives can be obtained in the same way with different right-hand sides. Since the coefficient matrix is the same, we just need to compute the SVD once.

The least squares interpolation scheme provides a bridge between the interface and the underlying grid. With the least squares interpolation scheme, it is possible to get a nearly second-order accurate gradient of a computed solution that is second-order accurate. This may not be true for other interpolation schemes.

An alternate approach is described in Chapter 5 to avoid surface derivatives of the jump conditions by transforming the interface problem into a different one with natural (homogeneous) jump conditions. We also refer the readers to [274] and the references therein for a numerical method to evaluate the surface Laplacian on a surface defined by a level set function.

4.4.2 The 10-point finite difference stencil at irregular grid points

To get a second-order discretization, we require the system of equations (4.43)–(4.52) to be satisfied. We can choose three more grid points in addition to the standard central 7-point finite difference stencil so that we have the same number of unknowns and equations for the finite difference coefficients $\{\gamma_m\}$. One way is to choose the three closest grid points from the cubic $(x_{i\pm1}, y_{j\pm1}, z_{k\pm1})$ to the orthogonal projection $\mathbf{X}_{ijk}^*$. The condition number of the coefficient matrix depends on the jump in β. The 10-point finite difference scheme usually works well for problems with small to modest jumps in the coefficient β. The stability of the method, however, is hard to establish except for the case when $[\beta] = 0$. Numerical results using this approach can be found in [160, 161].

4.4.3 The maximum principle preserving scheme in three dimensions

To guarantee the stability and preserve the discrete maximum principle, we also enforce the sign constraints

$$\begin{aligned} \gamma_m < 0 \quad &\text{if } (i_m, j_m, k_m) = (0,0,0), \\ \gamma_m \ge 0 \quad &\text{if } (i_m, j_m, k_m) \ne (0,0,0), \end{aligned} \tag{4.59}$$

[5] In MATLAB, one can use $\boldsymbol{\alpha} = pinv(A) * \mathbf{b}$ to solve a least squares problem $A\boldsymbol{\alpha} = \mathbf{b}$. In Lapack and Linpack, the *SVD* subroutine can be used to get the lease squares solution.

in addition to the linear system of equations (4.43)–(4.52). The sign constraints guarantee that the coefficient matrix of the system of the finite difference equations is an M-matrix.

To solve the linear system of equations (4.43)–(4.52) with the inequality constraints (4.59), we construct a quadratic optimization problem,

$$\min_{\gamma_m} \frac{1}{2} \sum_m (\gamma_m - d_m)^2 \tag{4.60}$$

subject to

$$\begin{aligned} &B\boldsymbol{\gamma} = \mathbf{b}, \quad \text{the system of (4.43)–(4.52)}, \\ &\gamma_m < 0 \quad \text{if } (i_m, j_m, k_m) = (0,0,0), \\ &\gamma_m \geq 0 \quad \text{if } (i_m, j_m, k_m) \neq (0,0,0), \end{aligned} \tag{4.61}$$

where $\boldsymbol{\gamma} = (\gamma_1, \gamma_2, \dots, \gamma_{n_s})^T$ is the vector composed of the coefficients of the finite difference scheme and $B\boldsymbol{\gamma} = \mathbf{b}$ denotes the equality constraints specified by (4.43)–(4.52). We also want to choose the $\{\gamma_m\}$'s in such a way that the finite difference scheme (4.27) becomes the standard central finite difference scheme if there is no interface or the coefficient β of the PDE is continuous across the interface. This can be done by choosing

$$\begin{aligned} d_m &= \frac{\beta_{i+i_m/2, j+j_m/2, k+k_m/2}}{h^2} \\ &\qquad \text{if } (i_m, j_m, k_m) \text{ is one of the six neighbors of } (0,0,0), \\ d_0 &= -\frac{1}{h^2} \sum_{m, m\neq 0} \beta_{i+i_m/2,\ j+j_m/2,\ k+k_m/2}, \\ d_m &= 0 \qquad\qquad \text{otherwise}, \end{aligned} \tag{4.62}$$

where the summation is over the six neighbors of the grid point (x_i, y_j, z_k) and $\beta_{i-1/2,j,k} = \beta(x_i - h_x/2, y_j, z_k)$, and so forth.

What the minimum n_s should be to guarantee the existence of the solution to the optimization problem is still an open theoretical problem. Similar to the maximum principle preserving scheme for two-dimensional problems, we can take all the grid points in the cube centered at (x_i, y_j, z_k), that is, $n_s = 27$. The discussions of the existence of the solution to the optimization problem and the error analysis are similar to that in two space dimensions, and thus are omitted. The maximum principle preserving scheme is second-order accurate, which has been confirmed by numerical tests; see §4.5 and [64, 65].

4.4.4 Solving the finite difference equations using an AMG solver

The size of the linear system of the finite difference equations is very large in three dimensions. For the variable coefficient β, an FFT-based fast solver generally is not applicable. Sparse matrix techniques, iterative methods, and multigrid approaches can be used to solve the linear system. The algebraic multigrid (AMG) method developed by the German National Research Center for Information Technology [227] is quite efficient for the linear system $QU = F$ obtained from the maximum principle preserving scheme in three dimensions. The AMG method is guaranteed to converge if the coefficient matrix Q satisfies one of the following conditions.

- Q is a positive/negative definite or semipositive/seminegative definite matrix, and the sum of the entries in most rows is zero.
- Q is of "essentially" positive type, i.e.,
  - the diagonal entries of Q must be positive/negative,
  - most of the off-diagonal entries of Q are nonpositive/nonnegative,
  - for each row, the sum of the entries in each row should be nonnegative/nonpositive.

The linear system of equations derived from the maximum principle preserving scheme is "essentially" a negative definite matrix (an M-matrix) and the AMG solver can be applied.

4.5 A numerical example for a three-dimensional elliptic interface problem

We show a grid refinement analysis for Example 3.1 in three dimensions. Almost everything is the same as in Example 3.1 except that $r = \sqrt{x^2 + y^2 + z^2}$ and $f(x, y, z) = 10r^2 + 6$ in three dimensions. In Table 4.1, we show the results for three different cases with $b = 1$, $b = 10$ (small jump in β), and $b = 1000$ (large jump in β). We see that the *average ratio* of two consecutive errors $\| E \|_\infty$ is close to 4, indicating second-order accuracy of the maximum principle preserving scheme.

When the interface is represented by a level set function, it is more flexible when dealing with complicated geometries. Table 4.2 shows the results of a grid refinement analysis for a multiconnected domain. The interface is the zero level set of the function

$$\varphi(x, y, z) = S_1(x, y, z) S_2(x, y, z),$$

where

$$S_1(x, y, z) = (x - 0.2)^2 + 2(y - 0.2)^2 + z^2 - 0.01,$$

$$S_2(x, y, z) = 3(x + 0.2)^2 + (y + 0.2)^2 + z^2 - 0.01.$$

Table 4.1. *A grid refinement analysis for Example* 3.1 *in three dimensions with* $M = N = L$.

	$b = 1$		$b = 10$		$b = 1000$	
N	$\| E_N \|_\infty$	Ratio	$\| E_N \|_\infty$	Ratio	$\| E_N \|_\infty$	Ratio
26	1.247×10^{-3}		1.525×10^{-3}		3.485×10^{-3}	
52	3.979×10^{-4}	3.134	5.240×10^{-4}	2.910	1.111×10^{-3}	3.137
104	9.592×10^{-5}	4.148	1.010×10^{-4}	5.188	1.605×10^{-4}	6.922

Table 4.2. *A grid refinement analysis for the multiconnected domain with $M = N = L$.*

N	$\beta^+ = 1,\ \beta^- = 1$		$\beta^+ = 1000,\ \beta^- = 1$	
	$\| E_N \|_\infty$	Ratio	$\| E_N \|_\infty$	Ratio
52	3.108×10^{-2}		2.032×10^{-2}	
104	6.758×10^{-3}	4.599	4.771×10^{-3}	4.259

Table 4.3. *Comparison of the CPU time (in seconds) of the SOR method and the AMG method with $M = N = L$.*

N	Example 3.1 in three dimensions ($b = 10^7$)		Multiconnected domain ($\beta^+ = 10^4,\ \beta^- = 1$)	
	SOR	AMG	SOR	AMG
20	0.21	1.57	0.06	0.83
40	27.51	25.56	29.62	13.89
80	1410.54	265.26	1464.57	265.84

The source term is

$$f(x, y, z) = \begin{cases} 6\beta^- e^{x+2y+z} & \text{in } \Omega^-, \\ -\beta^+(4\pi^2 \sin 2\pi x + \pi^2 \sin \pi y + 16\pi^2 \sin 4\pi z) & \text{in } \Omega^+. \end{cases}$$

The Dirichlet boundary condition is determined from the exact solution,

$$u(x, y, z) = \begin{cases} e^{x+2y+z} & \text{in } \Omega^-, \\ \sin 2\pi x + \sin \pi y + \sin 4\pi z & \text{in } \Omega^+. \end{cases} \tag{4.63}$$

The jump conditions are

$$[u] = \sin 2\pi x + \sin \pi y + \sin 4\pi z - e^{x+2y+z}$$

and $[\beta u_n] = [\beta u_\xi]$ from

$$\begin{pmatrix} [\beta u_\xi] \\ [\beta u_\eta] \\ [\beta u_\tau] \end{pmatrix} = A \begin{pmatrix} 2\pi\beta^+ \cos 2\pi x - \beta^- e^{x+2y+z} \\ \pi\beta^+ \cos \pi y - 2\beta^- e^{x+2y+z} \\ 4\pi\beta^+ \cos 4\pi z - \beta^- e^{x+2y+z} \end{pmatrix}$$

at a point $(x, y, z) \in \Gamma$, where A is the local coordinates transformation matrix defined in (4.4).

We tested two different cases, $\beta^- = \beta^+ = 1$ and $\beta^- = 1,\ \beta^+ = 1000$. Table 4.2 shows the results of a grid refinement analysis. Again second-order accuracy is observed.

In Table 4.3, we show a comparison of the CPU time (in seconds) of the SOR method and the AMG solver for the maximum principle preserving scheme on a Sun™ Ultra™ 10 computer. We see that the AMG solver is much faster than the SOR method when L, M, and N are large. For small problems, the AMG solver may be slower due to the setup time in the AMG solver. More numerical examples can be found in [64, 65].

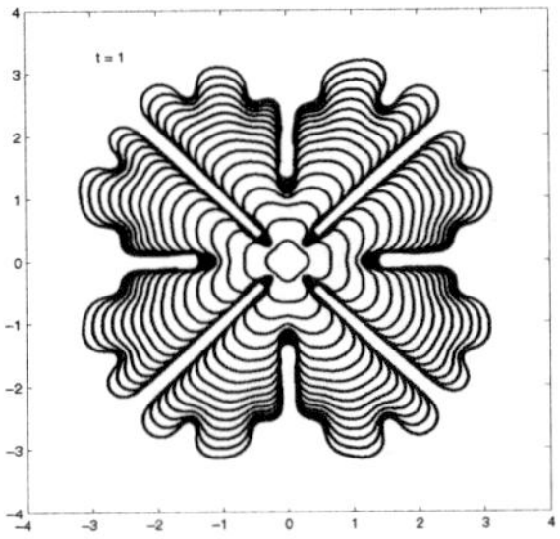

Chapter 5

Removing Source Singularities for Certain Interface Problems

From previous chapters we can see that for homogeneous jump conditions the correction terms in the finite difference equations using the IIM are simply zero. Such jump conditions are often called natural jump (or interface) conditions and have many applications. Physically, it means that there are no sources/sinks along the interface. The IIM is much simplified and easy to implement with natural jump conditions.

For nonhomogeneous jump conditions the IIM requires the first, and possibly the second, surface derivatives of the jump conditions in the correction terms. If the interface is represented by a cubic spline in two space dimensions, the surface derivatives are easy to compute; see [156, 160]. If the interface is represented by the zero level set of a function $\varphi(\mathbf{x})$, then we can use the approach for finding the surface derivatives that is described in [107] for two-dimensional problems and in [65, 64] for three-dimensional problems. The implementation, however, is not trivial for nonexperts in this area.

In [178], a special IIM that does not require the surface derivatives of the jump conditions was proposed for Poisson equations with only singular sources, that is, $[u]$ and $[u_n]$ are given. The new method is based on an extension of the known jump conditions along the normal lines of the interface using a level set function. The main idea is to construct a known function with the same jump conditions as that of the original interface problem. Thus, the difference between the solution to the original problem and the constructed function is smooth across the interface. The method described in [178] applies only to problems with a constant coefficient. In this chapter, we explain the method in Cartesian, polar, and spherical coordinates. We also describe how the strategy can be applied to general elliptic interface problems with discontinuous coefficients.

5.1 Eliminating source singularities using level set functions: The theory

We first introduce the idea of removing source singularities for the following Poisson interface problem:

$$\Delta u = f, \quad \mathbf{x} \in \Omega \setminus \Gamma, \qquad [u]_\Gamma = w, \qquad [u_n]_\Gamma = v \tag{5.1}$$

with a given boundary condition along $\partial\Omega$. We assume that $w \in C^2(\Gamma)$ and $v \in C^2(\Gamma)$.

Let the interface Γ be represented by the zero level set of a function $\varphi(\mathbf{x})$, that is,

$$\varphi(\mathbf{x}) \begin{cases} < 0 & \text{if } \mathbf{x} \in \Omega^-, \\ = 0 & \text{if } \mathbf{x} \in \Gamma, \\ > 0 & \text{if } \mathbf{x} \in \Omega^+. \end{cases} \tag{5.2}$$

We assume that $\varphi(\mathbf{x}) \in C^3(\Omega)$ and $|\nabla\varphi(\mathbf{x})| \neq 0$ in a neighborhood of the interface Γ.[6] The signed distance function will have these properties. In a neighborhood of the interface Γ, we define the extensions of $w(\mathbf{X})$ and $v(\mathbf{X})$ along the normal line (in both directions) as follows:

$$w_e(\mathbf{x}) = w_e(\mathbf{X} + \alpha\mathbf{n}) = w(\mathbf{X}), \qquad \mathbf{X} \in \Gamma, \tag{5.3}$$

$$v_e(\mathbf{x}) = v_e(\mathbf{X} + \alpha\mathbf{n}) = v(\mathbf{X}), \qquad \mathbf{X} \in \Gamma, \tag{5.4}$$

where $\mathbf{X}$ is a point on the interface Γ. In other words, we extend the jump conditions from $\mathbf{X}$ along the normal lines emitted from $\mathbf{X}$. The numerical methods to find approximate orthogonal projections are described in §5.2.2 and §4.3.2 for two and three dimensions, respectively. Here we assume that in a neighborhood of Γ (numerically it is within $|\alpha| \le 2h$ range) the normal lines do not intersect, where $\mathbf{n}$ is the unit normal direction pointing outward. We construct the following function based on the extension:

$$\tilde{u}(\mathbf{x}) = w_e(\mathbf{x}) + v_e(\mathbf{x}) \frac{\varphi(\mathbf{x})}{|\nabla\varphi(\mathbf{x})|}. \tag{5.5}$$

Note that $\tilde{u}(\mathbf{x}) \in C^2$ in the neighborhood of the interface Γ since we assume that w, v are in C^2 on Γ, and φ is in $C^3(\Omega)$ in the neighborhood of the interface Γ. Define also

$$\hat{u}(\mathbf{x}) = H(\varphi(\mathbf{x}))\tilde{u}(\mathbf{x}) = \begin{cases} 0 & \text{if } \varphi(\mathbf{x}) < 0, \\ \tilde{u}(\mathbf{x}) & \text{if } \varphi(\mathbf{x}) \ge 0 \end{cases} \tag{5.6}$$

in the same neighborhood in which $\tilde{u}(\mathbf{x})$ is well defined, where $H(\cdot)$ is the Heaviside function. We have the following theorem.

Theorem 5.1. *Let $u(\mathbf{x})$ be the solution of* (5.1) *and let $\hat{u}(\mathbf{x})$ be defined in* (5.6). *If we define $q(\mathbf{x}) = u(\mathbf{x}) - \hat{u}(\mathbf{x})$, then in the neighborhood of the interface, where $w_e(\mathbf{x})$ and $v_e(\mathbf{x})$ are well defined, the following are true:*

$$\Delta q(\mathbf{x}) = f(\mathbf{x}) - H(\varphi(\mathbf{x}))\, \Delta\hat{u}(\mathbf{x}), \qquad \mathbf{x} \in \Omega \setminus \Gamma, \tag{5.7}$$

$$[q]_\Gamma = 0, \qquad [q_\tau]_\Gamma = 0, \qquad [q_n]_\Gamma = 0, \tag{5.8}$$

where $\boldsymbol{\tau}$ is the unit tangential direction and $q_\tau = \nabla q \cdot \boldsymbol{\tau} = \frac{\partial q}{\partial \boldsymbol{\tau}}$. In other words, the new function $q(\mathbf{x})$ is a smooth function across the interface Γ and satisfies a new Poisson equation (5.7).

---

[6]In the algorithm implementation, $\varphi(\mathbf{x})$ just needs to be in $C^2(\Omega)$ in a small tube $d(\Gamma, \mathbf{x}) \le 2h$. Second-order accurate results are obtained when $\varphi(\mathbf{x}) \in C^2(\Omega)$.

Proof: If $\mathbf{x} \in \Omega^-$, then we have $\hat{u}(\mathbf{x}) = 0$, and thus $\Delta\hat{u}(\mathbf{x}) = 0$ and $H(\varphi(\mathbf{x})) = 0$. Therefore $\Delta q(\mathbf{x}) = \Delta u(\mathbf{x}) - 0 = f(\mathbf{x})$, and (5.7) is true. If $\mathbf{x} \in \Omega^+$, then $H(\varphi(\mathbf{x})) = 1$, and we have

$$\Delta q(\mathbf{x}) = \Delta u(\mathbf{x}) - \Delta\hat{u}(\mathbf{x}) = f(\mathbf{x}) - H(\varphi(\mathbf{x}))\Delta\hat{u}(\mathbf{x}).$$

What is left to prove are the jump conditions. Note that for any $\mathbf{X} \in \Gamma$, we have

$$[q]_{\mathbf{X}} = [u]_{\mathbf{X}} - [\hat{u}]_{\mathbf{X}} = w(\mathbf{X}) - \tilde{u}^+(\mathbf{X}) = w(\mathbf{X}) - w_e(\mathbf{X}) = 0.$$

Since w is differentiable along the interface, so is $w_e(\mathbf{X})$. Differentiating the expression above along the tangential direction, we get

$$\frac{\partial}{\partial\tau}[q]_{\mathbf{X}} = [q_\tau]_{\mathbf{X}} = \frac{\partial}{\partial\tau}[u] - \frac{\partial}{\partial\tau}w_e(\mathbf{X}) = 0.$$

To prove the last jump condition, we proceed with the following derivation:

$$\begin{aligned}\nabla\left(v_e(\mathbf{x})\,\frac{\varphi(\mathbf{x})}{|\nabla\varphi(\mathbf{x})|}\right)\cdot\mathbf{n} &= v_e(\mathbf{x})\,\nabla\frac{\varphi(\mathbf{x})}{|\nabla\varphi(\mathbf{x})|}\cdot\mathbf{n}\\ &= v_e(\mathbf{x})\left(\frac{\nabla\varphi(\mathbf{x})}{|\nabla\varphi(\mathbf{x})|} + \varphi(\mathbf{x})\,\nabla\frac{1}{|\nabla\varphi(\mathbf{x})|}\right)\cdot\mathbf{n};\end{aligned}$$

here we have used the fact that $v_e(\mathbf{x})$ is a constant along the normal line. Using the facts that $\mathbf{n} = \nabla\varphi/|\nabla\varphi|$, $\varphi = 0$ on Γ, and $\nabla w_e(\mathbf{x})\cdot\mathbf{n} = 0$, we conclude that

$$\begin{aligned}\hat{u}_n^+(\mathbf{X}) &= \nabla w_e(\mathbf{x})\cdot\mathbf{n} + v_e(\mathbf{X})\,\frac{\nabla\varphi(\mathbf{X})}{|\nabla\varphi(\mathbf{X})|}\cdot\mathbf{n}\\ &= v(\mathbf{X})\,\frac{\nabla\varphi(\mathbf{X})}{|\nabla\varphi(\mathbf{X})|}\cdot\frac{\nabla\varphi(\mathbf{X})}{|\nabla\varphi(\mathbf{X})|} = v(\mathbf{X})\end{aligned}$$

for any point $\mathbf{X}$ on the interface. Therefore we get $[q_n](\mathbf{X}) = [u_n](\mathbf{X}) - \hat{u}_n^+(\mathbf{X}) = 0$. □

Remark 5.1. *While $q(\mathbf{x})$ is smooth across the interface Γ, its second partial derivatives have finite jumps across the interface. This can also be observed from* (5.7)*, where the right-hand side is discontinuous across the interface.*

5.2 The finite difference scheme using the new formulation

Because $q(\mathbf{x})$ is a smooth function, that is, $q(\mathbf{x}) \in C^1(\Omega)$, it is easier to obtain second-order accurate finite difference methods for (5.7). Clearly, the crucial part is how to extend w and v along the normal lines accurately, and how far they should be extended.

5.2.1 The extension of jump conditions along the normal lines

For reasons that will be seen later, if the standard central 5-point finite difference is used, we need to extend the jump conditions to the nearby grid points, within a tube of width approximately 2h, from both sides of the interface Γ. For this purpose, we define

$$\begin{aligned} \varphi_{ij}^{max} &= \max\{\varphi_{i-1,j}, \varphi_{i,j}, \varphi_{i+1,j}, \varphi_{i,j-1}, \varphi_{i,j+1}\}, \\ \varphi_{ij}^{min} &= \min\{\varphi_{i-1,j}, \varphi_{i,j}, \varphi_{i+1,j}, \varphi_{i,j-1}, \varphi_{i,j+1}\}. \end{aligned} \tag{5.9}$$

As before, we call $\mathbf{x}_{ij}$ an *irregular grid point* in reference to the 5-point stencil if

$$\varphi_{ij}^{max}\varphi_{ij}^{min} \le 0. \tag{5.10}$$

Otherwise, a grid point is a regular grid point. We call a grid point $\mathbf{x}_{ij}$ a *subirregular grid point* in reference to a 5-point stencil if it is a regular grid point and one of its four neighbors is an irregular grid point.

Note that all irregular and subirregular grid points are located within two grid sizes from the interface. *Numerically, we need only extend the jump conditions to all irregular and subirregular grid points.* Given any such grid point $\mathbf{x} = \mathbf{x}_{ij}$, the normal extensions of the jump conditions are approximated by the jump conditions at the orthogonal projection of $\mathbf{x}_{ij}$ on the interface. Usually, the projections are approximated to third-order accuracy as explained below, as are the extensions of the jump conditions.

5.2.2 The orthogonal projections in Cartesian and polar coordinates in two dimensions

Assume that $\mathbf{x}$ is a grid point near the interface. Let $\mathbf{X}^*$ be the orthogonal projection of $\mathbf{x}$ on the interface. The orthogonal projection can be approximated by

$$\mathbf{X}^* = \mathbf{x} + \alpha \mathbf{p}, \tag{5.11}$$

where

$$\mathbf{p} = \begin{bmatrix} \varphi_x(\mathbf{x}) \\ \varphi_y(\mathbf{x}) \end{bmatrix} \tag{5.12}$$

in Cartesian coordinates and

$$\mathbf{p} = \begin{bmatrix} \varphi_r(\mathbf{x}) \\ \dfrac{\varphi_\theta(\mathbf{x})}{r^2} \end{bmatrix} \tag{5.13}$$

in polar coordinates. The scalar α is determined from the following quadratic equation:

$$\varphi(\mathbf{x}) + (\nabla\varphi(\mathbf{x}) \cdot \mathbf{p})\,\alpha + \frac{1}{2}\left(\mathbf{p}^T \mathrm{He}(\varphi(\mathbf{x}))\,\mathbf{p}\right)\alpha^2 = 0, \tag{5.14}$$

where

$$\mathbf{p}^T \mathrm{He}(\varphi)\,\mathbf{p} = \varphi_x^2\,\varphi_{xx} + 2\,\varphi_x\varphi_y\varphi_{xy} + \varphi_y^2\,\varphi_{yy} \tag{5.15}$$

in Cartesian coordinates and

$$\mathbf{p}^T \mathrm{He}(\varphi)\,\mathbf{p} = \varphi_r^2\,\varphi_{rr} + \frac{2}{r^2}\,\varphi_r\varphi_\theta\varphi_{r\theta} + \frac{1}{r^4}\,\varphi_\theta^2\,\varphi_{\theta\theta} \tag{5.16}$$

in polar coordinates. If $\varphi_{ij} < 0$, then α is chosen as positive; if $\varphi_{ij} > 0$, then α is chosen as negative; if $\varphi_{ij} = 0$, then $\alpha = 0$. The partial derivatives $\nabla\varphi(\mathbf{x})$ and the Hessian matrix $\mathrm{He}(\varphi)$ are computed at the grid point $\mathbf{x}$ using the standard centered 5-point finite difference formula. Using the method above, the computed projections have third-order accuracy; see Table 5.1.

5.2.3 The discretization strategy using the transformation

Since $q(\mathbf{x}) = u(\mathbf{x}) - \hat{u}(\mathbf{x})$ is a smooth function, we can apply the standard central 5-point finite difference scheme to (5.7) directly and obtain at least first-order accuracy. However, this is not practical because we do not know $\hat{u}(\mathbf{x})$ everywhere for at least two reasons: (1) the normal lines can intersect with each other away from the interface, which will make the extension ambiguous; (2) for a point that is far away from the interface, it is hard to determine the orthogonal projection on the interface accurately. Fortunately, we need only modify the finite difference equations at irregular grid points and extend the jump conditions to irregular and subirregular grid points. The justification is given below.

At a regular or subirregular grid point $\mathbf{x}_{ij}$ where $\varphi(\mathbf{x}_{ij}) > 0$, we have

$$\Delta_h \hat{u}(\mathbf{x}_{ij}) = H(\varphi(\mathbf{x}_{ij}))\,\Delta_h \hat{u}(\mathbf{x}_{ij}) = H(\varphi(\mathbf{x}_{ij}))\,\Delta_h \tilde{u}(\mathbf{x}_{ij}), \tag{5.17}$$

where Δ_h is the standard central 5-point finite difference operator. In a two dimensional Cartesian grid with $h_x = h_y = h$, Δ_h is

$$\Delta_h U_{ij} = \frac{U_{i-1,j} + U_{i+1,j} + U_{i,j-1} + U_{i,j+1} - 4U_{ij}}{h^2}, \tag{5.18}$$

while in polar coordinates, it is

$$\begin{aligned}\Delta_h U_{ij} = {} & \frac{U_{i-1,j} - 2U_{ij} + U_{i+1,j}}{(\Delta r)^2} + \frac{1}{r_i}\,\frac{U_{i+1,j} - U_{i-1,j}}{2\Delta r} \\ & + \frac{1}{r_i^2}\,\frac{U_{i,j-1} - 2U_{ij} + U_{i,j+1}}{(\Delta\theta)^2}.\end{aligned} \tag{5.19}$$

Note that in polar coordinates, we have a periodic boundary condition in j, that is, $U_{i,0} = U_{i,N}$. To avoid the singularity at $r = 0$, we often use a staggered grid such that $r_{-1/2} = -\Delta r/2$, $r_{1/2} = \Delta r/2$, and $\Delta r = r_{max}/(M-1/2)$; see [178]. Therefore, the finite difference scheme applied to (5.7),

$$\Delta_h q_{ij} = f_{ij} - H(\varphi_{ij})\Delta_h \tilde{u}_{ij}$$

or

$$\Delta_h \left(U_{ij} - \hat{u}_{ij}\right) = F_{ij} = f_{ij} - H(\varphi_{ij})\Delta_h \tilde{u}_{ij},$$

at a regular grid point (including a subirregular grid) is equivalent to the standard finite difference scheme[7]

$$\Delta_h U_{ij} = f_{ij}.$$

[7]Note that $\tilde{u}$ and $\hat{u}$ are defined using the known jump conditions and are not unknowns, so we still use the lowercase letters to represent them at the grid points.

It is straightforward that the above statement is also true for a regular grid point $\mathbf{x}_{ij}$, where $\varphi(\mathbf{x}_{ij}) < 0$, since both $H(\varphi_{ij})$ and $\Delta_h \hat{u}_{ij}$ are zero.

At an irregular grid point (x_i, y_j), however, we need to add a correction term C_{ij} to the standard 5-point finite difference scheme applied to (5.7),

$$\Delta_h \left(U_{ij} - \Delta_h \hat{u}_{ij}\right) = F_{ij} = f_{ij} - H(\varphi_{ij})\Delta_h \tilde{u}_{ij} + C_{ij}, \tag{5.20}$$

to treat the discontinuity in F_{ij}. Note that such a discontinuity exists even if f is continuous since $H(\varphi(\mathbf{x}))$ is discontinuous at Γ. The correction term is $C_{ij} = a_8[F]$ in (3.23) and has a closed form (5.23).

In the finite difference scheme above, we need to evaluate $\Delta_h \tilde{u}_{ij}$ and $\Delta_h \hat{u}_{ij}$ at an irregular grid point (x_i, y_j). Thus, information about $\tilde{u}_{ij}$ and $\hat{u}_{ij}$ is needed at the grid points in the central 5-point finite difference stencil as required by Δ_h. That is why we need only extend the jump conditions to the irregular and subirregular points.

5.2.4 An outline of the algorithm of removing source singularities

The algorithm for solving (5.1) using the new formulation is outlined below.

- Set up a grid.
- Label all the grid points as irregular, regular, or subirregular.
- Find the projections for irregular and subirregular grid points using (5.11)–(5.14).
- Form the extensions $\{\tilde{u}_{ij}\}$ and $\{\hat{u}_{ij}\}$ at irregular and subirregular grid points according to (5.5) and (5.6), respectively.
- Form the discrete equations

$$\Delta_h U_{ij} = \begin{cases} f_{ij} - H(\varphi_{ij})\Delta_h \tilde{u}_{ij} + \Delta_h \hat{u}_{ij} + C_{ij} & \text{if } \mathbf{x}_{ij} \text{ is irregular,} \\ f_{ij} & \text{otherwise.} \end{cases} \tag{5.21}$$

  The correction term C_{ij} is defined in (5.23).
- Apply a fast Poisson solver, for example, a solver from [2, 143], to solve the discrete system of (5.21).

5.2.5 A closed formula for the correction terms

Let $\mathbf{x}_{ij}$ be an irregular grid point, say in the "−" side of the interface. Below we discuss how to determine C_{ij} in (5.21). Without the correction term (i.e., $C_{ij} = 0$), the finite difference scheme (5.21) is first-order accurate. We denote

$$F_{ij} = f_{ij} - H(\varphi_{ij})\, \Delta_h \tilde{u}_{ij}, \tag{5.22}$$

which is computable. The correction term is $a_8[F]$ in (3.23) and can be simplified to

$$C_{ij} = \sum_{i_k, j_k} H(-\varphi_{i+i_k, j+j_k} \cdot \varphi_{ij})\, \gamma_{i+i_k, j+j_k} \left(\frac{\varphi_{i+i_k, j+j_k}}{|\nabla \varphi_{i+i_k, j+j_k}|} \right)^2 \frac{F_{i+i_k, j+j_k} - F_{ij}}{2}, \tag{5.23}$$

where $(i_k, j_k) = \{(-1,0), (1,0), (0,-1), (0,1)\}$, $H(\cdot)$ is the Heaviside function,[8] and $\gamma_{i+i_k, j+j_k}$ is the coefficient of the discrete Laplacian Δ_h in (5.18) or (5.19) corresponding to $U_{i+i_k, j+j_k}$. In Cartesian coordinates, these coefficients are simply $\gamma_{i\pm1, j\pm1} = 1/h^2$, while in polar coordinates, they are

$$\gamma_{i\pm1,j} = \frac{1}{(\Delta r)^2} \pm \frac{1}{2r_i \Delta r}, \qquad \gamma_{i,j\pm1} = \frac{1}{(\Delta\theta)^2}. \tag{5.24}$$

To show why we have the correction term (5.23), we introduce the following lemmas.

Lemma 5.2. *With the same notations and settings as in Theorem* 5.1. *Let* (ξ, η) *be the local coordinate system at a fixed point on the interface, and* $F(\mathbf{x}) = f(\mathbf{x}) - H(\varphi)\Delta\tilde{u}(\mathbf{x})$. *The following equality holds*

$$\left[\frac{\partial^2 q}{\partial \xi^2}\right] = [\Delta q] = \left[f - H(\varphi)\Delta\tilde{u}\right] = [F] \tag{5.25}$$

under the local coordinate system.

Proof: Since the Laplacian operator is invariant under orthogonal coordinates, we have

$$\Delta q = \frac{\partial^2 q}{\partial \xi^2} + \frac{\partial^2 q}{\partial \eta^2} = F.$$

From $[q] = 0$, $[q_\xi] = 0$, and $[q_\eta] = 0$ proved in Theorem 5.1, we can conclude that $[\frac{\partial^2 q}{\partial \eta^2}] = 0$, which is derived in §3.1 for the interface relations in the local coordinates. Therefore we conclude

$$[\Delta q] = [F] = \left[\frac{\partial^2 q}{\partial \xi^2}\right]. \qquad \square$$

The following lemma is the basis for the correction terms and will also be used in §5.2.6 for accurate evaluation of the partial derivatives (u_x, u_y).

Lemma 5.3. *Let* $\varphi(\mathbf{x}) \in C^3$ *in a neighborhood of the zero level set* $\varphi(\mathbf{x}) = 0$; *let* $\mathbf{x}_{ij}$ *be an irregular point with* $\varphi_{i-1,j} < 0$, $\varphi_{ij} < 0$, $\varphi_{i+1,j} > 0$, $\nabla\varphi_{i+1,j} = O(1)$; *and let* $\tilde{u}(\mathbf{x})$ *be well defined at the central* 5*-point finite difference stencil. The following relation is true:*

$$\begin{aligned} \delta_{xx}^h q(\mathbf{x}_{ij}) &\overset{def}{=} \frac{q(\mathbf{x}_{i-1,j}) - 2q(\mathbf{x}_{ij}) + q(\mathbf{x}_{i+1,j})}{h_x^2} \\ &= \frac{\partial^2 q}{\partial x^2}(\mathbf{x}_{ij}) + \gamma_{i+1,j}\left(\frac{\varphi_{i+1,j}}{|\nabla\varphi_{i+1,j}|}\right)^2 \frac{F_{i+1,j} - F_{ij}}{2} \\ &\quad + O\left(\varphi_{i+1,j}^3 / h_x^2\right). \end{aligned} \tag{5.26}$$

[8] In the discrete case, we define $H(\varphi) = 1$ if $\varphi \geq 0$.

Proof: Let $\mathbf{X}^*$ be the orthogonal projection of $\mathbf{x}_{i+1,j}$ on the interface. Expanding $\varphi(\mathbf{x})$ in a Taylor expansion at $\mathbf{X}^*$, we get

$$0 = \varphi(\mathbf{X}^*) = \varphi(\mathbf{x}_{i+1,j}) + \nabla\varphi(\mathbf{x}_{i+1,j}) \cdot \left(\mathbf{X}^* - \mathbf{x}_{i+1,j}\right) + O\left(\left|\mathbf{x}_{i+1,j} - \mathbf{X}^*\right|^2\right).$$

Since $\mathbf{X}^* - \mathbf{x}_{i+1,j}$ is parallel to the normal direction $\mathbf{n}$ $(= \nabla\varphi/|\nabla\varphi|)$ according to the definition of the projection, and $\varphi_{i+1,j} > 0$, we get

$$\left|\mathbf{x}_{i+1,j} - \mathbf{X}^*\right| = \frac{\varphi(\mathbf{x}_{i+1,j})}{|\nabla\varphi(\mathbf{x}_{i+1,j})|} + O\left(\left|\mathbf{x}_{i+1,j} - \mathbf{X}^*\right|^2\right). \tag{5.27}$$

Note that $\varphi(\mathbf{x})/|\nabla\varphi(\mathbf{x})|$ is known to be an approximation to the distance between $\mathbf{x}$ and the interface Γ. We expand $q(\mathbf{x}_{i+1,j})$ in a Taylor expansion at $\mathbf{X}^*$ to get

$$\begin{aligned} q(\mathbf{x}_{i+1,j}) &= q^+(\mathbf{X}^*) + \nabla q^+(\mathbf{X}^*) \cdot \left(\mathbf{x}_{i+1,j} - \mathbf{X}^*\right) \\ &\quad + \frac{1}{2}\left(\mathbf{x}_{i+1,j} - \mathbf{X}^*\right)^T \mathrm{He}(q^+(\mathbf{X}^*))\left(\mathbf{x}_{i+1,j} - \mathbf{X}^*\right) + O\left(\left|\mathbf{x}_{i+1,j} - \mathbf{X}^*\right|^3\right), \end{aligned}$$

where $\mathrm{He}(q^+(\mathbf{X}^*))$ is the Hessian matrix of $q(\mathbf{x})$ at $\mathbf{X}^*$. Since

$$\frac{\mathbf{X}^* - \mathbf{x}_{i+1,j}}{|\mathbf{X}^* - \mathbf{x}_{i+1,j}|} = -\mathbf{n} + O(h_x^2),$$

from the continuity condition of $q(\mathbf{x})$ and its derivative at $\mathbf{X}^*$, we get

$$\begin{aligned} q(\mathbf{x}_{i+1,j}) &= q^-(\mathbf{X}^*) - \frac{\partial q^-(\mathbf{X}^*)}{\partial n}\left|\mathbf{x}_{i+1,j} - \mathbf{X}^*\right| \\ &\quad + \frac{1}{2}\left(\mathbf{x}_{i+1,j} - \mathbf{X}^*\right)^T \left(\mathrm{He}(q^-(\mathbf{X}^*)) + \left[\mathrm{He}(q(\mathbf{X}^*))\right]\right)\left(\mathbf{x}_{i+1,j} - \mathbf{X}^*\right) \\ &\quad + O\left(\left|\mathbf{x}_{i+1,j} - \mathbf{X}^*\right|^3\right). \end{aligned}$$

Define a smooth extension of $q^-(\mathbf{x})$ from Ω^- into Ω^+ in the neighborhood of $\mathbf{X}^*$ as

$$\begin{aligned} q^-(\mathbf{x}) &= q^-(\mathbf{X}^*) - \frac{\partial q^-(\mathbf{X}^*)}{\partial n}\left|\mathbf{x} - \mathbf{X}^*\right| \\ &\quad + \frac{1}{2}\left(\mathbf{x}_{i+1,j} - \mathbf{X}^*\right)^T \mathrm{He}(q^-(\mathbf{X}^*))\left(\mathbf{x}_{i+1,j} - \mathbf{X}^*\right). \end{aligned}$$

Again from Theorem 5.1, we know that

$$[q_{\xi\eta}] = 0, \qquad [q_{\eta\eta}] = 0.$$

This leads to

$$\left(\mathbf{x}_{i+1,j} - \mathbf{X}^*\right)^T \left[\mathrm{He}(q(\mathbf{X}^*))\right]\left(\mathbf{x}_{i+1,j} - \mathbf{X}^*\right) = \left|\mathbf{x}_{i+1,j} - \mathbf{X}^*\right|^2 \left[\frac{\partial^2 q(\mathbf{X}^*)}{\partial \xi^2}\right]$$

and

$$q(\mathbf{x}_{i+1,j}) = q^-(x_{i+1,j}) + \frac{1}{2}\left|\mathbf{x}_{i+1,j} - \mathbf{X}^*\right|^2 \left[\frac{\partial^2 q(\mathbf{X}^*)}{\partial \xi^2}\right] + O\left(\left|\mathbf{x}_{i+1,j} - \mathbf{X}^*\right|^3\right).$$

Therefore, we have

$$\begin{aligned}
\delta^h_{xx} q(\mathbf{x}_{ij}) &= \frac{q(\mathbf{x}_{i-1,j}) - 2q(\mathbf{x}_{ij}) + q(\mathbf{x}_{i+1,j})}{h_x^2} \\
&= \frac{q(\mathbf{x}_{i-1,j}) - 2q(\mathbf{x}_{ij}) + q^-(\mathbf{x}_{i+1,j})}{h_x^2} \\
&\quad + \frac{1}{h_x^2}\left(\frac{1}{2}\left|\mathbf{x}_{i+1,j} - \mathbf{X}^*\right|^2 \left[\frac{\partial^2 q(\mathbf{x}^*)}{\partial \xi^2}\right] + O\left(\left|\mathbf{x}_{i+1,j} - \mathbf{X}^*\right|^3\right)\right).
\end{aligned}$$

Since $q^-(\mathbf{x}_{i+1,j})$ is the extension of $q^-(\mathbf{x})$ from Ω^- into Ω^+, the first term in the expression above is a second-order approximation to $\frac{\partial^2 q}{\partial \mathbf{x}^2}(\mathbf{x}_{ij})$. Therefore, the expression above can also be written as

$$\begin{aligned}
\delta^h_{xx} q(\mathbf{x}_{ij}) = \frac{\partial^2 q}{\partial x^2}(\mathbf{x}_{ij}) + \frac{1}{h_x^2}\Bigg(\frac{1}{2}\left|\mathbf{x}_{i+1,j} - \mathbf{X}^*\right|^2 \left[\frac{\partial^2 q(\mathbf{X}^*)}{\partial \xi^2}\right] \\
+ O\left(\left|\mathbf{x}_{i+1,j} - \mathbf{X}^*\right|^3\right)\Bigg).
\end{aligned} \tag{5.28}$$

Finally, from (5.27) and Lemma 5.2 we have

$$\begin{aligned}
\frac{1}{2}\left|\mathbf{x} - \mathbf{X}^*\right|^2 \left[\frac{\partial^2 q(\mathbf{X}^*)}{\partial \xi^2}\right] &= \frac{1}{2}\left(\frac{\varphi_{i+1,j}}{|\nabla \varphi_{i+1,j}|}\right)^2 \left[\frac{\partial^2 q(\mathbf{X}^*)}{\partial \xi^2}\right] + O\left(\left|\mathbf{x}_{i+1,j} - \mathbf{X}^*\right|^3\right) \\
&= \frac{1}{2}\left(\frac{\varphi_{i+1,j}}{|\nabla \varphi_{i+1,j}|}\right)^2 [F(\mathbf{X}^*)] + O\left(\left|\mathbf{x}_{i+1,j} - \mathbf{X}^*\right|^3\right) \\
&= \frac{1}{2}\left(\frac{\varphi_{i+1,j}}{|\nabla \varphi_{i+1,j}|}\right)^2 (F_{i+1,j} - F_{ij}) + O\left(\left|\mathbf{x}_{i+1,j} - \mathbf{X}^*\right|^3\right),
\end{aligned}$$

where we have used (5.25), the fact $\varphi_{i+1,j} = O(|\mathbf{x}_{i+1,j} - \mathbf{X}^*|)$, and $[F(\mathbf{X}^*)] = F_{i+1,j} - F_{ij} + O(|\mathbf{x}_{i+1,j} - \mathbf{X}^*|)$. Plugging the expression above into (5.28), we get the result of the lemma. □

The last term in (5.26) is the contribution to the correction term C_{ij} from $\mathbf{x}_{i+1,j}$ if it is on the opposite side of the interface from the grid point $\mathbf{x}_{ij}$. Similarly, we can show the contributions to the correction term from the other neighboring points if they are in the "+" side.

With the correction term C_{ij}, the finite difference scheme matches the differential equation up to second-order derivatives after the Taylor expansion at $\mathbf{X}^*$. As a result, the computed solution of (5.21) has global second-order accuracy in the infinity norm; see [166] for the analysis.

Since the left-hand side of the system of the finite difference equations (5.21) is the standard discrete Laplacian, the FFT-based fast Poisson solver [2, 143] can be applied directly. The main cost is the Poisson solver which typically requires about $MN\log(\max\{M, N\})$ operations. The cost in dealing with the interface is $O(N_I)$, where N_I is the total number of irregular and subirregular grid points. Usually, we have $N_I \sim \max\{M, N\}$ which is one dimension lower than the total number of grid points.

5.2.6 Computing the gradient using the new formulation

With the new formulation, we can also easily evaluate the gradient of the solution to (5.1). The first-order partial derivative can be approximated by

$$\frac{\partial U_{ij}}{\partial x} \approx \begin{cases} \dfrac{U_{i+1,j}-U_{i-1,j}}{2h_x} - \dfrac{\hat{u}_{i+1,j}-\hat{u}_{i-1,j}}{2h_x} \\ \qquad + H(\varphi_{ij})\dfrac{\tilde{u}_{i+1,j}-\tilde{u}_{i-1,j}}{2h_x} + C_{ij}^x & \text{if } \mathbf{x}_{ij} \text{ is irregular,} \\ \dfrac{U_{i+1,j}-U_{i-1,j}}{2h_x} & \text{otherwise,} \end{cases} \tag{5.29}$$

$$\frac{\partial U_{ij}}{\partial y} \approx \begin{cases} \dfrac{U_{i,j+1}-U_{i,j-1}}{2h_y} - \dfrac{\hat{u}_{i,j+1}-\hat{u}_{i,j-1}}{2h_y} \\ \qquad + H(\varphi_{ij})\dfrac{\tilde{u}_{i,j+1}-\tilde{u}_{i,j-1}}{2h_y} + C_{ij}^y & \text{if } \mathbf{x}_{ij} \text{ is irregular,} \\ \dfrac{U_{i,j+1}-U_{i,j-1}}{2h_y} & \text{otherwise,} \end{cases} \tag{5.30}$$

where the correction terms are

$$\begin{aligned} C_{ij}^x = {} & \frac{1}{2h_x} H(-\varphi_{i+1,j}\,\varphi_{ij}) \left(\frac{\varphi_{i+1,j}}{|\nabla\varphi_{i+1,j}|}\right)^2 \frac{F_{i+1,j}-F_{ij}}{2} \\ & - \frac{1}{2h_x} H(-\varphi_{i-1,j}\,\varphi_{ij}) \left(\frac{\varphi_{i-1,j}}{|\nabla\varphi_{i-1,j}|}\right)^2 \frac{F_{i-1,j}-F_{ij}}{2}, \end{aligned} \tag{5.31}$$

$$\begin{aligned} C_{ij}^y = {} & \frac{1}{2h_y} H(-\varphi_{i,j+1}\,\varphi_{ij}) \left(\frac{\varphi_{i,j+1}}{|\nabla\varphi_{i,j+1}|}\right)^2 \frac{F_{i,j+1}-F_{ij}}{2} \\ & - \frac{1}{2h_y} H(-\varphi_{i,j-1}\,\varphi_{ij}) \left(\frac{\varphi_{i,j-1}}{|\nabla\varphi_{i,j-1}|}\right)^2 \frac{F_{i,j-1}-F_{ij}}{2}. \end{aligned} \tag{5.32}$$

The correction term is zero at a regular grid point. The accuracy of the computed gradient from the computed solution at regular grid points is generally second order for a smooth solution. The correction term is nonzero and needed at irregular grid points to offset the discontinuity in the second-order derivatives of $q(\mathbf{x})$. The reasoning is the same as discussed in §5.2.5. Because $q(\mathbf{x})$ is a smooth function, the computed gradient is also second-order accurate.

Because the level set function is defined everywhere, or at least in a tube $|\varphi(\mathbf{x})| \le \delta$ that contains the interface, where $\delta \sim O(h)$ is a given width, we can naturally define the "normal derivative" at all grid points in the tube $|\varphi(\mathbf{x})| \le \delta$ as

$$\frac{\partial u}{\partial \mathbf{n}}(\mathbf{x}) = \nabla u(\mathbf{x}) \cdot \frac{\nabla \varphi(\mathbf{x})}{|\nabla \varphi(\mathbf{x})|}. \tag{5.33}$$

Thus we can get a second-order accurate "normal derivative" at grid points in the tube as well. The formula above is used in §5.2.7 to compute the normal derivative of the solution at irregular grid points.

5.2.7 An example of removing source singularities

We present an example of the numerical method discussed in this chapter using the polar coordinates. The interface is the zero level set of

$$\varphi(r,\theta) = r - (0.5 + 0.1\sin(4\theta + \pi)), \quad 0 \le r \le 1, \quad 0 \le \theta < 2\pi. \tag{5.34}$$

To check the order of accuracy of the method by removing source singularities, we use two nonlinear functions,

$$u(r,\theta) = \begin{cases} r^4 & \text{if } \varphi(r,\theta) < 0, \\ r^2 \sin\theta & \text{if } \varphi(r,\theta) \ge 0, \end{cases} \tag{5.35}$$

as the exact solution. Note that the solution depends on both r and θ. The source term excluding the interface Γ is

$$f(r,\theta) = \begin{cases} 16r^2 & \text{if } \varphi(r,\theta) < 0, \\ 3\sin\theta & \text{if } \varphi(r,\theta) \ge 0. \end{cases} \tag{5.36}$$

The Dirichlet boundary condition at $r_{max} = 1$, and the jump conditions in the solution and in the flux (the normal derivative) are computed from the exact solution at the projections.

Table 5.1 lists a grid refinement analysis of the computed results. The error of the computed projections is defined as

$$E_p = \max_k \left| r_e(\theta_k^*) - r_k^* \right|, \tag{5.37}$$

where (r_k^*, θ_k^*)'s are computed projections of all irregular grid points and $r_e(\theta^*)$ is the exact interface relation given by $r_e(\theta) = 0.5 + 0.1\sin(4\theta + \pi)$. The error in the solution E_u is defined as

$$E_u = \max_{\substack{0 \le i \le M \\ 0 \le j \le N}} \left| u(r_i,\theta_j) - U_{ij} \right|, \tag{5.38}$$

where U_{ij} is the computed solution at the grid points.

The errors of the normal derivatives are measured at two levels. $E_{u_n,g}$ is the error measured at all irregular grid points,

$$E_{u_n,g} = \max_{(r_i,\theta_j)\ \text{is irregular}} \left| u_{\mathbf{n}}(r_i,\theta_j) - (U_{\mathbf{n}})_{ij} \right|, \tag{5.39}$$

Table 5.1. *A grid refinement analysis of the numerical method based on removing source singularities. The parameters are chosen as $N = 2M$, where N and M are the number of grid lines in the r and θ directions, respectively.*

M	E_p	r_p	$E_{u_n,g}$	$r_{u_n,g}$	$E_{u_n,\Gamma}$	$r_{u_n,\Gamma}$	E_u	r_u
40	$1.91\ 10^{-4}$		$1.29\ 10^{-2}$		$9.04\ 10^{-3}$		$1.87\ 10^{-3}$	
80	$2.45\ 10^{-5}$	2.97	$3.05\ 10^{-3}$	2.08	$2.07\ 10^{-3}$	2.13	$3.38\ 10^{-4}$	2.47
160	$3.07\ 10^{-6}$	2.99	$6.14\ 10^{-4}$	2.31	$5.26\ 10^{-4}$	1.98	$7.42\ 10^{-5}$	2.19
320	$3.86\ 10^{-7}$	2.99	$1.59\ 10^{-4}$	1.95	$1.47\ 10^{-4}$	1.83	$1.78\ 10^{-5}$	2.06
640	$4.83\ 10^{-8}$	3.00	$3.76\ 10^{-5}$	2.08	$3.75\ 10^{-5}$	1.97	$4.24\ 10^{-6}$	2.07

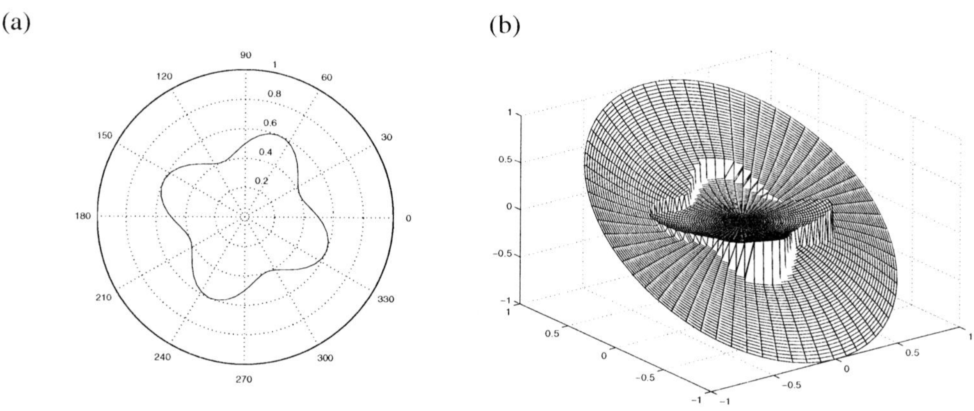

Figure 5.1. (a) *The domain $r = 1$ and the interface $r = 0.5 + 0.1\sin(4\theta + \pi)$ used in Table* 5.1. (b) *A mesh plot of the computed solution.*

while $E_{u_n,\Gamma}$ is the error measured at all projections of irregular grid points,

$$E_{u_n,\Gamma} = \max_{(r_k^*,\theta_k^*)} \left| u_{\boldsymbol{n}}(r_k^*, \theta_k^*) - (U_{\boldsymbol{n}})_{(r_k^*,\theta_k^*)} \right|, \tag{5.40}$$

where $(U_{\boldsymbol{n}})_{ij}$ and $(U_{\boldsymbol{n}})_{(r_k^*,\theta_k^*)}$ are the computed normal derivatives at grid points and at the projections, respectively. At a projection, we compare both $(U_{\boldsymbol{n}}^+)_{(r_k^*,\theta_k^*)}$ and $(U_{\boldsymbol{n}}^-)_{(r_k^*,\theta_k^*)}$ with the exact normal derivatives from each side of the interface. The convergence order $r_E(M)$ is defined as

$$r_E(M) = \frac{\log(E(M)/E(2M))}{\log 2}. \tag{5.41}$$

We see that the computed projections are *third*-order accurate, while the computed solution, the normal derivative at grid points, and the normal derivative at projections are all second-order accurate. We also plot the domain and the interface in Figure 5.1(a) and a mesh plot of a computed solution in Figure 5.1(b).

5.3 Removing source singularities for variable coefficients

In this section, we consider more general cases. For an elliptic interface problem

$$\nabla \cdot (\beta \nabla u) = f, \quad [u] = w, \quad [\beta u_n] = v, \tag{5.42}$$

if the coefficient β is discontinuous, then it is difficult, if not impossible, to transform the original problem into a new one with a smooth solution. However, using the idea described in the previous section in this chapter, we can transform the original problem into a new one with homogeneous jump conditions. This is significant for the IIM because it is not trivial to compute surface derivatives, especially in three dimensions.

Now we define

$$\tilde{u}(\mathbf{x}) = w_e(\mathbf{x}) + \frac{v_e(\mathbf{x})}{\beta^+(\mathbf{X}^*)} \frac{\varphi(\mathbf{x})}{|\nabla \varphi(\mathbf{x})|}, \tag{5.43}$$

where $\mathbf{X}^*$ is the orthogonal projection of $\mathbf{x}$ on the interface. Note that $\tilde{u}(\mathbf{x})$ is a smooth function provided that w and v are smooth functions defined along the interface Γ and $\varphi \in C^2(\Omega)$.

We define $\hat{u}$ in exactly the same way as in (5.6) and we let $q(\mathbf{x}) = u(\mathbf{x}) - \hat{u}(\mathbf{x})$. Then by following the proof process of Theorem 5.1, it is straightforward to show that

$$[q]_\Gamma = 0, \qquad [\beta q_n]_\Gamma = 0, \tag{5.44}$$

and

$$\nabla \cdot (\beta \nabla q) = f - H(\varphi)\, \nabla \cdot (\beta \nabla \tilde{u}), \quad \mathbf{x} \in \Omega - \Gamma. \tag{5.45}$$

Numerically, we need only modify the finite difference scheme at irregular grid points in reference to the standard 5-point stencil and the central finite difference scheme (3.14). At an irregular grid point, the finite difference scheme in two dimensions can be written as

$$\sum_k^{n_s} \gamma_k\, U_{i+i_k, j+j_k} = f_{ij} + \sum_k^{n_s} \gamma_k\, \hat{u}_{i+i_k, j+j_k} - H(\varphi_{ij}) L_h \tilde{u}_{ij} + C_{ij}, \tag{5.46}$$

where L_h is the *standard finite difference operator* used at regular grid points, for example, the central 5-point central finite difference scheme (3.14). The correction term

$$C_{ij} = a_8 \frac{[f] - L_h \tilde{u}_{ij}}{\beta^+} \tag{5.47}$$

involves only the jump in f and $[H(\varphi)\, \nabla \cdot (\beta \nabla \tilde{u})]$; see (3.19). Note that the correction term C_{ij} no longer needs any surface derivatives. This is the main motivation for this extension and the transformation for the interface problems.

Caution has to be taken when determining how far we should extend the jump conditions. When β is continuous and the standard 5-point central finite difference scheme is used, we need only extend the jump conditions to irregular and subirregular grid points that are within two grid sizes. When β is discontinuous, we need to use either a 6-point or 9-point stencil in the IIM. In this case, we need to extend the jump conditions to all the neighboring grid points of the finite difference stencil for the particular irregular master grid point. Specifically, if the compact central 9-point stencil is used, then we need to extend the jump conditions to irregular and subirregular grid points, plus to the neighboring grid points of subirregular grid points. Numerical experiments for (5.42) have confirmed the analysis here.

5.4 Orthogonal projections and extensions in spherical coordinates

The strategy of removing source singularities has been applied to Poisson equations with interfaces in the finite or the infinite domain in three dimensions [145] using the spherical coordinates. Most discussions are similar to that in two dimensions except for the orthogonal projections. In the following discussion, we use the same notation as we used in the previous sections of this chapter for problems in two space dimensions.

The spherical coordinates system is

$$\begin{aligned} x &= r\,\sin\phi\,\cos\theta, \\ y &= r\,\sin\phi\,\sin\theta, \qquad r \geq 0, \quad 0 \leq \theta < 2\pi, \quad 0 \leq \phi < \pi. \\ z &= r\,\cos\phi. \end{aligned}$$

Let $\mathbf{x}$ be a point near the interface Γ, and let $\mathbf{X}^*$ be the corresponding orthogonal projection of $\mathbf{x}$ on the interface. As before we write

$$\mathbf{X}^* = \mathbf{x} + \alpha\mathbf{p}, \tag{5.48}$$

where α is a scalar and $\mathbf{p}$ is a direction to be determined. If the direction $\mathbf{p}$ is known, α can be approximated by the solution of the quadratic equation

$$\varphi(\mathbf{x}) + (\nabla\varphi(\mathbf{x})\cdot\mathbf{p})\,\alpha + \frac{1}{2}\left(\mathbf{p}^T \mathrm{He}(\varphi(\mathbf{x}))\,\mathbf{p}\right)\alpha^2 = 0, \tag{5.49}$$

where $\nabla\varphi$ and the Hessian matrix $\mathrm{He}(\varphi)$ are evaluated at $\mathbf{x}$. The above equation is a third-order approximation to $\varphi(\mathbf{X}^*) = 0$. The key is knowing how to choose the right direction $\mathbf{p}$ in spherical coordinates to get an approximation of the orthogonal projection $\mathbf{X}^*$.

Let $[\Delta x, \Delta y, \Delta z\,]^T$ and $[\Delta r, \Delta\phi, \Delta\theta\,]^T$ be the first-order approximations of the vector $\mathbf{x} - \mathbf{X}^*$ in Cartesian and spherical coordinates, respectively. Then we have

$$\begin{aligned} \begin{bmatrix} \Delta x \\ \Delta y \\ \Delta z \end{bmatrix} &\approx \begin{bmatrix} \sin\phi\cos\theta & r\cos\phi\cos\theta & -r\sin\phi\sin\theta \\ \sin\phi\sin\theta & r\cos\phi\sin\theta & r\sin\phi\cos\theta \\ \cos\phi & -r\sin\phi & 0 \end{bmatrix} \begin{bmatrix} \Delta r \\ \Delta\phi \\ \Delta\theta \end{bmatrix} \\ &= A \begin{bmatrix} \Delta r \\ \Delta\phi \\ \Delta\theta \end{bmatrix} = A\Delta\bar{x}. \end{aligned}$$

The columns of the Jacobian matrix A above are orthogonal and the Euclidean norms are 1, r, and $r\sin\phi$, respectively. Therefore, the Euclidean distance between $\mathbf{x}$ and $\mathbf{X}^*$ is

$$\begin{aligned} d^2 &= \|\mathbf{x}-\mathbf{X}^*\|^2 \approx (\Delta x)^2 + (\Delta y)^2 + (\Delta z)^2 = (\Delta\bar{x})^T A^T A \Delta\bar{x} \\ &= (\Delta r)^2 + (r\,\Delta\phi)^2 + (r\sin\phi\Delta\theta)^2. \end{aligned}$$

Since $\varphi(r^*,\phi^*,\theta^*)=0$, we have $\varphi(r,\phi,\theta)=\varphi(r,\phi,\theta)-\varphi(r^*,\phi^*,\theta^*)$ and

$$\begin{aligned} \varphi(r,\phi,\theta) &= \varphi_r\,\Delta r + \varphi_\phi\,\Delta\phi + \varphi_\theta\,\Delta\theta \\ &= \varphi_r\,\Delta r + \frac{\varphi_\phi}{r}\,(r\,\Delta\phi) + \frac{\varphi_\theta}{r\,\sin\phi}\,(r\,\sin\phi\,\Delta\theta), \end{aligned} \tag{5.50}$$

where we have dropped the higher-order terms in the first line. Using the Cauchy–Schwarz inequality, we have

$$|\varphi| \le \sqrt{(\varphi_r)^2 + \left(\frac{\varphi_\phi}{r}\right)^2 + \left(\frac{\varphi_\theta}{r\,\sin\phi}\right)^2}\,\sqrt{(\Delta r)^2 + (r\,\Delta\phi)^2 + (r\,\sin\phi\,\Delta\theta)^2}.$$

In the limit case, the inequality above is

$$\frac{|\varphi|}{|\nabla\varphi|} \le \sqrt{(\Delta r)^2 + (r\,\Delta\phi)^2 + (r\,\sin\phi\,\Delta\theta)^2},$$

where the gradient vector of the level set function is

$$\nabla\varphi = \left(\varphi_r,\ \frac{1}{r}\varphi_\phi,\ \frac{1}{r\sin\phi}\,\varphi_\theta\right)^T. \tag{5.51}$$

The minimum of d can be reached if the equal sign is reached since $|\varphi|/|\nabla\varphi|$ is an approximation to the distance between (r,ϕ,θ) and its orthogonal projection (r^*,ϕ^*,θ^*). The equal sign can be reached if and only if the two vectors in the inner product (5.50) are collinear; i.e., there is a scalar α such that

$$\begin{aligned} r - r^* &= \alpha\,\varphi_r, \\ \phi - \phi^* &= \alpha\,\frac{\varphi_\phi}{r^2}, \\ \theta - \theta^* &= \alpha\,\frac{\varphi_\theta}{(r\,\sin\phi)^2}. \end{aligned}$$

Thus we conclude that the projection is along the direction

$$\mathbf{p} = \left[\varphi_r,\ \frac{\varphi_\phi}{r^2},\ \frac{\varphi_\theta}{(r\,\sin\phi)^2}\right]^T. \tag{5.52}$$

The undetermined scalar factor α can be determined approximately using (5.49).

The discrete finite difference scheme for (5.42) in three dimensions can be written as

$$\Delta_h U_{ijk} = \begin{cases} f_{ijk} & \text{if } \mathbf{x}_{ijk} \text{ is regular,} \\ f_{ijk} - H_h(\varphi_{ijk})\Delta_h\tilde{u}_{ijk} + \Delta_h\hat{u}_{ijk} + C_{ijk} & \text{otherwise,} \end{cases} \tag{5.53}$$

where C_{ijk} is a correction term needed to offset the discontinuities in the second-order partial derivatives of q, or the right-hand side of the expression above, Δ_h is the standard discrete Laplacian in spherical coordinates defined by

$$\begin{aligned}\Delta_h U_{ijk} &= \frac{U_{i+1,j,k} - 2U_{ijk} + U_{i-1,j,k}}{(\Delta r)^2} + \frac{2}{r_i}\frac{U_{i+1,j,k} - U_{i-1,j,k}}{2\Delta r} \\ &+ \frac{1}{r_i^2}\frac{U_{i,j+1,k} - 2U_{ijk} + U_{i,j-1,k}}{(\Delta\phi)^2} + \frac{\cot\phi_j}{r_i^2}\frac{U_{i,j+1,k} - U_{i,j-1,k}}{2\Delta\phi} \\ &+ \frac{1}{r_i^2 \sin^2\phi_j}\frac{U_{i,j,k+1} - 2U_{ijk} + U_{i,j,k-1}}{(\Delta\theta)^2},\end{aligned}$$

and $H_h(\varphi_{ijk})$ is the discrete Heaviside function,

$$H_h(\varphi_{ijk}) = \begin{cases} 1 & \text{if } \varphi_{ijk} \ge 0, \\ 0 & \text{if } \varphi_{ijk} < 0. \end{cases} \tag{5.54}$$

The correction term C_{ijk} has the closed form

$$C_{ijk} = \sum_{i_l, j_l, k_l} H_h(-\varphi_{ijk}\,\varphi_{I^*})\gamma_{I^*}\left(\frac{\varphi_{I^*}}{|\nabla\varphi_{I^*}|}\right)^2 \frac{F_{I^*} - F_{ijk}}{2}, \tag{5.55}$$

where

$$I^* = (i + i_l,\ j + j_l,\ k + k_l) \quad \text{abbreviated,}$$

$$(i_l, j_l, k_l) = \{(-1,0,0), (1,0,0), (0,-1,0), (0,1,0), (0,0,-1), (0,0,1)\},$$

and $\gamma_{i+i_l,j+j_l,k+k_l}$ are the coefficients of the discrete Laplacian in spherical coordinates,

$$\begin{aligned}\gamma_{i\pm1,j,k} &= \frac{1}{(\Delta r)^2} \pm \frac{1}{r_i \Delta r}, \qquad \gamma_{i,j\pm1,k} = \frac{1}{(r_i\,\Delta\phi)^2} \pm \frac{\cot\phi_j}{2r_i^2\,\Delta\phi}, \\ \gamma_{i,j,k\pm1} &= \frac{1}{(r_i \sin\phi_j \Delta\theta)^2}.\end{aligned} \tag{5.56}$$

The system of the finite difference equations (5.53) can be solved by a fast solver for spherical coordinates. We refer the reader to [145] for the fast Poisson solver, ways to deal with the infinite domain, and numerical examples.

The strategy of removing source singularities has also been applied to the immersed finite element methods (IFEMs) in §8.6.

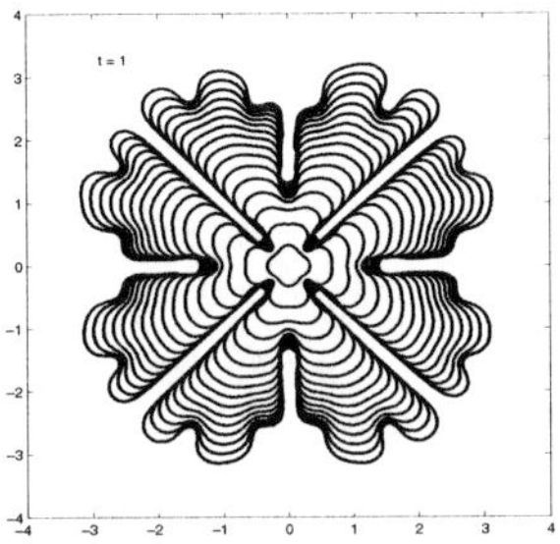

Chapter 6

Augmented Strategies

In this chapter, we explain how augmented strategies can be used to solve some interface problems and problems defined on irregular domains. The original idea of the augmented strategy for interface problems was proposed in [163] for elliptic interface problems with a piecewise but discontinuous coefficient. With a little modification, the augmented method developed in [163] was applied to generalized Helmholtz equations including Poisson equations on irregular domains, in [180]; see also §6.2. The augmented approach for the incompressible Stokes equations with discontinuous viscosity is presented in §10.4; see also [167].

There are at least two motivations for using augmented strategies. The first one is to get a faster algorithm compared to a direct discretization, particularly to take advantage of existing fast solvers. The second is that, for some interface problems, an augmented approach may be the only way to derive an accurate algorithm. This will be illustrated through the augmented IIM for incompressible Stokes equations, in which the jump conditions for the pressure and the velocity are coupled together. The augmented techniques enable us to decouple the jump conditions so that the idea of the IIM can be applied.

Using augmented strategies, some augmented variable $\mathbf{g}$ of codimension 1 will be introduced.[9] Once we know the augmented variable, it is relatively easy to solve the original problem. In discretization, the approximate solution (denoted by $\mathbf{U}$) to the original problem and the augmented variable $\mathbf{G}$ (discrete form of $\mathbf{g}$) together form a large linear system. If we eliminate $\mathbf{U}$ from the matrix vector equations, we will get the Schur complement system for the augmented variable, which is generally much smaller than that for $\mathbf{U}$. Therefore, we can use the GMRES iterative method [231] to solve the Schur complement for the augmented variable. In implementation, there is no need to explicitly form the matrices. The matrix-vector multiplication needed for the GMRES iteration includes two main steps: (1) solving the original problem, assuming the augmented variable is known; (2) finding the residual of the constraint using the computed approximate solution given the augmented variable. The constraint is often the jump condition or the boundary condition. In this chapter, we explain this technique for some interface problems and problems defined on irregular domains.

[9]There can be more than one augmented variable. In this case, we can put all the augmented variables together as a vector; see, for example, §10.4.

While an augmented approach for an interface or irregular domain problem has some similarities with an integral equation approach, or boundary element method, to find a source strength, the augmented methods described in this chapter have a few special features: (1) they do not need a Green's function; (2) there is no need to set up the system of equations; (3) they are applicable to general PDEs with or without source terms; (4) they do not depend on the boundary condition. On the other hand, we may have less knowledge about the condition of the Schur complement system and how to apply preconditioning techniques. The efficiency of the algorithm depends on the number of iterations of the GMRES method.

6.1 The augmented technique for elliptic interface problems

We start with the elliptic interface problem,

$$\nabla \cdot (\beta(x, y)\nabla u) = f(x, y), \quad x \in \Omega \setminus \Gamma, \tag{6.1a}$$

$$[u]_\Gamma = w, \qquad [\beta u_n]_\Gamma = v \tag{6.1b}$$

with a specified boundary condition on $\partial\Omega$. We assume that $\beta(x, y)$ has a constant value in each subdomain,

$$\beta(x, y) = \begin{cases} \beta^+ & \text{if } (x, y) \in \Omega^+, \\ \beta^- & \text{if } (x, y) \in \Omega^-. \end{cases} \tag{6.2}$$

If $\beta^+ = \beta^- = \beta$ is a constant, then we have a Poisson equations $\Delta u = f/\beta$ with source distributions along the interface. The finite difference method obtained from the IIM is the standard discrete Laplacian plus correction terms in the right-hand side as explained in Chapter 3. Or we can transform the interface problem to a problem with a smooth solution as explained in Chapter 5. Therefore, a fast Poisson solver can be used to solve the discrete system of equations if the boundary condition is also linear. If $\beta^+ \neq \beta^-$, we cannot separate the coefficient β from the flux jump condition. The motivation is to introduce an augmented variable so that we can take advantage of fast solvers for the Poisson equations with only singular sources.

6.1.1 The augmented variable for the elliptic interface problems

There is more than one way to introduce an augmented variable. Since the PDE (6.1) can be written as a Poisson equation in each subdomain separated by the interface after we divide β from the original equation excluding the interface Γ, it is a natural choice to introduce $[u_n]$ as the *augmented variable*.

Consider the solution set $u_g(x, y)$ of the following problem:

$$\begin{cases} \Delta u = \dfrac{f}{\beta^+} & \text{if } x \in \Omega^+, \\ \Delta u = \dfrac{f}{\beta^-} & \text{if } x \in \Omega^-, \end{cases} \tag{6.3a}$$

$$[u]_\Gamma = w, \qquad [u_n]_\Gamma = g \tag{6.3b}$$

with the same boundary condition on $\partial\Omega$ as in the original problem (6.1). The solution $u_g(x, y)$ is then a functional in g. Let the solution of (6.1) be $u^*(x, y)$, and define

$$g^* = [u_n^*]_\Gamma. \tag{6.4}$$

Then $u^*(x, y)$ satisfies the elliptic equation (6.3) and the jump conditions (6.3b) with $g \equiv g^*$. In other words, $u_{g^*}(x, y) \equiv u^*(x, y)$ and

$$\left[\beta \frac{\partial u_{g^*}}{\partial \boldsymbol{n}}\right]_\Gamma = v \tag{6.5}$$

is satisfied. This is the constraint for the choice of the augmented variable. Therefore, solving the original problem (6.1) is equivalent to finding the corresponding g such that $[\beta \frac{\partial u_g}{\partial \boldsymbol{n}}]_\Gamma = v$. Such a $u(x, y) = u_g(x, y)$ is the solution to (6.1). Notice that g is defined only along the interface, so it is one dimension lower than that of $u(x, y)$. If we are given $[u_{\boldsymbol{n}}] = g$, then it is easy to solve (6.3) since only correction terms at irregular grid points need to be added to the right-hand side of the finite difference equations. A fast Poisson solver such as the FFT can be used. The cost in solving (6.3) is just a little more than that in solving a regular Poisson equation on the rectangle Ω with a smooth solution. For a more general variable coefficient, the discussions are still valid, except we cannot use a fast Poisson solver because of the convection term $(\nabla\beta \cdot \nabla u)/\beta$. In this case, it is better to solve the original problem directly using the IIM described in Chapter 3.

6.1.2 The discrete system of equations in matrix-vector form

To use an augmented approach, the augmented variable is discretely defined at a set of selected points $\mathbf{X}_1, \mathbf{X}_2, \ldots, \mathbf{X}_{N_b}$ on the interface. These points can be the control points used in a spline interpolation, or the orthogonal projections of certain grid points on the interface.

Let $\mathbf{W} = [W_1, W_2, \ldots, W_{N_b}]^T$ and $\mathbf{G} = [G_1, G_2, \ldots, G_{N_b}]^T$ be the discrete values of the jump conditions (6.1b) at these points. Given $\mathbf{W}$ and $\mathbf{G}$, the discrete form of (6.3a) obtained from the IIM can be written as

$$\Delta_h U_{ij} = \frac{f_{ij}}{\beta_{ij}} + C_{ij}, \tag{6.6}$$

where C_{ij} is zero at a regular grid point with reference to the standard 5-point stencil, it is a nonzero correction term (from (3.23)) at an irregular grid point, and Δ_h is the discrete Laplacian operator

$$\Delta_h U_{ij} = \frac{U_{i+1,j} + U_{i-1,j} + U_{i,j+1} + U_{i,j-1} - 4U_{i,j}}{h^2}.$$

The correction term C_{ij} depends on $\{G_k\}$ and $\{W_k\}$ continuously; see §3.2 and the discussions there. Thus we can write

$$A\mathbf{U} + \mathcal{B}(\mathbf{W}, \mathbf{G}) = \mathbf{F}, \tag{6.7}$$

where $\mathbf{U}$ and $\mathbf{F}$ are the vectors formed by $\{U_{ij}\}$ and $\{F_{ij}\}$, and $A\mathbf{U} = \mathbf{F}$ is the discrete linear system of the finite difference equations for the Poisson equation when $\mathbf{W}$ and $\mathbf{G}$ are both zero vectors. In (6.7), $\mathcal{B}(\mathbf{W}, \mathbf{G})$ is a mapping from $\mathbf{W} = [W_1, W_2, \ldots, W_{n_b}]^T$ and $\mathbf{G} = [G_1, G_2, \ldots, G_{n_b}]^T$ to C_{ij} in (3.23) with v substituted by g. From expression (3.23) we know that $\mathcal{B}(\mathbf{W}, \mathbf{G})$ depends on the surface derivatives of w and g. In discretization, all the surface derivatives of the jump conditions are obtained by linear combinations of the values of $\mathbf{W}$ and $\mathbf{G}$ at $\{\mathbf{X}_k\}$. Therefore, $\mathcal{B}(\mathbf{W}, \mathbf{G})$ is a linear function of $\mathbf{W}$ and $\mathbf{G}$ and can be written as

$$\mathcal{B}(\mathbf{W}, \mathbf{G}) = B\mathbf{G} - B_1\mathbf{W},$$

where B and B_1 are two matrices with real entries. Thus (6.7) becomes

$$A\mathbf{U} + B\mathbf{G} = \mathbf{F} + B_1\mathbf{W} \stackrel{\text{def}}{=} \mathbf{F}_1. \tag{6.8}$$

The solution $\mathbf{U}$ of the equation above certainly depends on $\mathbf{G}$. We define the residual vector of the flux jump condition at $\{\mathbf{X}_k\}$ as

$$R(\mathbf{G}) = [\beta\mathbf{U}_n](\mathbf{G}) - \mathbf{V} = \beta^+\mathbf{U}_n^+(\mathbf{G}) - \beta^-\mathbf{U}_n^-(\mathbf{G}) - \mathbf{V}. \tag{6.9}$$

We are interested in finding a $\mathbf{G}^*$ such that $R(\mathbf{G}^*) = \mathbf{0}$, where the components of the vectors $\mathbf{U}_n^+$ and $\mathbf{U}_n^-$ are the discrete approximation of the normal derivative from each side of the interface at $\{\mathbf{X}_k\}$.

Once we know the solution $\mathbf{U}$ of the system (6.8) given an approximate $\mathbf{G}$, we can interpolate $\{U_{ij}\}$ linearly to get $\{U_n^{\pm}(\mathbf{X}_k)\}$, which is an approximation to the normal derivative from each side of the interface at $\{\mathbf{X}_k\}$, $1 \le k \le N_b$. The interpolation scheme is crucial to the success of the augmented algorithm and will be explained in detail in the next subsection. Since the interpolation is linear, we can write

$$\frac{\partial \mathbf{U}(\mathbf{G})^{\pm}}{\partial \boldsymbol{n}} = E^{\pm}\mathbf{U} + T^{\pm}\mathbf{G} + P^{\pm}\mathbf{V} + Q^{\pm}\mathbf{W}, \tag{6.10}$$

where E^+, E^-, T^+, T^-, P^+, P^-, Q^+, and Q^- are some sparse matrices determined by the interpolation scheme. The matrices are used only for theoretical purposes but are not actually constructed in practice. We need to choose a vector $\mathbf{G}$ such that the second interface condition $\beta^+\mathbf{U}_n^+ - \beta^-\mathbf{U}_n^- = \mathbf{V}$ is satisfied along the interface Γ. Therefore, we have the second matrix-vector equation

$$E\mathbf{U} + T\mathbf{G} - P\mathbf{V} - Q\mathbf{W} = 0, \tag{6.11}$$

where

$$E = \beta^+E^+ - \beta^-E^-, \qquad T = \beta^+T^+ - \beta^-T^-,$$

$$P = I + \beta^-P^- - \beta^+P^+, \qquad Q = \beta^-Q^- - \beta^+Q^+.$$

Once again, there is no need to generate the matrices E, T, P, and Q explicitly. If we put the two matrix-vector equations (6.8) and (6.11) together, we get

$$\begin{bmatrix} A & B \\ E & T \end{bmatrix} \begin{bmatrix} \mathbf{U} \\ \mathbf{G} \end{bmatrix} = \begin{bmatrix} \mathbf{F}_1 \\ P\,\mathbf{V} + Q\,\mathbf{W} \end{bmatrix}. \tag{6.12}$$

The solutions $\mathbf{U}$ and $\mathbf{G}$ are the discrete approximation of $u_{g^*}(x, y)$ and g^*, the solution of (6.3) which satisfies $[u_n] = g^*$ and $[\beta u_n(g^*)] = v$.

If we are trying to solve (6.12) directly, we can use the direct discretization described in Chapter 3. The main reason to use an augmented approach is to take advantage of fast Poisson solvers. Eliminating $\mathbf{U}$ from (6.12) gives a linear system for $\mathbf{G}$,

$$\begin{aligned} (T - EA^{-1}B)\mathbf{G} &= P\,\mathbf{V} + Q\,\mathbf{W} - EA^{-1}\mathbf{F}_1 \\ &\stackrel{\text{def}}{=} \mathbf{F}_2. \end{aligned} \tag{6.13}$$

This is an $N_b \times N_b$ system for $\mathbf{G}$, a much smaller linear system compared with the one for $\mathbf{U}$. Since we cannot guarantee the coefficient matrix of the Schur complement to be symmetric positive definite, the GMRES iterative method [231] is preferred.

Note that if β is constant ($[\beta] = 0$), then $E = \mathbf{0}$, $T = I$, $P = I/\beta$, and $Q = \mathbf{0}$. Thus the coefficient matrix of the Schur complement is the identity matrix. We recover the IIM for Poisson equations with only a singular source; see §3.4. For the general case, the analysis is given in §6.1.4.

The GMRES method requires only matrix-vector multiplication. We explain below how to evaluate the right-hand side $\mathbf{F}_2$ of the Schur complement and how to evaluate the matrix-vector multiplication needed by the GMRES iteration. We can see why we do not need to form the coefficient matrix $T - EA^{-1}B$ explicitly.

Evaluation of the right-hand side of the Schur complement

First, we set $\mathbf{G} = 0$ and solve the system (6.3), or (6.8) in the discrete form, to get $\mathbf{U}(\mathbf{0})$ which is $A^{-1}\mathbf{F}_1$ from (6.8). Note that the residual of the Schur complement for $\mathbf{G} = 0$ is

$$\begin{aligned} (T - EA^{-1}B)\,\mathbf{0} - \mathbf{F}_2 &= -\mathbf{F}_2 \\ &\stackrel{\text{from}(6.13)}{=} -\left(P\mathbf{V} + Q\mathbf{W} - EA^{-1}\mathbf{F}_1\right) \\ &\stackrel{\text{from}(6.8)}{=} -(P\mathbf{V} + Q\mathbf{W} - E\mathbf{U}(\mathbf{0})) \\ &\stackrel{\text{from}(6.9)\ \&\ (6.11)}{=} \beta^+ \frac{\partial \mathbf{U}(\mathbf{0})^+}{\partial \boldsymbol{n}} - \beta^- \frac{\partial \mathbf{U}(\mathbf{0})^-}{\partial \boldsymbol{n}} - \mathbf{V}, \end{aligned} \tag{6.14}$$

which gives the right-hand side of the Schur complement system with an opposite sign.

Matrix-vector multiplication of the Schur complement

The matrix-vector multiplication of the Schur complement system given $\mathbf{G}$ is obtained from the following two steps.

Step 1: Solve the coupled system (6.3), or (6.8) in the discrete form, to get $\mathbf{U}(\mathbf{G})$.

Step 2: Interpolate $\mathbf{U}(\mathbf{G})$ to get $R(\mathbf{U}(\mathbf{G}))$ defined in (6.9). Then the matrix-vector multiplication is

$$\begin{aligned}(T - EA^{-1}B)\mathbf{G} &= R(\mathbf{U}(\mathbf{G})) - R(\mathbf{U}(\mathbf{0})) \\ &= \beta^{+}\frac{\partial \mathbf{U}(\mathbf{G})^{+}}{\partial n} - \beta^{-}\frac{\partial \mathbf{U}(\mathbf{G})^{-}}{\partial n} - \mathbf{V} \\ &\quad - \left(\beta^{+}\frac{\partial \mathbf{U}(\mathbf{0})^{+}}{\partial n} - \beta^{-}\frac{\partial \mathbf{U}(\mathbf{0})^{-}}{\partial n} - \mathbf{V}\right) \\ &= \beta^{+}\frac{\partial \mathbf{U}(\mathbf{G})^{+}}{\partial n} - \beta^{-}\frac{\partial \mathbf{U}(\mathbf{G})^{-}}{\partial n} - \mathbf{V} + \mathbf{F}_2.\end{aligned} \tag{6.15}$$

This is because

$$\begin{aligned}(T - EA^{-1}B)\mathbf{G} &= T\mathbf{G} - EA^{-1}B\mathbf{G} \\ &\overset{\text{from(6.8)}}{=} T\mathbf{G} - E\left(A^{-1}\mathbf{F}_1 - \mathbf{U}(\mathbf{G})\right) \\ &= T\mathbf{G} + E\mathbf{U}(\mathbf{G}) + E\mathbf{U}(\mathbf{0}) \\ &= T\mathbf{G} + E\mathbf{U}(\mathbf{G}) - P\mathbf{V} - Q\mathbf{W} \\ &\quad - (T\mathbf{0} + E\mathbf{U}(\mathbf{0}) - P\mathbf{V} - Q\mathbf{W}) \\ &\overset{\text{from(6.9) \& (6.11)}}{=} \beta^{+}\frac{\partial \mathbf{U}(\mathbf{G})^{+}}{\partial n} - \beta^{-}\frac{\partial \mathbf{U}(\mathbf{G})^{-}}{\partial n} - \mathbf{V} \\ &\quad - \left(\beta^{+}\frac{\partial \mathbf{U}(\mathbf{0})^{+}}{\partial n} - \beta^{-}\frac{\partial \mathbf{U}(\mathbf{0})^{-}}{\partial n} - \mathbf{V}\right),\end{aligned}$$

from the equalities $A\mathbf{U} + B\mathbf{G} = \mathbf{F}_1$ and $\mathbf{U}(\mathbf{0}) = A^{-1}\mathbf{F}_1$.

Now we can see that a matrix-vector multiplication is equivalent to solving the system (6.3a), or (6.8) in the discrete form, to get $\mathbf{U}$, and then using an interpolation scheme to get the residual for the flux condition $[\beta\mathbf{U}_n]$ at the control points on the interface. We also see that the residual defined in (6.9) is actually the residual of the Schur complement.

6.1.3 The least squares interpolation scheme from a Cartesian grid to an interface

The interpolation scheme used to perform (6.10), particularly for evaluating $\mathbf{U}_n^{\pm}$, is crucial to the efficiency (accuracy and the number of iterations of the GMRES) of the augmented method. To reduce the number of iterations, it is important to couple the solutions on both sides of the interface through the jump conditions. We use the least squares interpolation scheme for interface problems, which was introduced in [163].

The least squares interpolation scheme discussed here is similar to that in (4.57) except that the interpolation scheme there is from an interface to the interface, while the interpolation here is from a grid function to its restriction and its derivatives on the interface.

The interpolation scheme for approximating $\mathbf{U}_n^-$ can be written as

$$\frac{\partial \mathbf{U}^-}{\partial \mathbf{n}}(\mathbf{X}) = \sum_{k=0}^{k_s-1} \gamma_k \, U_{i^*+i_k,\, j^*+j_k} - C, \tag{6.16}$$

where k_s is the number of grid points involved in the interpolation scheme, (x_{i^*}, y_{j^*}) is the closest grid point to $\mathbf{X}$, and C is a correction term. Below we discuss how to determine the coefficients $\{\gamma_k\}$ and the correction term C using the information from both sides of the interface. Note that $\{\gamma_k\}$ and C depend on $\mathbf{X}$, but for simplicity of notation, we omit the dependency.

The coefficients $\{\gamma_k\}$ are determined by minimizing the interpolation error of (6.16) when $U_{i^*+i_k,\, j^*+j_k}$ is substituted with the exact solution $u(x_{i^*+i_k}, y_{j^*+j_k})$. Using the local coordinates system (1.34) centered at the point $\mathbf{X}$, and denoting the local coordinates of $(x_{i^*+i_k}, y_{j^*+j_k})$ as (ξ_k, η_k), we have the following from the Taylor expansion at $\mathbf{X} = (X, Y)$ or $(0, 0)$ in the local coordinates:

$$\begin{aligned} u(x_{i^*+k}, y_{j^*+k}) &= u(\xi_k, \eta_k) = u^\pm + \xi_k u_\xi^\pm + \eta_k u_\eta^\pm \\ &+ \frac{1}{2}\xi_k^2 u_{\xi\xi}^\pm + \xi_k \eta_k u_{\xi\eta}^\pm + \frac{1}{2}\eta_k^2 u_{\eta\eta}^\pm + O(h^3), \end{aligned} \tag{6.17}$$

where the "+" or "−" sign is chosen depending on whether (ξ_k, η_k) lies on the "+" or "−" side of Γ, and $u^\pm$, $u_\xi^\pm$, $\ldots$, $u_{\eta\eta}^\pm$ are evaluated at the local coordinates $(0, 0)$ or $\mathbf{X} = (X, Y)$ in the original coordinates system. Note that we should have used something like $u(X, Y) = \bar{u}(0, 0)$ to distinguish the two coordinate systems; however, we omit the bars and use the same notation $u(X, Y) = u(0, 0)$ for simplicity.

We carry out this expansion for all the grid points involved in the interpolation scheme and plug (6.17) into (6.16). After collecting and arranging terms, we can write

$$\begin{aligned} \frac{\partial u^-}{\partial \boldsymbol{n}}(\mathbf{X}) &\sim a_1 \, u^- + a_2 \, u^+ + a_3 \, u_\xi^- + a_4 \, u_\xi^+ + a_5 \, u_\eta^- + a_6 u_\eta^+ \\ &+ a_7 \, u_{\xi\,\xi}^- + a_8 \, u_{\xi\,\xi}^+ + a_9 \, u_{\eta\,\eta}^- + a_{10} \, u_{\eta\,\eta}^+ + a_{11} \, u_{\xi\,\eta}^- - C, \end{aligned} \tag{6.18}$$

where the $\{a_i\}$'s are defined in (3.17). Note that $u^+ = u^- + w$ and $u_{\boldsymbol{n}}^+ = u_{\boldsymbol{n}}^- + g$ from (6.3b). From Theorem 3.1 we also have the following interface relations:

$$\begin{aligned} u_\eta^+ &= u_\eta^- + w', \\ u_{\xi\,\xi}^+ &= u_{\xi\,\xi}^- + g\,\chi'' - w'' + \left[\frac{f}{\beta}\right], \\ u_{\eta\,\eta}^+ &= u_{\eta\,\eta}^- - g\,\chi'' + w'', \\ u_{\xi\,\eta}^+ &= u_{\xi\,\eta}^- + g\,\chi'' + g'. \end{aligned} \tag{6.19}$$

Therefore, we can express all the quantities from the "+" side in (6.18) in terms of those from the "−" side and the known quantities. Thus, when $U_{i^*+i_k,\, j^*+j_k}$ is substituted for by the exact solution $u(x_{i^*+i_k}, y_{j^*+j_k})$, (6.16) can be written as

$$
\begin{aligned}
\frac{\partial u^-}{\partial \mathbf{n}}(\mathbf{X}) &\approx \sum_{k=0} \gamma_k \, u(x_{i^*+i_k}, y_{j^*+j_k}) \;-C \\
&= a_1\, u^- + a_2\, u^+ + a_3\, u_\xi^- + a_4\, u_\xi^+ + a_5\, u_\eta^- + a_6 u_\eta^+ + a_7\, u_{\xi\,\xi}^- + a_8\, u_{\xi\,\xi}^+ \\
&\quad + a_9\, u_{\eta\,\eta}^- + a_{10}\, u_{\eta\,\eta}^+ + a_{11}\, u_{\xi\,\eta}^- + a_{12}\, u_{\xi\,\eta}^+ - C \\
&= (a_1 + a_2) u^- + (a_3 + a_4) u_\xi^- + (a_5 + a_6) u_\eta^- + (a_7 + a_8) u_{\xi\xi}^- \\
&\quad + (a_9 + a_{10}) u_{\eta\eta}^- + (a_{11} + a_{12}) u_{\xi\eta}^- + a_2[u] + a_4[u_\xi] + a_6[u_\eta] \\
&\quad + a_8[u_{\xi\xi}] + a_{10}[u_{\xi\eta}] + a_{12}[u_{\xi\,\eta}] - C.
\end{aligned}
$$

To minimize the interpolation error, we should set the following linear system of equations for the coefficients $\{\gamma_k\}$ by matching the terms of $u^-, u_\xi^-, \ldots, u_{\xi\,\eta}^-$:

$$
\begin{aligned}
a_1 + a_2 &= 0, \qquad & a_3 + a_4 &= 1, \\
a_5 + a_6 &= 0, \qquad & a_7 + a_8 &= 0, \\
a_9 + a_{10} &= 0, \qquad & a_{11} + a_{12} &= 0.
\end{aligned}
\tag{6.20}
$$

If the linear system (6.20) has a solution, then we can obtain a second-order interpolation scheme for the normal derivative u_n^- by choosing an appropriate correction term C. From (3.17) and (6.20), we can see that the system of equations for the $\{\gamma_k\}$'s is independent of the jump conditions which means we can calculate $\{\gamma_k\}$ outside of the GMRES iteration.

If more than six different grid points ($n_s > 6$) in a neighborhood of $\mathbf{X}$ are used in the interpolation, we will have an underdetermined system of linear equations that has an infinite number of solutions. Often $9 \sim 16$ closest grid points to $\mathbf{X} = (X, Y)$ are chosen as the interpolation stencil. For stability, we solve (6.20) using the SVD.[10] The SVD algorithm can be found in many software packages, e.g., the *pinv* function in MATLAB and the SVD subroutine from Linpack and Lapack. The SVD solution has the smallest 2-norm among all feasible solutions,

$$
\sum_{k=0}^{k_s-1} \left(\gamma_k^*\right)^2 = \min_{\gamma_k} \sum_{k=1}^{k_s-1} \gamma_k^2 \qquad \text{subject to} \quad (6.20).
$$

For such a solution, the magnitude of γ_k^* is well under control, which is important to the stability of the entire algorithm.

Once the $\{\gamma_k\}$'s are computed, and thus the $\{a_k\}$'s, the correction term C is determined from the following:

$$
\begin{aligned}
C = a_2\, w + a_4\, g + a_6\, w' + a_8 \left(g\chi'' - w'' + \left[\frac{f}{\beta}\right]\right) \\
+ a_{10}\,[w'' - g\,\chi''] + a_{12}\,(w'\chi'' + g').
\end{aligned}
\tag{6.21}
$$

[10]Given a system of $Ax = b$, let the singular decomposition of A be $A = U\Sigma V^H$, where U and V are two unitary matrices, $\Sigma = \mathrm{diag}(D, \mathbf{0})$ with D being invertible. The SVD solution of $Ax = b$ is $x^* = V\Sigma^+ U^H b$, where $\Sigma^+ = \mathrm{diag}(D^{-1}, \mathbf{0})$.

If the interface is represented by a cubic spline interpolation, then the surface derivatives w', w'', g' are also computed from the spline interpolation using the arc-length parameter and the values of v and g at the control points. If the interface is represented by the zero set of a level set function, then the least squares interpolation is used to approximate the surface derivatives using their values at the orthogonal projections of irregular grid points; see, for example, §4.4.1.

We use the relation $u_n^+ = u_n^- + g$ to get

$$U_{\mathbf{n}}^+(\mathbf{X}) \simeq \sum_{k=0}^{k_s-1} \gamma_k \, U_{i^*+i_k,\, j^*+j_k} - C + g, \tag{6.22}$$

where $\{\gamma_k\}$'s are the coefficients of the interpolation scheme for U_n^-. In the next subsection, we will explain a modification of either (6.16) or (6.22), depending on the magnitude of β^- or β^+.

Remark 6.1. *The least squares interpolation from a grid function to an interface is second-order accurate with local support. It is robust in selecting interpolation points. The interpolation formulas* (6.16) *and* (6.22) *depend continuously on the location* $\mathbf{X}$ *and the grid point* (x_i, y_j) *as does the truncation error of these two interpolation schemes. In other words, the interpolation has a smooth error distribution. This is important in moving interface problems in which we do not want to introduce any nonphysical oscillations.*

The trade-off of the least squares interpolation is to solve an underdetermined $6 \times k_s$ linear system of equations (where $k_s \geq 6$) using an SVD. However, the size of the linear system is small and the coefficients can be predetermined before the GMRES iteration. The extra time needed in dealing with the interface is usually less that 5% of the total CPU time and the percentage decreases as the mesh sizes (M and N) increase.

An alternative is a one-sided interpolation in which one can use n_s grid points ($n_s \geq 6$) on the proper side of the interface, for example,

$$U_{\mathbf{n}}^- = \sum_{(x_{i^*+i_k}, y_{j^*+j_k}) \in \Omega^-}^{k_s-1} \gamma_k \, U_{i^*+i_k,\, j^*+j_k}. \tag{6.23}$$

This approach does not use the interface relations $[u] = w$, $[u_n] = g$ or those interface relations in (6.19). At least six grid points from each side have to be included in order for the approach to have a second-order interpolation. This may cause some difficulty if the curvature of the interface is large. The number of GMRES iterations using a one-sided interpolation is often much larger than that using an interpolation involving grid points from both sides.

6.1.4 Invertibility of the Schur complement system

Assume that we use the least squares interpolation formula (6.16) to compute u_n^-, and use (6.22) to compute u_n^+. Then the flux jump condition $[\beta u_n] - v = 0$ is approximated by

$$(\beta^+ - \beta^-) \sum_{k=0}^{k_s-1} \gamma_k \, u(x_{i^*+i_k}, y_{j^*+j_k}) + \beta^+(g - C) + \beta^- C - v = O(h^3).$$

In terms of g and its surface derivative, it is

$$(\beta^+ - \beta^-) \sum_{k=0}^{k_s-1} \gamma_k \, u(x_{i^*+i_k}, y_{j^*+j_k}) + \Big(\beta^+ - (\beta^+ - \beta^-)\,(a_4 + a_8\chi'' - a_{10}\chi'') \Big)\, g$$
$$- (\beta^+ - \beta^-) a_{12} g' - v - (\beta^+ - \beta^-)\bar{C} \approx 0,$$

where

$$\bar{C} = a_2 w + a_6\, w' - a_8 \left(w'' - \left[\frac{f}{\beta} \right] \right) + a_{10}\, w'' + a_{12}\, w' \chi''.$$

In the discrete form, it is the second equation in (6.12),

$$E\mathbf{U} + T\mathbf{G} = P\mathbf{V} + Q\mathbf{W}.$$

If $\beta^+ = \beta^- = \beta$, then we have the unique solution for $\mathbf{G}$, which is $\mathbf{G} = \mathbf{V}/\beta$. Assuming $\beta^+ \neq \beta^-$, the invertibility of the Schur complement system (6.13) is true through the following arguments.

Remark 6.2. *Assume that we use the least squares interpolation formula* (6.16) *to compute* U_n^-, *and use* (6.22) *to compute* U_n^+. *If* h *is small enough and* $\beta^- \neq \beta^+$, *then* $T - EA^{-1}B$ *is invertible.*

Proof: It is enough to consider the homogeneous case

$$w = 0, \quad f = 0, \quad v = 0, \quad \text{and } u(x, y)|_{\partial\Omega} = 0.$$

In this case, $\mathbf{F}_2 = -EA^{-1}\mathbf{F}_1 = \mathbf{0}$. If the remark is not true, then there exists $\mathbf{G}^* \neq \mathbf{0}$, $\|\mathbf{G}^*\|_2 = 1$ such that $(T - EA^{-1}B)\mathbf{G}^* = \mathbf{0}$. Let $\mathbf{U}^* = -A^{-1}B\mathbf{G}^*$, which is the discrete solution of (6.3a)–(6.3b), in which the flux jump condition $[u_n] = g^* = O(1)$ is a continuous interpolation of $\mathbf{G}^*$ along the interface. Let u^* be the solution to (6.3a)–(6.3b) with $[u_n] = g^*$. Since $(T - EA^{-1}B)\mathbf{G}^*$ is a second-order approximation to $\beta^+ u_n^{*+} - \beta^- u_n^{*-}$, we conclude that $[\beta u_n^*] = O(h^2)$. Since the solution of the interface problem depends on the jump conditions of $[u]$ and $[\beta u_n]$ continuously, the solution u^* to the interface problem, with $[u] = 0$ and $[\beta u_n] = O(h^2)$, has a magnitude at most $O(h)$, and thus so does u_n^{*-}. On the other hand, we have

$$g^* = [u_n^*] = \frac{1}{\beta^+}\big(- [\beta] u_n^{*-} + [\beta u_n] \big) = O(h),$$

which contradicts the fact that $g^* = O(1)$. Thus, this cannot happen if h is small enough. These arguments rely on the smoothness of g^*. □

6.1.5 A preconditioner for the Schur complement system

The convergence speed of the GMRES method depends on the condition number of the coefficient matrix and on the Krylov space generated by the initial guess. Preconditioning techniques are often used to accelerate the convergence. If we use (6.16) and (6.22) to compute $\mathbf{U}_n^{\pm}$, the number of iterations seems to grow linearly as the number of grid points increases. Since the coefficient matrix of the Schur complement system is not constructed explicitly, it is difficult to take advantages of existing preconditioners that depend on the structure of the coefficient matrix. A modification in computing $\mathbf{U}_n^{\pm}$ proposed in [163] seems to be an efficient preconditioner for the Schur complement.

If u_n^+ and u_n^- are exact, that is,

$$\beta^+ u_n^+ - \beta^- u_n^- - v = 0,$$

then we can solve u_n^- or u_n^+ in terms of v, β^-, β^+, and $[u_n]$. It is easy to get

$$u_n^- = \frac{v - \beta^+ [u_n]}{\beta^+ - \beta^-} \tag{6.24}$$

and

$$u_n^+ = \frac{v - \beta^- [u_n]}{\beta^+ - \beta^-}. \tag{6.25}$$

The idea is simple and intuitive. We use one of the formulas (6.16) or (6.22) obtained from the least squares interpolation to approximate u_n^- or u_n^+, and then use (6.25) or (6.24) to approximate u_n^+ or u_n^- to enforce the flux jump condition, where $[u_n]$ is a guess of the jump in the normal derivative and will be updated by the GMRES method. This is actually an acceleration process, or a preconditioner for the Schur complement system (6.13).

With this modification, the number of iterations for solving the Schur complement system seems to be independent of the mesh size h, and almost independent of the jump $[\beta]$ in the coefficient through our numerical tests.

Whether we use the pair (6.16), (6.25) or (6.22), (6.24) has only a marginal effect on the accuracy of the computed solutions and the number of iterations. The algorithm otherwise behaves the same, and the analysis in the next section seems to be true no matter what pair we choose.

The following criteria are recommended for choosing the desired pair:

$$\text{if} \quad \beta^+ < \beta^- : \qquad \begin{cases} U_n^+ \ \text{ is determined by (6.22)}, \\ U_n^- = \dfrac{V - \beta^+ G}{\beta^+ - \beta^-}; \end{cases}$$

$$\text{if} \quad \beta^+ > \beta^- : \qquad \begin{cases} U_n^- \ \text{ is determined by (6.16)}, \\ U_n^+ = \dfrac{V - \beta^- G}{\beta^+ - \beta^-}. \end{cases}$$

The augmented method for elliptic interface problem (6.1) is also called the *fast immersed interface method* or the *fast IIM*.

6.1.6 Numerical experiments and analysis of the fast IIM

We present some numerical analysis of the augmented method for the elliptic interface problem (6.1)–(6.2) through the following numerical experiments. More examples can be found in [163]. The fast Poisson solver is from Fishpack [2]. Almost all the simulations can be completed within seconds or minutes.

Example 6.1. The interface is given by

$$\begin{cases} X = \big(r_0 + 0.2\sin(\omega\theta)\big)\cos\theta + x_c, \\ Y = \big(r_0 + 0.2\sin(\omega\theta)\big)\sin\theta + y_c, \end{cases} \qquad 0 \le \theta < 2\pi, \tag{6.26}$$

in the domain $[-1, 1] \times [-1, 1]$.

Figure 6.1 shows three different interfaces with different parameters r_0, (x_c, y_c), and ω. These parameters enable us to test a variety of interfaces. Dirichlet boundary conditions, as well as the jump conditions $[u]$ and $[\beta u_n]$ along the interface, are determined from the exact solution

$$u(x,y) = \begin{cases} \dfrac{r^2}{\beta^-} & \text{if } (x,y) \in \Omega^-, \\[2ex] \dfrac{r^4 + C_0\log(2r)}{\beta^+} + C_1\left(\dfrac{r_0^2}{\beta^-} - \dfrac{r_0^4 + C_0\log(2r_0)}{\beta^+}\right) & \text{if } (x,y) \in \Omega^+, \end{cases}$$

where $r = \sqrt{x^2 + y^2}$. The source term can be determined accordingly,

$$f(x,y) = \begin{cases} \dfrac{4}{\beta^-} & \text{if } (x,y) \in \Omega^-, \\[2ex] \dfrac{16r^2}{\beta^+} & \text{if } (x,y) \in \Omega^+. \end{cases}$$

We provide numerical results for three typical cases below. The interface Γ is represented by periodic cubic splines with equally spaced $\Delta\theta$ in the discretization of (6.26). The total number of control points is N_b.

Case A. The interface is a circle centered at the origin; see the dashed line in Figure 6.1(a) with $r_0 = 0.5$, $(x_c, y_c) = (0, 0)$, and $\omega = 0$. With $C_1 = 1$, the solution is continuous everywhere, but u_n and βu_n are discontinuous across the circle. It is easy to verify that $[\beta u_n] = -0.7$ when we take $C_0 = -0.1$. Figure 6.2(a) is a plot of the solution $-u$ with $\beta^- = 1$ and $\beta^+ = 10$.

Case B. The interface is a five-pointed star with $r_0 = 0.5$, $x_c = y_c = 0.2/\sqrt{20}$, and $\omega = 5$; see the solid line in Figure 6.1(a). The center of the interface is shifted slightly to have a nonsymmetric solution relative to the grid so that the test problem is as general as possible. The interface is irregular but the curvature has modest magnitude. Since it is difficult to construct an exact solution that is continuous but nonsmooth across the interface, we simply set $C_1 = 0$ and $C_0 = -0.1$. Figure 6.2(b) is a plot of the solution $-u$ with $\beta^- = 1$ and $\beta^+ = 10$.

Case C. The interface is a twelve pointed star with $r_0 = 0.4$, $x_c = y_c = 0.2/\sqrt{20}$, and $\omega = 12$; see Figure 6.1(b). The magnitude of the curvature is large at some points on

(a) (b)

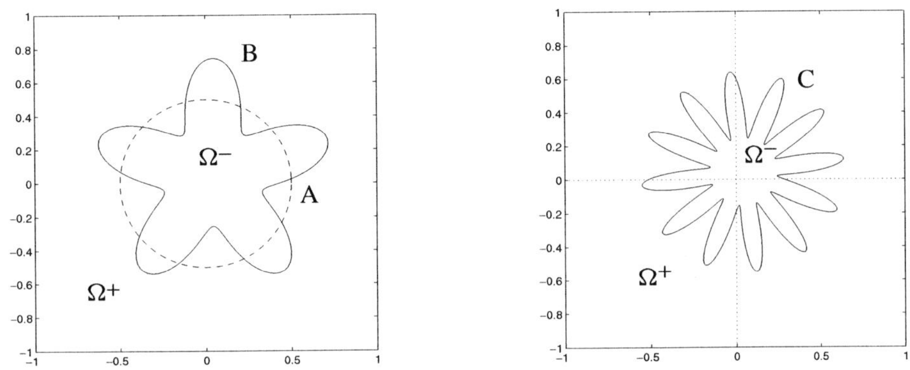

Figure 6.1. *Plots of different interfaces determined from* (6.26). (a) *Case* A *is the dashed line and Case* B *is the solid line.* (b) *A plot of the interface for Case* C.

(a) (b)

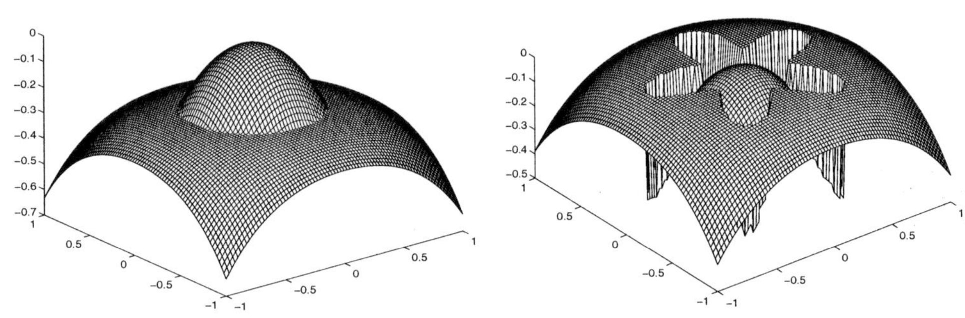

Figure 6.2. *Plots of* $-u(x, y)$ *of the computed solutions with* $\beta^+ = 10$ *and* $\beta^- = 1$. (a) *Case* A, *a circular interface where the solution is continuous but* $[\beta u_n] = -0.7$. (b) *Case* B, *an irregular interface where both the solution and the flux* $[\beta u_n]$ *are discontinuous.*

the interface and we must have enough control points to resolve it. The solution parameters are set to be the same as in Case B.

Figures 6.3 and 6.4 and Table 6.1 show plots and data from the computed solutions. In the table, E_1 is the relative error of the solution in the infinity norm; E_2 and E_3 are the relative errors of the normal derivative u_n^- and u_n^+, respectively, for example,

$$E_2 = \frac{\max\limits_{1 \le l \le n_b} \left| u_{\boldsymbol{n}}^-(\mathbf{X}_l) - \mathbf{U}_{\boldsymbol{n}\,l}^- \right|}{\max\limits_{1 \le l \le n_b} \left| u_{\boldsymbol{n}}^-(\mathbf{X}_l) \right|}; \tag{6.27}$$

r_i are the ratio of the two consecutive errors; and k is the number of GMRES iterations.

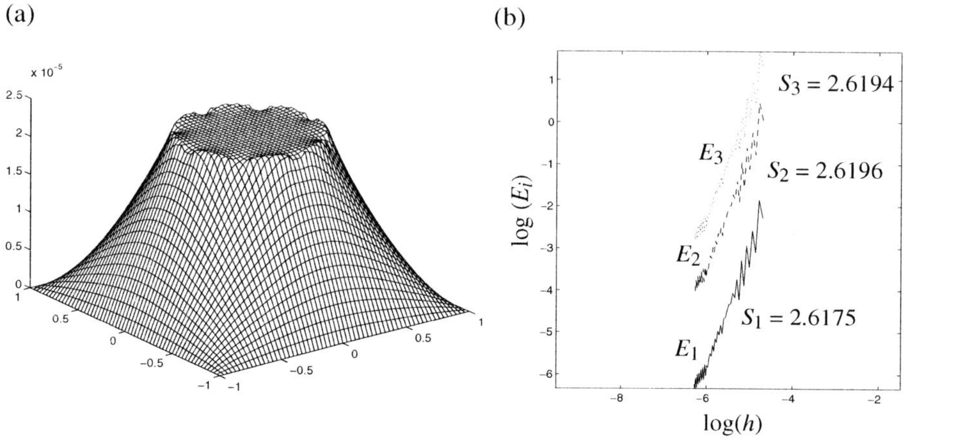

Figure 6.3. (a) *An error distribution for Case* A. *The largest error in magnitude is about* 2.5×10^{-6}. (b) *Errors* E_i *versus the mesh size* h *in* log-log *scale for Case* A *with* $M = N = N_b$, $\beta^- = 10^4$, *and* $\beta^+ = 1$.

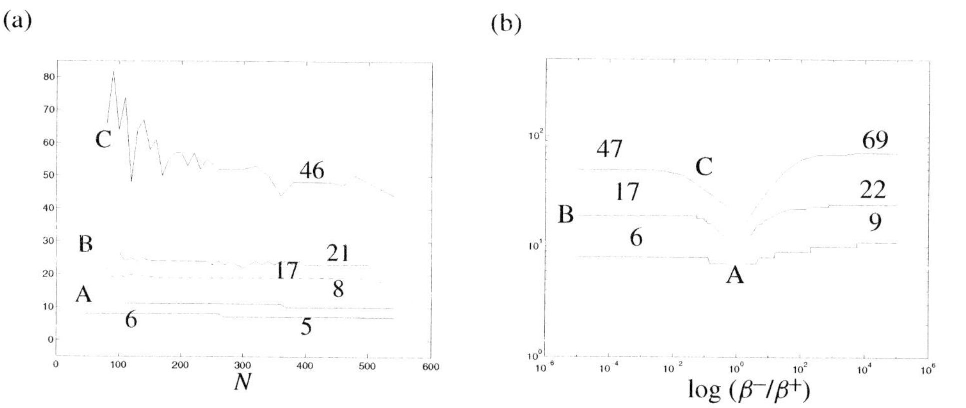

Figure 6.4. *The number of iterations for Example* 6.1 *with* $M = N = N_b$*:* (a) *versus the number of grid lines* M *in the* x*-direction* ($N = M$). *Case* A*: lower curve,* $\beta^- = 1$, $\beta^+ = 10^4$*; upper curve,* $\beta^- = 10^4$, $\beta^+ = 1$. *Case* B*: lower curve,* $\beta^- = 1$, $\beta^+ = 10^3$*; upper curve,* $\beta^- = 10^3$, $\beta^+ = 1$. *Case* C*:* $\beta^- = 1$, $\beta^+ = 100$*;* (b) *versus the ratio of jumps* β^-/β^+ *in* log-log *scale with* $M = N = N_b = 160$.

Convergence analysis

Table 6.1 shows the results of a grid refinement analysis for Case A with two very different values of the ratios β^-/β^+, $\beta^+ > \beta^-$. When $\beta^-/\beta^+ = 0.5$, the ratio r_i is very close to 4 indicating second-order convergence. With $\beta^- = 1$ and $\beta^+ = 10^4$, the error in the solution drops much more rapidly. This is because the solution in Ω^+ approaches a constant as β^+ becomes large, and it is quadratic in Ω^-. A second-order accurate method would give a highly accurate solution in both regions. So it is not surprising to see that the ratio r_1 is

Table 6.1. *A grid refinement analysis for Case A with $M = N = N_b$.*

N	β^+	β^-	E_1	E_2	E_3	r_1	r_2	r_3	k
40	2	1	$2.285\ 10^{-3}$	$2.23\ 10^{-3}$	$7.434\ 10^{-3}$				7
80	2	1	$5.225\ 10^{-4}$	$5.956\ 10^{-3}$	$1.987\ 10^{-2}$	4.37	3.74	3.74	7
160	2	1	$1.269\ 10^{-4}$	$1.827\ 10^{-4}$	$6.101\ 10^{-4}$	4.12	3.26	3.26	7
320	2	1	$2.988\ 10^{-5}$	$5.038\ 10^{-5}$	$1.678\ 10^{-4}$	4.25	3.63	3.64	7

N	β^+	β^-	E_1	E_2	E_3	r_1	r_2	r_3	k
40	10^4	1	$6.552\ 10^{-5}$	$6.331\ 10^{-4}$	$2.110\ 10^{-4}$				8
80	10^4	1	$7.847\ 10^{-6}$	$8.366\ 10^{-5}$	$2.785\ 10^{-5}$	8.35	7.57	7.58	8
160	10^4	1	$5.988\ 10^{-7}$	$9.192\ 10^{-7}$	$3.033\ 10^{-6}$	13.1	9.10	9.18	8
320	10^4	1	$5.859\ 10^{-8}$	$2.058\ 10^{-7}$	$6.887\ 10^{-7}$	10.2	4.47	4.40	7

much larger than 4. For the normal derivatives, we expect second-order accuracy again since $\beta^+ u_n^+$ is not quadratic and has a magnitude of $O(1)$. This agrees with the results r_2 and r_3 in Table 6.1.

In the opposite case, when $\beta^-/\beta^+ = 10^4$, $\beta^+ < \beta^-$, the solution is not quadratic, and we see the expected second-order accuracy. Figure 6.3(a) is a plot of the error distribution over the region. The error seems to change continuously even though the maximum error occurs on or near the interface.

For interface problems, the errors obtained from non-body-fitting grids usually do not decrease monotonically as we refine the grid unless the interface is aligned with one of the axes. It is more realistic to find the asymptotic convergence rate as the slope of the line fitting of the experimental data $(\log(h_i), \log(E_i))$. Figure 6.3(b) is a plot of the errors versus the mesh size h in log-log scale for the case $N_b = N$. The asymptotic convergence rate is about 2.62, indicating a second-order method. As h gets smaller, it is observed through our numerical tests that the curves for the errors become flatter indicating that the asymptotic convergence rate will approach 2. If the interface is represented by a set of particles, then the convergence rate may also be affected by the number of control points and their relative positions, unless the interface is well resolved. Often, we refine the Eulerian meshes M and N and Lagrangian mesh N_b simultaneously. We refer the readers to [163] for more detailed discussions; see also [198].

The number of iterations versus the mesh size h

Figure 6.4(a) shows the number of iterations versus the number of grid lines M in the x-direction ($N = M$) for Cases A, B, and C. It is not surprising to see that the number of iterations depends on the shape of the interface. The number of iterations required for Case C is larger than that for Cases A and B, but it is still almost independent of the mesh size h. For Case A, where the interface is a circle, the algorithm needs only $5 \sim 8$ iterations for all choices of the mesh size h for two extreme cases $\rho = 10^4$ and $\rho = 10^{-4}$. This is also true for different choices of the ratio $\rho = \beta^-/\beta^+$. In Figure 6.4(a), the lowest curve corresponds to Case A with $\beta^- = 1$ and $\beta^+ = 10^4$, and the lowest but the second curve corresponds to $\beta^- = 10^4$ and $\beta^+ = 1$. For Case B, the number of iterations required is about $17 \sim 21$ for $\rho = 10^{-3}$ and $\rho = 10^3$, respectively. For Case C, the most complicated interface, the number of iterations is about 46 with a reasonable number of control points on the interface for $\rho = 10^{-2}$ and $\rho = 10^2$, respectively.

The number of iterations versus the jump ratio $\rho = \beta^-/\beta^+$

Figure 6.4(b) is a plot of the number of iterations versus the jump ratio ρ in log-log scale with fixed mesh size $M = N = N_b = 160$. We set $\beta^- = 1$ when $\beta^-/\beta^+ < 1$ and $\beta^+ = 1$ when $\beta^-/\beta^+ > 1$. As ρ deviates from unity, we have a larger jump in the coefficient. The number of iterations increases proportionally with $|\log(\rho)|$ when ρ is close to unity but soon reaches a point after which the number of iterations remains constant. Such a point depends on the shape of the interface. For Case A, it requires only $5 \sim 6$ iterations at most for $\rho < 1$ and $7 \sim 8$ iterations for $\rho > 1$ in solving the Schur complement system using the GMRES method. For Case B, the numbers are $17 \sim 22$. For Case C, the most complicated interface in our examples, the numbers are $47 \sim 69$. As we mentioned in the previous paragraph (also see Figure 6.4(a)), for Case C, with only 160 control points we cannot express the complicated interface, Figure 6.1(b), very well. If we take more control points on the interface, then the number of iterations will be about 46. We refer the reader to [163] for further analysis about the effect of the number of control points.

6.2 The augmented method for generalized Helmholtz equations on irregular domains

The idea of the augmented approach for the elliptic interface problems described in the previous section can be used with little modification to solve generalized Helmholtz or Poisson equations of the form

$$\begin{aligned} \Delta u - \sigma u &= f(\mathbf{x}), \quad \mathbf{x} \in \Omega, \\ q(u, u_n) &= 0, \quad \mathbf{x} \in \partial\Omega, \end{aligned} \tag{6.28}$$

defined on an irregular domain Ω (interior or exterior), where $q(u, u_n)$ is either a Dirichlet or Neumann boundary condition along the boundary $\partial\Omega$. We assume that $\sigma \geq 0$, but the numerical methods described below should work also for $\sigma \leq 0$ with modest magnitude. For large negative σ, different methods are needed to deal with the wave-type equation.

To use an augmented approach, the domain Ω is embedded into a rectangle $R \supset \Omega$; the PDE and the source term are extended to the entire rectangle R,

$$\Delta u - \sigma u = \begin{cases} f & \text{if } \mathbf{x} \in \Omega, \\ 0 & \text{if } \mathbf{x} \in R - \Omega, \end{cases} \qquad q(u, u_n) = 0,$$

$$\begin{cases} [u] = g \text{ on } \partial\Omega, \\ [u_n] = 0 \text{ on } \partial\Omega, \\ u = 0 \text{ on } \partial R, \end{cases} \quad \text{or} \quad \begin{cases} [u] = 0 \text{ on } \partial\Omega, \\ [u_n] = g \text{ on } \partial\Omega, \\ u = 0 \text{ on } \partial R. \end{cases} \tag{6.29}$$

The solution u to the *interface problem* above is a functional of g, which is one dimension lower than that of u. We determine g such that the solution $u(g)$ satisfies the boundary condition $q(u, u_n) = 0$. Note that, given g, we can solve the above interface problem using the IIM (3.22) or (5.20) with a single call to a fast Poisson solver.

For the first option, $[u] = g$ and $[u_n] = 0$, we briefly describe the method below. For the second option, $[u] = 0$ and $[u_n] = g$, the process is similar to what has been discussed in §6.1.

For an interior Dirichlet boundary condition $u|_{\partial\Omega} = u_0(\mathbf{x})$, if we set $[u_n] = 0$ and $[u] = g$ as the augmented variable, then the condition that needs to be enforced is the boundary condition $u^-(g) = u_0(\mathbf{x})$, or $U^-(\mathbf{X}) - u_0(\mathbf{X}) = \mathbf{0}$, at all control points on the interface in the discrete form. Corresponding to (6.11), the second matrix-vector equation now takes the form

$$E\mathbf{U} + T\mathbf{G} - P\mathbf{U_0} = 0, \tag{6.30}$$

where $\mathbf{U_0}$ is the vector formed by the boundary values at control points, and E, T, and P are matrices that depend on the interpolation scheme for $U^-(\mathbf{X}) - u_0(\mathbf{X}) = \mathbf{0}$. We can use the same least squares interpolation scheme (6.16) to find the interpolation coefficients $\{\gamma_k\}$. We just need to change the first two equations in (6.20) to $a_1 + a_2 = 1$ and $a_3 + a_4 = 0$. That is, instead of approximating $U_n^-(\mathbf{X})$, we now approximate the solution $U^-(\mathbf{X})$ at the control points on the boundary. The rest of the procedure is exactly the same.

For an interior Neumann boundary condition $\frac{\partial u}{\partial n}|_{\partial\Omega} = u_1(\mathbf{x})$, where $u_1(\mathbf{x})$ is a given function, we can use exactly the same least squares interpolation (6.16) to compute $\frac{\partial u}{\partial n}|_{\partial\Omega} = \frac{\partial u^-}{\partial n}|_{\partial\Omega}$ to get the residual. The second matrix-vector equation takes the form

$$E\mathbf{U} + T\mathbf{G} - P\mathbf{U_1} = 0, \tag{6.31}$$

where $\mathbf{U_1}$ is the vector formed by the boundary values of $u_1(\mathbf{x})$ at the control points.

For an exterior problem within a rectangle (i.e., (6.28) holds for $x \in \Omega^c$) with a prescribed boundary condition along the boundary of the rectangle ∂R and the inside boundary Γ, we also extend the PDE to the entire domain and set $[u] = g$ and $[u_n]_\Gamma = 0$. We can use a standard finite difference method to deal with the boundary condition along ∂R. The augmented variable $[u] = g$ is defined along the inside irregular boundary Γ.

For Dirichlet boundary value problems, the Schur complement system is usually well-conditioned and there is no need for preconditioning the GMRES iteration. For Neumann boundary value problems, some preconditioning techniques can reduce the number of iterations of the GMRES method. For example, for an exterior Poisson equation with a Neumann boundary condition, in each iteration, we simply set the solution outside (the fictitious domain) to be zero, because it is the solution to the augmented problem. For pure Neumann problems, that is, $\sigma = 0$, the solution may not exist unless the compatibility condition is satisfied.

If the boundary is represented by the zero level set of a function $\varphi(\mathbf{x})$, then the unknowns of the augmented variable are defined at the orthogonal projections of irregular grid points from a particular side, say $\varphi \geq 0$, instead of the orthogonal projection of all irregular grid points from both sides of the boundary Γ. In this way, we can avoid clustered control points and an ill-conditioned Schur complement system.

The fast Helmholtz solvers on irregular domains in two dimensions with Dirichlet or Neumann boundary conditions using the augmented approach described here are available online via anonymous ftp; see [165]. The irregular boundary is allowed to be expressed implicitly by a level set function. The package was written in FORTRAN.

The application of the augmented approach for two-dimensional irregular domains can be found in [180] for simulating electromigration of voids in an integrated circuit; in [113] for simulating the motion of drops; in [177] for solving Navier–Stokes equations on irregular domains using the stream-vorticity formulation; in [125, 266] for microstructure evolution in the chemical vapor infiltration process; and in [41] for solving biharmonic equations on irregular domains. The augmented approach for generalized Helmholtz equations has also been developed in three dimensions and has been applied to an inverse problem of shape optimization in [64, 65].

There are other methods for elliptic PDEs defined on irregular domains using embedding or fictitious domain techniques. The earlier ones include the capacitance matrix method [221] and the integral equation approach [188, 198].

In [279], an immersed interface method for boundary value problems (IIMB) on irregular domains is developed, particularly for Dirichlet boundary conditions. It was shown in [279] that the IIMB is second-order accurate in the maximum norm and that the Schur complement system is well-conditioned. The IIMB was applied to underground water simulations using the stream-vorticity function in [279].

Fogelson and Keener [81] have developed an embedding method for Laplacian equations on irregular domains with a Neumann boundary condition in two and three dimensions. With careful selection of the stencils, the method is second-order accurate and produces a stable matrix (diagonally semidominant). Dumett and Keener [67] have also extended the embedding method to anisotropic elliptic boundary value problems on irregular domains in two space dimensions when $\beta(\mathbf{x})$ in (6.1) is an anisotropic matrix.

Most of the methods mentioned above have a very important feature in common, that is, fast Poisson solvers have been employed in these methods. We believe this is a criterion for the state of the art of an algorithm. The implementations of these methods, however, are not trivial.

One can also use a direct discretization at the intersections between the boundary and the axes for a Dirichlet boundary value problem; see, for example [202], the recent finite volume method [126], and the ghost fluid method (GFM) [183]. Second-order accuracy can still be achieved and the coefficient matrix of the finite difference equation is symmetric positive or negative definite. However, the finite difference coefficients at the stencils near the boundary are no longer constant, which makes it difficult to apply fast Poisson solvers. The advantage of such an approach is the simplicity of implementation. A global iterative method such as a preconditioning conjugate gradient method (PCG) may not be as efficient as a fast Poisson solver based on the FFT. A multigrid method is recommended for solving the resulting linear system of equations which may cost us the advantage of its simplicity. For Neumann boundary conditions, it is hard to get high-order accurate discretization using such an approach; see [202]. High-order discretization is possible if one uses a one-sided high-order extrapolation along each axis. However, the stability of such an algorithm cannot be guaranteed and the condition number of the resulting linear system of equations can be large. Generally, we do not recommend a direct discretization for PDEs whose boundary conditions involve partial derivatives.

6.2.1 An example of the augmented approach for Poisson equations on irregular domains

We provide one example with different boundary conditions to show the efficiency of the augmented algorithm for Poisson equations defined on irregular domains. We show second-order accuracy of the computed solution, and, more important, show that the number of iterations is nearly independent of the mesh size, except for a factor of $\log h$.

Example 6.2. We construct an exact solution,

$$\begin{aligned} u(x,y) &= -\frac{1}{2}\log r + r^2, \qquad r = \sqrt{x^2+y^2}, \\ f(x,y) &= 4. \end{aligned} \tag{6.32}$$

The exterior domain is composed of two pieces: the boundary $\partial\Omega_1$ of the rectangle, $-1 \le x, y \le 1$, and the boundary of $\partial\Omega_2$ of the ellipse

$$\frac{x^2}{a^2} + \frac{y^2}{b^2} = 1.$$

Case 1. The Dirichlet boundary condition on $\partial\Omega_1$ and the normal derivative boundary condition on $\partial\Omega_2$ are given using the exact solution. The first part of Table 6.2 shows a grid refinement analysis and other information, with $a = 0.5$ and $b = 0.4$.

In Table 6.2, N ($M = N$) is the number of grid lines in the x- and y-directions; E is the error of the computed solution in the maximum norm over all grid points inside the rectangle but outside the ellipse; k is the number of iterations of the GMRES method, which is also the number of the fast Poisson solver on the rectangular domain called. We observe

Table 6.2. *A grid refinement analysis with $a = 0.5$ and $b = 0.4, 0.15$ for Example 6.2. Dirichlet boundary conditions are prescribed on $\partial\Omega_1$, and Neumann boundary conditions are prescribed on $\partial\Omega_2$.*

N	a	b	E	r	k
40	0.5	0.4	$5.7116\ 10^{-4}$		16
80	0.5	0.4	$1.4591\ 10^{-4}$	3.9146	17
160	0.5	0.4	$3.5236\ 10^{-5}$	4.1408	19
320	0.5	0.4	$8.1638\ 10^{-6}$	4.3161	21

N	a	b	E	r	k
40	0.5	0.15	$4.4820\ 10^{-3}$		13
80	0.5	0.15	$1.1466\ 10^{-3}$	3.9089	15
160	0.5	0.15	$2.6159\ 10^{-4}$	4.3832	17
320	0.5	0.15	$6.7733\ 10^{-5}$	3.8621	20

Table 6.3. *A grid refinement analysis for an exterior problem in Example* 6.2 *with a Neumann boundary condition.*

N	a	b	E	r	k
40	0.5	0.15	$4.5064\ 10^{-3}$		14
80	0.5	0.15	$1.2529\ 10^{-3}$	3.5967	17
160	0.5	0.15	$3.3597\ 10^{-4}$	3.7292	19
320	0.5	0.15	$7.9409\ 10^{-5}$	4.2309	21

a second-order rate of convergence in the maximum norm. The number of iterations for the GMRES is proportional to $\log h$. Note that the error may not be reduced by a factor of 4 exactly but rather fluctuates, as explained earlier in this chapter and in [163].

Case 2. The normal derivative u_n is prescribed on $\partial\Omega_1$ using the exact solution. In this case, the solution is not unique and the compatibility condition has to be imposed in order to get a reasonably accurate solution. To get a unique solution, we specify the solution at one corner using the exact solution. In this way, we can measure the error of the computed solution. Table 6.3 shows the result of a grid refinement analysis. We have similar results as we analyzed for Case 1.

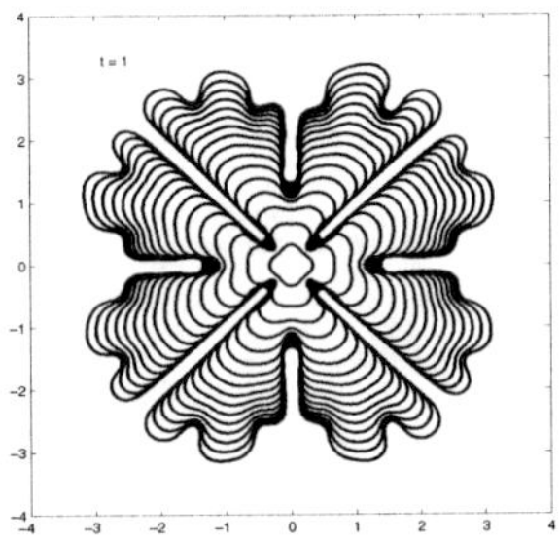

Chapter 7

The Fourth-Order IIM

In this chapter, we discuss an M-matrix finite difference scheme that has fourth-order accuracy for an elliptic interface problem of the form

$$-\nabla \cdot (\beta(\mathbf{x})\nabla u) + b(\mathbf{x}) \cdot \nabla u + \sigma(\mathbf{x})\, u = f(\mathbf{x}), \quad \mathbf{x} \in \Omega, \tag{7.1}$$

with a boundary condition

$$\frac{\partial u}{\partial \boldsymbol{n}} + \alpha(\mathbf{x})\, u(\mathbf{x}) = g(\mathbf{x}), \quad \mathbf{x} \in \Gamma, \tag{7.2}$$

where Ω is a bounded open domain in R^2 or R^3 with a smooth boundary Γ and $\boldsymbol{n}$ is the outward normal at the boundary Γ. Also, we discuss a fourth-order method for the heat equation

$$\frac{\partial}{\partial t} u(t,\mathbf{x}) = \nabla \cdot (\beta(\mathbf{x})\nabla u) + b(\mathbf{x}) \cdot \nabla u + \sigma(\mathbf{x})\, u, \quad (t,\mathbf{x}) \in R^+ \times \Omega, \tag{7.3}$$

with an initial condition $u(0,\mathbf{x}) = u_0(\mathbf{x})$ and the boundary condition (7.2).

We describe the finite difference method for (i) the general boundary condition (7.2) on an irregular domain, and (ii) variable coefficients of interface problems, i.e., piecewise constant media.

In §7.1 we describe the fourth-order finite difference method for one-dimensional boundary value problems and the basic idea of deriving high-order schemes based on the solution continuation across the interface and the boundary. In §7.2 we discuss the two-dimensional problems. A 9-point fourth-order compact finite difference scheme is introduced in §7.2.1.

The discussions in this chapter will proceed as follows.

- Section 7.2.2: A treatment of Neumann boundary conditions.
- Sections 7.2.3–7.2.4: Equations on irregular domains for two-dimensional problems.
- Section 7.2.5: Heat equations on irregular domains.
- Section 7.2.6: Variable coefficients.

- Section 7.2.7: Interface problems in two dimensions.
- Section 7.3.1: Equations on irregular domains for three-dimensional problems.
- Section 7.3.2: Interface problems in three dimensions.
- Section 7.4: Sparse subspace iterative methods.
- Section 7.5: Numerical results.

In §7.4 we also discuss the iteration method for solving the resulting system of linear equations based on the preconditioned sparse subspace iterative method.

7.1 Two-point boundary value problems

In this section we discuss the one-dimensional case to describe the basic idea for developing higher-order finite difference schemes. First we describe the higher-order finite difference schemes for the constant coefficient case. We then describe the fourth-order methods for (7.4) with the general boundary condition (7.5), smooth variable coefficients, and piecewise constant coefficients. Our derivation of the fourth-order methods also leads to the second-order method as an intermediate step. These methods are similar to those described in §2.1, but not identical. For example, the resulting matrix is symmetric for the approach in this section. Also, our description for the second-order methods in this section is meant to be paired with those for the fourth-order methods.

Consider the two-point boundary value problem

$$-\frac{d}{dx}\left(\beta(x)\,\frac{d}{dx}u\right) + \sigma(x)\,u = f(x), \qquad 0 < x < 1, \tag{7.4}$$

with boundary conditions

$$\begin{aligned} -a_0\,u_x(0) + b_0\,u(0) &= g_0, \\ a_1\,u_x(1) + b_0\,u(1) &= g_1. \end{aligned} \tag{7.5}$$

Let $x_i = i\,h,\ 0 \le i \le M$, be the uniform grid points with $h = 1/M$. Define the test function ϕ_h by

$$\phi_h(s) = \begin{cases} 1 - \dfrac{|s|}{h}, & s \in [-h, h], \\ 0 & \text{otherwise}. \end{cases} \tag{7.6}$$

Then

$$\begin{aligned} & -\int_{-h}^{h} \frac{d}{dx}\left(\beta(s+x_i)\,\frac{d}{dx}u(s+x_i)\right)\phi_h(s)ds \\ & = \frac{1}{h}\left(\int_{x_i}^{x_{i+1}} \beta(x)\frac{d}{dx}u(x)\,dx - \int_{x_{i-1}}^{x_i} \beta(x)\frac{d}{dx}u(x)\,dx\right). \end{aligned} \tag{7.7}$$

7.1.1 The constant coefficient case

If $\beta = 1$ and $\sigma = 0$, then from (7.7),

$$-\frac{u(x_{i+1}) - 2u(x_i) + u(x_{i-1})}{h^2} = \frac{1}{h}\int_{-h}^{h} f(s + x_i)\phi_h(s)\,ds. \tag{7.8}$$

Thus we obtain the standard central finite difference scheme

$$-\frac{U_{i+1} - 2U_i + U_{i-1}}{h^2} = \tilde{f}_i \tag{7.9}$$

with $U_0 = U_M = 0$ (for the homogeneous boundary condition). If

$$\tilde{f}_i = \frac{1}{h}\int_{-h}^{h} f(s + x_i)\phi_h(s)\,ds, \tag{7.10}$$

then $U_i = u(x_i)$ (exact solution). If we take $\tilde{f}_i = f(x_i)$, then

$$\frac{1}{h}\int_{-h}^{h} f(s + x_i)\phi_h(s)\,ds - f(x_i) = O(h^2),$$

and thus we obtain the second-order scheme

$$-\frac{U_{i+1} - 2U_i + U_{i-1}}{h^2} = F_i, \tag{7.11}$$

where $F_i = f_i - \sigma\, U_i$. If we let

$$\tilde{f}_i = \frac{f(x_{i-1}) + 10\, f(x_i) + f(x_{i+1})}{12}, \tag{7.12}$$

then

$$\frac{1}{h}\int_{-h}^{h} f(s + x_i)\phi_h(s)\,ds - F_i = O(h^4),$$

and thus we obtain the fourth-order scheme

$$-\frac{U_{i+1} - 2U_i + U_{i-1}}{h^2} = \frac{F_{i-1} + 10\, F_i + F_{i+1}}{12}. \tag{7.13}$$

We can deduce the finite difference scheme of any desired even order by choosing appropriate quadrature rule for evaluating

$$\frac{1}{h}\int_{-h}^{h} f(s + x_i)\phi_h(s)\,ds. \tag{7.14}$$

7.1.2 General boundary conditions

In this subsection we consider the general boundary condition of the form

$$u_x(1) + \alpha\, u(1) = g, \tag{7.15}$$

and the corresponding finite difference schemes. If we let $G = g - \alpha\, u(1)$ and $F = f - \sigma\, u$, then the problem is equivalent to

$$-u'' = F, \quad u'(1) = G. \tag{7.16}$$

At the boundary $x = 1$ ($i = M$), U_{M+1} is not an unknown in (7.11) and (7.13), and thus we must complete the equations by determining U_{M+1}. We use the continuation of the solution u to complete the finite difference schemes at $x = 1$. For the second-order scheme, note that

$$u_{M+1} = u(1+h) = u(1) + u'(1)\,h + \frac{h^2}{2}\,u''(1) + O(h^3).$$

Since $u'(1) = G$ and $u'' = -f(1)$,

$$u_{M+1} \sim u_M + G\,h - \frac{h^2}{2} F_M + O(h^3),$$

and by substituting this into (7.11) we obtain the second-order finite difference

$$\frac{U_n - U_{n-1}}{h^2} = \frac{1}{2} F_M + \frac{G}{h}. \tag{7.17}$$

For the fourth-order scheme, note that

$$u_{M+1} = u(1+h) = u(1) + u'(1)\,h + \frac{h^2}{2}\,u''(1) + \frac{h^3}{6}\,u'''(1) + \frac{h^4}{24}\,u''''(1) + O(h^5),$$
$$F_{M+1} = F(1+h) = F(1) + F'(1)\,h + \frac{h^2}{2}\,F''(1) + O(h^3).$$

Since $u''' = -F'$ and $u'''' = -F''$,

$$\frac{1}{h^2} u_{M+1} + \frac{1}{12} F_{M+1} = -\frac{5}{12} F(1) - \frac{h}{12} F'(1) + O(h^3),$$

and thus at $i = M$ ($x = 1$) the fourth-order scheme is

$$\frac{U_M - U_{M-1}}{h^2} = \frac{1}{12}(5F_M + F_{M-1}) + \frac{G}{h} - \frac{h}{12} F'(1), \tag{7.18}$$

where

$$F'(1) = f'(1) - \sigma\, G.$$

7.1.3 The smooth variable coefficient case

If we use the midpoint rule and

$$\frac{d}{dx} u(x_{i\pm 1/2}) \sim \pm \frac{u(x_{i\pm 1}) - u(x_i)}{h}, \quad \beta(x_{i\pm 1/2}) \sim \frac{\beta(x_{i\pm 1}) + \beta(x_i)}{2}$$

for approximating integrals in (7.7), then we obtain the standard central finite difference scheme

$$
\begin{aligned}
&-\frac{1}{h}\left(\beta_{i+1/2}\frac{U_{i+1}-U_i}{h}-\beta_{i-1/2}\frac{U_i-U_{i-1}}{h}\right)\\
&\quad+b(x_i)\frac{U_{i+1}-U_{i-1}}{2h}+\sigma(x_i)U_i=f(x_i),
\end{aligned}
\tag{7.19}
$$

where

$$\beta_{i\pm1/2}=\frac{\beta(x_{i\pm1})+\beta(x_i)}{2}.$$

Next we compute the truncation error of (7.19). Note that

$$(\beta\, u')'=\beta\, u''+\beta' u',$$

and let

$$
\begin{aligned}
B&=\beta\frac{u(x_{i+1})-2u(x_i)+u(x_{i-1})}{h^2}+\beta'\frac{u(x_{i+1})-u(x_{i-1})}{2h}\\
&=h^{-1}\left(\left(\beta+\frac{h}{2}\beta'\right)\frac{u(x_{i+1})-u(x_i)}{h}-\left(\beta-\frac{h}{2}\beta'\right)\frac{u(x_i)-u(x_{i-1})}{h}\right)
\end{aligned}
$$

and

$$A=h^{-1}\left(\beta_{i+1/2}\frac{u(x_{i+1})-u(x_i)}{h}-\beta_{i-1/2}\frac{u(x_i)-u(x_{i-1})}{h}\right).$$

Then we have

$$B-(\beta\, u')'=\frac{h^2}{12}\beta u''''+\frac{h^2}{6}\beta' u'''+O(h^4)$$

and

$$A-B=\frac{h^2}{4}\beta'' u''+\frac{h^2}{6}\beta''' u'+O(h^4).$$

Thus, we get

$$E_1=A-(\beta\, u')'=\frac{h^2}{12}(\beta u''''+2\beta' u'''+3\beta'' u''+2\beta''' u')+O(h^4).$$

Similarly, we have

$$E_2=-b_i\frac{u(x_i+h)-u(x_i-h)}{2h}+b\, u'=-\frac{h^2}{6}\, bu'''+O(h^4).$$

Since

$$
\begin{aligned}
(\sigma u-f)''&=(\beta u')'''-(bu')''\\
&=\beta u''''+3\beta' u'''+3\beta'' u''+\beta''' u'-(bu'''+2b' u''+b'' u'),
\end{aligned}
$$

we obtain the truncation error

$$
\begin{aligned}
E=E_1+E_2&=\frac{h^2}{12}(\beta u''''+2\beta' u'''+3\beta'' u''+2\beta''' u'-2bu''')\\
&=\frac{h^2}{12}(-f''+(\sigma u)''+(\beta'''+b'')u'-(\beta'+b)u'''+2b' u'').
\end{aligned}
\tag{7.20}
$$

Here,

$$(\beta''' + b'')u' - (\beta' + b)u''' = \left((\beta'' + b')u'\right)' - \left((\beta' + b)u''\right)' \tag{7.21}$$

and

$$(\beta' + b)u'' = \frac{\beta' + b}{\beta}(-(\beta' - b)u' + \sigma u - f). \tag{7.22}$$

Thus, eliminating the second-order error (7.20) from (7.19) using (7.21)–(7.22), we obtain the fourth-order 3-point finite difference scheme,

$$\begin{aligned} -\frac{1}{h}\left(\tilde{\beta}_{i+\frac{1}{2}}\frac{U_{i+1} - U_i}{h} - \tilde{\beta}_{i-\frac{1}{2}}\frac{U_i - U_{i-1}}{h}\right) + \gamma^k_{k-1}U_{i-1} + \frac{5\sigma_i}{6}U_i + \gamma^k_{i+1}U_{i+1} \\ = \alpha^i_{i-1}f_{i-1} + \frac{5}{6}f_i + \alpha^i_{i+1}f_{i+1}, \end{aligned} \tag{7.23}$$

where $\tilde{\beta}_{i\pm 1/2} = (\tilde{\beta}_i + \tilde{\beta}_{i\pm 1})/2 - h^2(b_{i+1} - b_{i-1})/12$ and

$$\begin{aligned} \tilde{\beta}_i &= \beta_i - \frac{h^2}{12}\left(\frac{1}{\beta_i}\left(\left(\frac{\beta_{i+1} - \beta_{i-1}}{2h}\right)^2 - b_i^2\right) + \frac{\beta_{i+1} - 2\beta_i + \beta_{i-1}}{h^2} + b_i\right), \\ \alpha^i_{i\pm 1} &= \frac{1}{12}\left(1 \mp h\frac{\beta' + b}{\beta}(x_{i\pm 1})\right), \\ \gamma^i_{i\pm 1} &= \frac{1}{12}\left(\sigma_{i\pm 1} \mp h\frac{c(\beta' + b)}{\beta}(x_{i\pm 1})\right). \end{aligned}$$

7.1.4 The piecewise constant coefficient case

Let β be a piecewise constant, i.e.,

$$\beta = \begin{cases} \beta^- & \text{on } [x_i, x^*) = I^-, \\ \beta^+ & \text{on } (x^*, x_{i+1}] = I^+. \end{cases} \tag{7.24}$$

We use the continuation method to complete the finite difference scheme at x_i,

$$\beta^- \frac{\hat{U}_1 - 2U_0 + U_{-1}}{h^2} = f_i, \tag{7.25}$$

where U_0 and U_{-1} denote approximate solution values at x_i and x_{i-1}, respectively, and the continuation value $\hat{U}_1$ is determined as follows. We represent u by the piecewise quadratic Taylor polynomial on each interval $I^\pm$,

$$u^\pm = u + u^\pm_x(x - x^*) + \frac{1}{2}u^\pm_{xx}(x - x^*)^2$$

with continuity conditions at x^*,

$$\beta^- u^-_x = \beta^+ u^+_x \quad \text{and} \quad \beta^- u^-_{xx} = \beta^+ u^+_{xx} = -f(x^*).$$

Thus, if u_1 denotes the solution value at x_{i+1} and if $h_1 = x^* - x_i$ and $h_2 = x_{i+1} - x^*$, then

$$\begin{cases} u_0 = u - u_x^- h_1 + \frac{1}{2} u_{xx}^- h_1^2, \\ u_1 = u + u_x^+ h_2 + \frac{1}{2} u_{xx}^+ h_2^2. \end{cases}$$

Given u_0, u_1, and $f(x^*)$, we solve this for (u, u_x^-) and obtain

$$\begin{aligned} \hat{u}_1 - u_0 &= u + u_x^- h_2 + \frac{1}{2} u_{xx}^- h_2^2 - u_0 \\ &= \frac{h}{h_1 + h_2 \beta^- / \beta^+} \left(u_1 - u_0 - \frac{h_1 h_2}{2} \left(\frac{1}{\beta^-} - \frac{1}{\beta^+} \right) f(x^*) \right). \end{aligned}$$

Substituting this into (7.25), we obtain the second-order symmetric difference,

$$\begin{aligned} &-\frac{1}{h} \left(\left(\frac{1}{\beta^-} \frac{h_1}{h} + \frac{1}{\beta^+} \frac{h_2}{h} \right)^{-1} \frac{U_{i+1} - U_i}{h} - \beta^- \frac{U_i - U_{i-1}}{h} \right) \\ &= \left(1 - \frac{\beta^+ - \beta^-}{2} \frac{h}{\beta^- h_2 + \beta^+ h_1} \frac{h_1 h_2}{h^2} \right) f_i. \end{aligned} \tag{7.26}$$

Here,

$$\left(\frac{1}{\beta^-} \frac{h_1}{h} + \frac{1}{\beta^+} \frac{h_2}{h} \right)^{-1} = \left(\int_{x_i}^{x_{i+1}} \frac{1}{\beta(x)} \, dx \right)^{-1}$$

is the harmonic average of β on $[x_i, x_{i+1}]$ and the resulting matrix is symmetric.

For the fourth-order method, we use

$$u^\pm = u + u_x^\pm (x - x^*) + \frac{1}{2} u_{xx}^\pm (x - x^*)^2 + \frac{1}{6} u_{xxx}^\pm (x - x^*)^3 + \frac{1}{24} u_{xxxx}^\pm (x - x^*)^4$$

with continuity conditions at x^*,

$$\beta^- u_x^- = \beta^+ u_x^+ \quad \text{and} \quad \beta^- (u^-)^{(k)} = \beta^+ (u^+)^{(k)}, \ 2 \le k \le 4.$$

Since

$$\beta^- u_{xx}^-(x_{i-1}) = -f_{i-1}, \ \beta^- u_{xx}^-(x_i) = -f_i, \ \beta^+ u_{xx}^+(x_{i+1}) = -f_{i+1},$$

then

$$\begin{aligned} C_1 &= \beta^- u_{xx}^- = \beta^+ u_{xx}^+ = f_i + \frac{f_{i+1} - f_{i-1}}{2h} h_1 + \frac{f_{i+1} - 2f_i + f_{i-1}}{2h^2} h_1^2, \\ C_2 &= \beta^- u_{xxx}^- = \beta^+ u_{xxx}^+ = \frac{f_{i+1} - f_{i-1}}{2h} + \frac{f_{i+1} - 2f_i + f_{i-1}}{h^2} h_1, \\ C_3 &= \beta^- u_{xxxx}^- = \beta^+ u_{xxxx}^+ = \frac{f_{i+1} - 2f_i + f_{i-1}}{h^2}. \end{aligned}$$

Using exactly the same arguments as above, we obtain the 3-point fourth-order finite difference equation,

$$-\frac{1}{h} \left(\left(\frac{1}{\beta^-} \frac{h_1}{h} + \frac{1}{\beta^+} \frac{h_2}{h} \right)^{-1} \frac{U_{i+1} - U_i}{h} - \beta^- \frac{U_i - U_{i-1}}{h} \right) = \tilde{f}_i, \tag{7.27}$$

where

$$\tilde{f}_i = f_i - (\beta^+ - \beta^-)\frac{1}{\beta^- h_2 + \beta^+ h_1}\frac{h_1 h_2}{h^2}\left(\frac{h}{2}C_1 + \frac{h_1^2 + h_2^2}{6}C_2 + \frac{h_1^3 + h_2^3}{24}C_3\right).$$

7.2 Two-dimensional cases

In this section we describe the fourth-order scheme for two-dimensional cases on irregular domains, and with interface and variable coefficients. First, we introduce the fourth-order compact finite difference scheme on which the method is based. We then describe the method for each case and for heat equations on two-dimensional domains.

7.2.1 The fourth-order compact central finite difference method

Consider the Poisson equation

$$-\Delta u = f.$$

Let $\psi_h(x, y) = \phi_h(x)\phi_h(y)$. Then we have

$$\begin{aligned}\iint (-\Delta u)\, \psi_h(x_i + x, y_j + y)\, dxdy \\ = \frac{1}{h}\int_{-h}^{h} (2u(x_i, y) - u(x_{i+1}, y) - u(x_{i-1}, y))\, \phi_h(y_j + y)\, dy \\ + \frac{1}{h}\int_{-h}^{h} \big(2u(x, y_j) - u(x, y_{j+1}) - u(x, y_{j-1})\big)\, \phi_h(x_i + x)\, dx.\end{aligned} \tag{7.28}$$

Note that

$$\frac{1}{h}\int_{-h}^{h} f(x)\phi_h(x)\, dx = f(0) + O(h^2)$$

and

$$\frac{1}{h}\int_{-h}^{h} f(x)\phi_h(x)\, dx = f(0) + \frac{1}{12}(f(h) - 2f(0) + f(-h)) + O(h^4).$$

Thus, using these quadrature formulas in (7.28), we obtain, respectively, the second-order central finite difference scheme,

$$\frac{4U_{i,j} - (U_{i-1,j} + U_{i,j-1} + U_{i,j+1} + U_{i+1,j})}{h^2} = f_{i,j} \tag{7.29}$$

and the fourth-order compact finite difference scheme (see Figure 7.1)

$$\begin{aligned}\bigg(\frac{3}{10}U_{i,j} - \frac{2}{3}(U_{i-1,j} + U_{i,j-1} + U_{i,j+1} + U_{i+1,j}) \\ -\frac{1}{6}(U_{i-1,j-1} + U_{i-1,j+1} + U_{i+1,j-1} + U_{i+1,j+1})\bigg)\frac{1}{h^2} \\ = \frac{2}{3}f_{i,j} + \frac{1}{12}\left(f_{i-1,j} + f_{i,j-1} + f_{i,j+1} + f_{i+1,j}\right).\end{aligned} \tag{7.30}$$

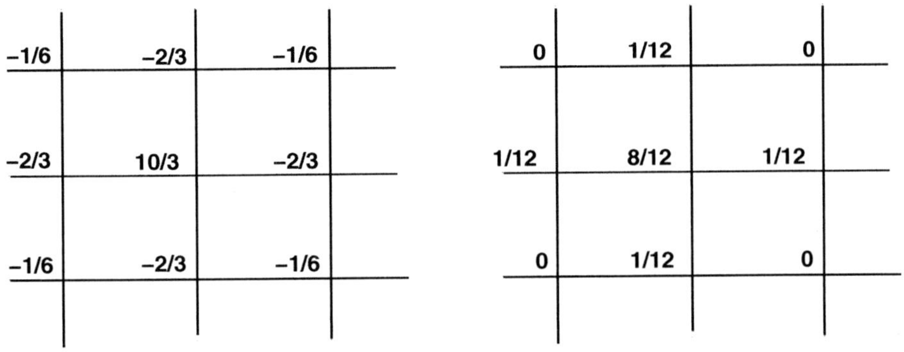

(a) Weights for solution. (b) Weights for source function.

Figure 7.1. *Weights on the compact 9-point stencil.*

Alternatively, we can derive (7.30) as follows. Note that the standard 5-point central finite difference scheme has the local truncation error

$$\frac{u(x_{i+1},y_j)+u(x_{i-1},y_j)+u(x_i,y_{j+1})+u(x_i,y_{j-1})-4u(x_i,y_j)}{h^2}-\Delta u_{xx}(\mathbf{x}_0)$$
$$=\frac{h^2}{12}\left(u_{xxxx}(\mathbf{x}_0)+u_{yyyy}(\mathbf{x}_0)\right)+O(h^4). \tag{7.31}$$

Differentiating the Poisson equation, we have

$$u_{xxxx}+u_{yyyy}=-\left(f_{xx}+f_{yy}\right)-2u_{xxyy}. \tag{7.32}$$

Here, u_{xxyy} can be approximated to second order by the standard 9-point finite difference scheme

$$h^2 u_{xxyy}\sim\frac{U_{i+1,j+1}-2U_{i,j+1}+U_{i-1,j+1}}{h^2}-2\frac{U_{i+1,j}-2U_{i,j}+U_{i-1,j}}{h^2}$$
$$+\frac{U_{i+1,j-1}-2U_{i,j-1}+U_{i-1,j-1}}{h^2}, \tag{7.33}$$

and f_{xx} and f_{yy} at $\mathbf{x}_0$ by the central finite difference

$$f_{xx}\sim\frac{f_{i+1,j}-2f_{i,j}+f_{i-1,j}}{h^2},\qquad f_{yy}\sim\frac{f_{i,j+1}-2f_{i,j}+f_{i,j-1}}{h^2}. \tag{7.34}$$

Substituting (7.33)–(7.34) into (7.31)–(7.32), we obtain the 9-point compact finite difference scheme as (7.30).

7.2.2 Neumann boundary conditions

In this section we discuss how to treat the Neumann boundary condition $u_n = g$ on a rectangular domain in the fourth-order manner. The fourth-order central finite difference

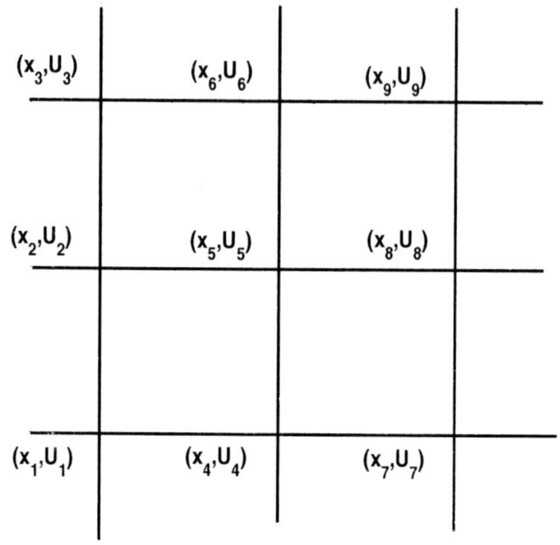

Figure 7.2. *9-point stencil for compact scheme.*

scheme for $-(u_{xx}+u_{yy})+\sigma\, u = f$ can be written as

$$\begin{aligned} C_0 = {} & \frac{1}{h^2}\left(\frac{10}{3}U_5 - \frac{2}{3}(U_2+U_4+U_6+U_8) - \frac{1}{6}(U_1+U_3+U_7+U_9)\right) \\ & - \frac{1}{12}\left(8F_5+F_2+F_4+F_6+F_8\right) = 0, \end{aligned} \tag{7.35}$$

where $F = f - \sigma u$, and we order the unknowns U and F in the 9-point neighborhood as in Figure 7.2.

If the nodal point (x_0, y_0) is on the boundary $x = 1$ (i.e., $x_0 = 1$), we need to complete the finite difference scheme (7.35), i.e., U_7, U_8, U_9, and F_8 are not unknowns. We determine them by the continuation of the solution and complete (7.35) as follows. Note that for smooth function ϕ,

$$\frac{\phi(h)+\phi(-h)}{2} = \phi(0) + \frac{h^2}{2}\,\phi''(0) + O(h^4). \tag{7.36}$$

Thus, by summing the Taylor series of u at (x_0, y_0) and $(x_0, y_0 \pm h)$ in x, we get

$$\begin{aligned} \frac{1}{6}u_7 + \frac{2}{3}u_8 + \frac{1}{6}u_9 = {} & \frac{1}{6}u_4 + \frac{2}{3}u_5 + \frac{1}{6}u_6 \\ & + \left(\frac{1}{6}\,g(y_0-h) + \frac{2}{3}\,g(y_0) + \frac{1}{6}\,g(y_0+h)\right)h \\ & + \frac{1}{2}\left(u_{xx}(x_0,y_0) + \frac{h^2}{12}\,u_{xxyy}(x_0,y_0)\right)h^2 \\ & + \frac{1}{6}\,u_{xxx}(x_0,y_0)h^3 + \frac{1}{24}\,u_{xxxx}(x_0,y_0)h^4 + O(h^5), \end{aligned} \tag{7.37}$$

where we used $u_x(x_0, y) = g(y)$, and from (7.36),

$$\frac{1}{6}u_{xx}(x_0, y_0 - h) + \frac{2}{3}u_{xx}(x_0, y_0) + \frac{1}{6}u_{xx}(x_0, y_0 + h)$$
$$= u_{xx}(x_0, y_0) + \frac{h^2}{12}u_{xxyy}(x_0, y_0) + O(h^4).$$

We now use the equation $-\Delta u = f$ to further reduce it to (7.42). Note that

$$u_{xx}(x_0, y_0) + u_{yy}(x_0, y_0) = -F(x_0, y_0) \tag{7.38}$$

and

$$\frac{u_6 - 2u_5 + u_4}{h^2} - u_{yy}(x_0, y_0) = \frac{h^2}{24}u_{yyyy}(x_0, y_0) + O(h^4). \tag{7.39}$$

Moreover, since $u_{xyy}(x_0, y_0) = g_{yy}(y_0)$, we have

$$u_{xxx}(x_0, y_0) + g_{yy}(y_0) = -F_x(x_0, y_0) \tag{7.40}$$

with

$$g_{yy}(y_0) = \frac{g(y_0 + h) - 2g(y_0) + g(y_0 - h)}{h^2} + O(h^2). \tag{7.41}$$

Substituting (7.38)–(7.41) into (7.37), we get

$$\frac{1}{6}u_7 + \frac{2}{3}u_8 + \frac{1}{6}u_9 = \frac{1}{6}u_4 + \frac{2}{3}u_5 + \frac{1}{6}u_6 - \frac{u_4 - 2u_5 + u_6}{2}$$
$$- \frac{h^2}{2}F(x_0, y_0) + g(y_0)h - \frac{h^3}{6}F_x(x_0, y_0)$$
$$+ \frac{h^4}{24}(u_{xxxx}(x_0, y_0) + 2\,u_{xxyy}(x_0, y_0) + u_{yyyy}(x_0, y_0)) + O(h^5),$$

where

$$u_{xxxx} + 2u_{xxyy} + u_{yyyy} = -(F_{xx} + F_{yy}).$$

Thus, we get

$$\begin{aligned}\frac{1}{6}u_7 + \frac{2}{3}u_8 + \frac{1}{6}u_9 &= -\frac{1}{3}u_4 + \frac{5}{3}u_5 - \frac{1}{3}u_6\\ &- \frac{h^2}{2}F(x_0, y_0) + g(y_0)h - \frac{h^3}{6}F_x(x_0, y_0) - \frac{h^4}{24}(F_{xx} + F_{yy}) + O(h^5).\end{aligned} \tag{7.42}$$

On the other hand, since

$$F_8 = F(x_0 + h, y_0) = F(x_0, y_0) + F_x(x_0, y_0)h + \frac{h^2}{2}F_{xx}(x_0, y_0) + O(h^3),$$

we have

$$\begin{aligned}\frac{h^2}{12}(F_8 - F_5) - \frac{h^4}{24}(F_{xx} + F_{yy}) &= \frac{h^2}{12}F_x(x_0, y_0)\\ &- \frac{h^2}{24}(F(x_0, y_0 + h) - 2F(x_0, y_0) + F(x_0, y_0 - h)) + O(h^5).\end{aligned} \tag{7.43}$$

Now, from (7.42)–(7.43), we can derive

$$\frac{1}{6}u_4+\frac{2}{3}u_5+\frac{1}{6}u_6+\frac{h^2}{12}F_8=-\frac{1}{3}u_4+\frac{5}{3}u_5-\frac{1}{3}u_6$$
$$-\frac{h^2}{24}(F_6+8F_5+F_4)+g(y_0)h-\frac{h^3}{12}F_x(x_0,y_0)+O(h^5).$$

Substituting this into (7.35), we obtain the finite difference scheme at the boundary,

$$\frac{1}{h^2}\left(\frac{5}{3}U_5-\frac{2}{3}U_2-\frac{1}{3}(U_4+U_6)-\frac{1}{6}(U_1+U_3)\right)+\frac{1}{24}(8F_5+2F_2+F_4+F_6)$$
$$=\frac{1}{h}g(y_0)-\frac{h}{12}F_x(x_0,y_0),$$

where

$$F_x(x_0,y_0)=f_x(x_0,y_0)-\sigma\, g(y_0).$$

At the corner (1, 1) we seek the symmetric finite difference scheme of the form

$$C_1=\frac{1}{h^2}\left(\frac{5}{6}U_5-\frac{1}{3}(U_2+U_4)-\frac{1}{6}U_1\right)-\frac{1}{24}(4F_5+F_2+F_4).$$

Using the fourth-order Taylor expansion of u at (1, 1), we have

$$\begin{aligned}
A &= C_1-C_0\\
&=\frac{2}{3h^2}\Bigg(2u_5+(u_x+u_y)h+\frac{h^2}{2}(u_{xx}+u_{yy})+\frac{h^3}{6}(u_{xxx}+u_{yyy})\\
&+\frac{h^4}{24}(u_{xxxx}+u_{yyyy})+\frac{1}{6h^2}2u_5+h^2(u_{xx}+u_{yy})+\frac{h^4}{12}(u_{xxxx}+6u_{xxyy}+u_{yyyy})\Bigg)\\
&+\frac{1}{3h^2}\Bigg(2u_5-h(u_x+u_y)+\frac{h^2}{2}(u_{xx}+u_{yy})-\frac{h^3}{6}(u_{xxx}+u_{yyy})\\
&+\frac{h^4}{24}(u_{xxxx}+u_{yyyy})+\frac{1}{6h^2}u_5+h(u_x+u_y)+\frac{h^2}{2}(u_{xx}+2u_{xy}+u_{yy})\\
&\qquad+\frac{h^3}{6}(u_{xxx}+3u_{xxy}+3u_{xyy}+u_{yyy})\\
&\qquad+\frac{h^4}{24}(u_{xxxx}+4u_{xxxy}+6u_{xxyy}+4u_{xyyy}+u_{yyyy})\Bigg)\\
&\qquad+\frac{3}{4}F_5+\frac{h^2}{16}(F_{xx}+F_{yy})+\frac{h}{24}(F_x+F_y)-\frac{5}{2h^2}U_5+O(h^3).
\end{aligned}$$

An elementary calculation shows that

$$A=\frac{1}{2h}(u_x+u_y)-\frac{h}{24}\left(u_{xxx}+u_{xxy}+u_{xyy}+u_{yyy}\right)$$
$$+\frac{1}{6}u_{xy}-\frac{h^2}{36}\left(u_{xxxy}+u_{xyyy}\right)+O(h^3),$$

where we used

$$u_{xx} + u_{yy} = -F, \quad u_{xxx} + u_{xxy} + u_{xyy} + u_{yyy} = -(F_x + F_y),$$
$$u_{xxxx} + 2u_{xxyy} + u_{yyyy} = -(F_{xx} + F_{yy}), \quad u_{xxxy} + u_{xyyy} = -F_{xy}.$$

Hence we obtain the difference scheme at $(1, 1)$,

$$\frac{1}{h^2}\left(\frac{5}{6}U_5 - \frac{1}{3}(U_2 + U_4) - \frac{1}{6}U_1\right) + \frac{1}{24}(4F_5 + F_2 + F_4)$$
$$= \frac{1}{2h}(g^1 + g^2) + \frac{g_y^1 + g_x^2}{12} - \frac{1}{24}\big(F_x(1,1) + F_y(1,1)\big) - \frac{h^2}{36}F_{xy}(1,1),$$

where

$$F_x = f_x - \sigma\, g^1, \quad F_y = f_y - \sigma\, g^2, \quad \text{and} \quad F_{xy} = f_{xy} - \frac{\sigma}{2}\left(g_y^1 + g_x^2\right).$$

7.2.3 The fourth-order method for Poisson equations on irregular domains

In this section we describe the fourth-order method for the Poisson equation $-\Delta u = f$ on an irregular domain Ω. First, we consider the Dirichlet boundary condition. The grid point $\mathbf{x}_0$ is *irregular* if the standard 9-point stencil contains grid points outside of $\bar{\Omega}$ on the irregular domain; see Figure 7.3.

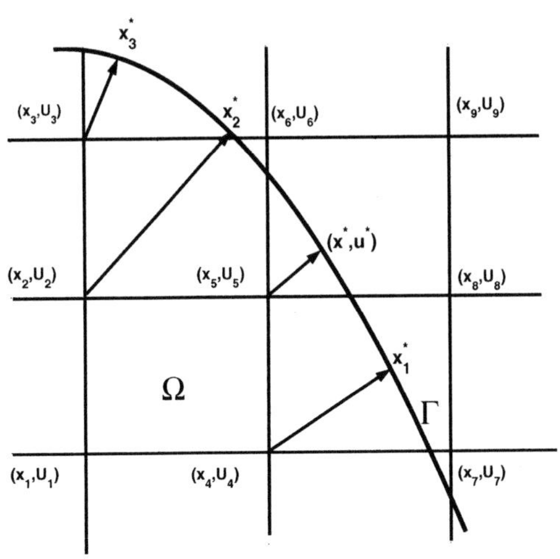

Figure 7.3. *A diagram of an irregular grid point* $\mathbf{x}_5$ *and its orthogonal projection* $\mathbf{x}_i^*$ *on the boundary* Γ.

At each regular point we apply the 9-point compact central finite difference scheme,

$$
\begin{aligned}
&\frac{1}{h^2}\left(\frac{10}{3}U_{i,j} - \frac{2}{3}(U_{i-1,j} + U_{i+1,j} + U_{i,j-1} + U_{i,j+1})\right.\\
&\qquad\left. - \frac{1}{6}(U_{i-1,j-1} + U_{i-1,j+1} + U_{i+1,j-1} + U_{i+,j+1})\right)\\
&= \frac{8}{12}f_{i,j} + \frac{1}{12}(f_{i-1,j} + f_{i+1,j} + f_{i,j-1} + f_{i,j+1}).
\end{aligned}
\tag{7.44}
$$

At each irregular grid point, in order to complete (7.44) we must define the unknowns $U_{i,j}$ outside grid points (x_i, y_j). To this end, we employ the continuation of the solution u from Ω to the outside of the domain Ω. The continuation procedure uses the multivariable Taylor expansion of the solution u at selected boundary points. The Taylor coefficients are then determined by incorporating the boundary conditions and the equation. This procedure can be described as follows.

Step 1: The fourth-order projection of $\mathbf{x}_0$ onto the boundary Γ and the formation of the local coordinate (ξ, η) are carried out by the fourth-order polynomial interpolation of ϕ [104] based on the standard 25-point grid points (see the details in §7.2.4). As a result there is a family of the projected points $\mathbf{x}_k^*$ on Γ, with the associated family of irregular grid points.

Step 2: We expand u in a neighborhood of $\mathbf{x}^*$ by the fourth-order Taylor polynomial in the local coordinate (ξ, η),

$$
u \sim \Phi(\xi, \eta) = \sum_{m=0}^{4} \sum_{\hat{i}+\hat{j}=m} \frac{C(m, \hat{i})u^{(m)}_{\hat{i},\hat{j}}}{m!} \xi^{\hat{i}}\eta^{\hat{j}}. \tag{7.45}
$$

We equate the equation and its differentiations up to the second order at $\mathbf{x}^*$,

$$
\begin{cases}
u_{\xi\xi} + u_{\eta\eta} = -f(\mathbf{x}^*),\\
u_{\xi\xi\xi} + u_{\xi\eta\eta} = -f_\xi(\mathbf{x}^*),\\
u_{\xi\xi\eta} + u_{\eta\eta\eta} = -f_\eta(\mathbf{x}^*),\\
u_{\xi\xi\xi\xi} + u_{\xi\xi\eta\eta} = -f_{\xi\xi}(\mathbf{x}^*),\\
u_{\xi\xi\xi\eta} + u_{\xi\eta\eta\eta} = -f_{\xi\eta}(\mathbf{x}^*),\\
u_{\xi\xi\eta\eta} + u_{\eta\eta\eta\eta} = -f_{\eta\eta}(\mathbf{x}^*).
\end{cases}
\tag{7.46}
$$

Thus, we select $X = (u_\xi, u_\eta, u_{\xi\eta}, u_{\eta\eta})$ and $(u_{\xi\eta\eta}, u_{\eta\eta\eta}, u_{\xi\xi\eta\eta}, u_{\xi\eta\eta\eta})$ as unknowns in R^8 by eliminating the others by (7.46). We use the equality $u(x_0) = u_{i,j}$ along with the seven additional Dirichlet boundary conditions $u(\mathbf{x}_i^*) = g(\mathbf{x}_i^*)$ at the closest projection points $\mathbf{x}_i^*$, $1 \le i \le 7$, to $\mathbf{x}^*$ on Γ, and it results in an 8×8 system $AX = b$ with (7.47)–(7.48) to determine the Taylor polynomial (7.45).

Step 3: Based on the Taylor polynomial (7.45), we complete (7.44) by setting

$$
U_{i+k,j+\ell} = \Phi(\xi_k, \eta_\ell),
$$

where (ξ_k, η_ℓ) is the corresponding local coordinate to the grid point $(x_{i+k}, y_{j+\ell})$ outside $\bar{\Omega}$.

In Step 2 we used the fact that the Laplacian is invariant under the rotation. Note that the Taylor polynomial (7.45) is completely determined as a linear function of $(u_{i,j}, f, g(\mathbf{x}^*), g(\mathbf{x}_i^*), 1 \le i \le 7)$. The finite difference equation at the grid point is obtained by replacing $10\,U_{i,j}/3$ and $8\,f_{i,j}/10$ in (7.44) by $h_{ij}U_{ij}$ and $\tilde{f}_{ij}$, respectively, while keeping the other terms unchanged. It is clear from (7.47)–(7.48) below that the weight $h_{i,j} > 0$ depends only on the local geometry of Γ and $\tilde{f}_{i,j}$ depends also on f and the boundary values $g(\mathbf{x}_i^*)$, $1 \le i \le 7$. The proposed algorithm works as follows: first, determine the projection point of $\mathbf{x}^*$ of each irregular grid point; second, evaluate the diagonal weights $h_{i,j}$ and $\tilde{f}_{i,j}$ at each irregular grid point. The resulting system matrix is symmetric and positive definite and has the same sparsity as the original compact finite difference scheme. Also, note that the local truncation error of (7.44) at the irregular grid point is of order $O(h^3)$.

The system (A, b) appearing in Step 2 is defined by

$$\begin{aligned} A_i = \big[\, &\xi_i \;\; \eta_i \;\; \xi_i\eta_i \;\; (\eta_i^2 - \xi_i^2)/2 \;\; (3\xi_i\eta_i^2 - \xi_i^3)/6 \;\; (\eta_i^3 - 3\xi_i^2\eta_i)/6 \\ &(\xi_i\eta_i^3 - \xi_i^3\eta_i)/6 \;\; (6\xi_i^2\eta_i^2 - \xi_i^4 - \eta_i^4)/24 \,\big] \in R^{1\times 8} \end{aligned} \tag{7.47}$$

for $0 \le i \le 8$, and

$$\begin{aligned} b_0 &= u_{i,j} - g(\mathbf{x}^*) - f(\mathbf{x}^*)\xi_0^2/2 - f_\xi\xi_0^3/6 - f_\eta\eta_0\xi_0^2/2 - f_{\xi\xi}\xi_0^4/24 \\ &\quad - f_{\xi\eta}\eta_0\xi_0^3/6 - f_{\eta\eta}\eta_0^4/24, \\ b_i &= g(\mathbf{x}_i^*) - g(\mathbf{x}^*) - f(\mathbf{x}^*)\xi_i^2/2 - f_\xi\xi_i^3/6 - f_\eta\eta_i\xi_i^2/2 - f_{\xi\xi}\xi_i^4/24 \\ &\quad - f_{\xi\eta}\eta_i\xi_i^3/6 - f_{\eta\eta}\eta_i^4/24 \end{aligned} \tag{7.48}$$

for $1 \le i \le 7$, where (ξ_0, η_0) is the local coordinate of $\mathbf{x}_0$ and (ξ_i, η_i) is the local coordinate for each $\mathbf{x}_i^*$, $1 \le i \le 7$.

Remark 7.1.

1. *If we are able to evaluate g_η accurately, then we may use the equality $u_\eta = g_\eta(\mathbf{x}^*)$ that leads to the option in which only the six boundary conditions $u(\mathbf{x}_i^*) = g(\mathbf{x}_i^*)$ (instead of seven) are equated, and thus the 7×7 system determines completely the Taylor polynomial* (7.45)*.*

2. *For the mixed boundary condition $u(\mathbf{x}) + \alpha\, u_{\boldsymbol{n}}(\mathbf{x}) = g(\mathbf{x})$, $\mathbf{x} \in \Gamma$, we have*

$$u(\mathbf{x}_i^*) + \alpha\, u_{\boldsymbol{n}}(\mathbf{x}_i^*) = g(\mathbf{x}_i^*)$$

at each projection point $\mathbf{x}_i^*$ and

$$u_{\boldsymbol{n}}(\mathbf{x}_i^*) = \boldsymbol{n}_i^* \cdot \boldsymbol{n}^* u_\xi + \boldsymbol{n}_i^* \cdot \boldsymbol{\tau}^* u_\eta,$$

where $(\boldsymbol{n}^*, \boldsymbol{\tau}^*)$ is the normal tangent vector pair at $\mathbf{x}^*$ and $\boldsymbol{n}_i^*$ is the normal vector at $\mathbf{x}_i^*$. In this case we select $X = (u^*, u_\eta, u_{\xi\eta}, u_{\eta\eta})$ and $(u_{\xi\eta\eta}, u_{\eta\eta\eta}, u_{\xi\xi\eta\eta}, u_{\xi\eta\eta\eta})$ as the unknowns for the Taylor polynomial (7.45)*. It leads to $AX = b$ (with (A, b) similar to* (7.47)–(7.48)*) and determines the Taylor polynomial* (7.45)*.*

3. *For the Helmholtz equation* $-\Delta u + \sigma u = f$, *we simply apply appropriate changes, i.e.,* $f_{i,j}$ *in* (7.29) *is replaced by* $F_{i,j} = f_{i,j} - \sigma\, u_{i,j}$ *and* $f(\mathbf{x})$ *in* (7.46) *is replaced by* $F(\mathbf{x}) = f(\mathbf{x}) - \sigma\, u(\mathbf{x})$.

7.2.4 Projections and a fourth-order polynomial interpolation

In this section we describe how to project an irregular point $\mathbf{x}_0$ onto $\Gamma = \{\mathbf{x} : \varphi(\mathbf{x}) = 0\}$ assuming $\varphi(\mathbf{x}_0) < 0$; i.e., Γ is the zero level set of φ and we assume the values of φ at the grid points are provided.

- We approximate the gradient φ at $\mathbf{x}_0$ by

$$
\begin{aligned}
\mathbf{d} &= \left(d_x, d_y\right), \qquad \text{where} \\
d_x &= \frac{\varphi_{i-2,j} - 8\varphi_{i-1,j} + 8\varphi_{i+1,j} - \varphi_{i+2,j}}{12h}, \\
d_y &= \frac{\varphi_{i,j-2} - 8\varphi_{i-1,j-1} + 8\varphi_{i,j+1} - \varphi_{i,j+2}}{12h},
\end{aligned}
\tag{7.49}
$$

  we let $s = |\mathbf{d}|$, and normalize $\mathbf{d}$ by $\mathbf{d} = \mathbf{d}/s$.

- We make the quadratic approximation of φ along the direction $\mathbf{d}$ by

$$F(\mathbf{x}_0 - \alpha\, \mathbf{d}) = \varphi_{i,j} - s\alpha + \frac{\alpha^2}{2}\mathbf{d}^T H \mathbf{d},$$

  where $H \in R^{2\times 2}$ is an approximate Hessian at $\mathbf{x}_0$ with

$$
\begin{aligned}
H_{11} &= \frac{\varphi_{i+1,j} - 2\varphi_{i,j} + \varphi_{i-1,j}}{h^2}, \qquad H_{22} = \frac{\varphi_{i,j+1} - 2\varphi_{i,j} + \varphi_{i,j-1}}{h^2}, \\
H_{12} = H_{21} &= \frac{\varphi_{i+1,j+1} - \varphi_{i-1,j+1} - \varphi_{i+1,j} + \varphi_{i-1,j-1}}{4h^2}.
\end{aligned}
$$

  Then we find a solution $\alpha_0 < 0$ for the quadratic equation $F(\mathbf{x}_0 - \alpha\, \mathbf{d}) = 0$.

- We use the fourth-order polynomial interpolation Ψ of the level set function φ at $\mathbf{x}_0$ based on its values at the standard 25-point grid points,

$$\Psi(x_1, x_2) = \sum_{m=0}^{4} \sum_{i+j=m} \frac{\varphi_{ij}^{(m)}}{m!} \left(\frac{x_1}{h}\right)^i \left(\frac{x_2}{h}\right)^j, \tag{7.50}$$

  where the coefficients $\varphi_{ij}^{(m)}$ are given below. We apply the Newton method to find the solution $\alpha^* < 0$ to $\Psi(\mathbf{x} - \alpha^*\, \mathbf{d}) = 0$ with initial guess α_0.

- Set $\mathbf{x}^* = \mathbf{x}_0 - \alpha^*\, \mathbf{d}$ and set up the local coordinates with the normal and the tangential vectors defined by

$$\boldsymbol{n} = \frac{\nabla\Psi(\mathbf{x}^*)}{|\nabla\Psi(\mathbf{x}^*)|}, \qquad \boldsymbol{\tau} = \frac{\left(\Psi_{\mathbf{x}_2}(\mathbf{x}^*),\, \Psi_{\mathbf{x}_1}(\mathbf{x}^*)\right)}{|\nabla\Psi(\mathbf{x}^*)|},$$

  respectively.

Let φ be the ordered vector consisting of the function values at the 25 standard grid points centered on $\mathbf{x}_0$. Then, we have

$$\varphi_{10}^{(1)} = \frac{1}{12}(\varphi_{i-2,j} - 8\varphi_{i-1,j} + 8\varphi_{i+1,j} - \varphi_{i+2,j}),$$

$$\varphi_{20}^{(2)} = \frac{1}{12}(-\varphi_{i-2,j} + 16\varphi_{i-1,j} - 30\varphi_{i,j} + 16\varphi_{i+1,j} - \varphi_{i+2,j}),$$

$$\varphi_{30}^{(3)} = \frac{1}{8}(-4\varphi_{i-2,j} + 8\varphi_{i-1,j} - 8\varphi_{i+1,j}4\varphi_{i+2,j}),$$

$$\varphi_{40}^{(4)} = \varphi_{i-2,j} - 4\varphi_{i-1,j} + 6\varphi_{i,j} - 4\varphi_{i+1,j} + \varphi_{i+2,j},$$

and

$$\varphi_{11}^{(2)} = [\ 1\ \ -8\ \ 0\ \ 8\ \ -1\ \ -8\ \ 64\ \ 0\ \ -64\ \ 8\ \ 0\ \ 0\ \ 0\ \ 0\ \ 0\ \ 8\ \ -64\ \ 0\ \ 64\ \ -8\ \ -1\ \ 8\ \ 0\ \ -8\ \ 1\ ]\ \varphi/144;$$

$$\varphi_{21}^{(3)} = [\ -1\ \ 8\ \ 0\ \ -8\ \ 1\ \ 16\ \ -128\ \ 0\ \ 128\ \ -16\ \ -30\ \ 240\ \ 0\ \ -240\ \ 30\ \ 16\ \ -128\ \ 0\ \ 128\ \ -16\ \ -1\ \ 8\ \ 0\ \ -8\ \ 1\ ]\ \varphi/144;$$

$$\varphi_{31}^{(4)} = [\ -4\ \ 32\ \ 0\ \ -32\ \ 4\ \ 8\ \ -64\ \ 0\ \ 64\ \ -8\ \ 0\ \ 0\ \ 0\ \ 0\ \ 0\ \ -8\ \ 64\ \ 0\ \ -64\ \ 8\ \ 4\ \ -32\ \ 0\ \ 32\ \ -4\ ]\ \varphi/96;$$

$$\varphi_{22}^{(4)} = [\ 1\ \ -16\ \ 30\ \ -16\ \ 1\ \ -16\ \ 256\ \ -480\ \ 256\ \ -16\ \ 30\ \ -480\ \ 900\ \ -480\ \ 30\ \ -16\ \ 256\ \ -480\ \ 256\ \ -16\ \ 1\ \ -16\ \ 30\ \ -16\ \ 1\ ]\ \varphi/144.$$

7.2.5 The fourth-order method for heat equations on irregular domains

In this section we describe the fourth-order scheme for the heat equation on an irregular domain Ω,

$$\frac{\partial}{\partial t}u(t,\mathbf{x}) = \Delta u, \quad u(0,\mathbf{x}) = u_0(\mathbf{x}), \quad \mathbf{x} \in \Omega, \tag{7.51}$$

with a boundary condition

$$u(t,\mathbf{x}) + \alpha\, u_n(t,\mathbf{x}) = g(t,\mathbf{x}), \quad \mathbf{x} \in \Gamma. \tag{7.52}$$

For clarity of our presentation, we describe only the method for the homogeneous (in time) Dirichlet boundary condition $u(t,\mathbf{x}) = g(\mathbf{x})$, $\mathbf{x} \in \Gamma$. But we indicate the specific changes needed for the general cases.

In order to develop the fourth-order method, we treat $-u_t$ as a forcing function f and the right-hand side of (7.1), and thus $f_{i,j}$, is replaced by $-(u_t)_{i,j}$; i.e.,

$$\begin{aligned}\frac{1}{h^2}\Big(\frac{10}{3}U_{i,j} &- \frac{2}{3}(U_{i-1,j} + U_{i+1,j} + U_{i,j-1} + U_{i,j+1}) \\ &- \frac{1}{6}(U_{i-1,j-1} + U_{i-1,j+1} + U_{i+1,j-1} + U_{i+,j+1})\Big) \\ &+ \frac{d}{dt}\left(\frac{8}{12}u_{i,j} + \frac{1}{12}(U_{i-1,j} + U_{i+1,j} + U_{i,j-1} + U_{i,j+1})\right) = 0,\end{aligned} \tag{7.53}$$

and then we apply the fourth-order method as in §7.2.3. Thus we need to be able to evaluate the vector $Y = (u_t(\mathbf{x}^*), u_{t\xi}(\mathbf{x}^*), u_{t\eta}(\mathbf{x}^*), u_{t\xi\xi}(\mathbf{x}^*), u_{t\xi\eta}(\mathbf{x}^*), u_{t\eta\eta}(\mathbf{x}^*))$ at $\mathbf{x}^*$ as the right-hand side of (7.46). To this end, we expand u_t in a neighborhood of the projected point $\mathbf{x}^* \in \Gamma$ by the second-order Taylor polynomial

$$u_t \sim u_{t\xi}(\mathbf{x}^*)\xi + \frac{1}{2}u_{t\xi\xi}(\mathbf{x}^*)\xi^2 + u_{t\xi\eta}\xi\eta + \frac{1}{2}u_{t\eta\eta}(\mathbf{x}^*)\eta^2, \tag{7.54}$$

where $u_t(\mathbf{x}^*) = u_{t\eta}(\mathbf{x}^*) = 0$ and

$$u_{t\xi\xi} + u_{t\eta\eta} = 0 \text{ at } \mathbf{x}^*. \tag{7.55}$$

As in Step 2 of §7.2.3 we determine (7.54) completely by

$$\begin{pmatrix} \xi_0 & \xi_0\eta_0 & 0.5(\eta_0^2 - \xi_0^2) \\ \xi_1 & \xi_1\eta_1 & 0.5(\eta_1^2 - \xi_1^2) \\ \xi_2 & \xi_2\eta_2 & 0.5(\eta_2 - \xi_2^2) \end{pmatrix} \begin{pmatrix} u_{t\xi} \\ u_{t\xi\eta} \\ u_{t\eta\eta} \end{pmatrix} = \begin{pmatrix} (U_0)_t \\ 0 \\ 0 \end{pmatrix}, \tag{7.56}$$

where (η_i, ξ_i), $i = 0, 1, 2$, are the local coordinates of $\mathbf{x}_0$, $\mathbf{x}_1^*$, $\mathbf{x}_2^*$, respectively. Thus the vector Y is determined as $Y = (U_0)_t\, V$ with $V \in R^7$. Now we know completely the right-hand side of (7.46) as Y and proceed to determine the fourth-order Taylor expansion (7.45) completely. In this way we complete (7.53) at each irregular grid point x_0 by substituting expressions (7.45) and (7.54) into (7.53) for the corresponding values outside the domain. The work proceeds as follows: first, solve (7.56) to determine the second-order Taylor expansion (7.54) for u_t; second, solve (7.47) to determine the fourth-order Taylor expansion (7.45) for u at $\mathbf{x}^*$. As a consequence at each irregular grid point, the diagonal entries $8/3$ and $8/12$ in (7.53) are modified accordingly, as is the right-hand side, due to the nonhomogeneous boundary condition $u = g(\mathbf{x})$ (otherwise, no changes are required). Thus the resulting system of equations is of the form

$$QU_t = -HU + \tilde{F}, \quad U(0) = u_0(\mathbf{x}_{i,j}), \tag{7.57}$$

where Q and H are symmetric sparse matrices. The positivity of the matrices Q and H is demonstrated and tested numerically for our test examples.

We demonstrate the applicability of the fourth-order method by the approximation of the eigenvalue problem for (7.51)–(7.52), i.e., finding the eigenpair $(\lambda, u) \in R \times L^2(\Omega)$ satisfying

$$\Delta u = \lambda u, \quad u|_\Gamma = 0. \tag{7.58}$$

We simply consider the generalized eigenvalue problem for (7.57),

$$\lambda\, QU = -HU, \tag{7.59}$$

and thus the eigenpairs (λ, U) approximate the ones for (7.58).

Remark 7.2. *For the Dirichlet nonhomogeneous condition $u(t, \mathbf{x}) = g(t, \mathbf{x})$, $\mathbf{x} \in \Gamma$, we have*

$$u_{t\xi\xi} + u_{t\eta\eta} = u_{tt} \quad \textit{at } \mathbf{x}^*.$$

In order to avoid the second-order derivatives in time, we do not equate (7.54) *as an equality condition for determining the second-order Taylor expansion for* u_t. *Instead we use the additional Dirichlet boundary condition* $u_t(\mathbf{x}_3^*) = g_t(\mathbf{x}_3^*)$, *where* $\mathbf{x}_3^*$ *is the next closest projection point to* $\mathbf{x}^*$. *Thus* (7.56) *is replaced by*

$$\begin{pmatrix} \xi_0 & \xi_0\eta_0 & \eta_0^2/2 & \xi_0^2/2 \\ \xi_1 & \xi_1\eta_1 & \eta_1^2/2 & \xi_1^2/2 \\ \xi_2 & \xi_2\eta_2 & \eta_2^2/2 & \xi_2^2/2 \\ \xi_3 & \xi_3\eta_3 & \eta_3^2/2 & \xi_3^2/2 \end{pmatrix} \begin{pmatrix} u_{t\xi} \\ u_{t\xi\eta} \\ u_{t\eta\eta} \\ u_{t\xi\xi} \end{pmatrix} = \begin{pmatrix} (U_0)_t - g_t(x^*) - g_{t\tau}(\mathbf{x}^*)\eta_0 \\ g_t(\mathbf{x}_1^*) - g_t(\mathbf{x}^*) - g_{t\tau}(\mathbf{x}^*)\eta_1 \\ g_t(\mathbf{x}_2^*) - g_t(\mathbf{x}^*) - g_{t\tau}(\mathbf{x}^*)\eta_2 \\ g_t(\mathbf{x}_3^*) - g_t(\mathbf{x}^*) - g_{t\tau}(\mathbf{x}^*)\eta_3 \end{pmatrix}.$$

Thus Y *is of the form* $Y = (U_0)_t\, V + \tilde{V}(t)$, *where* $\tilde{V} \in R^7$ *depends on* $(g_t(\mathbf{x}^*), g_t(\mathbf{x}_i^*),\ 1 \le i \le 3)$. *The step to determine the fourth-order Taylor expansion for* u *remains the same, and the resulting system has the form*

$$QU_t = -HU + \tilde{F}_1(t) + \frac{d}{dt}\tilde{F}_2(t),$$

where only the diagonal entry of Q *at each irregular grid point is modified and depends on the geometry of the boundary* Γ, *and the corresponding right-hand sides* $\tilde{F}_1(t)$, $\tilde{F}_2(t)$ *also depend on the boundary values* $g(t, \mathbf{x})$.

For the mixed boundary condition $u_{\mathbf{n}} + \alpha\, u = g$, *we use exactly the same procedures as in* §7.2.3 *to determine the second-order Taylor expansion of* u_t *in a neighborhood of* $\mathbf{x}^*$. *For example, if* g *is time invariant, the second option uses*

$$\tilde{A}\left(u_t^*, u_{t\eta}, u_{t\xi\eta}, u_{t\eta\eta}\right)^T = ((u_0)_t \;\; 0 \;\; 0 \;\; 0)^T,$$

where the matrix $\tilde{A} \in R^{4\times 4}$ *is given by*

$$\begin{pmatrix} 1-\alpha\xi_0 & \eta_0 & \xi_0\eta_0 & 0.5(\eta_0^2-\xi_0^2) \\ \alpha(1-\alpha\xi_1)-\alpha a_1 & b_1+\alpha\eta_1 & (a_1\eta_1+b_1\xi_1)+\alpha\xi_1\eta_1 & b_1\eta_1-a_1\xi_1+\frac{\alpha}{2}(\eta_1^2-\xi_1^2) \\ \alpha(1-\alpha\xi_2)-\alpha a_2 & b_2+\alpha\eta_2 & (a_2\eta_2+b_2\xi_2)+\alpha\xi_2\eta_2 & b_2\eta_2-a_2\xi_2+\frac{\alpha}{2}(\eta_2^2-\xi_2^2) \\ \alpha(1-\alpha\xi_3)-\alpha a_3 & b_3+\alpha\eta_3 & (a_3\eta_3+b_3\xi_3)+\alpha\xi_3\eta_3 & b_3\eta_3-a_3\xi_3+\frac{\alpha}{2}(\eta_3^2-\xi_3^2) \end{pmatrix}.$$

7.2.6 The fourth-order method for PDEs with variable coefficient on irregular domains

In this section, we consider the variable elliptic equation on a rectangular domain of the form

$$-\nabla\cdot(\beta\,\nabla u) + b\cdot\nabla u + \sigma\, u = f. \tag{7.60}$$

Or, equivalently,

$$\beta\left(u_{xx}+u_{yy}\right) + (\beta_x - b_1)\,u_x + \left(\beta_y - b_2\right)u_y - \sigma\, u + f = 0. \tag{7.61}$$

At this point, we are describing the method for a rectangular domain with Dirichlet boundary conditions. With our method in §7.2.3, the cases for general boundary conditions and irregular domains can be treated.

Motivated by (7.31)–(7.34), we consider the standard 9-point finite difference scheme of the form

$$\sum_{k=-1}^{i}\sum_{\ell=-1}^{1} \gamma_{k,\ell}\, u_{i+k,j+\ell} + \alpha_1 D_x f + \alpha_2 D_y f + \alpha_3 D_{xx} f + \alpha_4 D_{xy} f + \alpha_5 D_{yy} f, \quad (7.62)$$

with $(i, j) \in (i_0, j_0) + S_9$, to approximate (7.60) at each node (x_i, y_j). Here, S_9 denotes the index set of the standard compact 9-point stencils ordered in $(-1,-1)$, $(-1,0)$, $(-1,1)$, $(0,-1)$, $(0,0)$, $(0,1)$, $(1,-1)$, $(1,0)$, $(1,1)$, and

$$\begin{aligned} D_x f &= \frac{f_{i+1,j} - f_{i-1,j}}{2h}, \quad D_y f = \frac{f_{i,j+1} - f_{i,j-1}}{2h}, \\ D_{xx} f &= \frac{f_{i+1,j} - 2f_{i,j} + f_{i-1,j}}{h^2}, \quad D_{yy} f = \frac{f_{i,j+1} - 2f_{i,j} + f_{i,j-1}}{h^2}, \\ D_{xy} f &= \frac{f_{i+1,j+1} - f_{i+1,j-1} - f_{i-1,j+1} + f_{i-1,j-1}}{4h^2}. \end{aligned} \quad (7.63)$$

From (7.61),

$$f_{i+k,j+\ell} = f(x_i, y_j) = -\beta\,(u_{xx} + u_{yy}) - (\beta_x - b_1)\,u_x - (\beta_y - b_2)\,u_y + \sigma\, u \quad (7.64)$$

for each (k, ℓ) at node $\mathbf{x}_0 = (x_i, y_j)$, and with this we determine the weight (γ, α) so that the finite difference scheme (7.62) expresses the equation up to the fourth order in the Taylor series expansion. This procedure is equivalent to expressing the correction term in (7.31) by using the differentiations of (7.60) and the right-hand side as in (7.32)–(7.34) for the Poisson equation. In particular, we seek for the scheme (7.62) to have the M-property [19] of finite difference schemes. That is, we demand that $\gamma_k \geq 0$ for $k \neq 5$ in (7.62).

Now, we proceed as follows. We first expand u in a neighborhood of each node x_0 by the fourth-order Taylor polynomial (assuming, without loss of generality, that $\mathbf{x}_0 = (0, 0)$),

$$u \sim \sum_{m=0}^{4} \sum_{\hat{i}+\hat{j}=m} \frac{C(m,\hat{i})\,u^{(m)}_{\hat{i},\hat{j}}}{m!}\, h_x^{\hat{i}} h_y^{\hat{j}}. \quad (7.65)$$

Let $Y = u^{(m)}_{\hat{i},\hat{j}} \in R^{15}$ denote the Taylor vector of unknown coefficients. By (7.65), we describe the approximations of the solutions on the 9-point stencil by

$$u_{i+k,j+\ell} \sim T_{k,\ell} \cdot Y. \quad (7.66)$$

Substituting (7.65) into (7.64), we then describe the source function $f_{i,j}$ at each grid point of the 9-point stencil as a linear functional of Y. Thus, by (7.63) we obtain

$$(D_x f, D_y f, D_{xx} f, D_{xy} f, D_{yy} f)^T = SY, \quad S \in R^{5\times 15}, \quad (7.67)$$

and hence

$$\sum_{k,\ell} \gamma_{k.\ell}\, u_{i+k,j+\ell} + \alpha_1 D_x f + \alpha_2 D_y f + \alpha_3 D_x x f + \alpha_4 D_{xy} f + \alpha_5 D_{yy} f$$
$$= \left(\sum_{k,\ell} \gamma_{k,\ell}\, T_{k,\ell} + \sum_i \alpha_i\, S_i \right) \cdot Y. \tag{7.68}$$

If we equate (7.60) at $\mathbf{x}_0 = (x_i, y_j)$, then

$$\left(\sum_{k,\ell} \gamma_{k,\ell}\, T_{k,\ell} + \sum_i \alpha_i\, S_i \right) \cdot Y$$
$$= \Big((-\sigma, \beta_x - b_1, \beta_y - b_2, \beta, 0, \beta(\mathbf{x}_0)\,, 0, \ldots, 0 \Big) \quad \text{in } R^{15}. \tag{7.69}$$

We now formulate the following constrained minimization over (γ, α):

$$\min \frac{1}{2} \left| \left(\sum_{k,\ell} \gamma_{k,\ell}\, T_{k,\ell} + \sum_i \alpha_i\, S_i \right)_{7:15} \right|^2_{R^9}$$
$$\text{subject to} \quad \left(\sum_{k,\ell} \gamma_{k.\ell}\, T_{k,\ell} + \sum_i \alpha_i\, S_i \right)_{1:6} \tag{7.70}$$
$$= \big(-\sigma, \beta_x - b_1, \beta_y - b_2, \beta, 0, \beta \big)\, (x_i, y_j),$$
$$\gamma_{k,\ell} \geq 0, \quad (k, \ell) \neq (0, 0).$$

The solution pair (γ, α) to (7.70) determines the weight in the finite difference scheme (7.62). The constrained minimization (7.70) is formulated so that the resulting finite difference scheme leads to an M-matrix and the first six relations of (7.69) are equated and then the remainder (higher-order terms) are equated in the least squares sense.

Remark 7.3.

1. *If the solution u is C^5 in a neighborhood of $\mathbf{x}_0$, then the truncation error of* (7.65) *is of order $O(h^5)$.*

2. *The β_x and β_y at the grid points can be evaluated by the appropriate finite difference scheme, say, $\beta_x(x_i, y_j) \simeq (\beta(x_{i+1}, y_j) - \beta(x_{i+1}, y_j))/2h$, to avoid differentiating β, b, and σ explicitly.*

3. *In order to obtain an* M*-matrix difference through constrained minimization* (7.70), *we must have the CFL-like condition,* $\max |b_{i,j}| \leq h\, \min \beta_{i,j}$.

7.2.7 The fourth-order method for interface problems

In this section we discuss the method for (7.1) with β, b, and σ being piecewise constants. For the case with a material interface, a level set function φ is used to describe the discontinuity in β over the domain Ω such that

$$\Omega^- = \{\mathbf{x} \in \Omega : \varphi(\mathbf{x}) < 0\},\ \Omega^+ = \{\mathbf{x} \in \Omega : \varphi(\mathbf{x}) > 0\},\ \Gamma = \{\mathbf{x} \in \Omega : \varphi(\mathbf{x}) = 0\}$$

and

$$\beta(\mathbf{x}) = \begin{cases} \beta^+, & \mathbf{x} \in \Omega^+, \\ \beta^-, & \mathbf{x} \in \Omega^-, \end{cases} \tag{7.71}$$

where $\beta^\pm$ are positive constants. The zero level set Γ of φ defines the material interface Γ. The discontinuities in b and σ are defined likewise. Since the problem with the material discontinuity describes an elliptic interface problem, we consider the following homogeneous continuity conditions:

$$[u]_\Gamma = 0, \quad \left[\beta \frac{\partial u}{\partial \boldsymbol{n}}\right]_\Gamma = 0. \tag{7.72}$$

For clarity of presentation we describe the method for the case when $b = 0$. The grid point $\mathbf{x}_0 = (x_i, y_j)$ is *irregular* if the standard 9-point stencil at $\mathbf{x}_0$ contains grid points from different subdomains, say $\Omega^\pm$. Again we seek the finite difference scheme of the form (7.62) at the irregular point. That is, at an irregular grid point, we apply the continuation method and proceed as in the following steps.

Step 1: The fourth-order projection of each irregular $\mathbf{x}_0$ onto the interface Γ is carried out by the fourth-order polynomial interpolation of φ based on the standard 25-point grid points described in §7.2.4. As a result there is a family of the projected points $\mathbf{x}_k^*$ on Γ with the associated family of irregular grid points. Also, the local coordinate system at $\mathbf{x}^*$ is formed at $\mathbf{x}^*$ in terms of the normal and tangential coordinates of the interface Γ.

Step 2: We expand u in a neighborhood of $\mathbf{x}^*$ by the fourth-order Taylor polynomial in the local coordinates (ξ, η),

$$u \sim \sum_{m=0}^{4} \sum_{\hat{i}+\hat{j}=m} \frac{C(m,\hat{i})u_{\hat{i},\hat{j}}^{(m)}}{m!} \xi^{\hat{i}} \eta^{\hat{j}} \tag{7.73}$$

in both domains Ω^+ and Ω^- with a total of 30 unknowns. We use the interface conditions (7.79)–(7.82), in total 15, to determine the Taylor vector $Y^+ = ((u^+)_{\hat{i},\hat{j}}^{(m)},\ 0 \le m \le 4) \in R^{15}$ as the linear function of $Y = Y^- = ((u^-)_{\hat{i},\hat{j}}^{(m)},\ 0 \le m \le 4) \in R^{15}$. That is, we form the transformation matrix P such that $Y^+ = PY^-$.

Step 3: Since $Y^+ = PY$, it follows from (7.73) that there exists a vector T_k such that

$$u(\xi_i, \eta_j) = T_k \cdot Y \quad \text{for } k \in S_9,\ \ (i,j) \in (i_0, j_0) + k, \tag{7.74}$$

where (ξ_i, η_j) are the local coordinates of (x_i, y_j) of the standard compact 9-point neighborhood, and thus

$$\gamma_k u(x_i, y_j) = \gamma_k T_k \cdot Y.$$

Note also, from (7.73), that each

$$f_{i,j} = f(x_i, y_j) = -\beta^-(u_{\xi\xi} + u_{\eta\eta})(\xi_i, \eta_j) + \sigma^- u(\xi_i, \eta_i)$$

is expressed as a linear functional of Y and therefore from (7.63),

$$(D_x f, D_y f, D_{xx} f, D_{xy} f, D_{yy} f)^t = SY, \quad S \in R^{5\times 15}. \tag{7.75}$$

Hence from (7.74)–(7.75), we have

$$\begin{aligned}&\sum_{k\in S_9} \gamma_k U_{i,j} + \alpha_1 D_x f + \alpha_2 D_y f + \alpha_3 D_{xx} f + \alpha_4 D_{xy} f + \alpha_5 D_{yy} f \\ &= \left(\sum_k \gamma_k T_k + \sum_i \alpha_i S_i\right) \cdot Y.\end{aligned} \tag{7.76}$$

Equating $-\beta^- (u_{\xi\xi} + u_{\eta\eta}) + \sigma^- u = f$ at $\mathbf{x}_0$ with approximation (7.73), we have

$$\sum_k \gamma_k T_k + \sum_i \alpha_i S_i = z \quad \text{in } R^{15}, \tag{7.77}$$

where $z \in R^{15}$ is defined by

$$z \cdot Y = \left(-\beta^- (u^-_{\xi\xi} + u^-_{\eta\eta}) + \sigma^- u^-\right)(\xi_0, \eta_0),$$

with (ξ_0, η_0) being the local coordinate of $\mathbf{x}_0$. We therefore formulate the following constrained minimization to determine the weight (γ, α) to obtain a fourth-order method on the compact 9-point stencil for the interface problem at $\mathbf{x}_0$:

$$\begin{aligned}&\min \quad \frac{1}{2}\left|\left(\sum_k \gamma_k T_k + \sum_i \alpha_i S_i - z\right)_{7:15}\right|^2_{R^9} \\ &\quad \text{subject to} \\ &\left(\sum_k \gamma_k T_k + \sum_i \alpha_i S_i\right)_{1:6} - z_{1:6} = 0 \quad \text{and} \quad \gamma_k \geq 0, \ k \neq 5.\end{aligned} \tag{7.78}$$

The interface conditions in Step 2 are as follows. From the continuity conditions

$$u^+(\mathbf{x}^*) = u^-(\mathbf{x}^*), \quad u^+_\eta = u^-_\eta, \quad \beta^+ u^+_\xi = \beta^- u^-_\xi. \tag{7.79}$$

Since the Laplace operator is invariant under the rotation,

$$\beta^+ (u^+_{\xi\xi} + u^+_{\eta\eta}) - \sigma^+ u^+ = \beta^- (u^-_{\xi\xi} + u^-_{\eta\eta}) - \sigma^- u^-, \tag{7.80}$$

and differentiating this in the local coordinates,

$$\begin{cases} \beta^+ (u^+_{\xi_x i\xi} + u^+_{\xi\eta\eta}) - \sigma^+ u^+_\xi = \beta^- (u^-_{\xi\xi_x i} + u^-_{\xi\eta\eta}) - \sigma^- u^-_\xi, \\ \beta^+ (u^+_{\xi\xi\eta} + u^+_{\eta\eta\eta}) - \sigma^+ u^+_\eta = \beta^- (u^-_{\xi\xi\eta} + u^-_{\eta\eta\eta}) - \sigma^- u^-_\eta, \\ \beta^+ (u^+_{\xi\xi\xi\xi} + u^+_{\xi\xi\eta\eta}) - \sigma^+ u^+_{\xi\xi} = \beta^- (u^-_{\xi\xi\xi\xi} + u^-_{\xi\xi\eta\eta}) - \sigma^- u^-_{\xi\xi}, \\ \beta^+ (u^+_{\xi\xi\xi\eta} + u^+_{\xi\eta\eta\eta}) - \sigma^+ u^+_{\xi\eta} = \beta^- (u^-_{\xi\xi\xi\eta} + u^-_{\xi\eta\eta\eta}) - \sigma^- u^-_{\xi\eta}, \\ \beta^+ (u^+_{\xi\xi\eta\eta} + u^+_{\eta\eta\eta\eta}) - \sigma^+ u^+_{\eta\eta} = \beta^- (u^-_{\xi\xi\eta\eta} + u^-_{\eta\eta\eta\eta}) - \sigma^- u^-_{\eta\eta}. \end{cases} \tag{7.81}$$

Two additional conditions are obtained by taking the tangential derivative of the continuity condition (7.79),

$$\begin{aligned} u^+_{\eta\eta} + \kappa\, u^+_{\xi} &= u^-_{\eta\eta} + \kappa\, u^-_{\xi}\,, \\ \beta^+\,(u^+_{\xi\eta} - \kappa\, u^+_{\eta}) &= \beta^-\,(u^-_{\xi\eta} - \kappa\, u^-_{\eta}), \end{aligned} \tag{7.82}$$

where κ is the curvature of Γ at $\mathbf{x}^*$. Lastly, additional continuity conditions are obtained by equating the continuity condition (3.2) at $\mathbf{x}_1^*$ and $\mathbf{x}_2^*$,

$$u^+(\mathbf{x}_i^*) = u^-(\mathbf{x}_i^*), \quad \beta^+ u^+_{\boldsymbol{n}}(\mathbf{x}_i^*) = \beta^- u^-_{\boldsymbol{n}}(\mathbf{x}_i^*), \quad i = 1, 2, \tag{7.83}$$

where $\mathbf{x}_1^*$ and $\mathbf{x}_2^*$ are the two closest projected points to $\mathbf{x}^*$ on Γ.

Remark 7.4.

1. *If the solution u is piecewise C^5 in a neighborhood of $\mathbf{x}^*$, then the truncation error of* (7.79) *is of order $O(h^5)$.*

2. *Step 2 is an extension of the second-order IIM in §3.5 to the fourth-order finite difference and it allows us to extend the solution u across the interface (from Ω^- to Ω^+).*

3. *Adaptation of the method to include the piecewise constant convection term, $b \cdot \nabla u$, is straightforward.*

4. *When β^+ and β^- are variables, the corresponding interface conditions* (7.82) *require the derivatives of $\beta^\pm$. Instead of* (7.82) *we may use* (7.83) *at an additional point $\mathbf{x}_3^*$. Otherwise, the method can be extended to the variable case without much difficulty following the procedure described in §7.2.6.*

5. *The curvature κ is given by*

$$\kappa = \frac{\varphi_y^2 \varphi_{xx} - 2\varphi_x\varphi_y\varphi_{xy} + \varphi_x^2 \varphi_{yy}}{(\varphi_x^2 + \varphi_y^2)^{3/2}}$$

and can be evaluated at $\mathbf{x}^*$ using the 25-point interpolation of φ in §7.2.4.

6. *The constrained minimization* (7.78) *is formulated so that the resulting finite difference scheme yields an* M*-matrix by equating* (7.76) *in the least squares sense.*

7.2.8 The fourth-order method for heat equations with interfaces

In this section we consider the heat equation

$$\frac{\partial u}{\partial t} = \nabla \cdot (\beta\, \nabla u(t, \mathbf{x})) - b(\mathbf{x})\nabla u - \sigma\, u(t, \mathbf{x}), \qquad u(0, \mathbf{x}) = u_0(\mathbf{x}). \tag{7.84}$$

We discuss the case with $b = 0$ and the piecewise constant diffusion coefficient, i.e.,

$$\frac{\partial u}{\partial t} = \beta^{\pm}\, \Delta u(t, \mathbf{x}) - \sigma^{\pm}\, u(t, \mathbf{x}) \qquad \text{on each } \Omega^{\pm}, \tag{7.85}$$

but the extension to the general case is straightforward. The fourth-order finite difference scheme at an irregular point $\mathbf{x}_0$ is developed by treating $-u_t$ as a forcing function f in (7.1). Note that $-u_t$ satisfies the proper interface condition as follows from (7.87)–(7.89). We first expand u_t as the piecewise second-order Taylor expansion in a neighborhood of the projected point $\mathbf{x}^*$ (see Step 1 in §7.2.3),

$$\frac{\partial}{\partial t}u^{\pm} \sim \dot{u}^{\pm} + \dot{u}^{\pm}_{\xi}\xi + \dot{u}^{\pm}_{\eta}\eta + \frac{1}{2}\dot{u}^{\pm}_{\xi\xi}\xi^2 + \dot{u}^{\pm}_{\xi\eta}\xi\eta + \frac{1}{2}\dot{u}^{\pm}_{\xi\xi}\eta^2. \tag{7.86}$$

Let $X^{\pm} = (\dot{u}^{\pm}, \dot{u}^{\pm}_{xi}, \dot{u}^{\pm}_{\eta}, \dot{u}^{\pm}_{\xi\xi}\dot{u}^{\pm}_{\xi\eta}, \dot{u}^{\pm}_{\xi\xi}) \in R^6$ be the Taylor vectors for $u^{\pm}_t$. Then we relate them as $X^+ = \tilde{P}X,\ \tilde{P} \in R^{6\times 6}$ $(X = X^-)$, by the following interface conditions at $\mathbf{x}^*$:

$$\dot{u}^+ = \dot{u}^-, \quad \dot{u}^+_{\eta} = \dot{u}^-_{\eta}, \quad \beta^+\dot{u}^+_{\xi} = \beta^-\dot{u}^-_{\xi}, \tag{7.87}$$

$$\begin{aligned} &\dot{u}^+_{\eta\eta} + \kappa\,\dot{u}^+_{\xi} = \dot{u}^-_{\eta\eta} + \kappa\,\dot{u}^-_{\xi}, \\ &\beta^+\,(\dot{u}^+_{\xi\eta} - \kappa\,\dot{u}^+_{\eta}) = \beta^-\,(\dot{u}^-_{\xi\eta} - \kappa\,\dot{u}^-_{\eta}), \end{aligned} \tag{7.88}$$

and

$$\beta^+\,(\dot{u}^+_{\xi\xi} + \dot{u}^+_{\eta\eta}) - \sigma^+\,\dot{u}^+ = \beta^-\,(\dot{u}^-_{\xi_x i} + \dot{u}^-_{\eta\eta}) - \sigma^-\,\dot{u}^-. \tag{7.89}$$

Differentiating (7.85) in local coordinates (ξ, η), we have

$$\begin{aligned} &\beta^{\pm}\,(u^{\pm}_{\xi_x i\xi} + u^{\pm}_{\xi\eta\eta}) - \sigma^{\pm}\,u^{\pm}_{\xi} = \dot{u}^{\pm}_{\xi}, \\ &\beta^{\pm}\,(u^{\pm}_{\xi\xi\eta} + u^{\pm}_{\eta\eta\eta}) - \sigma^{\pm}\,u^{\pm}_{\eta} = \dot{u}^{\pm}_{\eta}, \\ &\beta^{\pm}\,(u^{\pm}_{\xi\xi\xi\xi} + u^{\pm}_{\xi\xi\eta\eta}) - \sigma^{\pm}\,u^{\pm}_{\xi\xi} = \dot{u}^{\pm}_{\xi\xi}, \\ &\beta^{\pm}\,(u^{\pm}_{\xi\xi\xi\eta} + u^{\pm}_{\xi\eta\eta\eta}) - \sigma^{\pm}\,u^{\pm}_{\xi\eta} = \dot{u}^{\pm}_{\xi\eta}, \\ &\beta^{\pm}\,(u^{\pm}_{\xi\xi\eta\eta} + u^{\pm}_{\eta\eta\eta\eta}) - \sigma^{\pm}\,u^{\pm}_{\eta\eta} = \dot{u}^{\pm}_{\eta\eta}. \end{aligned} \tag{7.90}$$

Thus, (7.81) (in which we assume f is smooth) should be replaced by (7.87)–(7.90). With this change we proceed with Steps 2 and 3 in §7.2.7 and determine (γ, α) by the constrained optimization (7.78) at $\mathbf{x}_0$. Then with $f = -u^-_t$ and by (7.86),

$$f - (\alpha_1\,D_x f + \alpha_2\,D_y f + \alpha_3\,D_{xx} f + \alpha_4 D_{xy} f + \alpha_5\,D_{yy} f) = \beta \cdot X \tag{7.91}$$

for $\beta \in R^6$. It follows from (7.86) that there exists a vector $\tilde{T}_k$ such that

$$\left(\frac{\partial u^-}{\partial t}\right)_k = \frac{\partial u}{\partial t}(\xi_i, \eta_j) = \tilde{T}_k \cdot X \quad \text{for } k \in S_9,\ \ (i, j) \in (i_0, j_0) + k. \tag{7.92}$$

We now further reduce the equation to (7.42). Here (ξ_i, η_j) is the local coordinate of (x_i, y_j) of the standard 9-point neighborhood. Now, we select the weight q_k such that

$$\sum q_k \left(\frac{\partial u^-}{\partial t}\right)_k = \sum q_k\,\tilde{T}_k \cdot X = \beta \cdot X,$$

i.e., $\tilde{T}q = \beta$ with $\tilde{T} \in R^{6\times 9}$, and we demand $q_k \geq 0$ and $\sum_{k\neq 5} q_k \leq q_5$. We formulate the constraint optimization for $\{q_k\}$,

$$\begin{aligned} &\min \frac{1}{2}\left|\left(\sum_k q_k \tilde{T}_k - \beta\right)_{4:6}\right|^2_{R^3} \\ &\text{subject to } \left(\sum_k q_k \tilde{T}_k - \beta\right)_{1:3} = 0 \quad \text{and} \quad \gamma_k \geq 0,\ \sum_{k\neq 5} q_k \leq q_5. \end{aligned} \tag{7.93}$$

Now, we obtain the finite difference scheme for (7.85) with (7.72) at the irregular point,

$$\sum_{k\in S_5} q_k \frac{d}{dt} U_{(i,j)+k} = -\sum_{k\in S_9} \gamma_k\, U_{(i+j)+k}. \tag{7.94}$$

We demonstrate the applicability of the fourth-order method (7.94) for the corresponding eigenvalue problem for (7.85) with (7.72), i.e., finding the eigenpair $(\lambda, u) \in R \times L^2(\Omega)$ satisfying (7.72) and

$$-\nabla \cdot (\beta\, \nabla u) = \lambda u, \quad u|_{\partial\Omega} = 0. \tag{7.95}$$

The corresponding approximation method reduces to the generalized eigenvalue problem for (7.94),

$$\lambda\, QU = -HU, \tag{7.96}$$

where Q, H are matrices that represent the left-hand side and the right-hand side of (7.94), respectively. Thus the eigenpairs (λ, U) approximate the ones for (7.95).

7.3 The fourth-order methods for three-dimensional cases

In this section we discuss the three-dimensional cases. We highlight the necessary steps to extend the method in R^2 to R^3.

7.3.1 The fourth-order scheme for problems on irregular domains in three dimensions

We use the compact 19-point central finite difference at $\mathbf{x}_{10} = \mathbf{x}_{i,j,k}$ (see Figure 7.4)

$$\begin{aligned} &\frac{1}{h^2}\Bigg(4U_{i,j} - \frac{1}{3}(U_{i-1,j,k} + U_{i+1,j,k} + U_{i,j-1,k} + U_{i,j+1,k} + U_{i,j,k-1} + U_{i,j,k+1}) \\ &\quad - \frac{1}{6}(U_{i+1,j+1,k} + U_{i+1,j-1,k} + U_{i-1,j+1,k} + U_{i-1,j-1,k} + U_{i,j+1,k+1} + U_{i,j+1,k-1} \\ &\quad + U_{i,j-1,k+1} + U_{i,j-1,k-1} + U_{i+1,j,k+1} + U_{i-1,j,k+1} + U_{i+1,j,k-1} + U_{i-1,j,j-1})\Bigg) \\ &= \frac{1}{2} f_{i,j} + \frac{1}{12}(f_{i-1,j,k} + f_{i+1,j,k} + f_{i,j-1,k} + f_{i,j+1,k} + f_{i,j,k-1} + f_{i,j,k+1}). \end{aligned} \tag{7.97}$$

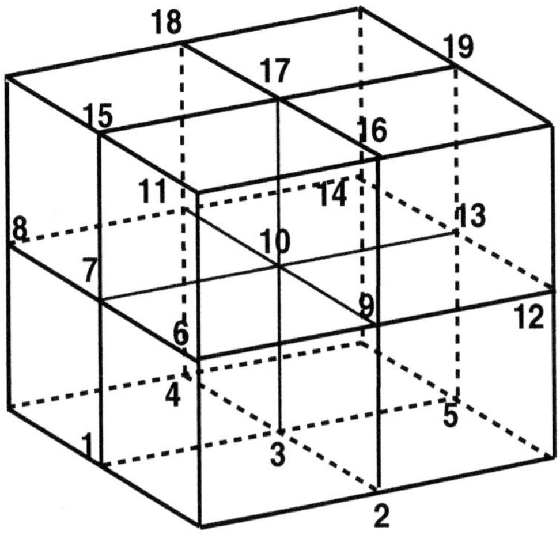

Figure 7.4. *The* 19-*point stencil for three-dimensional fourth-order compact scheme.*

If $\mathbf{x}_0$ is an irregular grid point, then we project $\mathbf{x}_0$ to the boundary Γ by the corresponding formula to (7.50) as in §7.2.4. Then we set the local coordinate system (ξ, η, τ) as

$$\xi = \frac{\mathbf{d}}{|\mathbf{d}|}, \quad \eta = \frac{(d_2, -d_1, 0)}{|\mathbf{d}|},$$
$$\tau = \frac{(d_3 d_1, d_3 d_2, -d_1^2 - d_2^2)}{\sqrt{(d_3 d_1)^2 + (d_3 d_2)^2 + (d_1^2 + d_2^2)^2}}$$

if $\sqrt{d_1^2 + d_2^2} \geq \sqrt{d_2^2 + d_3^2}$; otherwise

$$\xi = \frac{\mathbf{d}}{|\mathbf{d}|}, \quad \tau = \frac{(d_3, 0, -d_2)}{|\mathbf{d}|},$$
$$\eta = \frac{(-d_1 d_2, d_1^2 + d_3^2, -d_2 d_3)}{\sqrt{(d_1 d_2)^2 + (d_1^2 + d_3^2)^2 + (d_2 d_3)^2}}.$$

Here (η, τ) stands for the tangential coordinate and $\mathbf{d}$ is the approximated gradient at the projected point $\mathbf{x}^*$. Then, the boundary $\Gamma = \{\mathbf{x} \in R^3 : \phi(\mathbf{x}) = 0\}$ is described as the graph $\xi = \chi(\eta, \tau)$ in a neighborhood of $\mathbf{x}^*$, and the curvatures at $\mathbf{x}^*$ are given by

$$\chi_{\eta\eta}(\mathbf{x}^*) = \frac{\boldsymbol{\eta}^T H \boldsymbol{\eta}}{|\mathbf{d}|}, \quad \chi_{\eta\tau}(\mathbf{x}^*) = \frac{\boldsymbol{\eta}^T H \boldsymbol{\tau}}{|\mathbf{d}|}, \quad \chi_{\tau\tau}(\mathbf{x}^*) = \frac{\boldsymbol{\tau}^T H \boldsymbol{\tau}}{|\mathbf{d}|},$$

where H is the approximated Hessian of ϕ at $\mathbf{x}^*$.

We expand the solution u in a neighborhood of $\mathbf{x}^*$ by

$$u \sim \Phi(\xi, \eta, \tau) = \sum_{m=0}^{4} \sum_{\hat{i}+\hat{j}+\hat{k}=m} \frac{\hat{i}!\,\hat{j}!\,\hat{k}!\,u^{(m)}_{\hat{i},\hat{j},\hat{k}}}{m!} \frac{}{m!} \xi^{\hat{i}} \eta^{\hat{j}} \tau^{\hat{k}}. \tag{7.98}$$

Among the 35 Taylor coefficients, $u^* = g(\mathbf{x}^*)$ is known from the boundary condition, and thus we have a total of 34 unknowns denoted by X in (7.98), i.e., all derivatives. Let $\mathbf{x}_i^*$, $i = 1, \dots, 23$, be the 23 closest projected points to $\mathbf{x}^*$ on Γ. We use the following 34 conditions to obtain the complete system for the unknowns:

$$u(x_0) = u_{i,j}, \quad u(\mathbf{x}_i^*) = g(\mathbf{x}_i^*), \quad 1 \le i \le 23,$$

$$\begin{cases} u_{\xi\xi} + u_{\eta\eta} + u_{\tau\tau} = -f(\mathbf{x}^*), \\ u_{\xi\xi\xi} + u_{\xi\eta\eta} + u_{\xi\tau\tau} = -f_\xi(\mathbf{x}^*), \\ u_{\xi\xi\eta} + u_{\eta\eta\eta} + u_{\eta\tau\tau} = -f_\eta(\mathbf{x}^*), \\ u_{\xi\xi\tau} + u_{\eta\eta\tau} + u_{\tau\tau\tau} = -f_\tau(\mathbf{x}^*), \\ u_{\xi\xi\xi\xi} + u_{\xi\xi\eta\eta} + u_{\xi\xi\tau\tau} = -f_{\xi\xi}(\mathbf{x}^*), \\ u_{\xi\xi\eta\eta} + u_{\eta\eta\eta\eta} + u_{\eta\eta\tau\tau} = -f_{\eta\eta}(\mathbf{x}^*), \\ u_{\xi\xi\tau\tau} + u_{\eta\eta\tau\tau} + u_{\tau\tau\tau\tau} = -f_{\tau\tau}(\mathbf{x}^*), \\ u_{\xi\xi\xi\eta} + u_{\xi\eta\eta\eta} + u_{\xi\eta\tau\tau} = -f_{\xi\eta}(\mathbf{x}^*), \\ u_{\xi\xi\eta\tau} + u_{\eta\eta\eta\tau} + u_{\eta\tau\tau\tau} = -f_{\eta\tau}(\mathbf{x}^*), \\ u_{\xi\xi\xi\tau} + u_{\xi\eta\eta\tau} + u_{\xi,\tau\tau\tau} = -f_{\eta\tau}(\mathbf{x}^*). \end{cases} \tag{7.99}$$

System (7.99) is a linear equation $AX = b$ in R^{34}. Thus, all the unknowns in (7.98) are determined as a linear function of $(u_{i,j}, f(\mathbf{x}^*), g(\mathbf{x}^*), g(\mathbf{x}_i^*), 1 \le i \le 23)$. Now, we complete (7.97) at the irregular grid point by expressing the value $u_{i\pm1,j\pm1,k\pm1}$ corresponding to the grid points outside of Ω by (7.98). Again as in §7.2.3 for the two-dimensional case, we modify only the diagonal entry $h_{i,j,k}$ and the corresponding right-hand side $\tilde{f}_{i,j,k}$ at each irregular point.

Remark 7.5.

1. *For the second option in which we eliminate (u_η, u_τ) by the tangential derivatives $(g_\eta(\mathbf{x}^*), g_\tau(\mathbf{x}^*))$, we use the 21 additional boundary conditions at $\mathbf{x}_i^*$, $1 \le i \le 21$.*
2. *For the mixed boundary condition (7.2), we use the following equality:*

$$u_\xi = g(\mathbf{x}^*) - \alpha\, u^*,$$

and use

$$\frac{\partial u}{\partial \boldsymbol{n}}(\mathbf{x}_i^*) = g(\mathbf{x}_i^*) - \alpha\, u(\mathbf{x}_i^*), \quad 1 \le i \le 23.$$

7.3.2 The fourth-order scheme for three-dimensional interface problems

We seek the finite difference scheme of the form

$$\sum_{i_x=-1}^{i} \sum_{i_y=-1}^{1} \sum_{i_z=-1}^{1} \gamma_{i_x,i_y,i_z} u_{i+i_x,j+i_y,k+i_z} + \alpha_1 D_x f + \alpha_2 D_y f + \alpha_3 D_z f \\ + \alpha_4 D_{xx} f + \alpha_5 D_{yy} f + \alpha_6 D_{zz} f + \alpha_7 D_{xy} f + \alpha_8 D_{yz} f + \alpha_9 D_{zx} f, \tag{7.100}$$

where $D_{\ldots}$ denotes the central finite difference approximation of the second derivatives as in (7.63). A grid point $\mathbf{x}_0 = (x_i, y_j, z_k)$ is *irregular* if the standard 19-point stencil for the compact fourth-order scheme (7.97) at $\mathbf{x}_0$ contains grid points from different subdomains, say $\Omega^{\pm}$. At the regular point we use (7.97). At each irregular point we proceed as in Steps 1–3 in §7.2.7. We first project each irregular point $\mathbf{x}_0$ onto the interface Γ by the fourth-order projection method, and thus we have a family of projection points $\mathbf{x}^*$ on Γ. We then expand the solution u in a neighborhood of $\mathbf{x}^*$ by

$$u^{\pm} \sim \Phi(\xi, \eta, \tau) = \sum_{m=0}^{4} \sum_{\hat{i}+\hat{j}+\hat{k}=m} \frac{\hat{i}!\,\hat{j}!\,\hat{k}!}{m!} \frac{(u^{\pm})^{(m)}_{\hat{i},\hat{j},\hat{k}}}{m!} \xi^{\hat{i}} \eta^{\hat{j}} \tau^{\hat{k}}. \tag{7.101}$$

We use the following interface conditions, (7.102)–(7.104), to relate the Taylor expansion $u^{\pm}$ across the interface Γ:

$$\begin{aligned}
& u^+ = u^-, \quad \beta^+ u_\xi = \beta^- u_\xi, \quad u^+_\eta = u^-_\eta, \quad u^+_\tau = u^-_\tau, \\
& u^+_{\eta\eta} + \chi_{\eta\eta} u^+_\xi = u^-_{\eta\eta} + \chi_{\eta\eta} u^-_\xi, \quad u^+_{\eta\tau} + \chi_{\eta\tau} u^+_\xi = u^-_{\eta\eta} + \chi_{\eta\tau} u^-_\xi, \\
& u^+_{\tau\tau} + \chi_{\tau\tau} u^+_\xi = u^-_{\tau\tau} + \chi_{\tau\tau} u^-_\xi, \\
& \beta^+ (u^+_{\xi\eta} - \chi_{\eta\eta} u^+_\eta - \chi_{\eta\tau} u^+_\tau) = \beta^- (u^-_{\xi\eta} - \chi_{\eta\eta} u^-_\eta - \chi_{\eta\tau} u^-_\tau), \\
& \beta^+ (u^+_{\xi\tau} - \chi_{\eta\tau} u^+_\eta - \chi_{\tau\tau} u^+_\tau) = \beta^- (u^-_{\xi\tau} - \chi_{\eta\tau} u^-_\eta - \chi_{\tau\tau} u^-_\tau).
\end{aligned} \tag{7.102}$$

$$\begin{cases}
\beta^+(u^+_{\xi\xi} + u^+_{\eta\eta} + u^+_{\tau\tau}) = \beta^-(u^-_{\xi\xi} + u^-_{\eta\eta} + u^-_{\tau\tau}), \\
\beta^+(u^+_{\xi\xi\xi} + u^+_{\xi\eta\eta} + u^+_{\xi\tau\tau}) = \beta^-(u^-_{\xi\xi\xi} + u^-_{\xi\eta\eta} + u^-_{\xi\tau\tau}), \\
\beta^+(u^+_{\xi\xi\eta} + u^+_{\eta\eta\eta} + u^+_{\eta\tau\tau}) = \beta^-(u^-_{\xi\xi\eta} + u^-_{\eta\eta\eta} + u^-_{\eta\tau\tau}), \\
\beta^+(u^+_{\xi\xi\tau} + u^+_{\eta\eta\tau} + u^+_{\tau\tau\tau}) = \beta^-(u^-_{\xi\xi\tau} + u^-_{\eta\eta\tau} + u^-_{\tau\tau\tau}), \\
\beta^+(u^+_{\xi\xi\xi\xi} + u^+_{\xi\xi\eta\eta} + u^+_{\xi\xi\tau\tau}) = \beta^-(u^-_{\xi\xi\xi\xi} + u^-_{\xi\xi\eta\eta} + u^-_{\xi\xi\tau\tau}), \\
\beta^+(u^+_{\xi\xi\eta\eta} + u^+_{\eta\eta\eta\eta} + u^+_{\eta\eta\tau\tau}) = \beta^-(u^-_{\xi\xi\eta\eta} + u^-_{\eta\eta\eta\eta} + u^-_{\eta\eta\tau\tau}), \\
\beta^+(u^+_{\xi\xi\tau\tau} + u^+_{\eta\eta\tau\tau} + u^+_{\tau\tau\tau\tau}) = \beta^-(u^-_{\xi\xi\tau\tau} + u^-_{\eta\eta\tau\tau} + u^-_{\tau\tau\tau\tau}), \\
\beta^+(u^+_{\xi\xi\xi\eta} + u^+_{\xi\eta\eta\eta} + u^+_{\xi\eta\tau\tau}) = \beta^-(u^-_{\xi\xi\xi\eta} + u^-_{\xi\eta\eta\eta} + u^-_{\xi\eta\tau\tau}), \\
\beta^+(u^+_{\xi\xi\eta\tau} + u^+_{\eta\eta\eta\tau} + u^+_{\eta\tau\tau\tau}) = \beta^-(u^-_{\xi\xi\eta\tau} + u^-_{\eta\eta\eta\tau} + u^-_{\eta\tau\tau\tau}), \\
\beta^+(u^+_{\xi\xi\xi\tau} + u^+_{\xi\eta\eta\tau} + u^+_{\xi\tau\tau\tau}) = \beta^-(u^-_{\xi\xi\xi\tau} + u^-_{\xi\eta\eta\tau} + u^-_{\xi\tau\tau\tau}).
\end{cases} \tag{7.103}$$

Let $\mathbf{x}^*_i$, $1 \le i \le 7$, be the seven closest projected points to $\mathbf{x}^*$ on Γ, and then we equate

$$\begin{aligned}
& \frac{\partial}{\partial \eta} u^+(\mathbf{x}^*_i) = \frac{\partial}{\partial \eta} u^-(\mathbf{x}^*_i), \quad \frac{\partial}{\partial \tau} u^+(\mathbf{x}^*_i) = \frac{\partial}{\partial \tau} u^-(\mathbf{x}^*_i), \; i = 1, \ldots, 4, 5, \\
& \beta^+ \frac{\partial}{\partial \boldsymbol{n}} u^+(\mathbf{x}^*_i) = \beta^- \frac{\partial}{\partial \boldsymbol{n}} u^-(\mathbf{x}^*_i), \qquad i = 1, \ldots, 7.
\end{aligned} \tag{7.104}$$

That is, using the interface conditions (7.102)–(7.104), we express the Taylor vector $Y^+ = ((u^+)^{(m)}_{\hat{i},\hat{j},\hat{k}},\ 0 \le m \le 4) \in R^{35}$ as a linear function of $Y = Y^- = ((u^-)^{(m)}_{\hat{i},\hat{j},\hat{k}}\ 0 \le m \le 4) \in$

R^{35}, i.e., $Y^+ = PY^-$. Since $Y^+ = PY$, it follows from (7.100) that there exists a vector T_{i_x,i_y,i_z} such that

$$u(i_x, i_y, i_z) = T_{i_x,i_y,i_z} \cdot Y, \tag{7.105}$$

where $(\xi_{i_x}, \eta_{i_y}, \tau_{i_z})$ is the local coordinate of $(x_{i+i_x}, y_{j+i_y}, z_{k+i_z})$ of the standard 19-point neighborhood. Since

$$\begin{aligned} f_{i+i_x,j+i_y,k+i_z} &= f(x_{i+i_x}, y_{j+i_y}, z_{k+i_z}) \\ &= \left(-\beta^- (u_{\xi\xi} + u_{\eta\eta} + u_{\tau,\tau}) + \sigma^- u\right)(\xi_{i_x}, \eta_{i_y}, \tau_{i_z}), \end{aligned}$$

from (7.101) there exists a matrix $S \in R^{9\times 35}$ such that

$$(D_x f, D_y f, D_z f, D_{xx} f, D_{xy} f, D_{zz} f, D_{xy} f, D_{yz} f, D_{zx} f)^T = SY. \tag{7.106}$$

Hence, from (7.104)–(7.106),

$$\begin{aligned} &\sum_{i_x=-1}^{i} \sum_{i_y=-1}^{1} \sum_{i_z=-1}^{1} \gamma_{i_x,i_y,i_z} u_{i+i_x,j+i_y,k+i_z} + \alpha_1 D_x f + \alpha_2 D_y f + \alpha_3 D_z f \\ &\quad + \alpha_4 D_{xx} f \alpha_5 D_{yy} f + \alpha_6 D_{zz} f + \alpha_7 D_{xy} f + \alpha_8 D_{yz} f + \alpha_9 D_{zx} f \\ &= \left(\sum\nolimits_{i_x,i_y,i_z} \gamma_{i_x,i_y,i_z} T_{i_x,i_y,i_z} + \sum\nolimits_i \alpha_i S_i\right) \cdot Y. \end{aligned}$$

Equating $-\beta^- (u_{\xi\xi} + u_{\eta\eta} + u_{\tau\tau}) + \sigma^- u = f$ at $\mathbf{x}_0$ with approximation (7.100), we have

$$\begin{aligned} &\left(\sum\nolimits_{i_x,i_y,i_z} \gamma_{i_x,i_y,i_z} T_{i_x,i_y,i_z} + \sum\nolimits_i \alpha_i S_i\right) \\ &\qquad = \left(\sigma^-, 0, 0, 0, -\beta^-, -\beta^-, -\beta^-, 0, 0, 0), 0, \ldots, 0\right) \end{aligned} \tag{7.107}$$

in R^{35}. We now formulate the following constrained minimization to determine the weight (γ, α) at $\mathbf{x}_0$:

$$\begin{aligned} &\min \frac{1}{2} \left\| \left(\sum_{i_x,i_y,i_z} \gamma_{i_x,i_y,i_z} T_{i_x,i_y,i_z} + \sum_i \alpha_i S_i \right)_{11:35} \right\|_{R^9}^2 \\ &\text{subject to} \left(\sum_{i_x,i_y,i_z} \gamma_{i_x,i_y,i_z} T_{I_x,i_y,i_z} + \sum_i \alpha_i S_i \right)_{1:10} \\ &\qquad = \left(-\sigma^-, 0, 0, 0, \beta^-, \beta^-, \beta^-, 0, 0, 0 \right), \\ &\qquad \gamma_{i_x,i_y,i_z} \geq 0, \quad (i_x, i_y, i_z) \neq (0, 0, 0). \end{aligned} \tag{7.108}$$

7.4 The preconditioned subspace iteration method

The resulting linear system of equations via the IIM, in general, leads to a coefficient matrix whose rows differ only from that obtained from the standard discretization for the homogeneous region near the interface. For the homogeneous problem on rectangular domains, we can have a direct solver based on the FFT or other fast solvers. If we use this direct solver as a preconditioner for the interface problem, we can take advantage of the sparse subspace reduction [140, 120]. The approach can drastically reduce computational costs for the preconditioned GMRES method.

We now present the preconditioned sparse subspace iteration method for $Ax = b$. A can be singular, in which case a solution exists only if b belongs to the subspace range(A). The system can be solved by

$$AB^{-1}v = b, \tag{7.109}$$

where B is a right preconditioner. After we have the vector v, the solution vector x of the original problem is given by

$$x = B^{-1}v. \tag{7.110}$$

Another possibility would be to consider the left preconditioned system $B^{-1}Au = B^{-1}b$ or to change the Euclidean inner product in the iterative method to the B-inner product. In all these cases we can take advantage of the sparsity of arising subspaces in a manner similar to the one considered in the following.

The sparse subspace is

$$X = \text{range}(A - B) \cap \text{range}(A). \tag{7.111}$$

That is, the ith component x_i of x on X can be nonzero only if the ith row of the matrices $A - B$ and A are both nonzero. From the sparsity point of view, it is advantageous to make the dimension of X as small as possible. However, this should not be done in a way which increases the computational cost of the multiplication of a vector by B^{-1} or which deteriorates the conditioning of AB^{-1}, since this increases the number of iterations.

Note that if $\hat{v} = v - b$, then

$$AB^{-1}\hat{v} = b - AB^{-1}b = -(A - B)B^{-1}b = \hat{b} \in X,$$

where $AB^{-1} = (A - B)B^{-1} + I$. Thus $\hat{v} \in X$ satisfies

$$(A - B)B^{-1}\hat{v} + \hat{v} = \hat{b}. \tag{7.112}$$

This is the reduced equation in the sparse subspace X. Moreover, the Krylov subspace

$$span\left\{r, AB^{-1}r, \ldots, (AB^{-1})^{m-1}\right\} \subset X$$

if $r \in X$. Any iterative method based on the Krylov subspace for the solution of $AB^{-1}v = b$ generates a sequence of approximate solutions v^k in the subspace X, provided that $v^0 = b$. Hence, all required operations are carried out in the subspace X; i.e., we summarize the method in the outline below.

An outline of the preconditioned subspace iteration method

1. Determine $X = \text{range(A)} \cap \text{range }(A - B)$.

2. Set $v_0 = b_0$ and compute $r^0 = (A - B)B^{-1}b_0$.

3. Apply Krylov subspace iterates in X until convergence for the residual vectors generated by
$$b^k = (A - B)B^{-1}b^{k-1} + b^{k-1}, \quad b^0 = r^0 \in X.$$

The basic operations of the type

$$(A - B)B^{-1}x, \quad x \in X, \tag{7.113}$$

performed during the iteration require the solution $B^{-1}x$ on the range of $(A - B)^T$. The dimension of this range is usually of the same order as the dimension of X. For a Poisson equation, we use the matrix B corresponding to the compact fourth-order finite difference scheme on the extended problem on a cube. Then, $U = B^{-1}f,\ f \in X$, is carried out by the FFT as follows:

$$\hat{f}_{\hat{i}\hat{j}\hat{k}} = \sum f_{ijk} \sin(\hat{i}\pi\ x_i)\sin(\hat{j}\pi\ y_j)\sin(\hat{k}\pi\ z_k),$$
$$U_{ijk} = \sum \Lambda^{-1}_{\hat{i}\hat{j}\hat{k}} \hat{f}_{\hat{i}\hat{j}\hat{k}} \sin(i\pi\ x_{\hat{i}})\sin(j\pi\ y_{\hat{j}})\sin(k\pi\ z_{\hat{k}}),$$

where $\Lambda_{ijk} = \lambda_i + \lambda_j + \lambda_k$ with $\lambda_k = \sin^2(k\pi)$. This uses three-dimensional Fourier sine forward and inverse transforms and requires order $O(N^3\,log(N))$ operations and order $O(N^3)$ storage. We can take advantage of the sparsity by observing that, for each $\hat{k}$,

$$\begin{aligned} &B_2 + \lambda_{\hat{k}}\hat{U}_{\ldots\hat{k}} = F_{\ldots\hat{k}}, \qquad F_{\ldots\hat{k}} = \sum_k f_{\ldots k}\sin(\hat{k}\pi\ z_k), \\ &U = \sum_{\hat{k}} U_{\ldots\hat{k}}\sin(\hat{k}\pi\ z_k), \end{aligned} \tag{7.114}$$

where B_2 is the corresponding two-dimensional discrete Poisson matrix. For each $\hat{k}$, vector $F_{\ldots\hat{k}}$ is evaluated by order $O(N)$ operations since if $f \in X$, then $f_{i,j,k}$ are nonzero order $O(1)$ k's for each fixed (i, j). Similarly, for $S = (A - B)U$,

$$S_{ijk} = S_{ijk} + \sum_{\bar{i},\bar{j},\bar{k}} C_{(i,j,k),(\bar{i},\bar{j},\bar{k})}\, U_{\bar{i}\bar{j}\hat{k}} \sin(\hat{k}\pi\ z_{\bar{k}})$$

for all $(i, j, k) \in X$ and $\hat{k} = 1, 2, \ldots, N$, where $C = A - B$ and $C_{(i,j,k),(\bar{i},\bar{j},\bar{k})}$ is nonzero with order $O(1)$ for each $(i, j, k) \in X$. Thus, in this way we require only $O(N^2)$ storage and N two dimensional elliptic solvers. In summary, we can reduce the computational cost and the storage requirement of these solutions considerably.

7.4.1 The irregular domain case

We are solving the Poisson equation on an irregular domain Ω which can be embedded into a larger, rectangular domain Π. The aim is to solve more efficiently an extended problem in Π. We use the notation $A_{11}u_1 = f_1$ for the system of linear equations obtained using the fourth-order method in Ω. Furthermore, we denote the matrix corresponding to the compact fourth-order finite difference scheme on the extended problem in Π by B. Here we use FFT to solve problems with B. Then, it is natural to use B as a preconditioner. The dimension of B is larger than the dimension of A_{11}. We introduce a compatible block presentation of B as

$$B = \begin{pmatrix} B_{11} & B_{12} \\ B_{21} & B_{22} \end{pmatrix}, \tag{7.115}$$

where the matrix block B_{11} corresponds to A_{11}. Now we extend A_{11} to A in (7.109). The most trivial way is by the zero extension, that is,

$$A = \begin{pmatrix} A_{11} & 0 \\ 0 & 0 \end{pmatrix}. \tag{7.116}$$

The matrix A is singular, but this does not cause any difficulties with our sparse subspace method. In general, the trivial extension by zeros does not lead to good conditioning for AB^{-1}. Thus, we use the following two forms of extensions:

$$A = \begin{pmatrix} A_{11} & B_{12} \\ 0 & A_{22} \end{pmatrix} \quad \text{and} \quad A = \begin{pmatrix} A_{11} & 0 \\ B_{21} & A_{22} \end{pmatrix}. \tag{7.117}$$

They are called the upper extension and the lower extension, respectively. Then, for the upper and lower extensions, the subspace X is given by the ranges of the matrices

$$\begin{pmatrix} A_{11} - B_{11} & 0 \\ -B_{21} & A_{22} - B_{22} \end{pmatrix} \quad \text{and} \quad \begin{pmatrix} A_{11} - B_{11} & -B_{12} \\ 0 & A_{22} - B_{22} \end{pmatrix}, \tag{7.118}$$

respectively. Here the original problem corresponding to A_{11} is a Dirichlet boundary value problem. We consider the extension with $A_{22} = B_{22} + D$. If $A_{11} = B_{11}$, then the choice $D = -B_{21}B_{11}^{-1}B_{12}$ would lead to perfect conditioning, but it is not computationally feasible. Choosing A_{22}, which corresponds to a Neumann boundary value problem, leads to good conditioning and thus small computational cost; e.g., see [26].

7.4.2 The interface case

We consider the solution of the equation

$$-\nabla \cdot \beta \nabla u = f \quad \text{in } \Pi \setminus \Gamma, \qquad [u] = \left[\beta \frac{\partial u}{\partial \boldsymbol{n}}\right] = 0 \quad \text{on } \Gamma, \qquad u = 0 \quad \text{on } \partial\Pi, \tag{7.119}$$

where $[\cdot]$ denotes the jump. The coefficient β is piecewise constant. Thus, scaling the equation by $1/\beta$ on each subdomain, we obtain the Poisson equation with the interface condition on Γ. We discretize the scaled equation with the second-order and fourth-order accurate IIMs. The preconditioner B is the Laplace equation discretized using the standard compact 9-point finite difference on $\Omega = \Pi$. Figure 7.5 shows the grid points associated with the subspace X. Table 7.1 gives the dimension n of the problem, the dimension m of the subspace X, and the number k of GMRES iterations to reduce the norm of the residual by the factor 10^{-6} for three different grids.

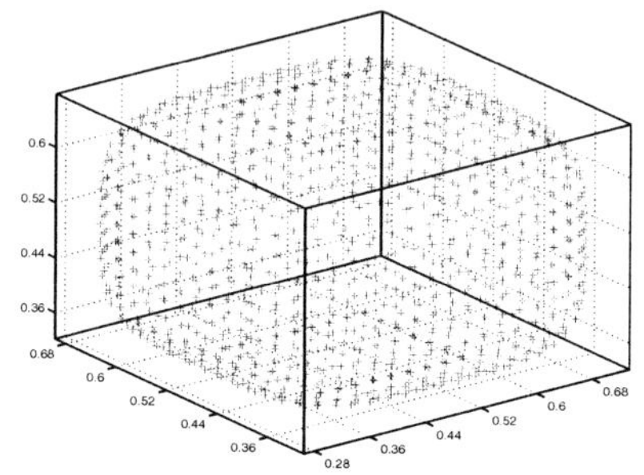

Figure 7.5. *The interface problem and the grid points associated with X.*

Table 7.1. *The dimensions and the number of GMRES iterations for three different grids.*

N	m	k
$39 \times 39 \times 39$	1226	25
$59 \times 59 \times 59$	2842	32
$79 \times 79 \times 79$	5186	41

7.5 Numerical experiments

In this section we present numerical results that demonstrate the feasibility and the applicability of the fourth-order method.

7.5.1 The irregular domain case

We have done a number of numerical tests which confirm the order of accuracy of our proposed fourth-order methods for different geometries and different boundary conditions. All our computations are done using the sparse matrix routines in MATLAB for two-dimensional domains. For the three-dimensional case we apply the iterative method described in §7.4. In our numerical tests, we have tested the method for several nonrectangular domains and present the selected results on domains with boundary geometries as shown in Figures 7.6(a) and 7.6(b). In Figure 7.6, the "+"s indicate the irregular grid points and the " · "s represent the projected points which collectively approximate the boundary geometry. We tested the schemes in §7.2.3 by using the following examples.

Example 7.5.1(a). In the first example, the exact solution is

$$u = \sin(x^2 + y^2) \quad \text{with} \quad f = -4\cos(x^2 + y^2) + 4(x^2 + y^2)\sin(x^2 + y^2). \tag{7.120}$$

Example 7.5.1(b). In the second example, we have

$$\begin{aligned} &u = \exp(-((x - 0.5)^2 + (y - 0.5)^2)) \\ &\text{with } f = (4 - (1 - 2x)^2 - (1 - 2y)^2)\exp(-((x - 0.5)^2 + (y - 0.5)^2)). \end{aligned} \tag{7.121}$$

The errors $e(h)$ are measured in the ℓ_2 and ℓ^∞ norms of the difference between the finite difference solutions and the exact solutions of the differential equation. The order of convergence r is given by

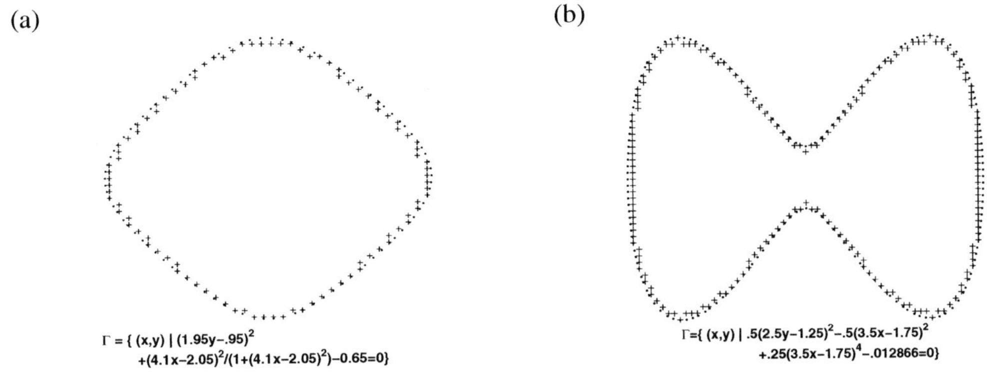

Figure 7.6. *Nonrectangular boundary geometries. (a) A diamond-shaped geometry. (b) A butterfly-shaped geometry.*

$$e(h) = Ch^r$$

for some C, where the step size or grid spacing is given as $h = 1/N$. Therefore, for a constant C, the order of accuracy r is calculated by

$$r = \frac{\log\left(e(h_1)/e(h_2)\right)}{\log\left(h_1/h_2\right)} \tag{7.122}$$

for any two errors due to grid spacings h_1 and h_2 [245, 159, 85].

Examples for Dirichlet boundary value problems

Among the test examples we conducted, the fourth-order scheme produced machine accuracy for the case when the solution u is a fourth-degree polynomial for all tested geometries. We present the results of the tests for the fourth-order scheme with Dirichlet boundary conditions in Table 7.2. The right-hand side of the table shows the results for the second option where we avoid the calculation of tangential derivatives at $\mathbf{x}^*$, while the left-hand side shows the results for the first option where g_n is computed analytically (see Remark 7.1). In the first option, the curvature at the projected points $\mathbf{x}^*$ is computed as described in item 5 of Remark 7.4 in §7.2.7. We use a least squares procedure [85, 159] to better determine the order of convergences of the methods as indicated in (7.122) since C may not be uniformly a constant but may depend on the closeness of the irregular grid points to the boundary (i.e., C depends on h).

Table 7.2. *A grid refinement analysis using* ℓ^∞ *for Example* 7.5.1.

Fig.		Example 7.5.1(a)	
	N	Error $\|e\|_{\ell^\infty}$	Order r
7.6(a)	16	1.013-4	
	32	6.924-6	4.13
	64	1.688-7	5.36

Fig.		Example 7.5.1(a)	
	N	Error $\|e\|_{\ell^\infty}$	Order r
7.6(b)	16	1.068-4	
	32	1.250-6	6.42
	64	3.576-8	5.13

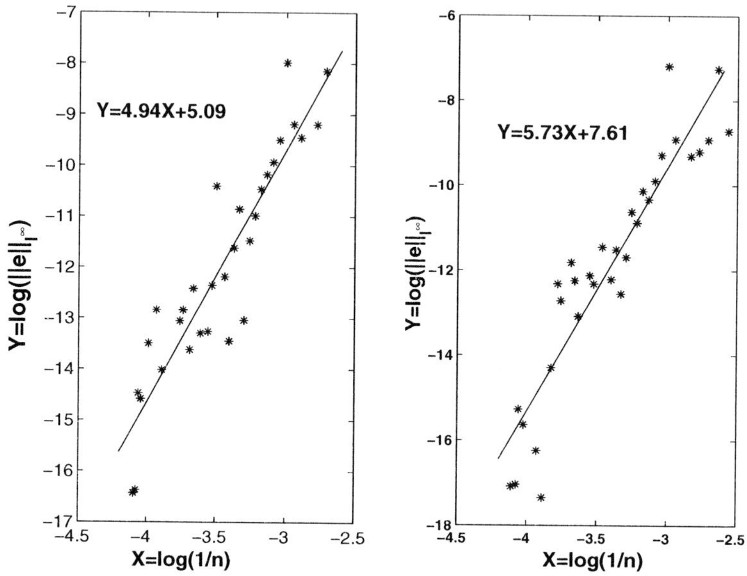

Figure 7.7. *Least squares estimates of the order of convergence for the first and second options shown in Table 7.2.*

In Figure 7.7, the left graph represents the order of convergence determination for the first option with Dirichlet boundary conditions, while the right graph is for the second option. For the first option results reported in the left plot of Figure 7.7, we plotted the errors for 31 different values of h with $N = 15, 16, 18\text{–}35, 37, 39, 40, 42, 43, 49, 51, 54, 57\text{–}60$. Then, we determined the error rate r from (7.122) by

$$\log(\|e\|_{\ell^\infty}) = 4.94\log(1/N) + 5.09.$$

For the second option results reported in the right plot of Figure 7.7, we plotted the errors for 32 different values of h with $N =$ 13–17,19–30, 32, 34, 35, 38–40, 43, 44, 46, 49, 51, 56, 58, 59, 61. The error rate is therefore determined by

$$\log(\|e\|_{\ell^\infty}) = 5.73\log(1/N) + 7.61.$$

Examples for Neumann boundary value problems

The results for the Neumann boundary conditions are shown in Table 7.3 for the fourth-order method and Table 7.4 for the second-order method.

In Figure 7.8, the left graph is for the fourth-order method while the right graph is for the second-order method with Neumann boundary conditions. In the least squares determination of the rate for the fourth-order method, we tried selecting values of h with $N =$ 14–20, 23, 24, 27, 29, 30, 32, 34, 36, 38–41, 43, 47, 49–52, 57, 60, 61, 63. As a result, we obtain the rate as

$$\log(\|e\|_{\ell^\infty}) = 4.07\log(1/N) + 4.74,$$

as shown in the left graph in Figure 7.8.

Table 7.3. *A grid refinement analysis for the fourth-order method with Neumann boundary conditions.*

Fig.		Example 7.5.1(b)	
	N	Error $\|e\|_{\ell_2}$	Order r
7.6(a)	16	4.355-2	
	32	2.832-3	3.94
	64	1.186-4	4.58

Fig.		Example 7.5.1(b)	
	N	Error $\|e\|_{\ell^\infty}$	Order r
7.6(b)	16	5.559-3	
	32	1.715-4	5.03
	64	4.799-6	5.16

Table 7.4. *A grid refinement analysis for the second-order method with Neumann boundary conditions.*

Fig.		Example 7.5.1(b)	
	N	Error $\|e\|_{\ell_2}$	Order r
7.6(a)	37	1.056-2	
	74	2.495-3	2.08
	148	4.529-4	2.45

Fig.		Example 7.5.1(b)	
	N	Error $\|e\|_{\ell^\infty}$	Order r
7.6(b)	37	3.639-2	
	74	8.41-3	2.11
	148	4.990-4	2.44

For the second-order method with Neumann boundary conditions, the different values of h have N=16, 17, 21, 24, 25, 29, 32, 34, 36, 38, 45, 46, 51, 62, 67, 71, 75, 81, 84, 92, 93, 97, 101, 136, 143, 148, 165, 178, 180, 186, 188, 199, 200. We therefore obtained the rate as

$$\log(\|e\|_{\ell^\infty}) = 2.02\log(1/N) + 2.49,$$

as shown in the right graph in Figure 7.8.

7.5.2 Examples for eigenvalues and eigenfunctions in a circular domain

Next, we validate the scheme in §7.2.5 for the heat equation on irregular domains by solving the associated eigenvalue problem (7.58) for the circular domain and compare the computed eigenvalues with exact eigenvalues. To this end, we use the scheme to solve the associated eigenvalue problem with a circular domain Ω of radius a centered at (0, 0) for homogeneous Dirichlet boundary conditions. Thus we have

$$\begin{aligned} -\Delta u &= \lambda u \quad \text{for} \quad r^2 = x^2 + y^2 < a^2, \\ u &= 0 \quad \text{for} \quad r^2 = x^2 + y^2 = a^2. \end{aligned} \tag{7.123}$$

Through a polar coordinate transformation, we obtain the eigenpair solution as

$$\left(\lambda_m, \quad J_m(\sqrt{\lambda} r) \begin{array}{c} \cos m\theta \\ \sin m\theta \end{array}\right), \qquad m = 0, 1, 2, \ldots,$$

where $J_m(\cdot)$ is the Bessel function of the first kind [218], with index m, and the eigenvalue λ_m satisfies

$$J_m(\sqrt{\lambda} a) = 0. \tag{7.124}$$

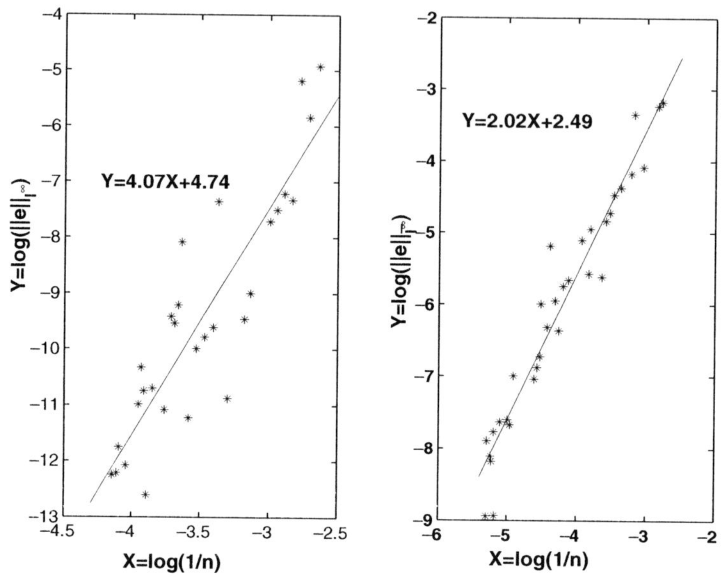

Figure 7.8. *Least squares estimates of orders shown in Tables* 7.4 *and* 7.3

Thus, with X_m^i representing the ith root of the Bessel function of index m, we have

$$\lambda_m^i = \left[\frac{X_m^i}{a}\right]^2, \quad \text{i=1,2,} \ldots . \tag{7.125}$$

In Table 7.5 we compare the estimates from the fourth-order method for different N's with the exact eigenvalues as calculated by (7.125). The "exact" zeros of the Bessel functions X_m are obtained from Maple® software. The last four columns in Table 7.5 show the difference between the estimates and the exact eigenvalues. The order of convergence of our proposed fourth-order method is confirmed to be 4 by employing (7.122). In the following tables, $nnz(\cdot)$ refers to the number of nonzero elements in the stiffness matrix h and the mass matrix q, respectively. $Iter$ refers to the number of iterations for the MATLAB function $eigs$ to converge with default tolerance.

In Table 7.6, we show the efficiency of the fourth-order method and illustrate the complexity level and the conditioning of the resulting system matrix for the fourth-order method. Specifically, we use the second-order central finite difference scheme for estimating the eigenvalues as the benchmark. The comparisons are made through the number of iterations and the CPU time required to solve the problem by employing the MATLAB function eig. The function $eigs$ is an iterative routine for finding a few eigenvalues and eigenvectors of a square matrix based on an implicitly restarted Arnoldi iteration algorithm [151, 152].

In Table 7.5, the first row indicates the size of the matrices which are selected to be almost equal for both methods. Table 7.5 shows that the complexity level for our fourth-order method is comparable to the one for the benchmark.

Next, in Table 7.7 we compare the fourth-order method with our second-order method and with the second-order finite element method through the MATLAB pde toolbox.The

Table 7.5. *Comparison of the estimates of the first 20 eigenvalues from our fourth-order method with the exact eigenvalues for a circular domain.*

λ_m^i	Exact eigenvalue	The numbers in () indicate grid points in Ω at that level $\lvert\lambda_{\text{exact}} - \lambda_{\text{estimate}}\rvert$			
	λ	$N = 21(52)$	$N = 41(216)$	$N = 81(840)$	$N = 121(1884)$
λ_0^1	140.71012075	0.00344560	0.00033383	0.00022202	0.00007944
λ_1^1	357.22556306	0.41984796	0.02957536	0.00063390	0.00004888
λ_1^1	357.22556306	0.41984796	0.02957536	0.00063390	0.00004888
λ_1^2	641.71816124	5.52882510	0.28363512	0.01667297	0.00040774
λ_1^2	641.71816124	1.80972635	0.14485952	0.02413351	0.00435833
λ_0^2	741.39324437	6.04472129	0.51548679	0.03500361	0.00839482
λ_3^1	990.42495908	7.32499973	0.14629899	0.01974972	0.01035190
λ_3^1	990.42495908	7.32499973	0.14629899	0.01974972	0.01035190
λ_1^2	1197.52935089	22.14275733	1.74116827	0.11284935	0.02544642
λ_1^2	1197.52935089	22.14275733	1.74116827	0.11284935	0.02544641
λ_4^1	1401.04479083	28.50252033	1.22523225	0.02972824	0.02677208
λ_4^1	1401.04479083	15.58313267	0.42389398	0.06932715	0.02740201
λ_2^2	1723.84425593	110.35200645	7.16848208	0.47841299	0.08676289
λ_2^2	1723.84425593	19.19016496	1.48474518	0.04257484	0.02086816
λ_0^3	1822.06829174	74.19996634	5.90117617	0.38859898	0.08757072
λ_5^1	1871.99339011	59.86580361	1.18701162	0.08585961	0.05461152
λ_5^1	1871.99339011	59.86580361	1.18701162	0.08585961	0.05461153
λ_3^2	2318.18911299	208.09760052	9.59792708	0.52775690	0.09968339
λ_3^2	2318.18911299	208.09760052	9.59792708	0.52775690	0.09968339
λ_6^1	2402.09908703	146.57829637	3.411546756	0.05850641	0.02598372

Table 7.6. *First* 10 *eigenvalues for square and circular domains.*

	λ of square domain by 2nd order central finite difference scheme			λ of circular domain by 4th order method		
	225^2	900^2	2025^2	216^2	840^2	1884^2
nnz(h)	1,065	4,380	9,945	1,788	7,244	16,480
nnz(q)	$\mathbf{I}_{225}$	$\mathbf{I}_{900}$	$\mathbf{I}_{2025}$	1,016	4,072	9,220
Iter	9	10	10	9	9	8
cputime	0.11	0.32	0.64	0.26	0.44	0.81

Table 7.7. *First* 10 *eigenvalues for a circular domain computed with eigs.*

	$\lvert\lambda_{\text{exact}} - \lambda_{\text{estimate}}\rvert$ at that level					
	MATLAB via *pde* tool		2nd order method		4th order method	
			N=124	N=250	N=29	N=40
	1985^2	8065^2	1993^2	8069^2	112×112	213×213
	0.090905	0.022700	0.079863	0.019737	0.001991	0.000492
	0.681314	0.170351	0.516020	0.127375	0.100726	0.020078
	0.654829	0.163679	0.515830	0.127375	0.100724	0.020078
	2.120571	0.530002	2.024477	0.500254	0.998014	0.216279
	1.886004	0.471609	1.302557	0.321602	0.177470	0.066793
	2.899047	0.723868	2.226507	0.549263	1.457512	0.399027
	4.894667	1.224993	3.961473	0.978608	1.400833	0.255192
	4.577428	1.144803	3.961224	0.978608	1.400758	0.255192
	7.788358	1.951316	5.801800	1.432623	5.413804	1.490994
	7.396741	1.852561	5.801164	1.432623	5.413771	1.490994
nnz(h)	13633	55937	9761	39941	892	1757
nnz(q)	1985	8065	$\mathbf{I}_{1993}$	$\mathbf{I}_{8069}$	512	997
Iter	6	7	8	10	7	9
cputime	0.58	2.91	0.51	2.71	0.17	0.25

comparison is done so that our fourth-order method yields similar levels of accuracy. The columns in Table 7.7 show the absolute error for the eigenvalue estimates. The comparison is described in terms of the CPU time and number of iterations *Iter* for the *eigs* iterative procedure. By that, the efficiency and effectiveness of the fourth-order method are demonstrated.

For the sake of completeness, the *pde* toolbox uses a similar method (MATLAB function *pdeeig* [280]) to compute the eigenvalues in a selected interval.

7.5.3 Results for the variable coefficient case

We consider the variable elliptic problem

$$-\nabla \cdot (\beta(x, y)\nabla u) = f \tag{7.126}$$

with smooth variable coefficients over the entire rectangular domain for the case when β is highly oscillating. In particular, we present the results for β given by

$$\beta(x, y) = 1 + 0.7\sin(20x + 30y), \quad \text{i.e.,} \quad 0.3 \le \beta \le 1.7.$$

Also, we are looking at a problem that has a highly oscillating source function f for β in (7.126). Combining the proposed method with the irregular domain treatment to deal with the general boundary value problem on the irregular domains Ω [117] is straightforward, and therefore we only present results for rectangular domains $[0, 1] \times [0, 1]$ in two dimensions.

We consider the following four exact solutions:

$$\text{(a)}\quad u(x,y) = \sin(2(x-0.5)^2 + (y-0.5)^2 - 0.4811), \tag{7.127}$$

$$\text{(b)}\quad u(x,y) = (1.95y-0.95)^2 + \frac{(4.1x-2.05)^2}{1+(4.1x-2.05)^2} - 0.65, \tag{7.128}$$

$$\text{(c)}\quad u(x,y) = (x-0.5)^4 + 2(y-0.5)^4 - 0.01, \tag{7.129}$$

$$\text{(d)}\quad u(x,y) = \exp((x-0.5)^2 + 2(y-0.5)^2) - 0.9, \tag{7.130}$$

with their corresponding Dirichlet boundary values on the square domain $\Omega = [0,1]\times[0,1]$. The source function f in each case is given by

$$f(x,y) = -(\beta u_{xx} + \beta u_{yy} + \beta_x u_x + \beta_y u_y).$$

The results of the experiments for examples (7.127)–(7.130) are presented in Table 7.8, where $\|e\|_{\ell^\infty}$ represents the absolute maximum error between the exact solution and the computed solution. The results in Table 7.8 are for the case when β_x and β_y are determined analytically. In Table 7.9, we show the results for the case when β_x and β_y are computed from β using the second-order central finite difference scheme in (7.70) at the (i,j)th grid point.

Next we investigate how our proposed method handles significant fluctuations in amplitude of the source functions in terms of the absolute errors. First, consider variations in β given by

$$\beta(x,y) = 1 + k\sin(20x + 30y), \tag{7.131}$$

Table 7.8. *A grid refinement analysis for the fourth-order* M-*matrix* 9-*point compact scheme for the elliptic equation with variable coefficients and a Dirichlet condition on the rectangular domain with analytic β_x and β_y.*

	Ex. 7.5.3(a)		Ex. 7.5.3(b)		Ex. 7.5.3(c)		Ex. 7.5.3(d)	
N	$\|e\|_{\ell^\infty}$	r	$\|e\|_{\ell^\infty}$	r	$\|e\|_{\ell^\infty}$	r	$\|e\|_{\ell^\infty}$	r
15	6.579-6		1.049-3		1.802-6		3.649-6	
30	2.968-7	4.09	4.986-5	4.419	1.033-7	4.12	5.278-8	6.11
60	1.748-8	4.47	2.514-6	4.061	2.082-9	5.63	2.082-9	4.66

Table 7.9. *A grid refinement analysis for the fourth-order* M-*matrix* 9-*point compact scheme for the elliptic equation with variable coefficients and a Dirichlet condition on the rectangular domain with second-order central finite difference approximations of β_x and β_y.*

	Ex. 7.5.3(a)		Ex. 7.5.3(b)		Ex. 7.5.3(c)		Ex. 7.5.3(d)	
N	$\|e\|_{\ell^\infty}$	r	$\|e\|_{\ell^\infty}$	r	$\|e\|_{\ell^\infty}$	r	$\|e\|_{\ell^\infty}$	r
15	5.599-6		8.413-4		3.223-6		1.360-6	
30	2.987-7	4.23	4.248-5	4.31	1.263-8	7.99	5.120-8	4.73
60	1.755-8	4.09	2.507-6	4.08	3.561-10	5.15	2.129-9	4.59

Table 7.10. *A grid refinement analysis for Example* 7.5.3(a) *with analytic* β_x *and* β_y.

	$k = 0.80$		$k = 0.85$		$k = 0.90$		$k = 0.95$	
N	$\|e\|_{l^\infty}$	r	$\|e\|_{l^\infty}$	r	$\|e\|_{l^\infty}$	r	$\|e\|_{l^\infty}$	r
15	8.345-6		1.485-5		1.763-4		1.269-4	
30	2.913-7	4.84	2.843-7	5.71	3.212-7	9.10	5.115-7	7.95
60	1.663-8	4.13	1.604-8	4.15	1.522-8	4.40	1.490-8	5.10

Table 7.11. *A grid refinement analysis for Example* 7.5.3(a) *with central finite difference approximations to* β_x *and* β_y.

	$k = 0.80$		$k = 0.85$		$k = 0.90$		$k = 0.95$	
N	$\|e\|_{l^\infty}$	r	$\|e\|_{l^\infty}$	r	$\|e\|_{l^\infty}$	r	$\|e\|_{l^\infty}$	r
15	6.033-6		6.534-6		7.907-6		1.211-5	
30	2.952-7	4.35	2.925-7	4.48	2.861-7	4.79	3.733-7	5.02
60	1.668-8	4.15	1.605-8	4.19	1.517-8	4.24	1.363-8	4.78

Table 7.12. *A grid refinement analysis for Example* 7.5.3(d) *with central finite difference approximations to* β_x *and* β_y.

	$a = 20, b = -10$		$a = -20, b = 25$		$a = 20, b = 15$		$a = 30, b = 20$	
N	$\|e\|_{l^\infty}$	r	$\|e\|_{l^\infty}$	r	$\|e\|_{l^\infty}$	r	$\|e\|_{l^\infty}$	r
25	2.520-3		6.691-3		2.497-3		1.365-2	
50	1.542-4	4.03	4.007-4	4.06	1.422-4	4.13	8.118-4	4.07
100	9.449-6	4.03	2.456-5	4.03	8.665-6	4.04	4.99-5	4.02

when β_x and β_y are determined analytically and k varies from 0.8 to 0.95 in Table 7.10. Table 7.11 shows the results when β_x and β_y are computed from β by a second-order central finite difference scheme.

Second, we consider fluctuations induced mainly by high frequencies in the solution in Example 7.5.3(d) and for $k = 0.8$ in (7.44) . Varying a and b, we present the results in Table 7.12 for the case when β_x and β_y are computed by the central finite difference scheme:

$$u(x, y) = \sin(a(x - 0.5)^2 + b(y - 0.5)^2 - 0.4811). \tag{7.132}$$

Next, we use a grid refinement analysis to confirm the order of our proposed method for three different source functions in Table 7.13; we calculate the orders r according to (see [85, 159, 160]),

$$\frac{\|u_{4N} - u_N\|_{l^\infty}}{\|u_{4N} - u_{2N}\|_{l^\infty}} = 2^r + 1. \tag{7.133}$$

Thus, we have

$$r = \log_2\left(\frac{\|u_{4N} - u_N\|_{l^\infty}}{\|u_{4N} - u_{2N}\|_{l^\infty}} - 1\right).$$

Table 7.13. *A grid refinement analysis.*

$\|u_N - u_{2N}\|_{l^\infty}$	$f = 1$	$f = \sin(30xy)$	$f = \exp(xy)$
$\|u_{100} - u_{25}\|_{l^\infty}$	6.107-3	4.468-3	6.482-3
$\|u_{100} - u_{50}\|_{l^\infty}$	2.927-4	2.027-4	3.109-4
Order r	4.31	4.39	4.31

7.5.4 Results for the interface problem

Now for the interface problem (7.1) with (7.71)–(7.72), we consider the exact solution given in the form

$$u(x, y) = \frac{\varphi(x, y)}{\beta} \quad \text{in } \Omega^{\pm}, \tag{7.134}$$

where β is a piecewise constant function given by (7.71) and $\varphi(x, y)$ is the level set function such that the interface is given by

$$\Gamma = \{(x, y) \mid \varphi(x, y) = 0\} \in \Omega.$$

The first example we present on the interface problem is a square domain with a butterfly-shaped interface geometry, illustrated in Figure 7.9(a). In all the interface geometries, the "$*$"s represent the boundary or the projected points and "$\cdot$"s and "$\circ$"s represent the irregular grid points on both sides of the interface according to the standard compact 9-point stencil scheme. The exact solution is given by (7.134) such that

$$\text{(a)} \qquad \begin{aligned} \varphi(x, y) &= 3.125(y - 0.5)^2 - 6.125(x - 0.5)^2 \\ &\quad + 0.25\,(3.5(x - 0.5))^4 - 0.1 + 0.087134. \end{aligned}$$

With the step size $h = 1/N$ and running the method for $N = 33$–38, 42, 43, 46, 45, 48, 49, 50, 52, 54, 55, 57, 59, 60, 65, 67, 69, we determine the error rate according to

$$\log |e(h)|_{l^\infty} = r \log h + \log C$$

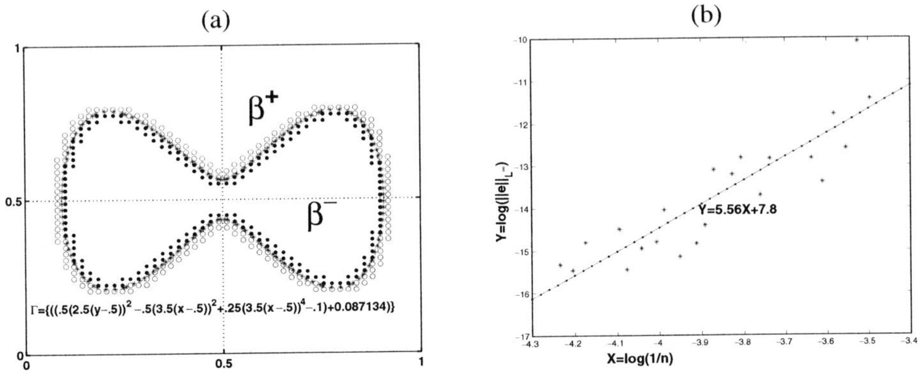

Figure 7.9. *A butterfly-shaped geometry and a grid refinement analysis. (a) A butterfly-shaped interface. (b) The plot of rate estimate using least squares fitting.*

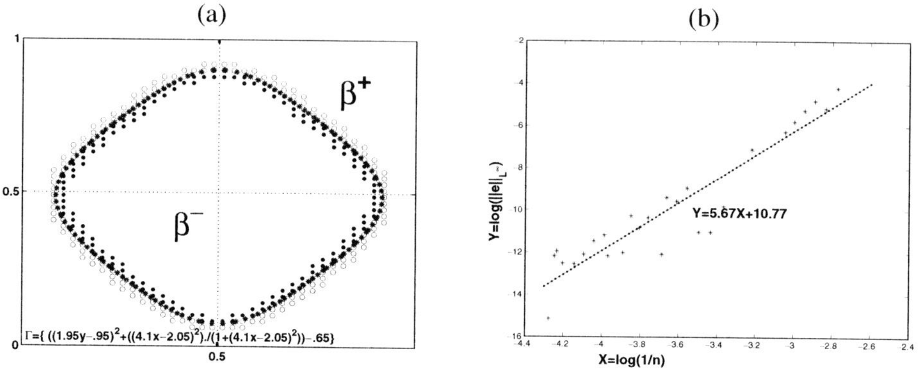

Figure 7.10. *A diamond-shaped interface geometry and an error analysis. (a) A diamond-shaped interface. (b) A grid refinement analysis using the least squares fitting.*

by

$$\log |e(h)|_{l^\infty} = 5.56 \log (1/N) + 7.8$$

as in Figure 7.9(b) when $|\beta^+ - \beta^-| = 99$.

In the second example, we consider an exact solution where φ is given by

$$\text{(b)} \quad \varphi(x, y) = (1.95y - 0.95)^2 + \frac{(4.1x - 2.05)^2}{1 + (4.1x - 2.05)^2} - 0.65,$$

and repeating the experiment as before for N =16–21, 25, 29, 31, 33, 35, 37, 39, 40, 43, 45, 47, 49, 53, 54, 57, 60, 63, 67, 69, 70, 72, the least squares estimate of the rate is determined by

$$\log |e(h)|_{l^\infty} = 5.67 \log (1/N) + 10.77$$

as in Figure 7.10.

In our next experiment, we consider the exact solution such that

$$\text{(c)} \qquad \varphi(x, y) = (x - 0.5)^4 + 2(y - 0.5)^4 - 0.01.$$

Running the experiment with $|\beta^+ - \beta^-| = 99$ for N =18, 20, 21, 25, 24, 29, 31, 33, 35, 37, 39, 41, 44, 46, 49, 54, 58, 61, 65, 66, 68, we determine the rate by

$$\log |e(h)|_{l^\infty} = 4.17 \log (1/N) - 0.4$$

as indicated in Figure 7.11(b).

Consider one more example where φ is given by

$$\text{(d)} \qquad \varphi(x, y) = \sin(2(x - 0.5)^4 + (y - 0.5)^2 - 0.03),$$

whose interface is as shown in Figure 7.12(a). Running the experiment for N =20–22, 25, 28, 31,35, 36, 40–42, 45, 46, 49–51, 55, 56, 60, 61, 65, 66, 70–72, we determine the accuracy rate by

$$\log |e(h)|_{l^\infty} = 5.06 \log (1/N) + 3.23$$

as shown in Figure 7.12(b) with $|\beta^+ - \beta^-| = 99$.

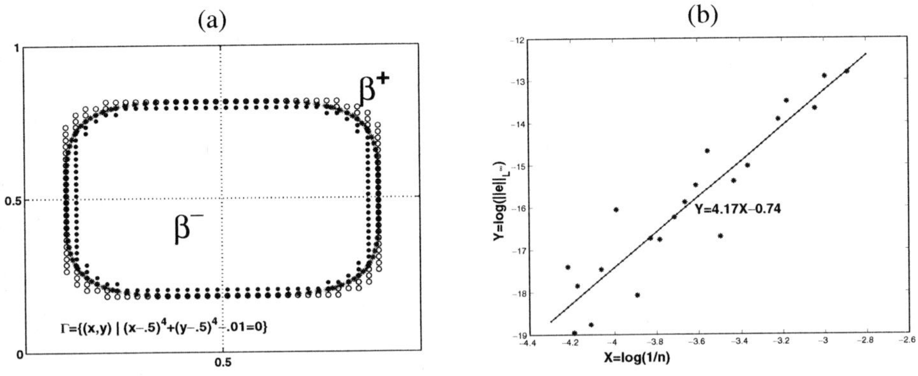

Figure 7.11. *A square-shaped interface geometry. (a) A square-shaped interface. (b) A grid refinement analysis using the least squares fitting.*

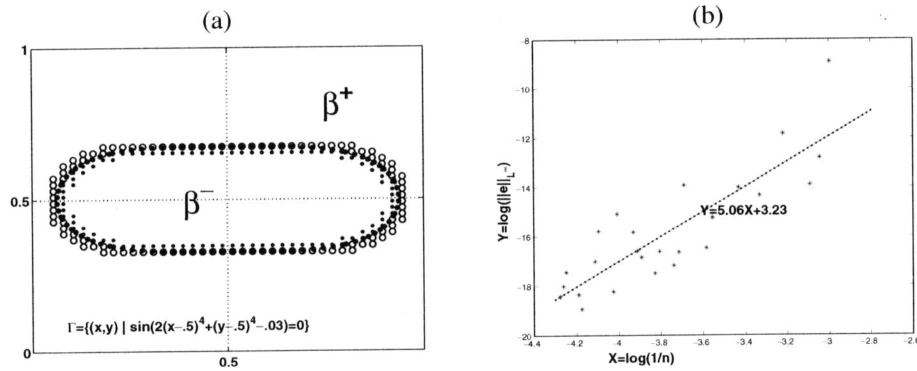

Figure 7.12. *An oval-shaped interface geometry. (a) A domain with an oval-shaped interface. (b) A grid refinement analysis using the least squares fitting.*

Next, we use a grid refinement analysis to estimate the order of the method according to (7.133); see also [85, 159, 160]. Using the source functions $f = 1$, $f = \sin 30xy$, and $f = \exp(xy)$ with $|\beta^+ - \beta^-| = 9$, the order of the method (r) for the interface problem is computed for 23 different values of h with N = 20–26, 28–33, 35–39, 44–48. The averages r for the source functions are, respectively, determined as $r = 4.02$, $r = 3.98$, and $r = 4.13$ as illustrated in Figure 7.13.

7.5.5 An eigenvalue problem with an interface

We tested the method for heat equations with interface in §7.2.8 by solving the corresponding eigenvalue problem

$$-\nabla \cdot (\beta(x)\, \nabla u) = \lambda\, u \tag{7.135}$$

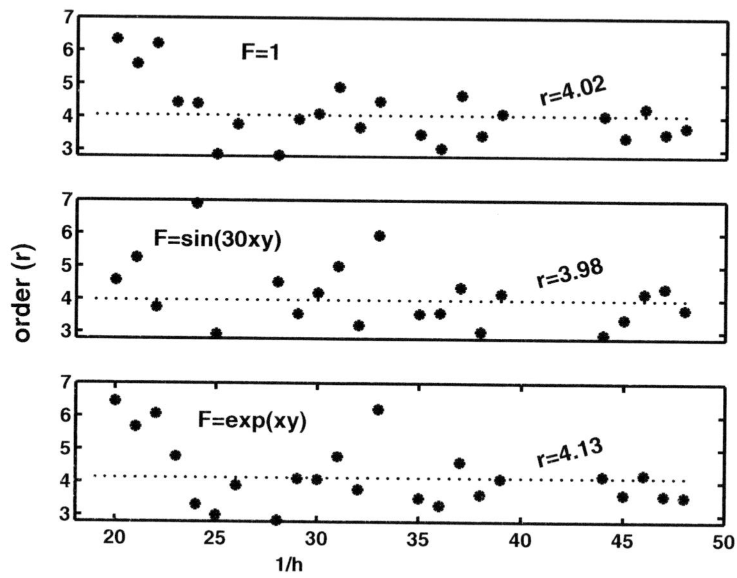

Figure 7.13. *Rate determination using grid refinement analysis for three source functions.*

with $u = 0$ on $\partial\Omega$ for $\Omega = (0, 1) \times (0, 1)$, where β is piecewise constant. Computation of eigenvalue problems provides an important test for the validity of the method. Our method leads to the approximating eigenvalue problem

$$HU = \lambda QU, \tag{7.136}$$

where H and Q are the stiffness and mass matrices that are formed by the proposed fourth-order scheme. That is, at each irregular point the entries γ of H and q of Q are determined by (7.78) and (7.93), respectively. We report the results for (7.135) with $\beta^- = 1, \beta^+ = 10$, and the interface Γ is defined by

$$\varphi = 2(x - 0.5)^2 + (y - 0.5)^2 - 0.0411.$$

We computed the first 10 eigenvalues of (7.95) for $N = 30, 60, 90$, using the MATLAB routine *eigs*. The function *eigs* is an iterative routine for finding a few eigenvalues and eigenvectors of a square matrix based on an implicitly restarted Arnoldi iteration algorithm [152]. It converged in six iterations with default tolerance for all N.

Table 7.14 shows that

$$\text{Relative error} = \frac{|\lambda_{90} - \lambda_{60}|}{\lambda_{90}}, \quad \text{Rate} = \frac{\lambda_{90} - \lambda_{60}}{\lambda_{60} - \lambda_{30}}$$

for each eigenvalue. If the method is fourth-order accurate, then the corresponding rate is estimated as 5.35E-2.

Table 7.14. *A grid refinement analysis for the eigenvalue problem with an interface.*

Eigenvalues	Relative error	Rate
1st	2.359E-5	3.541E-2
2nd	5.551E-6	8.627E-2
3rd	1.714E-5	6.193E-2
4th	7.460E-6	7.663E-2
5th	2.040E-5	4.265E-2
6th	1.294E-5	3.339E-2
7th	3.687E-6	8.253E-2
8th	1.608E-5	4.897E-2
9th	7.751E-6	2.209E-2
10th	1.452E-6	3.874E-2

7.6 The well-posedness and the convergence rate

As we stated in §7.2.3, the resulting matrices $-L_h$ are symmetric for our proposed methods for Poisson's equation on irregular domains. The positivity and diagonal dominance of the matrices are confirmed through our extensive numerical tests.

We generated the random geometry by selecting randomly distributed projection points $\mathbf{x}^*$ in the local square depicted in Figure 7.3 and then generated the random graphs for the boundary Γ with the local coordinate determined as

$$\xi = \chi(\eta) = \frac{\kappa}{2}\eta^2 + \frac{\kappa_1}{6}\eta^3 + \frac{\kappa_2}{24}\eta^4,$$

where $\kappa = \chi''(0)$, $\kappa_1 = \chi'''(0)$, and $\kappa_2 = \chi''''(0)$, were selected randomly in the range $[-5, 5]$. Then, we evaluated the corresponding diagonal entry $c_{i,j}$. All the tests conducted in this manner show that the resulting matrix is diagonally dominant.

Next, we examine this analytically for the two-dimensional Dirichlet boundary condition case for the second-order scheme. To determine the diagonal entry $c_{i,j}$, we let $f = g = 0$. As in (7.56), for the second-order scheme, the corresponding method to (7.45), (7.47)–(7.48)is given by

$$u \sim u + u_\xi \xi + u_\tau \tau + \frac{1}{2}u_{\xi\xi}\xi^2 + u_{\xi\eta}\xi\eta + +\frac{1}{2}u_{\eta\eta}\eta^2$$

and

$$\begin{pmatrix} \xi_0 & \xi_0\eta_0 & 0.5(\eta_0^2 - \xi_0^2) \\ \xi_1 & \xi_1\eta_1 & 0.5(\eta_1^2 - \xi_1)^2 \\ \xi_2 & \xi_2\eta_1 & 0.5(\eta_2^2 - \xi_2)^2 \end{pmatrix} \begin{pmatrix} u_\xi \\ u_{\xi\eta} \\ u_{\eta\eta} \end{pmatrix} = \begin{pmatrix} U_0 \\ 0 \\ 0 \end{pmatrix}, \tag{7.137}$$

where we used $u_\eta = 0$. We have $\eta_0 = 0$ and the asymptotic relations

$$\eta_1 + \eta_2 = 0 + O(h^2), \quad \xi_i = \frac{\kappa}{2}\eta_i^2 + O(h^3), \quad i = 1, 2. \tag{7.138}$$

By (7.138), adding the last two equations of (7.137), we get

$$\xi_1 \eta_1 u_{\xi\eta} = O(h^3),$$

and thus $u_{\xi\eta} = O(1)$ and (7.137) reduces to

$$\begin{pmatrix} \xi_0 & -0.5\xi_0^2 \\ \xi_1 & 0.5\eta_1^2 \end{pmatrix} \begin{pmatrix} u_\xi \\ u_{\eta\eta} \end{pmatrix} = \begin{pmatrix} U_0 \\ 0 \end{pmatrix} + \begin{pmatrix} 0 \\ O(h^3) \end{pmatrix}.$$

Thus

$$u_\xi = \frac{U_0}{\xi_0} + O(h^2), \quad u_{\eta\eta} = -\frac{2\xi_1}{\eta_1^2}\frac{U_0}{\xi_0} + O(h^2).$$

For the case of Figure 7.3,

$$c_{i,j} = 4 - \frac{\xi_3 + \xi_4}{\xi_0} + \left(\frac{\eta_3^2 - \xi_3^2}{\eta_1^2} + \frac{\eta_4^2 - \xi_4^2}{\eta_1^2} \right) \frac{\xi_1}{\xi_0} + O(h^2).$$

Since $\frac{\xi_3+\xi_4}{\xi_0} < 0$, it follows from (8.1) that

$$c_{i,j} \sim 4 - \frac{\xi_3 + \xi_4}{\xi_0} > 4.$$

7.6.1 Convergence rate

Since the central finite difference scheme and the fourth-order compact finite difference scheme (2.1) are diagonally dominant and the set P of all regular points are connected by their neighbors, thus the discrete maximum principle holds: if

$$(-L_h U)(p) \geq 0 \quad \forall p \in P,$$

then

$$\max_{p \in P} U_p \leq \max U_q, \quad q \in Q,$$

where Q is the set of all irregular grid points.

Consider the comparison function Φ,

$$\Phi(p) = \frac{1}{4} |x_p - x_0|^2.$$

Then

$$(-L_h \Phi)(p) = 1 \quad \forall p \in P. \tag{7.139}$$

Define the error e by $e_p = \bar{U}_p - U_p$, with $\bar{U}_p = u(x_p)$, and the exact solution evaluated at the node x_p and the truncation error T by

$$T = (-L_h \bar{U}) - F \quad \text{since} \quad -L_h U = F.$$

Let $\bar{T} = \max_{p \in P} |T_p|$. Then

$$(-L_h(\bar{T}\,\Phi + e))(p) = \bar{T} - T_p \geq 0 \quad \forall p \in P.$$

It thus follows from the maximum principle and (7.139) that

$$\max_{p\in P} e_p \le \max_{p\in P} (\bar{T}\,\Phi + e_p) \le \max_{q\in Q} (\bar{T}\,\Phi_q + e_q). \tag{7.140}$$

At an irregular grid point $q \in Q$, we have

$$c_{qq}\, e_q \le \sum_{p\in P} c_{qp}\, e_p + \sum_{q'\in Q} c_{qq'}\, e_{q'} + h^2\, |T_q|, \quad q \in Q.$$

If the maximum of e_q over Q is attained at q_0, we have

$$c_{q_0q_0}\, e_{q_0} \le \left(\sum_{p\in P} c_{q_0 p}\right) \max_{p\in P} e_p + \left(\sum_{q'\in Q} c_{q_0q'}\right) e_{q_0} + h^2 \max_{q\in Q} |T_q|.$$

Thus,

$$\frac{\sum_{p\in P} c_{q_0 p}}{c_{q_0q_0} - \sum_{q'\neq q_0} c_{q_0q'}} \le \rho < 1 \tag{7.141}$$

for all $q_0 \in Q$. For example, for the second-order scheme in the two-dimensional domain we have the two cases,

$$c_q\, e_q \le \max_{p\in P} e_p + 2 \max_{q\in Q} e_q + h^2\, |T_q|, \quad q \in Q_1,$$

and

$$c_q\, e_q \le 2 \max_{p\in P} e_p + h^2\, |T_q|, \quad q \in Q_2.$$

Thus,

$$\rho = \max\left(\frac{1}{c_q - 2}, \frac{2}{c_q - 1}\right) < 1 \quad (c_q > 3).$$

Then, it follows from (7.140)–(7.141) that

$$(1 - \rho) \max_{q\in Q} e_q \le \bar{T} \max_{q\in Q} \Phi_q + h^2 \max_{q\in Q} |T_q|.$$

If we apply exactly the same arguments for $\bar{T}\,\Phi - e_p$, we obtain the same bound for $-e_p$. Hence we conclude the error estimates

$$(1 - \rho) \max_{p\in P} |e_p| \le \bar{T} \max_{q\in Q} \Phi_q + h^2 \max_{q\in Q} |T_q|$$

and

$$\max_{q\in Q} |e_q| \le \rho \max_{p\in P} |e_p| + h^2 \max_{q\in Q} |T_q|.$$

For example, in the case of the second-order scheme, $|T_p| \le C_1\, h^2,\ p \in P$, and $|T_q| \le C_2\, h,\ q \in Q$, assuming $u \in C^4(\bar{\Omega})$. The proposed second-order method is of order 2 provided the diagonally dominant condition $\rho < 1$.

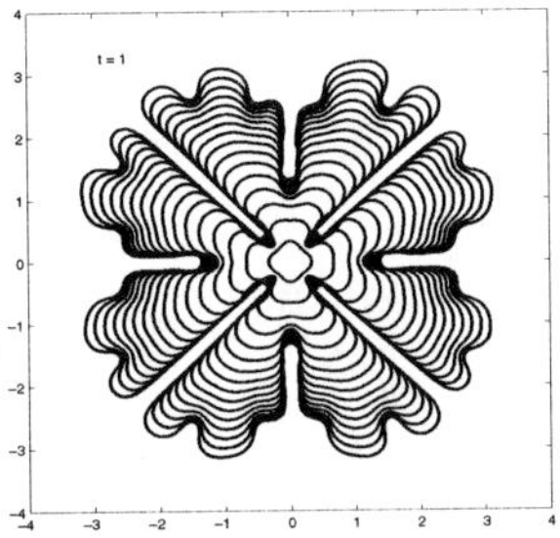

Chapter 8

The Immersed Finite Element Methods

In previous chapters, we have described the immersed interface method (IIM) using a finite difference discretization. Sometimes numerical methods using a finite element formulation may be preferred for various reasons such as theoretical analysis, personal background and preferences, available resources and linear solvers, etc. If a finite element method is applied to a self-adjoint elliptic PDE, the resulting linear system of equations is symmetric positive or negative definite. The Sobolev space theory provides strong theoretical foundations for convergence analysis for finite element methods.

It is well known that a second-order accurate approximation to the solution of an interface problem can be generated by the Galerkin finite element method with the standard linear basis functions if the triangulation is aligned with the interface, that is, a body-fitted mesh is used; see, for example, [11, 30, 39, 44, 100, 273]. Applications of such methods can be found in [43, 45, 204, 272] and many others. Mortar finite element methods for elliptic interface problems are discussed in [111]. Finite volume methods for interface Maxwell system have been developed in [50, 51]. An alternative to body-fitted meshes is locally fitted mesh techniques in which the mesh is almost Cartesian except near the interface, where the mesh is modified; see [21, 25, 26].

However, it may be difficult and time consuming to generate a body-fitted mesh for an interface problem in which the interface separates the solution domain into pieces or problems with complicated geometries. Such difficulty may become even more severe for moving interface problems, because a new grid has to be generated at each time step, or every other time step. A number of efficient software packages and methods based on Cartesian grids such as the FFT, the level set method, and others may not be applied directly to a body-fitted mesh. Few publications can be found in the literature about using body-fitted meshes for moving interface problems that have topological changes such as merging and splitting.

In this chapter, we present the immersed finite element methods (IFEMs) for elliptic interface problems based on Cartesian triangulations in one and two space dimensions. The purpose is to combine the advantages of the simple structure of Cartesian grids with a finite element formulation to develop accurate and stable numerical methods for interface problems. The triangulations in these methods do not need to fit the interfaces. The basis functions in these methods are constructed to satisfy the jump conditions either exactly or approximately. For two-dimensional problems, both nonconforming and conforming finite element spaces are considered.

The IIM using a finite element formulation was proposed and analyzed for one-dimensional interface problems in [164]. The nonconforming and conforming IFEMs for two-dimensional problems were proposed and analyzed in [170]. The nonconforming finite volume method for interface problems was proposed and analyzed in [75]. Additional error estimates and applications of the nonconforming IFEM can be found in [169]. The IFEM for a three-dimensional problem with an application is discussed in [133]. A quadratic immersed interface finite element method has been developed and analyzed in [37] based on rectangular partitions.

8.1 The IFEM for one-dimensional interface problems

The IFEM for one-dimensional interface problems was proposed in [164]. To explain the idea, we will use conventional notations of finite element methods and consider the model problem,

$$-(\beta(x)\, u')' + \sigma(x)u(x) = f(x), \qquad 0 \le x \le 1, \tag{8.1a}$$

$$u(0) = 0, \qquad u(1) = 0. \tag{8.1b}$$

We assume that $\beta(x) \ge \beta_0 > 0$ is a piecewise continuous function that may have a finite jump at an interface α. The solution of (8.1a) is typically nonsmooth at the interface if $\beta(x)$ has a finite jump there.

If the finite element method with the standard linear basis is used for (8.1) with the presence of interfaces, a second-order accurate solution can still be obtained if the interfaces lie on the grid points. This can be proved strictly in one-dimensional space. If any of the interfaces is not a grid point, then the solution obtained from the finite element method is only first-order accurate in the infinity norm; see Figure 8.1 for an illustration.

8.1.1 New basis functions satisfying the jump conditions

Define the standard bilinear form,

$$\begin{gathered} a(u,v) = \int_0^1 \Big(\beta(x)u'(x)v'(x) + \sigma(x)u(x)v(x)\Big)\, dx, \\ u(x),\, v(x) \in H_0^1(0,1), \end{gathered} \tag{8.2}$$

where $H_0^1(0,1)$ is the Sobolev space. The solution of the differential equation, $u(x) \in H_0^1(0,1)$, is also the solution of the following variational problem:

$$a(u,v) = (f,v) = \int_0^1 f(x)v(x)dx \quad \forall v \in H_0^1(0,1). \tag{8.3}$$

Without loss of generality, we assume that there is only one interface α in the interval $(0,1)$ and $f(x)$ is piecewise continuous and bounded. Assuming that $f(x) \in L^2$, we see that

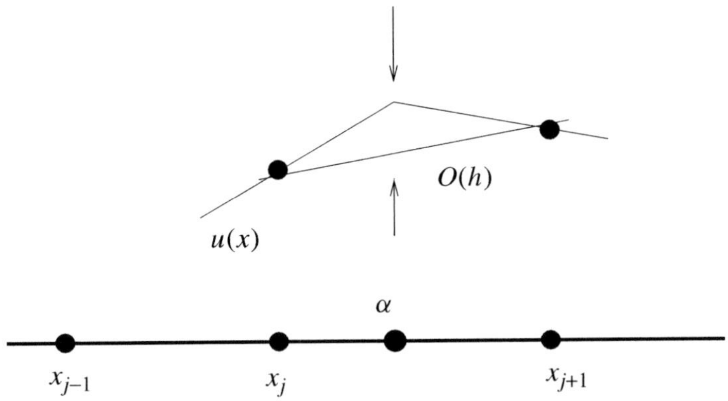

Figure 8.1. *A diagram showing why a finite element method cannot be second-order accurate in the infinity norm if the interface is not a grid point.*

$u(x) \in H_0^1$. Integration by parts over the separated intervals $(0, \alpha)$ and $(\alpha, 1)$ yields

$$0 = \int_0^{\alpha} \{-(\beta u')' + \sigma u - f\}v + \beta^- u_x^- v^-$$
$$+ \int_{\alpha}^{1} \{-(\beta u')' + \sigma u - f\}v - \beta^+ u_x^+ v^+$$

for a testing function $v(x) \in C^{\infty}$, $v(0) = v(1) = 0$. Since $v(x) \in H_0^1$ is arbitrary, it follows that the differential equation holds in each interval and that

$$[u] = u^+ - u^- = 0, \qquad [\beta u_x] = \beta^+ u_x^+ - \beta^- u_x^- = 0. \tag{8.4}$$

This gives another way to derive the flux jump condition. These relations indicate that the discontinuity in the coefficient $\beta(x)$ does not cause any difficulty for the theoretical analysis of the finite element method and the weak solution will satisfy the jump conditions (8.4). If the solution itself has a jump, then we can use the technique described in Chapter 5 to transform the problem into the form of (8.1).

Now we discuss how to construct the basis function. For simplicity, we use a uniform grid $x_i = ih$, $i = 0, 1, \dots, M$, with $x_0 = 0$, $x_M = 1$, and $h = 1/M$. The standard linear basis function satisfies

$$\phi_i(x_k) = \begin{cases} 1 & \text{if } i = k, \\ 0 & \text{otherwise.} \end{cases} \tag{8.5}$$

The solution $u_h(x)$ is a specific linear combination of the basis function from the finite-dimensional space V_h,

$$V_h = \left\{ v_h : \ v_h = \sum_{i=1}^{M-1} \eta_i \phi_i(x) \right\}, \tag{8.6}$$

and it satisfies

$$a(u_h, v_h) = (f, v_h) \quad \forall v_h \in V_h. \tag{8.7}$$

If an interface is not one of the grid points x_i, usually the solution u_h is only a first-order approximation to the exact solution in the infinite norm; see Figure 8.1. The problem is that some basis functions which have nonzero support near the interface do not satisfy the natural jump condition (8.4).

The idea of the IFEM is to modify the basis functions in such a way that natural jump conditions are satisfied, that is,

$$\phi_i(x_k) = \begin{cases} 1 & \text{if } i = k, \\ 0 & \text{otherwise,} \end{cases} \tag{8.8a}$$

$$[\phi_i] = 0, \qquad [\beta\phi_i'] = 0. \tag{8.8b}$$

Obviously, if $x_j < \alpha < x_{j+1}$, then only ϕ_j and ϕ_{j+1} need to be changed to satisfy the second jump condition. Using an undetermined coefficient method, we can conclude that

$$\phi_j(x) = \begin{cases} 0, & 0 \le x < x_{j-1}, \\ \dfrac{x - x_{j-1}}{h}, & x_{j-1} \le x < x_j, \\ \dfrac{x_j - x}{D} + 1, & x_j \le x < \alpha, \\ \dfrac{\rho\,(x_{j+1} - x)}{D}, & \alpha \le x < x_{j+1}, \\ 0, & x_{j+1} \le x \le 1, \end{cases} \tag{8.9}$$

where

$$\rho = \frac{\beta^-}{\beta^+}, \qquad D = h - \frac{\beta^+ - \beta^-}{\beta^+}(x_{j+1} - \alpha), \tag{8.10}$$

and

$$\phi_{j+1}(x) = \begin{cases} 0, & 0 \le x < x_j, \\ \dfrac{x - x_j}{D}, & x_j \le x < \alpha, \\ \dfrac{\rho\,(x - x_{j+1})}{D} + 1, & \alpha \le x < x_{j+1}, \\ \dfrac{x_{j+2} - x}{h}, & x_{j+1} \le x \le x_{j+2}, \\ 0, & x_{j+2} \le x \le 1. \end{cases} \tag{8.11}$$

Figure 8.2 shows several plots of the modified basis functions $\phi_j(x)$ and $\phi_{j+1}(x)$ and some neighboring basis functions that are the standard hat functions. At the interface, we can see clearly the kink in the basis functions that reflect the natural jump conditions. Since the basis functions are built on a uniform Cartesian mesh with the natural jump conditions, we call the spanned finite element space an *immersed finite element (IFE) space*. The corresponding Galerkin finite element method is the IFEM.

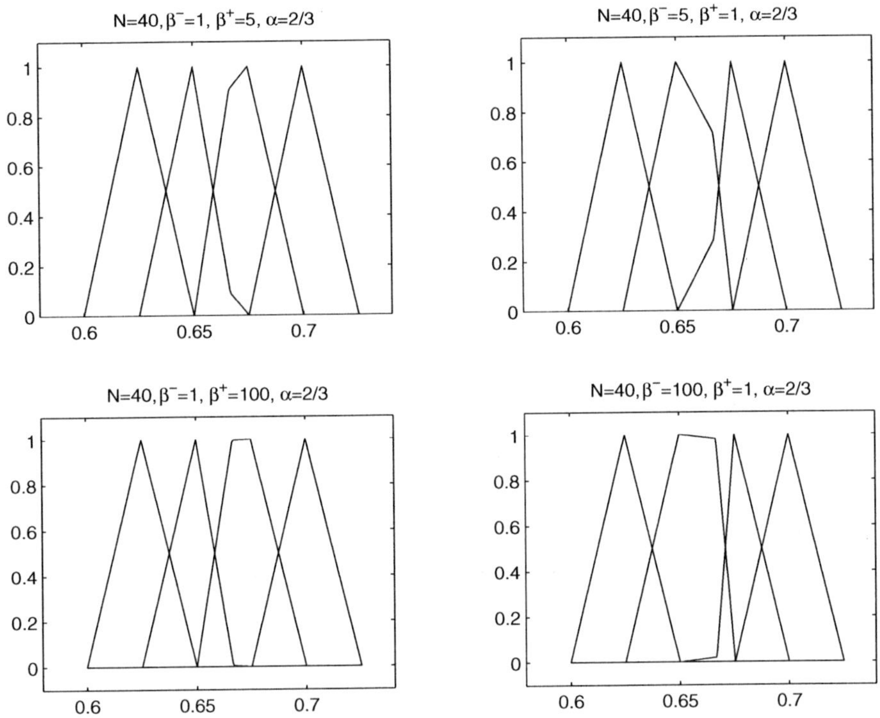

Figure 8.2. *Plots of some basis function near the interface with different β^- and β^+. The interface is $\alpha = 2/3$. Top left: $\beta^- = 1$, $\beta^+ = 5$. Top right: $\beta^- = 5$, $\beta^+ = 1$. Bottom left: $\beta^- = 1$, $\beta^+ = 100$. Bottom right: $\beta^- = 100$, $\beta^+ = 1$.*

8.1.2 The interpolation functions in the one-dimensional IFEM space

In the following subsections, we discuss the error analysis for the IFEM for a one-dimensional problem. For the sake of a clean and concise proof, we derive the theoretical analysis for the simple model,

$$
\begin{aligned}
& -\beta(x)\, u'' = f(x), \qquad f(x) \in C[0,1], \qquad 0 < x < 1, \\
& u(0) = 0, \qquad u(1) = 0, \qquad \text{with} \\
& \beta(x) = \begin{cases} \beta^- & \text{if } 0 \le x < \alpha, \\ \beta^+ & \text{if } \alpha < x \le 1, \end{cases}
\end{aligned}
\tag{8.12}
$$

where β^- and β^+ are two constants. The solution $u(x) \in H_0^1$ satisfies the natural jump conditions at α. If the value of the solution at α is known, say u_α, then the problem is equivalent to the following two separated problems:

$$\begin{cases} -\beta^- u'' = f(x), & 0 < x < \alpha, \\ u(0) = 0, & u(\alpha) = u_\alpha, \end{cases} \quad \text{and} \quad \begin{cases} -\beta^+ u'' = f(x), & \alpha < x < 1, \\ u(\alpha) = u_\alpha, & u(1) = 0. \end{cases}$$

Therefore, from the regularity theory we know that $u(x) \in C^2((0,\alpha) \cup (\alpha, 1))$ in each subdomain and u_{xx}^- and u_{xx}^+ are finite. We define

$$||u''||_\infty = \max\left\{ \sup_{0<x<\alpha} |u_{xx}|, \sup_{\alpha<x<1} |u_{xx}| \right\}, \tag{8.13}$$

which is bounded.

As in the standard finite element method analysis, an interpolation function of the solution plays an important role in the error analysis. In this subsection, we define a piecewise linear function in the space V_h which also interpolates $u(x)$ at the node points. Assuming that $x_j \le \alpha < x_{j+1}$, we define an interpolant of $u(x)$ as follows:

$$u_I(x) = \begin{cases} \dfrac{x_{i+1} - x}{h} u(x_i) + \dfrac{x - x_i}{h} u(x_{i+1}), & i \ne j, \ x_i \le x \le x_{i+1}, \\ u(x_j) + \kappa (x - u(x_j)), & x_j \le x < \alpha, \\ u(x_{j+1}) + \kappa\, \rho\, (x - x_{j+1}), & \alpha \le x \le x_{j+1}, \end{cases} \tag{8.14}$$

where

$$\rho = \frac{\beta^-}{\beta^+}, \qquad \kappa = \frac{u(x_{j+1}) - u(x_j)}{\alpha - x_j - \rho(\alpha - x_{j+1})}. \tag{8.15}$$

It is easy to verify that

$$\begin{aligned} &u_I(x_i) = u(x_i), \qquad i = 0, 1, \ldots, M-1, \\ &[u_I] = 0, \qquad [\beta u_I'] = 0, \end{aligned} \tag{8.16}$$

and hence $u_I(x) \in V_h$. Before giving an error bound for $||u_I(x) - u(x)||_\infty$, we need the following lemma which gives the error estimates for the first derivative of $u_I(x)$ approximating $u'(x)$.

Lemma 8.1. *Given the boundary value problem* (8.12), *assume that* $f(x)$ *in* (8.12) *is continuous in* $(0, \alpha)$ *and* $(\alpha, 1)$ *and bounded. Let* $u(x)$ *be the solution of* (8.12). *Given* $u_I(x)$ *as defined in* (8.14), *the following inequalities hold:*

$$\alpha - x_j - \rho(\alpha - x_{j+1}) \ge \min\left\{ \frac{h}{2}, \frac{h}{2}\rho \right\}, \tag{8.17a}$$

$$|\kappa - u_x^-| \le C\, ||u''||_\infty\, h, \tag{8.17b}$$

$$|\rho\,\kappa - u_x^+| \le C\, ||u''||_\infty\, h, \tag{8.17c}$$

where

$$C = \frac{2\max\{1, \rho\}}{\min\{1, \rho\}}. \tag{8.18}$$

Proof: It is obvious that

$$\begin{cases} \alpha - x_j - \rho(\alpha - x_{j+1}) \geq \rho(x_{j+1} - \alpha) \geq \dfrac{h}{2}\rho & \text{if } \alpha - x_j \leq \dfrac{h}{2}, \\ \alpha - x_j - \rho(\alpha - x_{j+1}) \geq \alpha - x_j \geq \dfrac{h}{2} & \text{if } \alpha - x_j > \dfrac{h}{2}, \end{cases}$$

which concludes the first inequality. Using the Taylor expansion about α, we have

$$\begin{aligned} |\kappa - u_x^-| &= \left| \frac{u(x_{j+1}) - u(x_j)}{\alpha - x_j - \rho(\alpha - x_{j+1})} - u_x^- \right| \\ &= \left| \frac{u^+ + u_x^+(x_{j+1} - \alpha) + \frac{u_{xx}(\xi_1)}{2}(x_{j+1} - \alpha)^2}{\alpha - x_j - \rho(\alpha - x_{j+1})} \right. \\ &\quad \left. - \frac{u^- + u_x^+(x_j - \alpha) + \frac{u_{xx}(\xi_2)}{2}(x_j - \alpha)^2}{\alpha - x_j - \rho(\alpha - x_{j+1})} - u_x^- \right|, \end{aligned}$$

where $\xi_1 \in (\alpha, x_{j+1})$ and $\xi_2 \in (x_j, \alpha)$. With the jump conditions $u^+ = u^-$ and $u_x^+ = \rho u_x^-$, the expression above is simplified to

$$\begin{aligned} |\kappa - u_x^-| &= \frac{\left| \frac{u_{xx}(\xi_2)}{2}(x_{j+1} - \alpha)^2 \right| + \left| \frac{u_{xx}(\xi_2)}{2}(x_j - \alpha)^2 \right|}{\alpha - x_j - \rho(\alpha - x_{j+1})} \\ &\leq \frac{1}{\min\{\frac{1}{2},\ \frac{\rho}{2}\}} \|u''\|_\infty h, \end{aligned}$$

which implies the second inequality (8.17b). At last we have the following:

$$|\rho\kappa - u_x^+| = \rho|\kappa - u_x^-| \leq C\|u''\|_\infty h.$$

This completes the proof of the lemma. We are now ready to prove the following theorem on the accuracy of the interpolating function $u_I(x)$. □

Theorem 8.2. *With the same assumptions and conditions as in Lemma* 8.1, *we have the error estimate,*

$$\|u(x) - u_I(x)\|_\infty \leq \bar{C}h^2\|u''\|_\infty, \tag{8.19}$$

where

$$\bar{C} = \frac{2\max\{1,\ \rho\}}{\min\{1,\ \rho\}} + \frac{3}{2} = C + \frac{3}{2}.$$

Proof: Again we assume that $\alpha \in [x_j, x_{j+1})$ for some integer $0 < j \leq M - 1$. For any $x \in [x_i, x_{i+1}]$ which does not contain the interface α, from the standard interpolation theory, we know that

$$|u(x) - u_I(x)| \leq \frac{h^2}{8}\|u''\|_\infty \leq \bar{C}h^2\|u''\|_\infty.$$

If $x_j \le x \le \alpha$, then

$$\begin{aligned} u(x) &= u(x_j) + u'(x_j)(x - x_j) + \frac{(x - x_j)^2}{2} u''(\xi_1) \\ &= u(x_j) + u_x^-(x - x_j) + u''(\xi_2)(x - \alpha)(x - x_j) + \frac{(x - x_j)^2}{2} u''(\xi_1), \end{aligned}$$

where $\xi_1 \in (x_j, \alpha)$ and $\xi_2 \in (x, \alpha)$ (from the intermediate value theorem). Thus, using the bound in (8.17b) and the fact that $|x - \alpha| \le h$ and $|x - x_j| \le h$, we have

$$\begin{aligned} |u(x) - u_I(x)| &= |(u_x^- - \kappa)(x - x_j)| + \left| u''(\xi_2)(x - \alpha)(x - x_j) + \frac{(x - x_j)^2}{2} u''(\xi_1) \right| \\ &\le C h^2 ||u''||_\infty + \frac{3}{2} h^2 ||u''||_\infty \\ &\le \bar{C} h^2 ||u''||_\infty. \end{aligned}$$

The proof is similar if $\alpha < x \le x_{j+1}$. □

8.1.3 The convergence analysis for the one-dimensional IFEM

Before we prove that the approximate solution obtained from the IFEM is a second-order approximation to the exact solution in the infinite norm, we need to prove the following lemma.

Lemma 8.3. *With the same assumptions and conditions as in Lemma* 8.1, *the following equality is true:*

$$\int_{x_i}^{x_{i+1}} \beta(x)(u(x) - u_I(x))' v_h'(x) dx = 0 \quad \forall v_h \in V_h, \quad 0 \le i \le M - 1. \tag{8.20}$$

Proof: If $\alpha \notin [x_i, x_{i+1})$, then $\beta(x)$ and $v_h'(x)$ are two constants, thus

$$\begin{aligned} \int_{x_i}^{x_{i+1}} \beta(x)(u(x) - u_I(x))' v_h'(x) dx &= \beta(x_{i+\frac{1}{2}})\, v_h'(x_{i+\frac{1}{2}}) \int_{x_i}^{x_{i+1}} (u(x) - u_I(x))'\, dx \\ &= \beta(x_{i+\frac{1}{2}})\, v_h'(x_{i+\frac{1}{2}})\, (u(x) - u_I(x))|_{x_i}^{x_{i+1}} \\ &= 0, \end{aligned}$$

where $x_{i+\frac{1}{2}} = x_i + \frac{h}{2}$. On the other hand, if $\alpha \in [x_j, x_{j+1})$, then

$$\begin{aligned} &\int_{x_j}^{x_{j+1}} \beta(x)(u(x) - u_I(x))' v_h'(x) dx \\ &= \beta^- v_h'(\xi_1) \int_{x_j}^{\alpha} (u(x) - u_I(x))' dx + \beta^+ v_h'(\xi_2) \int_{\alpha}^{x_{j+1}} (u(x) - u_I(x))' dx \\ &= \beta^- v_h'(\xi_1)(u^- - u_I^-) - \beta^+ v_h'(\xi_2)(u^+ - u_I^+) \\ &= 0, \end{aligned}$$

where ξ_1 and ξ_2 are any two points in the intervals (x_j, α) and (α, x_{j+1}), respectively. In the derivation above, we have used the natural jump condition (8.4) for the basis function v_h and the continuity conditions for $u(x)$ and $u_I(x)$ at α,

$$\beta^- v_h'(\xi_1) = \beta^+ v_h'(\xi_2), \quad u^+ = u^-, \quad u_I^+ = u_I^-.$$

Below is the main theorem of the convergence for the modified finite element method. □

Theorem 8.4. *Assume that the conditions in Lemma* 8.1 *hold and β is piecewise constant. Let $u_h(x)$ be the solution obtained from the IFEM with the modified basis functions. Then $u_h(x) = u_I(x)$, and the following error estimate holds:*

$$||u(x) - u_h(x)||_\infty \le C||u''||_\infty h^2, \tag{8.21}$$

where

$$C = \frac{2\,\max\{1,\ \rho\}}{\min\{1,\ \rho\}} + \frac{3}{2}. \tag{8.22}$$

Proof: For any $v_h \in V_h$, we have

$$\begin{aligned} a(u - u_I, v_h) &= a(u - u_h + u_h - u_I, v_h) \\ &= a(u - u_h, v_h) + a(u_h - u_I, v_h) \\ &= a(u_h - u_I, v_h). \end{aligned}$$

From the definition of $a(u, v)$ and Lemma 8.3, we know that

$$\begin{aligned} a(u - u_I, v_h) &= \int_0^1 \beta(x)(u - u_I)' v_h' dx \\ &= \sum_{i=0}^{M-1} \int_{x_i}^{x_{i+1}} \beta(x)(u(x) - u_I(x))' v_h'(x) dx \\ &= 0. \end{aligned}$$

Taking $v_h = u_h - u_I \in V_h$, we conclude that $a(u_h - u_I, u_h - u_I) = 0$, which implies that $u_h(x) \equiv u_I(x)$ since $u_h - u_I$ is continuous in $[0,\ 1]$. Thus we get

$$\begin{aligned} |u(x) - u_h(x)| &\le |u(x) - u_I(x)| + |u_I(x) - u_h(x)| \\ &\le |u(x) - u_I(x)| \\ &\le C||u''||_\infty h^2. \end{aligned}$$

□

8.1.4 A numerical example of one-dimensional IFEM

We present an example to verify the theoretical analysis for a one-dimensional IFEM. The integrations are evaluated using Simpson's rule.

Example 8.1. The differential equation is

$$(\beta u_x)_x = 12x^2, \quad 0 \le x \le 1, \quad \beta = \begin{cases} \beta^- & \text{if } x < \alpha, \\ \beta^+ & \text{if } x > \alpha, \end{cases}$$

$$u(0) = 0, \quad u(1) = 1/\beta^+ + (1/\beta^- - 1/\beta^+)\alpha^4.$$

The exact solution is

$$u(x) = \begin{cases} x^4/\beta^- & \text{if } x < \alpha, \\ x^4/\beta^+ + (1/\beta^- - 1/\beta^+)\alpha^4 & \text{if } x > \alpha, \end{cases}$$

where the parameters β^- and β^+ are two constants.

Since Simpson's rule has degree of precision "three" and $f(x)$ is quadratic, there are no errors in computing $\int \phi_i(x) f(x) dx$ and $\int \phi_i'(x)\ \phi_j'(x) dx$. We expect the solution obtained using the IFEM to be the same as the interpolating function defined in (8.14), provided there are no round-off errors involved. In other words, the computed solution agrees with the exact solution at the grid points and is second-order accurate at other points. Numerical experiments have confirmed the theoretical analysis; see Figure 8.3. The infinity norm of the computed solution at grid points is between 6×10^{-15} and 3×10^{-13} when double precision is used. At other points, the finite element method solution is defined as

$$u_h(x) = \sum_{i=0}^{M} U_i \phi_i(x), \tag{8.23}$$

where U_i is the computed solution at the grid point x_i. The error E_M decreases by a factor of 4 if we reduce the mesh size by half. Table 8.1 shows a grid refinement analysis in the infinity norm at two different points that are not part of the grid points. In the first case, $\beta^- = 1$, $\beta^+ = 100$, and the interface is $\alpha = 2/3$ which is not a grid point and the error is the largest in magnitude near $x = \alpha$. We see that the computed solution at the interface itself has average second-order accuracy. In the second case, the interface is $\alpha = 0.5$ which is a grid point. The solution at the interface α is accurate to machine precision up to a factor of the condition number of the discrete linear system. Table 8.1(b) shows a grid refinement analysis at $x = \alpha + 1/3$ which is not a grid point. We see that the error is reduced by a factor of 4. Notice that, for interface problems, the error constant which is $O(1)$ may not approach a constant; it depends on the relative position of the interface and the underlying grid. This is the case in the third column of Table 8.1(a). By the second-order accuracy, we actually mean the average convergence rate of the solution; the reader is referred to the previous chapter and to [110, 163] for more information on the error analysis. For Table 8.1(b), since the interface is a grid point, the error constant will indeed approach a fixed number.

Figure 8.3(a) is the plot of the solution with $M = 40$. There is no difference between the computed and the exact solutions at the grid points. The differences in other places are too small to be visible. Figure 8.3(b) is the plot of the error in the entire interval. We see that the errors are zero at grid points and $O(h^2)$ at other points.

(a) (b)

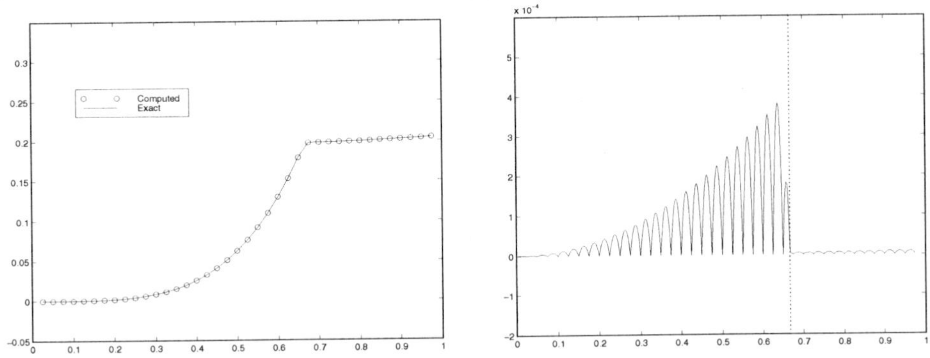

Figure 8.3. *A comparison of the computed solution using IFEM and the exact solution when* $M = 40$, $\beta^- = 1$, $\beta^+ = 100$, *and* $\alpha = 2/3$. (a) *The solid line is the exact solution while the "o"s are the computed solutions at the grid points.* (b) *The error plot of the computed solution. The largest error in magnitude is* 4×10^{-4}.

Table 8.1. *A grid refinement analysis of the IFEM for Example* 8.1 *with* $\beta^- = 1$, $\beta^+ = 100$. (a) *The error of the computed solution evaluated at the interface* $\alpha = 2/3$. (b) *The error of the solution evaluated at* $x = 0.5 + 1/3, \alpha = 0.5$.

(a)

M	E_M	E_M/E_{2M}	E_M/E_{4M}
20	4.4312×10^{-5}		
40	5.4822×10^{-6}	8.0829	
80	2.7347×10^{-6}	2.0047	16.2038
160	3.4478×10^{-7}	7.9318	15.9010
320	1.7038×10^{-7}	2.0235	16.0503
640	2.1582×10^{-8}	7.8948	15.9752

(b)

M	E_M	E_M/E_{2M}
20	2.2844×10^{-5}	
40	5.8259×10^{-6}	3.9211
80	1.4420×10^{-6}	4.0403
160	3.6229×10^{-7}	3.9801
320	9.0347×10^{-8}	4.0100
640	2.2615×10^{-8}	3.9950

8.2 The weak form of two-dimensional elliptic interface problems

Now we consider the following two-dimensional model problem using conventional finite element notations:

$$\begin{aligned} -\nabla \cdot (\beta \nabla u) &= f(\mathbf{x}), \quad (x, y) \in \Omega, \\ u|_{\partial\Omega} &= 0, \end{aligned} \tag{8.24}$$

in a domain Ω with an immersed interface Γ, where $f \in L^2(\Omega)$ is a bounded function; see Figure 1.6 for an illustration. The natural jump conditions of (8.24) are

$$[u] = 0, \qquad \left[\beta \frac{\partial u}{\partial \boldsymbol{n}}\right] = Q. \tag{8.25}$$

Note that we use Q to represent the flux since v has been used as a testing function.

To derive the weak formulation of the interface problem, we multiply both sides of the first equation in (8.24) by a test function $v(x, y) \in H_0^1(\Omega)$ and integrate over the domains Ω^+ and Ω^- respectively. Since $f \in L^2(\Omega)$, we have

$$\iint_{\Omega} fv\, dxdy = \iint_{\Omega^+} fv\, dxdy + \iint_{\Omega^-} fv\, dxdy. \tag{8.26}$$

Applying Green's theorem in the domain Ω^+, outside of the closed interface Γ, we get

$$\begin{aligned} &-\int_{\partial\Omega} \beta v \frac{\partial u}{\partial \boldsymbol{n}} ds - \int_{\Gamma} \beta^+ v^+ \frac{\partial u^+}{\partial \boldsymbol{n}^+} ds + \iint_{\Omega^+} \beta \nabla u \cdot \nabla v\, dxdy \\ &= \iint_{\Omega^+} fv\, dxdy, \end{aligned} \tag{8.27}$$

where $\mathbf{n}^+$ and $\mathbf{n}^- = \mathbf{n}$ are the unit normal directions of the interface Γ pointing outward and inward, respectively. Similarly, we get the following relation from the inside of the interface Ω^-:

$$-\int_{\Gamma} \beta^- v^- \frac{\partial u^-}{\partial \boldsymbol{n}^-} ds + \iint_{\Omega^-} \beta \nabla u \cdot \nabla v\, dxdy = \iint_{\Omega^-} fv\, dxdy. \tag{8.28}$$

Since

$$-\int_{\Gamma} \beta^+ v^+ \frac{\partial u^+}{\partial \boldsymbol{n}^+} ds = \int_{\Gamma} \beta^+ v^+ \frac{\partial u^+}{\partial \boldsymbol{n}^-} ds = \int_{\Gamma} \beta^+ v^+ \frac{\partial u^+}{\partial \boldsymbol{n}} ds,$$

by applying the zero boundary condition $v|_{\partial\Omega} = 0$ and adding (8.27) and (8.28) together, we get

$$\int_{\Gamma} \beta^+ v^+ \frac{\partial u^+}{\partial \boldsymbol{n}} ds - \int_{\Gamma} \beta^- v^- \frac{\partial u^-}{\partial \boldsymbol{n}} ds + \iint_{\Omega} \beta \nabla u \cdot \nabla v\, dxdy = \iint_{\Omega} fv\, dxdy.$$

Thus, we obtain the weak form for the interface problem,

$$\iint_{\Omega} \beta(x, y) \nabla u \cdot \nabla v\, dxdy = \iint_{\Omega} fv\, dxdy - \int_{\Gamma} vQ\, ds, \tag{8.29}$$

and the flux jump condition $[\beta u_n] = Q$ since $v(\mathbf{x}) \in H_0^1$ is arbitrary. The weak form does allow discontinuities in the coefficient and the normal derivatives of the solution. The existence of the weak solution is discussed in [30, 44]. Theoretically, the weak form is the same as those discussed in most standard textbooks on finite element methods; see [127] for example.

8.3 A nonconforming IFE space and analysis

Following the idea of the IFEM for one-dimensional problems, we describe a finite element space whose basis functions are piecewise linear functions satisfying the natural (homogeneous) jump conditions.

A Cartesian grid is used to form a uniform triangular partition $\mathcal{T}_h$ with mesh size h on Ω such that each element $T \in \mathcal{T}_h$ is a triangle constructed of two legs and one of the diagonals in a subrectangle. We call an element $T \in \mathcal{T}_h$ an interface element if the interface Γ passes through the interior of T (see Figure 8.4 for a typical geometric configuration); otherwise we call T a noninterface element. We assume that the interface meets the edges of an interface element at no more than two intersections.[11] Such an assumption is reasonable if h is small, and is guaranteed if the interface is expressed (or approximately expressed) in terms of the zero level set of the signed distance function of the interface. As is common practice, we approximate the interface in T by a line segment connecting the intersections of the interface and the edges of the triangles; for example, the line segment $\overline{DE}$ in Figure 8.4. The line segment divides T into two parts T^+ and T^- with $T = T^+ \cup T^- \cup \overline{DE}$. There is a small region in T,

$$T_r = T - \Omega^+ \cap T^+ - \Omega^- \cap T^-, \tag{8.30}$$

whose area is of order $O(h^3)$. This indicates that the interface is perturbed in a magnitude of $O(h^2)$. From [44] and the discussions later in this section, such a perturbation will affect only the solution, and the interpolation function to an order of h^2.

As usual, we want to construct local basis functions on each element T of the partition $\mathcal{T}_h$. For a noninterface element $T \in \mathcal{T}_h$, we simply use the standard linear shape functions on T, and use $S_h(T)$ to denote the linear spaces spanned by the three nodal basis functions on T. Attention is needed only for interface elements, which will be discussed in the following subsection.

8.3.1 Local basis functions on an interface element

For simplicity of discussion, we assume that β is a piecewise constant. Without loss of generality, we consider a reference interface element T, whose geometric configuration is given in Figure 8.4, in which the curve between points D and E is part of the interface. The basis function in a general interface element can then be defined through the usual affine transformation. We assume that the coordinates at A, B, C, D, and E are

$$(0, h), \qquad (0, 0), \qquad (h, 0), \qquad (0, y_1), \qquad (h - y_2, y_2), \tag{8.31}$$

[11] If one of the edges is part of the interface, then the element is a noninterface element.

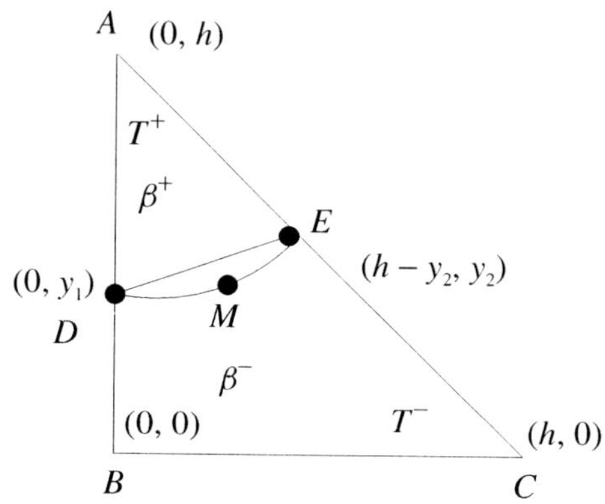

Figure 8.4. *A typical triangle element with an interface cutting through it. The curve between D and E is part of the interface curve Γ which is approximated by the line segment $\overline{DE}$. In this diagram, T is the triangle $\triangle ABC$, $T^+ = \triangle ADE$, $T^- = T - T^+$, and T_r is the region enclosed by the $\overline{DE}$ and the arc DME.*

respectively, with the restriction

$$0 \le y_1 \le h, \quad 0 \le y_2 < h. \tag{8.32}$$

Once the values at vertices A, B, and C of the element T are specified, we construct the following piecewise linear function:

$$u(\mathbf{x}) = \begin{cases} u^+(\mathbf{x}) = a_0 + a_1 x + a_2(y - h) & \text{if } \mathbf{x} = (x, y) \in T^+, \\ u^-(\mathbf{x}) = b_0 + b_1 x + b_2 y & \text{if } \mathbf{x} = (x, y) \in T^-, \end{cases} \tag{8.33a}$$

$$u^+(D) = u^-(D), \quad u^+(E) = u^-(E), \quad \beta^+ \frac{\partial u^+}{\partial \boldsymbol{n}} = \beta^- \frac{\partial u^-}{\partial \boldsymbol{n}}, \tag{8.33b}$$

where $\mathbf{n}$ is the unit normal direction of the line segment $\overline{DE}$. This is a piecewise linear function in T that satisfies the natural jump conditions along $\overline{DE}$. Intuitively, there are six constraints and six parameters, so we can expect that the solution exists and is unique as confirmed in the following theorem.

Theorem 8.5. *Given a right triangle ABC as indicated in Figure 8.4. The piecewise linear function $u(\mathbf{x})$ defined by (8.33a) and (8.33b) is uniquely determined by $u(A)$, $u(B)$, and $u(C)$.*

Proof: Let $\mathbf{x} = (x, y)^T$. Because u^+ and u^- are linear functions, we have

$$u(\mathbf{x}) = \begin{cases} u^+(x, y) = u(A) + a_1 x + a_2(y - h), & \mathbf{x} \in T^+, \\ u^-(x, y) = u(B) + \dfrac{u(C) - u(B)}{h} x + b_2 y, & \mathbf{x} \in T^-. \end{cases} \tag{8.34}$$

From the continuity condition at D and E, we have two equations,

$$a_2(y_1 - h) - b_2 y_1 = u(B) - u(A), \tag{8.35}$$

$$a_1(h - y_2) + a_2(y_2 - h) - b_2 y_2 = u(B) - u(A) + b_1(h - y_2), \tag{8.36}$$

where

$$b_1 = \frac{u(C) - u(B)}{h}. \tag{8.37}$$

The third equation is from the flux jump condition,

$$a_1\alpha - a_2 + \rho b_2 = \rho\alpha b_1, \tag{8.38}$$

where $\rho = \beta^- / \beta^+$, and we have used the fact that the normal direction of the line segment is $(\alpha, \ -1)$ with $\alpha = (y_2 - y_1)/(h - y_2)$. The coefficient matrix of the linear system for the unknowns a_1, a_2, and b_2 is

$$\mathcal{A} = \begin{bmatrix} 0 & y_1 - h & -y_1 \\ h - y_2 & y_2 - h & -y_2 \\ \alpha & -1 & \rho \end{bmatrix}. \tag{8.39}$$

Evaluating the determinant of the matrix above, using the relation $h - y_1 = (h - y_2)(1+\alpha)$, we obtain the following after some manipulations:

$$\begin{aligned} \det(\mathcal{A}) &= -(y_1 - h)y_2\alpha + (h - y_2)y_1 + \alpha(y_2 - h)y_1 - \rho(h - y_2)(y_1 - h) \\ &= (h - y_2)y_2\alpha(1 + \alpha) + (h - y_2)y_1 - \alpha(h - y_2)\,(y_2 - \alpha(h - y_2)) \\ &\quad + \rho(h - y_2)^2(1 + \alpha) \\ &= (h - y_2)\left(y_1 + h\alpha^2 + \rho(h - y_2)(1 + \alpha)\right) > 0. \end{aligned}$$

Thus there is a unique solution to the linear system (8.35), (8.36), and (8.38). □

We introduce a local finite element space on each element T of the partition T_h as follows:

$$S_h(T) = \begin{cases} \{u(\mathbf{x}) \mid u(\mathbf{x}) \text{ is linear on } T\} & \text{if } T \text{ is a noninterface element,} \\ \{u(\mathbf{x}) \mid u(\mathbf{x}) \text{ is defined by (8.33)}\} & \text{if } T \text{ is an interface element.} \end{cases}$$

It is well known that the dimension of $S_h(T)$ is 3 if T is a noninterface element. When T is an interface element, $S_h(T)$ contains three basis functions whose values at one of the vertices of T are unity, and zero at the other two vertices. Furthermore, Theorem 8.5 tells us that any function in $S_h(T)$ is a linear combination of these three basis functions. Therefore the dimension of $S_h(T)$ is also 3 even if T is an interface element.

8.3.2 The nonconforming IFE space

To describe the finite element space on the whole domain Ω, we let Ω' be the union of all interface elements. We define the IFE space $S_h(\Omega)$ as a set of functions such that

$$S_h(\Omega) = \left\{\phi(\mathbf{x}) \mid \forall\, T \in \mathcal{T}_h,\ \ \phi|_T \in S_h(T),\ \ \phi|_{\Omega\backslash\Omega'} \in H^1(\Omega\backslash\Omega')\right\}. \tag{8.40}$$

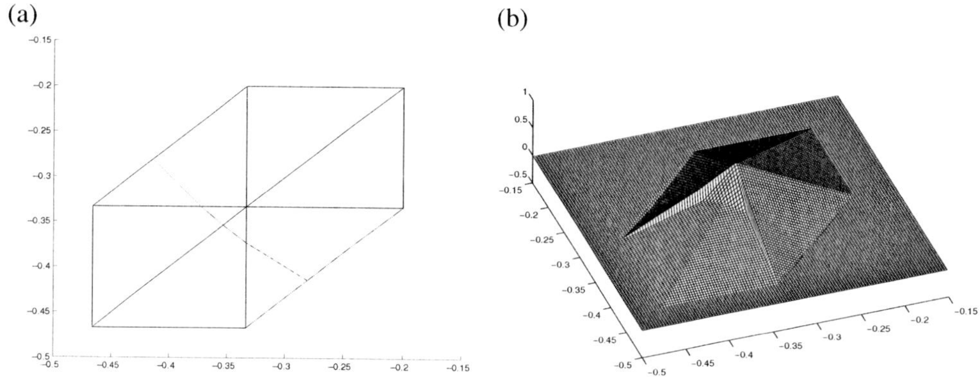

Figure 8.5. (a) *A standard domain of six triangles with an interface cutting through it.* (b) *A global basis function on its support in the nonconforming IFE space. The basis function has a small jump across some edges.*

Note that this finite element space is formed by piecewise linear functions defined according to the partition $\mathcal{T}_h$ and the interface, but the partition does not have to align with the interface. Part of the interface can be immersed in some elements of $\mathcal{T}_h$, and this is the reason we call $S_h(\Omega)$ an IFE space. On the other hand, the IFE space is rather similar to the usual linear finite element space defined by the partition $\mathcal{T}_h$. First, they are exactly the same on every noninterface element. Second, they have the same dimension.

Figure 8.5(b) shows a typical basis function of $S_h(\Omega)$ on its support with an interface cutting through its nonzero support region. If $\beta(\mathbf{x})$ is continuous (i.e., $[\beta] = 0$), then the IFE space becomes the usual linear finite element space. However, for a discontinuous $\beta(\mathbf{x})$, the IFE space is more sophisticated than the usual finite element space since the jump conditions across the interface are satisfied to a certain extent. In this case, the IFE space is similar to a nonconforming finite element space in the way that the basis functions may not be continuous across the edges of the elements in $\mathcal{T}_h$. Hence the IFE space is a nonconforming finite element space.

The dimension of the nonconforming IFE space is the number of interior points for the Dirichlet problem. The basis function centered at a node is defined as

$$\phi_i(\mathbf{x}_j) = \begin{cases} 1 & \text{if } i = j, \\ 0 & \text{otherwise}, \end{cases} \qquad \left[\beta \frac{\partial \phi_i}{\partial \boldsymbol{n}} \right]_{\bar{\Gamma}} = 0, \qquad \phi_i|_{\partial\Omega} = 0, \tag{8.41}$$

and ϕ_i is continuous in each element T except some edges if $\mathbf{x}_i$ is a vertex of one or several interface triangles; see Figure 8.5. We use $\bar{\Gamma}$ to denote the union of the line segment used to approximate the interface.

8.3.3 Approximation properties of the nonconforming IFE space

Given a function $u(\mathbf{x})$ which is continuous on the entire domain and satisfies the natural jump conditions, its interpolant in the IFE space $S_h(\Omega)$ is defined as the function $u_I(\mathbf{x}) \in S_h(\Omega)$

such that

$$u_I(\mathbf{x}) = u(\mathbf{x}) \quad \text{if } \mathbf{x} \text{ is a node of } \mathcal{T}_h.$$

Since $u_I(\mathbf{x})$ is the usual linear function on each *noninterface* element, we have the following standard error estimate [52]:

$$\|u_I - u\|_{0,T} + h\,\|I_h u - u\|_{1,T} \le C_1 h^2\,\|u\|_{2,T}\,,$$

where $\|\cdot\|_{s,\mathcal{B}}$ is the norm of the Sobolev space $H^s(\mathcal{B})$ defined in a set $\mathcal{B}$, and C_1 is a constant.[12] Similar error estimates can also be given in the norm of the space $W^{s,p}(\mathcal{B})$. When T is an interface element, each partial derivative of u_I on T is a piecewise constant function consisting of two values, $\partial u_I^+/\partial x$ and $\partial u_I^-/\partial x$, or $\partial u_I^+/\partial y$ and $\partial u_I^-/\partial y$, where u_I^i are the restrictions of u_I on T^i, $i = +, -$. The following theorem provides error estimates for u_I^i.

Theorem 8.6. *Let $T \in \mathcal{T}_h$ be an interface element. Let $u(\mathbf{x})$ be a continuous function such that its restriction $u^i = u|_{T^i}$ on T^i, $i = +, -$, is twice differentiable in each* subdomain *$\Omega^+ \cap T$ and $\Omega^- \cap T$, and satisfies the natural jump conditions* (8.25). *Then we have the following error estimates:*

$$\left\| \frac{\partial u_I}{\partial x} - \frac{\partial u}{\partial x} \right\|_{\infty, T\setminus T_r} \le \left(22\,\rho_{cond}^2 + 1\right)\,||\mathcal{D}^2 u||_{\infty,T} h, \tag{8.42}$$

$$\left\| \frac{\partial u_I}{\partial y} - \frac{\partial u}{\partial y} \right\|_{\infty, T\setminus T_r} \le \left(22\,\rho_{cond}^2 + 1\right)\,||\mathcal{D}^2 u||_{\infty,T} h, \tag{8.43}$$

where

$$\rho_{cond} = \frac{\rho_{max}}{\rho_{min}}, \qquad \rho_{max} = \max_{\mathbf{x}\in\Omega} \left\{ \rho,\ \frac{1}{\rho} \right\}, \qquad \rho_{min} = \min_{\mathbf{x}\in\Omega} \left\{ \rho,\ \frac{1}{\rho} \right\}, \tag{8.44}$$

and

$$||\mathcal{D}^2 u||_{\infty,T} = \max_{i=+,-} \left\{\, ||u_{xx}^i||_{\infty,T\cap\Omega^i} + 2\,||u_{xy}^i||_{\infty,T\cap\Omega^i} + ||u_{yy}^i||_{\infty,T\cap\Omega^i} \right\}. \tag{8.45}$$

The proof is omitted here because it is long and technical. We refer the reader to [170] for the proof. With the theorem above and the Taylor expansion, we can derive an error estimate for u_I itself given in the following theorem.

Theorem 8.7. *Let $T \in \mathcal{T}_h$ be an interface element. Let $u(\mathbf{x})$ be a continuous function such that its restriction $u^i = u|_{T^i}$ on T^i, $i = +, -$, is twice differentiable in each* subdomain *$\Omega^+ \cap T$ and $\Omega^- \cap T$, and satisfies the natural jump conditions* (8.25). *Then we have the following inequality:*

$$|u(\mathbf{x}) - u_I(\mathbf{x})| \le \begin{cases} C_1 \tilde{h}\, h & \text{if } \mathbf{x} \in T\setminus T_r, \\ C_2 h^2 & \text{if } \mathbf{x} \in T_r, \end{cases} \tag{8.46}$$

[12] We use C_1 instead of C because we have already used C in Figure 8.4.

where $\tilde{h}$ is the shortest distance between $\mathbf{x}$ *and the vertices of T that are on the same side of the interface as* $\mathbf{x}$,

$$C_1 \le \left(1 + 44\,\rho_{cond}^2\right) \|\mathcal{D}^2 u\|_{\infty,T}, \tag{8.47}$$

and C_2 is some constant.

Proof: Without loss of generality, we still use Figure 8.4 to illustrate our proof and assume that point B is the vertex closest to $\mathbf{x}$. If $\mathbf{x} \in T^-\backslash T_r$, then we have

$$\begin{aligned} u_I^-(\mathbf{x}) &= u_I(B) + \frac{\partial u_I}{\partial x}^-(x - x_B) + \frac{\partial u_I}{\partial y}^-(y - y_B) \\ &= u(B) + \frac{\partial u^-}{\partial x}(M)(x - x_B) + \frac{\partial u^-}{\partial y}(M)(y - y_B) + R_1 \\ &= u(\mathbf{x}) + R_1 + R_2, \end{aligned}$$

where x_B and y_B are the coordinates of point B,

$$\begin{aligned} |R_1| &\le 2\max\left\{ \left| \frac{\partial u_I^-}{\partial x} - \frac{\partial u^-(M)}{\partial x} \right|, \left| \frac{\partial u_I^-}{\partial y} - \frac{\partial u^-(M)}{\partial y} \right| \right\} h \\ &\le 44\,\rho_{cond}^2\,\|\mathcal{D}^2 u\|_{\infty,T}\,\tilde{h}\,h \end{aligned}$$

from (8.42)–(8.43), and

$$|R_2| \le \|\mathcal{D}^2 u\|_{\infty,T}\,\tilde{h}\,h$$

from the Taylor expansion. A similar result holds for $\mathbf{x} \in T^+\backslash T_r$.

If $\mathbf{x} \in T_r$, say $\mathbf{x} \in T \cap T^-$, but $\mathbf{x} \in \Omega^+$, for example, then we take the *closest* point $R \in T$ to $\mathbf{x}$ from the line segment. We know that $\|\mathbf{x} - R\| \approx h^2$. Using the triangle inequality, we obtain

$$\begin{aligned} |u_I(\mathbf{x}) - u(\mathbf{x})| &\le |u_I^-(\mathbf{x}) - u_I^-(R)| + |u_I^+(R) - u(\mathbf{x})| \\ &\le \|\mathcal{D}^2 u\|_{\infty,T}\|\mathbf{x} - R\| + |u_I^+(R) - u^+(R)| + |u^+(R) - u^+(\mathbf{x})| \\ &\le \|\mathcal{D}^2 u\|_{\infty,T}\|\mathbf{x} - R\| + C_1\tilde{h}\,h + \|\mathcal{D}^2 u\|_{\infty,T}\|\mathbf{x} - R\| \\ &\approx h^2. \end{aligned}$$

In the proof, we have used the fact that $u_I^+(R) = u_I^-(R)$, the continuity condition of $u_I(\mathbf{x})$ and $u(\mathbf{x})$, and the first error estimate in (8.46) of this theorem, which has already been proved. □

Note that the intersections of the interface and the edges of the triangles are not in T_r, so they satisfy the first inequality in (8.46).

Remark 8.1. *Although we have the error estimate for the interpolation functions for the nonconforming finite element method in terms of piecewise $C^2(\Omega)$ space, the convergence analysis for the FE solution is not straightforward for the nonconforming IFE space. Some recent results are given in* [169] *in the usual Sobolev space. Our result indicates that the nonconforming IFE space has an approximation capability similar to that of the standard conforming linear finite space based on body-fitting partitions.*

8.3.4 A nonconforming IFEM

It is obvious that the finite element space $S_h(\Omega)$ discussed above is not in the space to which the solution of the interface problem belongs. A function ϕ of $S_h(\Omega)$ is continuous in the union of noninterface triangles but may be discontinuous on the edges of interface triangles. Therefore the finite element method based on $S_h(\Omega)$ is a nonconforming one. For the interface problem, we define its nonconforming IFE solution as a function $u_h \in S_{h0}(\Omega)$ satisfying

$$a_h(u_h, v_h) = \iint_\Omega f v_h dxdy \qquad \forall v_h \in S_{h0}(\Omega), \tag{8.48}$$

where $S_{h0}(\Omega) = \{\phi \in S_h(\Omega) \mid \phi|_{\partial\Omega} = 0\}$, and

$$a_h(u, v) = \sum_{T\in\mathcal{T}_h} \iint_{T\in\mathcal{T}_h} \beta\nabla u \cdot \nabla v \, dxdy. \tag{8.49}$$

Remark 8.2. *If the flux jump is not homogeneous* (i.e., $[\beta u_{\mathbf{n}}] = Q \neq 0$ *in* (8.24)), *we can use the finite element method solution* $u_h \in S_{h0}(\Omega)$ *that satisfies the derived weak form* $a_h(u_h, v_h) = \iint_\Omega f v_h dxdy - \int_\Gamma v_h q ds$. *The solution is likely to be first-order accurate in the infinity norm. A better approach is to use the transformation technique discussed in Chapter* 5 *to get an interface problem with homogeneous jump conditions. Then we can use the nonconforming finite element method.*

The nonconforming IFEM is simple, easy to implement, and has an algebraic system similar to that of the Galerkin finite element method based on the standard finite element space. In particular, the partition of the IFE space does not have to be restricted by the geometry of the interface. The basis functions of the IFE space satisfy the natural jump conditions, which enables us to obtain sharp solutions near the interface. The same idea can be applied to treat three-dimensional problems; see [133].

8.4 A conforming IFE space and analysis

While the nonconforming IFEM performs better than the standard finite element method for interface problems, it does not always seem to be second-order accurate in the infinite norm. Note that, for regular boundary value problems, the standard conforming finite element method using the piecewise linear basis functions has second-order convergence in the infinity norm. However, the requirements of the continuity and the jump relations (8.25) with a non-body-fitting mesh turn out to be rather difficult to satisfy simultaneously with the same local support. One of the key ideas of the conforming IFEM is to enlarge the support of some basis functions so that the continuity condition can be maintained.

Let us examine the IFE space again to see why we need to enlarge the support of some basis functions. The nonconforming basis functions have the same compact support as the standard linear basis functions. However, it may be impossible to construct linear basis functions with the same compactness that satisfy

$$\phi_i(\mathbf{x}_j) = \begin{cases} 1 & \text{if } i = j, \\ 0 & \text{otherwise} \end{cases} \tag{8.50}$$

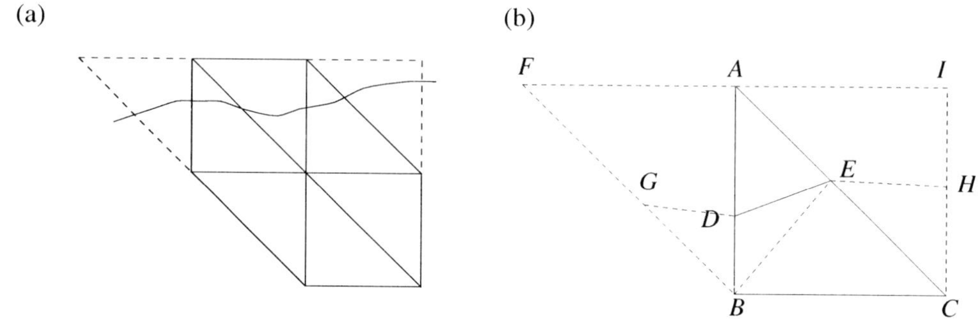

Figure 8.6. (a) *An extended region of support of the conforming basis function.* (b) *A diagram for constructing the basis function on* $\triangle ABC$.

in a conforming IFE space. To see this, let us consider an interface element $\triangle ABC$ as in Figure 8.6(b) in which the line $\overline{DE}$ is part of the interface. Let $u(\mathbf{x})$ be a function satisfying the natural jump conditions on the interface and have the following values:

$$u(C) = h, \qquad u(A) = u(B) = 0.$$

It is likely that $u(D) = O(h)$ for an interface problem. The interpolation function has the form

$$u_I(x, y) = \sum u_j \phi_j(\mathbf{x}),$$

where ϕ_j are conforming basis functions. Then we must have $u_I(D) = \phi_C(D) = 0$ since $u(A) = u(B) = 0$ and all the basis functions that are not centered at A and B are zero on the entire line segment $\overline{AB}$. Hence, $|u_I(D) - u(D)| = O(h)$ and the approximation is only first-order accurate.

Intuitively, it is not very difficult to approximate any piecewise twice differentiable function to second order by piecewise polynomials. The challenge is how to simultaneously maintain the continuity along the edges and the jump conditions along the interface. A simple idea is to average the values of nonconforming basis functions with the same values at nodal points to keep the continuity.

8.4.1 The conforming local basis functions on an interface element

We describe a procedure to construct basis functions in a typical interface element $\triangle ABC$ sketched in Figure 8.6(b) such that they can be used to form a conforming IFE space. We assume that the interface meets edges of this element at D and E. The key idea is to make sure that some of the local basis functions in two adjacent interface elements, such as $\triangle ABC$ and $\triangle AFB$, can take the same value at the interface point on their common edge, such as point D.

We use the standard five-dimensional Euclidean vector e_i (whose ith entry is unity while the other entries are zero) to assign values of a local basis function $\psi_i(x, y)$ at the vertices A, B, C, F, and I, and this basis function is piecewise constructed as follows.

P1. Use the values at the vertices A, B, C, F, and I to construct the three nonconforming IFE basis functions defined on the elements $\triangle ABC$, $\triangle AFB$, and $\triangle ACI$, respectively.

P2. Assign a value to the point D as the average (or a certain weighted average) of the values at this point of the nonconforming IFE functions defined on the elements ΔABC and ΔAFB formed in **P1**.

P3. Similarly, assign a value to the point E as the average (or a certain weighted average) of values at this point of the nonconforming IFE functions defined on the elements ΔABC and ΔACI formed in **P1**.

P4. Partition the element ΔABC into three subtriangles by an auxiliary line, say line segment $\overline{BE}$ or $\overline{DC}$, such that at least one of the acute angles, or the supplementary angle if an angle is bigger than $\pi/2$, of the triangle formed by the auxiliary line is bigger than or equal to $\pi/4$.

P5. Define the basis function ψ_i to be the piecewise linear function in the three subtriangles determined by the values at the points A, B, C, D, and E.

As in §8.3.1, we define a local finite element space on each element T of the partition $\mathcal{T}_h$ as follows:

$$S_h(T) = \begin{cases} \{u(\mathbf{x}) \mid u(\mathbf{x}) \text{ is linear on } T\} & \text{if } T \text{ is a noninterface element,} \\ \text{span}\,\{\psi_i(\mathbf{x}),\ 1 \le i \le 5 \mid \psi_i(\mathbf{x}) \text{ is defined by } \textbf{P1–P5}\} & \text{otherwise.} \end{cases}$$

8.4.2 A conforming IFE space

For the ith vertex in the partition $\mathcal{T}_h$, we let $\phi_i(\mathbf{x})$ be the continuous piecewise linear function satisfying (8.50) and $\phi_i|_T \in S_h(T)$ for any element $T \in \mathcal{T}_h$. Then we let the new IFE space $S_h(\Omega)$ be a set of functions such that

$$S_h(\Omega) = \text{span}\,\{\phi_i(\mathbf{x})\}. \tag{8.51}$$

Because all the basis functions are continuous and in H^1, the IFE space $S_h(\Omega)$ is conforming. Also, this conforming IFE space has the same dimension as the nonconforming IFE space and the standard linear finite element space defined on the partition $\mathcal{T}_h$.

The basis function ϕ_i of $S_h(\Omega)$ centered at the ith node has a nonzero support on the six surrounding triangles if the interface does not cut through any of these triangles. If the coefficient is continuous, i.e., $\rho \equiv 1$, then this basis function becomes the standard linear basis function. If the interface cuts through any of the surrounding triangles, then, by the definition of $S_h(\Omega)$, the support of this basis function is extended to two more triangles along the direction of the interface (see Figure 8.6(a)), where the support of the basis function includes the triangles marked by dashed lines. As a consequence, the corresponding finite difference scheme will generally have a nonstandard 9-point stencil.

8.4.3 Approximation properties of the conforming IFE space

Given a piecewise smooth function $u(\mathbf{x})$ satisfying the jump conditions (8.25) along a smooth interface, it has been shown in [170] that its interpolation function $u_I(x, y)$ in the conforming

IFE space, using the values of $u(\mathbf{x})$ at vertices, can approximate $u(\mathbf{x})$ to second order, and its first derivatives can approximate those of $u(\mathbf{x})$ to first-order accuracy in the maximum norm almost everywhere. We assume that the values of the basis functions at intersections, for example, points D and E in Figure 8.6(b), are simple *averages* of the nonconforming interpolation functions in the two neighborhood triangles. From this point of view, the conforming interpolation function is obtained by perturbing the values of the nonconforming interpolation functions at intersections. From Theorem 8.7, such perturbations are bounded by $C_1\tilde{h}h$, where $\tilde{h}$ is the shortest distance from the intersection points, such as D and E, to the vertices in interface element, such as B and A, in Figure 8.6(b). The following lemma shows that the perturbations in the first derivatives between two interpolation functions are of order h.

Lemma 8.8. *Assume that* (i) $T \in \mathcal{T}_h$ *is an interface element;* (ii) $u(\mathbf{x})$ *is a continuous function whose restriction* $u^i = u|_{T^i}$ *on* T^i, $i = +, -$, *is twice differentiable in each* subdomain $\Omega^+ \cap T$ *and* $\Omega^- \cap T$, *and satisfies the natural jump conditions* (8.25)*;* (iii) $\bar{u}_I$ *and* u_I *are the interpolation functions of* u *in the nonconforming and conforming IFE spaces, respectively. Then*

$$\left\| \frac{\partial u_I}{\partial x} - \frac{\partial \bar{u}_I}{\partial x} \right\|_{\infty,T} \le \left(4+\sqrt{2}\right) C_1 h, \tag{8.52}$$

$$\left\| \frac{\partial u_I}{\partial y} - \frac{\partial \bar{u}_I}{\partial y} \right\|_{\infty,T} \le \left(4+\sqrt{2}\right) C_1 h, \tag{8.53}$$

where C_1 *is given in Theorem* 8.6.

The results are trivial in every noninterface element since $\bar{u}_I$ and u_I are identical. The proof for an interface element is long and technical; we omit it here and refer the reader to [170] for the proof.

From Lemma 8.8, the following error estimates for the interpolation function in the conforming IFE space can be derived.

Theorem 8.9. *Let* $T \in \mathcal{T}_h$ *be an interface element, and let* $u(\mathbf{x})$ *be a continuous function whose restriction* $u^i = u|_{T^i}$ *on* T^i, $i = +, -$, *is twice differentiable in each* subdomain $\Omega^+ \cap T$ *and* $\Omega^- \cap T$, *and satisfies the natural jump conditions* (8.25). *Then we have the following error estimates:*

$$\left\| \frac{\partial u_I}{\partial x} - \frac{\partial u}{\partial x} \right\|_{\infty,T\setminus T_r} \le C_2 h, \tag{8.54}$$

$$\left\| \frac{\partial u_I}{\partial y} - \frac{\partial u}{\partial y} \right\|_{\infty,T\setminus T_r} \le C_2 h, \tag{8.55}$$

where

$$C_2 \le C_1\left(4+\sqrt{2}\right) + \left(22\,\rho_{cond}^2 + 1\right) ||\mathcal{D}^2 u||_{\infty,T} \le 7C_1.$$

Proof: Denote again the interpolation function using the nonconforming IFE space as $\bar{u}_I$; then

$$\left\| \frac{\partial u_I}{\partial x} - \frac{\partial u}{\partial x} \right\|_{\infty, T\backslash T_r} \leq \left\| \frac{\partial u_I}{\partial x} - \frac{\partial \bar{u}_I}{\partial x} \right\|_{\infty, T\backslash T_r} + \left\| \frac{\partial \bar{u}_I}{\partial x} - \frac{\partial u}{\partial x} \right\|_{\infty, T\backslash T_r}$$

$$\leq C_1 \left(4 + \sqrt{2}\right) h + \left(22\,\rho_{cond}^2 + 1\right) ||\mathcal{D}^2 u||_{\infty, T} h$$

from Lemma 8.8 and Theorem 8.6. A similar proof can be done for $\partial u_I / \partial y$. □

From this theorem and the proof of Theorem 8.7, it is straightforward to get the following theorem for an error estimate of the interpolation function.

Theorem 8.10. *Assume that* (i) $T \in \mathcal{T}_h$ *is an interface element;* (ii) $u(\mathbf{x})$ *is a continuous function whose restriction* $u^i = u|_{T^i}$ *on* T^i, $i = +, -$, *is twice differentiable in each* subdomain $\Omega^+ \cap T$ *and* $\Omega^- \cap T$, *and satisfies the natural jump conditions* (8.25)*;* (iii) u_I *is the interpolation function of* u *in the conforming IFE spaces. Then the following inequality holds:*

$$|u(\mathbf{x}) - u_I(\mathbf{x})| \leq \begin{cases} C_3 \tilde{h}\, h & \text{if } \mathbf{x} \in T \backslash T_r, \\ C_4 h^2 & \text{if } \mathbf{x} \in T_r, \end{cases} \tag{8.56}$$

where C_4 *is a constant,* $\tilde{h}$ *is the shortest distance between* $\mathbf{x}$ *and the vertices of* T *that are on the same side of the interface as* $\mathbf{x}$, *and*

$$C_3 \leq C_2 + ||\mathcal{D}^2 u||_{\infty, T}. \tag{8.57}$$

Remark 8.3. *The interpolation errors actually depend on the jump in the coefficient, the mesh size h, and the geometry. The error generally is not a monotonous function of h, because the error depends on the relative position of the interface and the underlying grid; see Figure* 8.9.

We now define the conforming IFE solution to the interface problem as a function $u_h \in S_{h0}(\Omega)$ such that

$$\iint_\Omega \beta(x, y) \nabla u_h \nabla v_h \, dxdy = \iint_\Omega f\, v_h \, dxdy \quad \forall v_h \in S_{h0}(\Omega),$$

and again, we let $S_{h0}(\Omega) = \{\phi \in S_h(\Omega) \mid \phi|_{\partial\Omega} = 0\}$. For this conforming IFE solution, we can obtain an error estimate in the energy norm given in the following theorem.

Theorem 8.11. *Let* u *be the solution of* (8.24), *and let* u_h *be the conforming IFE solution. If* u *is in* $H_0^1(\Omega)$ *and is* piecewise *twice differentiable on each* subdomain Ω^i, $i = +, -$, *then we have the following error estimate:*

$$\|u - u_h\|_{1,\Omega} \leq C_5 h, \tag{8.58}$$

where C_5 *is a constant independent of* h.

Proof: Since u, u_h, and the IFE finite-dimensional space all belong to $H_0^1(\Omega)$, then, from the standard finite element method theory, u_h is the best solution in the IFE space in the H^1 norm. Therefore, we have

$$\begin{aligned} \|u - u_h\|_{1,\Omega} &\le \tilde{C}\, \|u - u_I\|_{1,\Omega} \\ &\le \tilde{C}\left(\|u - u_I\|_{1,\sum T \backslash T_r} + \|u - u_I\|_{1,\sum T_r} \right), \end{aligned}$$

where $u_I \in H^1$ is the interpolation function of u in the conforming IFE space. $\sum T\backslash T_r$ is the union of the mismatched region of the line segments and the interface as shown in Figure 8.4. From Theorem 8.9, we know that $u - u_I$ and its first derivatives are of order $O(h^2)$ and $O(h)$, respectively, in the maximum norm on $T\backslash T_r$ of an element T; therefore, $u - u_I$ should be of order $O(h)$ in the H^1 norm on the unions of these regions as well. On each T_r, $u - u_I$ and its first derivatives are of orders $O(h^2)$ and $O(1)$. However, with the interface being approximated by the line segment on each element, the area of each T_r is of order $O(h^3)$. Since the interface is one dimension lower than the solution domain, we also conclude that

$$\|u - u_I\|_{1,\sum T_r} \approx h,$$

which leads to the result of this theorem. □

Remark 8.4. *For many practical interface problems, the solutions are indeed* piecewise *smooth. Generally, if the source term $f(x, y) \in L^2(\Omega)$ is also γth-Hölder* piecewise *continuous for $\gamma > 0$, then the solution $u(\mathbf{x})$ is* piecewise *twice differentiable; see* [73].

8.5 A numerical example and analysis for IFEMs

We present a nontrivial example for the standard Galerkin finite element method using the nonconforming and conforming IFE spaces. In this example, we consider the boundary value problem defined by (8.24) with a Dirichlet boundary condition.

Example 8.2. The computational domain is the rectangle $-1 \le x,\ y \le 1$, and the interface is a circle centered at the origin with radius r_0. The boundary condition and the source term f are determined from the exact solution,

$$u(x, y) = \begin{cases} \dfrac{r^\alpha}{\beta^-} & \text{if } r \le r_0, \\[2ex] \dfrac{r^\alpha}{\beta^+} + \left(\dfrac{1}{\beta^-} - \dfrac{1}{\beta^+} \right) r_0^\alpha & \text{otherwise}, \end{cases} \tag{8.59}$$

where $r = \sqrt{x^2 + y^2}$ and $\alpha = 3$. Notice that the exact solution satisfies the natural jump conditions (8.25).

The error estimates for the interpolation functions obtained in §8.3.3 indicate that the finite element solution in the IFE spaces has a second-order approximation capability. Hence we naturally expect that the IFE solutions are second-order accurate in the L^2 norm. Since the large errors occur near or at the interface which is one dimension lower than that

(a) (b)

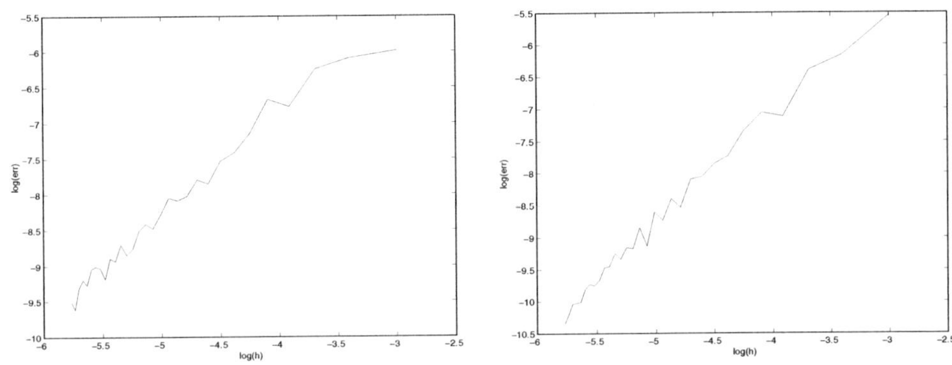

Figure 8.7. *The error plots of the finite element solutions obtained from the nonconforming IFE space in the maximum norm versus the mesh size h in* log-log *scale with $r_0 = \pi/6.28$. (a) $\beta^- = 1$, $\beta^+ = 1000$; the linear regression analysis gives $||u - u_h||_\infty \approx 0.64657\, h^{1.56459}$. (b) $\beta^- = 1000$, $\beta^+ = 1$; the linear regression analysis gives $||u - u_h||_\infty \approx 2.79434\, h^{1.94833}$.*

of the solution domain, we present only the errors in the maximum norm in Figure 8.7 for nonconforming IFEM, in which the IFE solutions u_h are found with various grid sizes h. The involved linear algebraic system has a structure similar to that in the Galerkin method with the usual linear finite element space. The jump in the coefficient of these tests is taken as $\rho = \beta^-/\beta^+ = 1 : 1000$ or $\rho = 1000 : 1$, quite a large ratio. As we mentioned before, the errors in the numerical solutions generally do not decrease monotonically for interface problems. Therefore, we use the linear regression analysis (the least squares fitting) to find the asymptotic convergence rate. In this way, we notice the second-order convergence for one ratio, $\|u - u_h\|_\infty \sim h^2$, and superlinear convergence for the other, $\|u - u_h\|_\infty \sim h^{1.565}$, where u is the exact solution of the boundary value problem. Similar behavior is observed for other examples. The magnitude of the errors with a 160×160 grid is about 10^{-4} for both ratios.

8.5.1 Numerical results for the conforming IFEM

Now we present the numerical results for the same boundary value problem for the conforming IFEM. We also report the error of the interpolation function that is important in applying the finite element theory and is useful in deriving the error estimate for the maximum norm.

Figure 8.8(a) plots the errors between the exact solution and its interpolation functions in the conforming IFE space $S_h(\Omega)$ with the jump ratio $\rho = \beta^-/\beta^+ = 1 : 1000$ and various partition sizes h. Figure 8.8(b) is the plot of the error in the x partial derivative of the interpolation function. We obtained similar results with other ratios and partial derivatives. Thus, this example confirmed the error analysis for the interpolation function. Note that the magnitude of the interpolation error is about 10^{-4} for the solution and 10^{-2} for the x partial derivative in a typical 160×160 grid.

(a) (b)

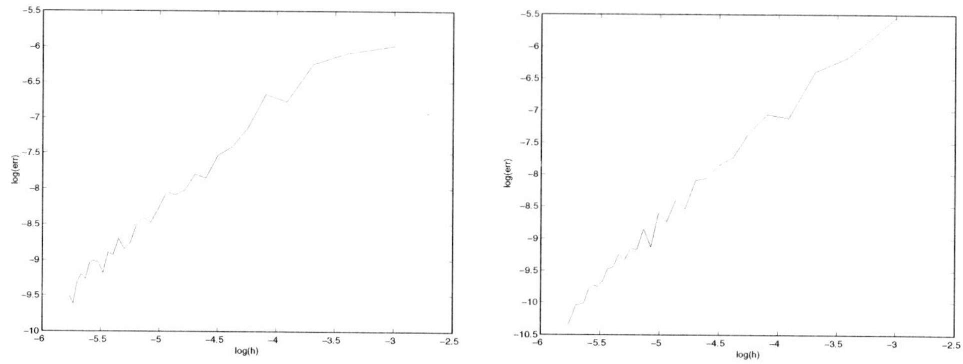

Figure 8.8. *The interpolation errors in the maximum norm versus the mesh size h for conforming basis functions in* log-log *scale with* $r_0 = \pi/6.28$, $\beta^- = 1$, *and* $\beta^+ = 1000$. (a) *The linear regression analysis gives* $||u - u_I||_\infty \approx 3.22816\, h^{2.06743}$. (b) *The error in the partial derivative* $\partial u/\partial x$ *excluding the region* $\sum T_r$; *the linear regression analysis gives* $||(u - u_I)_x||_{\infty, \sum T \backslash T_r} \approx 2.89806\, h^{0.96056}$.

(a) (b)

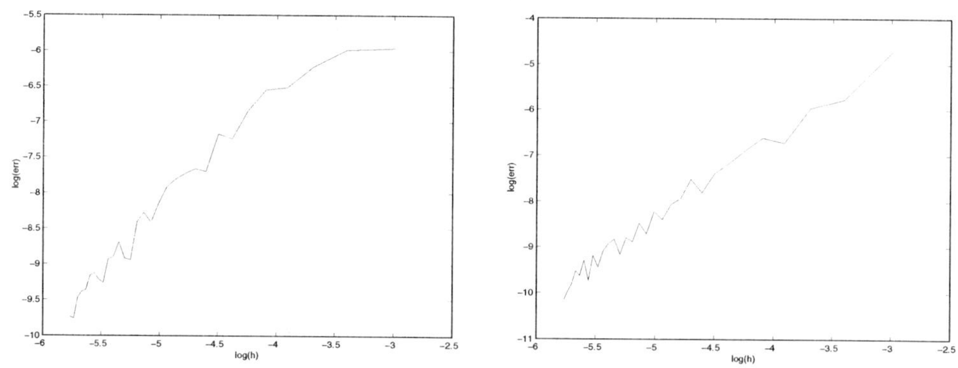

Figure 8.9. *Errors of finite element solutions obtained from the conforming basis function in the maximum norm versus the mesh size h in* log-log *scale with* $r_0 = \pi/6.28$. (a) $\beta^- = 1$, $\beta^+ = 1000$; *the linear regression analysis gives* $||u - u_h||_\infty \approx 6.85126\, h^{2.01002}$. (b) $\beta^- = 1000$, $\beta^- = 1$; *the linear regression analysis gives* $||u - u_h||_\infty \approx 5.65703\, h^{2.01542}$.

Figure 8.9 plots the errors in the maximum norm of the conforming IFE solutions u_h from $S_h(\Omega)$ with various h for two different ratios. The linear regression analysis shows that the data in Figure 8.9 obey

$$||u - u_h||_\infty \approx 6.85126\ h^{2.01002}, \qquad \rho = 1 : 1000,$$
$$||u - u_h||_\infty \approx 5.65703\ h^{2.01542}, \qquad \rho = 1000 : 1,$$

which suggests that the conforming IFE solution has a second-order convergence rate in the maximum norm.

8.5.2 A comparison with the finite element method with added nodes

As a slightly different method between a uniform Cartesian mesh and a body-fitted mesh, a natural approach is to add the intersections of the edges of the triangles and the interface as additional nodal points. Specifically, the triangulation is generated as follows.

1. We first generate a Cartesian triangulation composed of the right triangles over Ω.
2. We keep all the elements over the noninterface triangles unchanged.
3. For each interface triangle, we break it into three small triangles *in the same way* as we did in step **P4** in §8.4.1; see also Figure 8.6. Therefore the breakup satisfies the same maximum angle condition as we did earlier for the conforming IFEM.

The standard Galerkin finite element method with the usual linear basis functions is then applied to this triangulation. This method is called the finite element method using a Cartesian grid with added nodes, or FEMCGAN. The computational complexity of this approach is about the same as the conforming finite element method. Below we list some features of the two finite element methods.

- The convergence result of Theorem 8.11 is also valid for the FEMCGAN. However, this is guaranteed only with the choice of the maximum angles described here and in [170].
- In the FEMCGAN approach, all the intersections between the interface and the edges of Cartesian triangles are the added nodal points. However, in the IFEMs, either nonconforming or conforming, those intersections are not part of the nodal points. Therefore, the linear system of equations from the IFE approach will be of order $O(1/h)$ smaller in dimension compared with that from the FEMCGAN approach.
- More important, some linear solvers based on Cartesian grids can be applied to the nonconforming or conforming IFEMs but not to the FEMCGAN approach. In some applications, we are interested only in the solution at the grid points; there is no need to recover the solution at the points of the intersections.
- The FEMCGAN space contains the IFE space, so we can expect the energy norm of the error to be smaller than that obtained from the IFEM; see Table 8.2.

In Table 8.2, we show the results of the errors in $L^2(\Omega)$ and energy norms of the FEMCGAN approach and the IFEM for the same example. We can see clearly from the table that the two methods are comparable. Both methods give second-order accurate results in the $L^2(\Omega)$ norm and first-order accuracy in the energy norm. The linear regression analysis is conducted for the convergence in the L^∞ norm for both methods. The comparison results are listed in Table 8.3.

Again, these numerical results indicate that these two methods perform comparably. The discussions in this chapter can be modified for almost any arbitrary grid that is not necessarily aligned with interfaces. The methods based on the Cartesian grids can be easily used as finite difference methods. While the conforming IFEM becomes a little

Table 8.2. *Comparisons of errors of the FEMCGAN and the conforming IFEM, where $e_0(h)$ and $e_a(h)$ are errors of a numerical solution in the $L^2(\Omega)$ and energy norms, respectively. The example is the same as the example in §8.5 for the case when $\beta^- = 1$, $\beta^+ = 1000$.*

	The FEMCGAN			
h	$e_0(h)$	Ratio	$e_a(h)$	Ratio
1/20	5.5479×10^{-4}		3.0085×10^{-2}	
1/40	1.4040×10^{-4}	3.9516	1.5376×10^{-2}	1.9566
1/80	3.5525×10^{-5}	3.9520	7.7803×10^{-3}	1.9762
1/160	9.1518×10^{-6}	3.8817	3.9160×10^{-3}	1.9868
	The conforming IFEM			
1/20	7.7184×10^{-4}		3.4742×10^{-2}	
1/40	1.9050×10^{-4}	4.0516	1.7136×10^{-2}	2.0275
1/80	4.5729×10^{-5}	4.1659	8.4975×10^{-3}	2.0165
1/160	1.0596×10^{-5}	4.3158	4.1195×10^{-3}	2.0627

Table 8.3. *Comparison of errors of the FEMCGAN and the IFEMs using linear regression analysis, where e_0, e_1, e_a, and e_∞ are errors in the $L^2(\Omega)$, $H^1(\Omega)$, energy norms, and $L^\infty(\Omega)$, respectively.*

	e_0	e_1	e_a	e_∞
IFEM	$0.208\, h^{1.9803}$	$0.588\, h^{0.9902}$	$0.604\, h^{0.9923}$	$0.142\, h^{1.8562}$
FEMCGAN	$0.774\, h^{2.2106}$	$0.669\, h^{1.0142}$	$0.924\, h^{1.0666}$	$1.701\, h^{2.0100}$

more complicated in terms of programming due to the extension of the support of the basis functions, the simple structure of a Cartesian triangulation should offset the increased complexity. More important, the IFEM can be incorporated into other Cartesian grids–based methods and packages, for examples, LeVeque's Clawpack [153] and Berger's AMR package [18], to solve interface problems.

8.6 IFEM for problems with nonhomogeneous jump conditions

The IFEM discussed in this chapter works well for interface problems with natural jump conditions. When the flux has a jump at the interface, that is, $[\beta u_n]_\Gamma = Q$, the weak form for $-\nabla \cdot (\beta \nabla u) = f$ is

$$\int_\Omega \beta u_x v_x dx = \int_\Omega f v dx - Q \tag{8.60}$$

in one dimension, and

$$\iint_{\Omega} \beta \nabla u \cdot \nabla v dxdy = \iint_{\Omega} f v dxdy - \int_{\Gamma} Q v ds \tag{8.61}$$

in two dimensions. Note that we use Q to represent the flux since v has been used as a testing function. However, if the IFEM discussed in previous sections can be applied to the weak form directly, the numerical result is first-order accurate in the maximum norm at best. This is because the basis functions do not satisfy the flux jump condition. To gain insight on this, consider the simplest case in which $\beta \equiv 1$, $f \equiv 0$. The finite element method using the weak form is equivalent to the finite difference method,

$$-\frac{U_{i-1} - 2U_i + U_{i+1}}{h^2} = \begin{cases} -\dfrac{Q}{h^2} & \text{if } i = j,\, j+1,\, x_j \le \alpha < x_{j+1}, \\ 0 & \text{otherwise.} \end{cases} \tag{8.62}$$

The right-hand side of the linear system of equations above can be regarded as a discrete delta function applied to $Q\delta(x-\alpha)$. Obviously, the discrete delta function does not satisfy the moments equation described in [23, 258] and is no better than the discrete hat function (1.19) and the discrete cosine function (1.20). It is obvious that such a method generally does not yield second-order accurate results in the maximum norm.

In order to make the IFEM work for interface problems with nonhomogeneous jump conditions, we can transform the interface problem into a new one with homogeneous jump conditions using the strategy described in Chapter 5. Then we can apply the IFEM to the transformed interface problem. A different approach, which is likely to be between first- and second-order accurate, is given in [105].

Now we assume that both the solution and the flux have jumps as described by $[u] = w$ and $[\beta u_n] = Q$. Following the notations from Chapter 5, we can transform the interface problem into

$$-\nabla \cdot (\beta \nabla q) = f + H(\varphi)\, \nabla \cdot (\beta \nabla \tilde{u}), \qquad \mathbf{x} \in \Omega \setminus \Gamma, \tag{8.63}$$

with $[q] = 0$ and $[\beta q_n] = 0$, where $q = u - \hat{u}$, φ, and $\tilde{u}$ have the same meaning as those in Chapter 5. The weak form in terms of u is

$$\begin{aligned} \iint_{\Omega} \beta \nabla u \cdot \nabla v \, dxdy = \iint_{\Omega} f\, v \, dxdy + \iint_{\Omega} \beta \nabla \hat{u} \cdot \nabla v \, dxdy \\ + \iint_{\Omega} H(\varphi)\, \nabla \cdot (\beta \nabla \tilde{u})\, v \, dxdy. \end{aligned} \tag{8.64}$$

At a noninterface triangle, using the IFEM, the last two terms cancel each other to an order $O(h^2)$. Thus, it is only at interface triangles that the stiffness matrix and the local vector need to be modified. We show an example in one dimension using this approach.

Example 8.3. Consider the two-point boundary value problem

$$-u''(x) = \begin{cases} 4\sin(2x) & \text{if } x < \alpha, \\ \dfrac{1}{x^2} & \text{if } x > \alpha, \end{cases} \qquad x \in (0,\alpha) \cup (\alpha, 1),$$

$$u(0) = 0, \quad u(1) = 0, \quad [u]_{x=\alpha} = \log(\alpha) - \sin(2\alpha), \quad [u_x]_{x=\alpha} = \frac{1}{\alpha} - 2\cos(2\alpha).$$

Table 8.4. *A grid refinement analysis of the IFEM for Example* 8.3 *in which both the solution and the flux have a nonzero jump at* $\alpha = 1/3$.

M	E_M	E_M/E_{2M}
20	1.2000×10^{-3}	
40	2.9334×10^{-4}	4.0908
80	7.5255×10^{-5}	3.8979
160	1.8745×10^{-5}	4.0147
320	4.7166×10^{-6}	4.9743
640	1.1781×10^{-6}	4.0036

The exact solution is

$$u(x) = \begin{cases} \sin(2x) & \text{if } x < \alpha, \\ \log(x) & \text{if } x > \alpha. \end{cases}$$

Both the solution and the derivative have a finite jump at α.

Table 8.4 shows a grid refinement analysis result using the IFEM applied to the transformed problem (8.64) with a modified basis function. Second-order accuracy in the maximum norm measured at the grid points is achieved.

Instead of a piecewise linear conforming finite element space, a piecewise quadratic conforming IFE space has been developed in [93]. One of the advantages of the piecewise quadratic conforming IFE space is that it does not need to enforce the angle constraint and it is potentially useful for high-order, say piecewise $H^2(\Omega)$, spaces. The method has been coupled with the removing source singularity technique to deal with nonhomogeneous jump conditions.

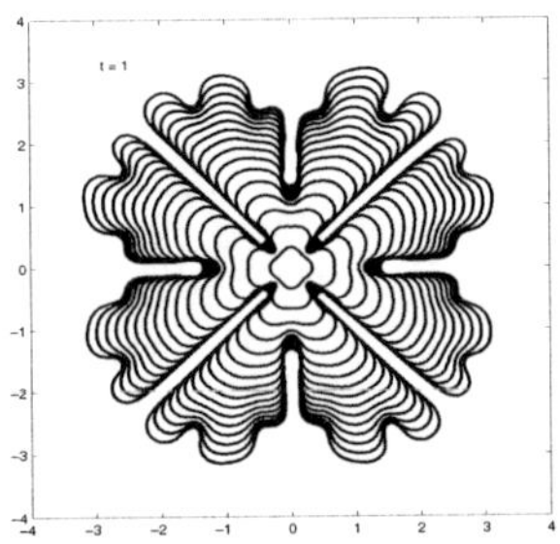

Chapter 9

The IIM for Parabolic Interface Problems

The IIM for parabolic interface problems with applications has been developed in [6, 160, 162, 172, 173, 175, 177, 181]. In this chapter, we explain the method for one-dimensional elliptic interface problems with fixed and moving interfaces, the alternative directional implicit (ADI) method for heat equations with a fixed interface, and the IIM for diffusion and advection equations with a fixed interface. The IIM for Stokes and Navier–Stokes equations with interfaces is explained in the next chapter.

9.1 The IIM for one-dimensional heat equations with fixed interfaces

Consider the model problem,

$$\begin{gathered} u_t(x,t) = (\beta(x,t)\,u_x)_x - \sigma(x,t)\,u(x,t) - f(x,t) + v(t)\delta(x-\alpha), \\ 0 \le x \le 1, \qquad 0 < \alpha < 1, \qquad t \ge 0, \end{gathered} \tag{9.1}$$

with specified boundary and initial conditions. We assume that $\beta(x,t)$, $\sigma(x,t)$, and $f(x,t)$ are bounded but may have a finite discontinuity at the interface α. From the equation we can conclude that

$$[\beta u_x] = v(t). \tag{9.2}$$

We also specify

$$[u] = w(t). \tag{9.3}$$

With the IIM, the standard Crank–Nicolson scheme, which is unconditionally stable, is used at regular grid points. The finite difference scheme from time level t^n to t^{n+1} has the following generic form:

$$\begin{aligned} \frac{U_i^{n+1} - U_i^n}{\Delta t} = \frac{1}{2}\, &\left\{\gamma_{i,1}^n U_{i-1}^n + \gamma_{i,2}^n U_i^n + \gamma_{i,3}^n U_{i+1}^n - \sigma_i^n\, U_i^n - f_i^n + C_i^n \right. \\ &\left. + \gamma_{i,1}^{n+1} U_{i-1}^{n+1} + \gamma_{i,2}^{n+1} U_i^{n+1} + \gamma_{i,3}^{n+1} U_{i+1}^{n+1} - \sigma_i^{n+1}\, U_i^{n+1} - f_i^{n+1} + C_i^{n+1} \right\}, \end{aligned}$$

where Δt is the time step and the ratio $\Delta t/h$ is a constant, $\sigma_i^n = \sigma(x_i, t^n)$, and so on. At regular grid points for which $\alpha \notin (x_{i-1}, x_{i+1})$, we have the standard finite difference coefficients,

$$\begin{aligned} \gamma_{i,1}^n &= \frac{\beta_{i-\frac{1}{2}}^n}{h^2}, \qquad \gamma_{i,2}^n = -\frac{\left(\beta_{i-\frac{1}{2}}^n + \beta_{i+\frac{1}{2}}^n\right)}{h^2}, \\ \gamma_{i,3}^n &= \frac{\beta_{i+\frac{1}{2}}^n}{h^2}, \qquad C_i^n = 0, \end{aligned} \tag{9.4}$$

where $\beta_{i-1/2}^n = \beta(x_{i-1/2}, t^n)$, and so on. Since the interface is fixed, the derivation for the finite difference scheme is just slightly different from that in Chapter 2. So we omit the details and give the results directly.

Suppose $x_j \le \alpha < x_{j+1}$; then x_j and x_{j+1} are two irregular grid points. In this case the coefficients $\gamma_{j,1}^n$, $\gamma_{j,2}^n$, and $\gamma_{j,3}^n$ satisfy the following system of equations:

$$\gamma_{j,1}^n + \gamma_{j,2}^n + \left(1 + \frac{(x_{j+1}-\alpha)^2}{2\,(\beta^+)^n}\,[\sigma]^n\right)\gamma_{j,3}^n = 0,$$

$$\begin{aligned} &(x_{j-1}-\alpha)\,\gamma_{j,1}^n + (x_j-\alpha)\,\gamma_{j,2}^n + \left\{ \frac{(\beta^-)^n}{(\beta^+)^n}\,(x_{j+1}-\alpha) \right. \\ &\qquad \left. + \left(\frac{(\beta_x^-)^n}{(\beta^+)^n} - \frac{(\beta^-)^n(\beta_x^+)^n}{\{(\beta^+)^n\}^2} \right) \frac{(x_{j+1}-\alpha)^2}{2} \right\} \gamma_{j,3}^n = (\beta_x^-)^n, \end{aligned}$$

$$\frac{(x_{j-1}-\alpha)^2}{2}\gamma_{j,1}^n + \frac{(x_j-\alpha)^2}{2}\gamma_{j,2}^n + \frac{(x_{j+1}-\alpha)^2\,(\beta^-)^n}{2\,(\beta^+)^n}\,\gamma_{j,3}^n = (\beta^-)^n,$$

where

$$[\sigma]^n = \left[\sigma(\alpha, x^n)\right], \qquad (\beta^-)^n = \beta(\alpha^-, t^n), \qquad (\beta^+)^n = \beta(\alpha^+, t^n),$$

and so on; see (2.14) for a comparison. The correction term C_j^n is

$$\begin{aligned} C_j^n &= \gamma_{j,3}^n \left\{ w^n + (x_{j+1}-\alpha)\,\frac{v^n}{(\beta^+)^n} \right\} \\ &\quad - \gamma_{j,3}^n\,\frac{(x_{j+1}-\alpha)^2}{2} \left\{ \frac{(\beta_x^+)^n\,v^n}{\{(\beta^+)^n\}^2} - \frac{(\sigma^+)^n\,w^n + \dfrac{dw}{dt}(t^n) + [f]^n}{(\beta^+)^n} \right\}. \end{aligned}$$

Notice that now there is an extra term $\frac{dw(t^n)}{dt}$ compared with the correction term (2.15) in the general one-dimensional elliptic problem due to the $[u_t]$ term. Similarly, at the grid point

x_{j+1}, the coefficients $\gamma_{j+1,1}^n$, $\gamma_{j+1,2}^n$, and $\gamma_{j+1,3}^n$ satisfy the following system of equations:

$$\left(1-\frac{(x_j-\alpha)^2}{2\,(\beta^-)^n}\,[\sigma]^n\right)\gamma_{j+1,1}^n+\gamma_{j+1,2}^n+\gamma_{j+1,3}^n=0,$$

$$\left\{\frac{(\beta^+)^n}{(\beta^-)^n}(x_j-\alpha)+\left(\frac{(\beta_x^+)^n}{(\beta^-)^n}-\frac{(\beta_x^-)^n\,(\beta^+)^n}{\{(\beta^-)^n\}^2}\right)\frac{(x_j-\alpha)^2}{2}\right\}\gamma_{j+1,1}^n$$
$$+\,(x_{j+1}-\alpha)\,\gamma_{j+1,2}^n+(x_{j+2}-\alpha)\,\gamma_{j+1,3}^n=\left(\beta_x^+\right)^n,$$

$$\frac{(x_j-\alpha)^2}{2}\frac{(\beta^+)^n}{(\beta^-)^n}\gamma_{j+1,1}^n+\frac{(x_{j+1}-\alpha)^2}{2}\gamma_{j+1,2}^n+\frac{(x_{j+2}-\alpha)^2}{2}\gamma_{j+1,3}^n=(\beta^+)^n;$$

see (2.16) for a comparison. The correction term now is

$$C_{j+1}^n=\gamma_{j+1,1}^n\left\{-w^n-(x_j-\alpha)\frac{v^n}{(\beta^-)^n}\right\}$$
$$-\,\gamma_{j+1,1}^n\frac{(x_j-\alpha)^2}{2}\left\{\frac{(\beta_x^-)^n\,v^n}{\{-(\beta^-)^n\}^2}+\frac{(\sigma^-)^n\,w^n+dw/dt\,(t^n)+[f]^n}{(\beta^-)^n}\right\}.$$

Note that the formulas are exactly the same for the time level $n+1$. The method is unconditionally stable if $\sigma\ge 0$ regardless of the jumps, provided that $\beta(x,t)$ has the same sign across the interface.

9.2 The IIM for one-dimensional moving interface problems

In this section, we discuss the IIM for the one-dimensional moving interface problem,

$$\begin{aligned}&u_t+\lambda u u_x=(\beta u_x)_x-f(x,t), && x\in(0,\,\alpha)\cup(\alpha,\,1),\\&\frac{d\alpha}{dt}=g\left(t,\alpha;\,u^-,u^+,u_x^-,u_x^+\right), && t>0,\end{aligned}\tag{9.5}$$

with an initial condition and a prescribed boundary condition at $x=0$ and $x=1$, where $\beta(x,t)>0$ and g are given functions. As before, the source term $f(x,t)$ may be discontinuous or have a delta function singularity at the interface $\alpha(t)$. It is reasonable to assume that the solution is piecewise smooth and discontinuities can occur only at the interface $\alpha(t)$.

The interface $\alpha(t)$ divides the solution domain into two parts: $0\le x<\alpha(t)$ and $\alpha(t)<x\le 1$. The solution in each domain $[0,\alpha(t))$ and $(\alpha(t),1]$ is assumed to be smooth, but coupled with the solution on the other side by interface conditions (or internal boundary conditions) that usually take one of the following forms.

Case 1: The solution on the interface is given. One example is the classical Stefan model for one-dimensional solidification problems. The temperature at the melting/freezing interface is given by the melting temperature, that is, $u(\alpha,t)=u_0$ is known.

Various approaches have been used to solve the Stefan problem and other *linear* free or moving interface problems numerically; see, for example, [9, 38, 59, 78, 83, 84, 114, 175, 204, 226, 271]; also see [42, 236] which use the level set method. Compared with Case 2 discussed below, the Stefan problem is easier to solve because the value of the solution on the interface is known. However, a few numerical methods are second-order accurate in the maximum norm for both the solution and the interface. Several methods involve some transformations for either the differential equations or the coordinate system, which complicates the problem in some way. The IIM proposed in [162] is simple, stable, and is second-order accurate for both the solution u and the interface $\alpha(t)$ simultaneously for more general equations.

Case 2: The jump conditions of the form

$$[u] = w(t), \qquad [\beta u_x] = v(t) \tag{9.6}$$

are given. This is a one-dimensional model for the immersed boundary method formulation with a more general equation for the motion. The problem can be written as a single equation without using the jump conditions:

$$u_t + \lambda u u_x = (\beta u_x)_x - v(t)\,\delta(x-\alpha(t)) - \hat{C}(t)\,\delta'(x-\alpha(t)) - f(x,t)$$

for some function $\hat{C}(t)$.

Beyer and LeVeque [23] studied various one-dimensional moving interface problems for the heat equation assuming a priori knowledge of the interface. In their approach a discrete delta function is carefully selected and some correction terms are added if necessary to get second-order accuracy. Wiegmann and Bube [269] applied the IIM for certain one-dimensional nonlinear problem with a *fixed* interface. However, for the moving interface problem (9.5), the interface is *unknown* and *moving*, and the discrete difference scheme is a nonlinear system of equations involving the solution and the interface.

Case 2 (with $\lambda = 0$) is also a model of the heat conduction with an interface between two different materials. In this case u is the temperature, and hence is continuous, i.e., $w(t) \equiv 0$ in (9.6). The net heat flux across the interface is $v(t)$ in the second jump condition in (9.6). Again, in this case we do not know the value of the solution on the interface but only the jump conditions.

For many classical Stefan problems, the motion of the interface is proportional to the flux across the interface,

$$\frac{d\alpha}{dt} = \sigma(t)\,[\beta u_x], \qquad u(\alpha, t) = u_0, \tag{9.7}$$

where u_0 is the known temperature at the interface. This type of problems fits both Case 1 and Case 2.

9.2.1 The modified Crank–Nicholson scheme

Given a uniform grid, set

$$x_i = ih, \qquad i = 0, 1, \ldots, M, \quad x_0 = 0, \quad x_M = 1.$$

Let Δt be the time step size and let the ratio $\Delta t/h$ be a constant so that we can write $O(\Delta t)$ as $O(h)$ or vice versa. Using the Crank–Nicolson scheme, the semidiscrete difference scheme for (9.5) can be written in the following general form:

$$\begin{aligned}&\frac{U_i^{n+1}-U_i^n}{\Delta t}-Q_i^{n+\frac{1}{2}}+\frac{\lambda}{2}\left(U_i^nU_{x,i}^n+U_i^{n+1}U_{x,i}^{n+1}\right)\\&=\frac{1}{2}\left((\beta U_x)_{x,i}^n+(\beta U_x)_{x,i}^{n+1}\right)-\frac{1}{2}\left(f_i^n+f_i^{n+1}\right),\end{aligned}\tag{9.8}$$

where $U_{x,i}^n$ and $(\beta U_x)_{x,i}^n$ are discrete analogues of u_x and $(\beta u_x)_x$ at (x_i,t^n), and $Q_i^{n+\frac{1}{2}}$ is a correction term needed when α crosses the grid line $x=x_i$ at some time between t^n and t^{n+1}. We will discuss how to determine $Q_i^{n+\frac{1}{2}}$ in the next subsection. For simplicity, we will drop the superscript n in the discussion of the spatial discretization if there is no confusion. At a grid point x_i, which is away from the interface (i.e., $\alpha\notin[x_{i-1},x_{i+1}]$), the classic 3-point central finite difference discretizations are

$$U_{x,i}=\frac{U_{i+1}-U_{i-1}}{2h},\tag{9.9}$$

$$(\beta U_x)_{x,i}=\frac{\beta_{i-\frac{1}{2}}U_{i-1}-\left(\beta_{i-\frac{1}{2}}+\beta_{i+\frac{1}{2}}\right)U_i+\beta_{i+\frac{1}{2}}U_{i+1}}{h^2},\tag{9.10}$$

where $\beta_{i+\frac{1}{2}}=\beta(x_i+\frac{h}{2},:)$. In §9.2.3 we will discuss how to discretize u_x and $(\beta u_x)_x$ when $\alpha\in[x_{j-1},x_{j+1})$ for Cases 1 and 2.

The interface location is determined by the trapezoidal method applied to the second equation in (9.5),

$$\frac{\alpha^{n+1}-\alpha^n}{\Delta t}=\frac{1}{2}\left(g^n+g^{n+1}\right),\tag{9.11}$$

where $g^n=g(t^n,\alpha^n;\,u^{-,n},\,u^{+,n},\,u_x^{-,n},\,u_x^{+,n})$ and α^n, $u^{\pm,n}$, $u_x^{\pm,n}$ are the approximations to $\alpha(t^n)$, $u(\alpha^\pm,t^n)$, and $u_x(\alpha^\pm,t^n)$, respectively. The same is true for the time level $n+1$. The discretizations (9.8) and (9.11) are second-order accurate and fully implicit. The core of the algorithm at a time level t^n consists of the following.

- Determine $Q_j^{n+\frac{1}{2}}$ if the interface crosses the grid line $x=x_j$ from time t^n to time t^{n+1}.
- Derive the finite difference approximations for u_x and $(\beta u_x)_x$ at the two grid points closest to the interface.
- Compute the interface quantities $u^\pm$, $u_x^\pm$, $[u_t]$, etc.
- Solve the nonlinear system of equations for the approximate solution $\{U_i^{n+1}\}$ and the approximate interface location α^{n+1}.

Away from the interface, the local truncation errors for the finite difference scheme are $O(h^2)$. But at a few grid points near the interface, we allow the local truncation errors to be $O(h)$ based on the fact that the local truncation error of a finite difference scheme on a boundary can be one order lower than those of interior points without affecting global second-order accuracy.

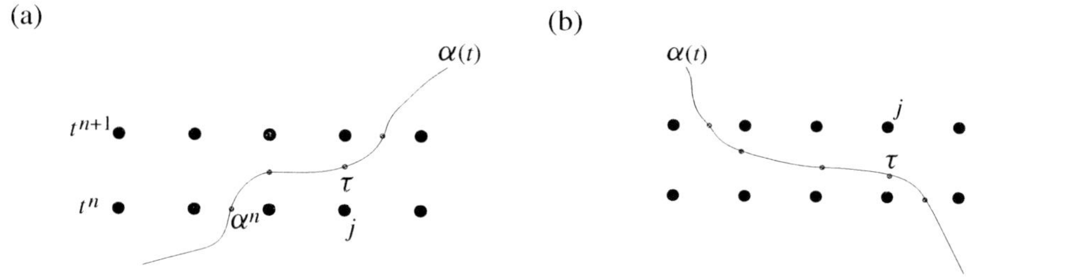

Figure 9.1. *A diagram of an interface crossing a grid line.* (a) $\alpha(t)$ *increases with time.* (b) $\alpha(t)$ *decreases with time.*

9.2.2 Dealing with grid crossing

If there is no grid crossing at a grid point x_i from time t^n to time t^{n+1}, that is, (x_i, t^n) and (x_i, t^{n+1}) are on the same side of the interface $\alpha(t)$, then $Q_i^{n+\frac{1}{2}} = 0$ and

$$\frac{u(x_i, t^{n+1}) - u(x_i, t^n)}{\Delta t} = \frac{1}{2}\Big(u_t(x_i, t^{n+1}) + u_t(x_i, t^n)\Big) + O((\Delta t)^2). \tag{9.12}$$

However, if the interface crosses the grid line $x = x_j$ at some time $\tau,\ t^n < \tau < t^{n+1}$, such that $x_j = \alpha(\tau)$[13] (see Figure 9.1), then the time derivative of u may have a finite jump at $t = \tau$. In this case, even though we can approximate the x-derivatives of u well at each time level (see §9.2.3), the standard Crank–Nicolson scheme needs to be corrected to guarantee second-order accuracy. This is done by choosing a correction term $Q_j^{n+\frac{1}{2}}$ based on the following theorem.

Theorem 9.1. *Let* $u(x,t)$ *be the solution of* (9.5). *Suppose that the equation* $\alpha(t) = x_j$ *has a unique solution* τ *in the interval* $t^n < t < t^{n+1}$. *If we choose*

$$Q_j^{n+\frac{1}{2}} = \frac{[u]_{;\tau}}{\Delta t} + \frac{1}{\Delta t}\left(t^{n+\frac{1}{2}} - \tau\right)[u_t]_{;\tau}, \tag{9.13}$$

then

$$\begin{aligned} &\frac{u(x_j, t^{n+1}) - u(x_j, t^n)}{\Delta t} - Q_j^{n+\frac{1}{2}} \\ &\qquad = \frac{1}{2}\left(u_t(x_j, t^{n+1}) + u_t(x_j, t^n)\right) + O(\Delta t). \end{aligned} \tag{9.14}$$

Proof: We expand $u(x_j, t^n)$ and $u(x_j, t^{n+1})$ in Taylor series at approximately time τ from each side of the interface to get

$$u(x_j, t^n) = u(x_j, \tau^-) + (t^n - \tau)\, u_t(x_j, \tau^-) + O((\Delta t)^2),$$

$$\begin{aligned} u(x_j, t^{n+1}) &= u(x_j, \tau^+) + (t^{n+1} - \tau)\, u_t(x_j, \tau^+) + O(\Delta t^2) \\ &= u(x_j, \tau^-) + [u]_{;\tau} + (t^{n+1} - \tau)\, u_t(x_j, \tau^-) + (t^{n+1} - \tau)\, [u_t]_{;\tau} + O((\Delta t)^2). \end{aligned}$$

[13]The crossing time τ really depends on the grid index j as well as the time index n; see Figure 9.1. To simplify the notation, τ will be used to indicate the crossing time, without explicitly showing its dependence on j and n.

Combining the two expressions above gives

$$u(x_j, t^{n+1}) - u(x_j, t^n) = \Delta t\, u_t(x_j, \tau^-) + [u_t]_{;\tau} \\ + (t^{n+1} - \tau)\,[u_t]_{;\tau} + O\big((\Delta t)^2\big). \tag{9.15}$$

On the other hand, we also have

$$u_t(x_j, t^n) = u_t(x_j, \tau^-) + O(\Delta t),$$

$$u_t(x_j, t^{n+1}) = u_t(x_j, \tau^-) + [u_t]_{;\tau} + O(\Delta t).$$

Thus it follows that

$$\Delta t\, u_t(x_j, \tau^-) = \frac{\Delta t}{2}\big(u_t(x_j, t^n) + u_t(x_j, t^{n+1})\big) - \frac{\Delta t}{2}[u_t]_{;\tau} + O\big((\Delta t)^2\big). \tag{9.16}$$

Substituting (9.16) into (9.15) gives

$$\frac{u(x_j, t^{n+1}) - u(x_j, t^n)}{\Delta t} = \frac{1}{2}\big(u_t(x_j, t^n) + u_t(x_j, t^{n+1})\big) + \frac{[u]_{;\tau}}{\Delta t} \\ + \frac{t^{n+\frac{1}{2}} - \tau}{\Delta t}\,[u_t]_{;\tau} + O(\Delta t). \tag{9.17}$$

This is equivalent to (9.14). □

We know $[u]_{;\tau}$ from the jump conditions. However, to compute $Q_j^{n+\frac{1}{2}}$, we also need to find the crossing time τ and the jump $[u_t]_{;\tau}$. We first discuss how to find τ if it exists. We will discuss how to approximate $[u_t]_{;\tau}$ in §9.2.4. Using the Crank–Nicolson formula twice, we get

$$\frac{\alpha^\tau - \alpha^n}{\tau - t^n} = \frac{1}{2}\left(g^n + g^\tau\right),$$

$$\frac{\alpha^{n+1} - \alpha^\tau}{t^{n+1} - \tau} = \frac{1}{2}\left(g^\tau + g^{n+1}\right).$$

Eliminating g^τ from the two expressions above and using $\alpha^\tau = x_j$, we get

$$\frac{x_j - \alpha^n}{\tau - t^n} - \frac{\alpha^{n+1} - x_j}{t^{n+1} - \tau} = \frac{1}{2}\left(g^n - g^{n+1}\right). \tag{9.18}$$

This is the equation for the crossing time τ and it is coupled with (9.8), (9.11), and (9.13).

Note that the discussion above is still valid even if the interface crosses several grid points during one time step. However, it would be better to control the time step so that the interface crosses only one grid point during one time step. This will give a smaller error constant.

9.2.3 The discretizations of u_x and $(\beta u_x)_x$ near the interface

As before, only the finite difference equations at the closest grid points from the left and the right of the interface need to be modified at each time level t^n or t^{n+1}. The discretization apparently depends on interface conditions and will be discussed separately in this section.

Case 1: The solution on the interface is known. Let the solution on the interface be

$$u\,(\alpha(t), t) = r(t), \tag{9.19}$$

where $r(t)$ is a given function. Since we know the value of the solution on the interface, we could discretize u_x and u_{xx} using a one-sided interpolation. For example, if $x_j \le \alpha^n < x_{j+1}$, then

$$u^n_{x,j} \approx U^n_x(j-1,\ j,\ \alpha^n, x_j),$$

where

$$\begin{aligned} U_x(j-1,\ j,\ \alpha, x) = {} & \frac{x_j + \alpha - 2x}{(x_{j-1} - \alpha)\,h}\, U_{j-1} + \frac{x_{j-1} + \alpha - 2x}{(\alpha - x_j)\,h}\, U_j \\ & + \frac{x_{j-1} + x_j - 2x}{(x_j - \alpha)(\alpha - x_{j-1})}\, r(t). \end{aligned} \tag{9.20}$$

This is a second-order approximation to $u_x(x_j, t)$. However, notice that $\alpha(t)$ changes with time, as does $x_j - \alpha$. If $|x_j - \alpha|$ becomes too small, the magnitudes of the coefficients in the interpolation (9.20) become very large and sometimes even blow up. An intuitive fix would be

$$u^n_{x,j} \approx U^n_x(j-2,\ j-1,\ \alpha^n, x_j),$$

when $|x_j - \alpha|$ is small. A more robust fix is by the linear combination of those two above,

$$\begin{aligned} u^n_{x,j} \approx {} & \frac{\alpha^n - x_j}{h}\, U^n_x(j-1,\ j,\ \alpha^n, x_j) \\ & + \frac{x_{j+1} - \alpha^n}{h}\, U^n_x(j-2,\ j-1,\ \alpha^n, x_j). \end{aligned} \tag{9.21}$$

This approach is the simplified version of the least squares interpolation in one dimension. There are several advantages of this robust approach. First of all, the interpolation is still second-order accurate. Second, if we rewrite (9.21) as

$$u^n_{x,j} \approx \sum_{j=j-2}^{j} \gamma^n_j\, U^n_j + \gamma^n_\alpha\, r(t^n),$$

then the magnitudes of the coefficients γ^n_j and γ^n_α will be always of order $O(1/h)$. Furthermore, the truncation error in such an interpolation varies smoothly, which gives better stability.

In the same manner, we use the following interpolation to discretize $u^n_{x,j+1}$:

$$\begin{aligned} u^n_{x,j+1} \approx {} & \frac{\alpha^n - x_j}{h}\, U^n_x(j+2,\ j+3,\ \alpha^n, x_{j+1}) \\ & + \frac{x_{j+1} - \alpha^n}{h}\, U^n_x(j+1,\ j+2,\ \alpha^n, x_{j+1}). \end{aligned} \tag{9.22}$$

For the second-order derivative u_{xx}, the corresponding one-sided finite difference approximation for u_{xx} is

$$\begin{aligned} U_{xx}(j-1,\ j,\ \alpha,x) = & \frac{2}{(\alpha - x_{j-1})\,h} U_{j-1} + \frac{2}{(\alpha - x_j)\,h} U_j \\ & + \frac{2}{(x_{j-1}-\alpha)(x_j-\alpha)} r(t). \end{aligned} \tag{9.23}$$

Using the formula above, we get the following robust finite difference approximations for $u^n_{xx,j}$:

$$\begin{aligned} u^n_{xx,j} \approx & \frac{\alpha^n - x_j}{h} U^n_{xx}(j-1,\ j,\ \alpha^n, x_j) \\ & + \frac{x_{j+1}-\alpha^n}{h} U^n_{xx}(j-2,\ j-1,\ \alpha^n, x_j), \end{aligned} \tag{9.24}$$

where U^n_{j-2}, U^n_{j-1}, and U^n_j are used. Similarly, we have

$$\begin{aligned} u^n_{xx,j+1} \approx & \frac{\alpha^n - x_j}{h} U^n_{xx}(j+2,\ j+3,\ \alpha^n, x_{j+1}) \\ & + \frac{x_{j+1}-\alpha^n}{h} U^n_{xx}(j+1,\ j+2,\ \alpha^n, x_{j+1}), \end{aligned} \tag{9.25}$$

where U^n_j, U^n_{j+1}, and U^n_{j+2} are used.

With (9.21)–(9.22) and (9.24)–(9.25), we can determine the explicit terms in (9.8) that involve the first- and second-order derivatives.

At the time level $n+1$, we do not wish to increase the finite difference stencil so that we can solve the nonlinear system of equations more efficiently. Note that if x_j is close to α^{n+1}, we simply set

$$u^{n+1}_j = r(t^{n+1}) \quad \text{if} \quad |x_j - \alpha^{n+1}| \le h^2, \tag{9.26}$$

without affecting second-order accuracy. If $|x_j - \alpha^{n+1}| > h^2$, we use (9.23) to approximate u_{xx} and use (9.20) to approximate u_x. By doing so we will have a tridiagonal system for the linearized equations.

Case 2: The jump conditions are known. Suppose we know the jump conditions (9.6) across the interface, i.e., $[u] = w(t)$ and $[\beta u_x] = v(t)$ are given. The finite difference equations at the two irregular grid points x_j and x_{j+1} are based on the following theorems, Theorems 9.2 and 9.3, as follows.

Theorem 9.2. *Let $u(x,t)$ be the solution of* (9.5) *with jump conditions* (9.6). *If $x_j \le \alpha(t) < x_{j+1}$, then*

$$\begin{bmatrix} u(\alpha^-,t) \\ u_x(\alpha^-,t) \\ u_{xx}(\alpha^-,t) \end{bmatrix} = S \begin{bmatrix} u(x_{j-1},t) \\ u(x_j,t) \\ u(x_{j+1},t) \end{bmatrix} - \begin{bmatrix} C_{j,1} \\ C_{j,2} \\ C_{j,3} \end{bmatrix} + \begin{bmatrix} O(h^3) \\ O(h^2) \\ O(h) \end{bmatrix}, \tag{9.27}$$

where

$$C_{j,k} = s_{k3}\left\{w + \frac{v}{\beta^+}(x_{j+1}-\alpha) + \frac{(x_{j+1}-\alpha)^2}{2\beta^+}\left(\frac{dw}{dt} + [f] - \frac{v}{\beta^+}\left(g - \lambda u^+ + \beta_x^+\right)\right)\right\}, \tag{9.28}$$

$k = 1, 2, 3$, $S = \{s_{kj}\} = A^{-1}$, *and*

$$A^T = \{a_{kj}\}^T = \begin{bmatrix} 1 & 1 & 1 \\ x_{j-1}-\alpha & x_j - \alpha & a_{32} \\ \dfrac{(x_{j-1}-\alpha)^2}{2} & \dfrac{(x_j-\alpha)^2}{2} & \dfrac{\rho\,(x_{j+}-\alpha)^2}{2} \end{bmatrix}.$$

The element a_{32} is given by

$$a_{32} = \rho(x_{j+1}-\alpha) + \frac{(x_{j+1}-\alpha)^2}{2\beta^+}\left(g\,(1-\rho) + \lambda(u^+\rho - u^-) + \beta_x^- - \rho\beta_x^+\right),$$

where $\rho = \beta^-/\beta^+$.

The proof of this theorem can be found in [162]. Using this theorem we get a discretized form of $(\beta u_x)_x - \lambda u u_x$ at the grid point x_j, $x_j \le \alpha < x_{j+1}$, using the grid points from both sides of the interface,

$$\begin{aligned}(\beta u_x)_{x,j} - \lambda u_j u_{x,j} &= \beta^- u_{xx}^- + \beta_x^- u_x^- - \lambda u^- u_x^- + O(h)\\ &\approx \left(\beta^- s_{31} + s_{21}(\beta_x^- - \lambda U^-)\right)U_{j-1} + \left(\beta^- s_{32} + s_{22}(\beta_x^- - \lambda u^-)\right)U_j \\ &\quad + \left(\beta^- s_{33} + s_{23}(\beta_x^- - \lambda U^-)\right)U_{j+1} - \left(\beta^- C_{j,3} + C_{j,2}(\beta_x^- - \lambda U^-)\right).\end{aligned} \tag{9.29}$$

An important feature of the discretization above is that we can still use a 3-point stencil, and thus the discretization is valid for any location α. The theorem above also gives an interpolation formula which can be used to compute the values of u^- and u_x^- that are needed to approximate g and the frozen term $\lambda u u_x$.

For the grid point x_{j+1}, $x_j \le \alpha < x_{j+1}$, there exists a similar formula that we state as follows.

Theorem 9.3. *Let $u(x,t)$ be the solution of* (9.5) *with jump conditions* (9.6). *If* $x_j \le \alpha(t) < x_{j+1}$, *then*

$$\begin{bmatrix} u(\alpha^+,t) \\ u_x(\alpha^+,t) \\ u_{xx}(\alpha^+,t)\end{bmatrix} = \tilde{S}\begin{bmatrix} u(x_j,t) \\ u(x_{j+1},t) \\ u(x_{j+2},t)\end{bmatrix} - \begin{bmatrix} C_{j+1,1} \\ C_{j+1,2} \\ C_{j+1,3}\end{bmatrix} + \begin{bmatrix} O(h^3) \\ O(h^2) \\ O(h)\end{bmatrix}, \tag{9.30}$$

where

$$C_{j+1,k} = \tilde{s}_{k1}\left\{-[u] - \frac{[\beta u_x]}{\beta^-}(x_j-\alpha) - \frac{(x_j-\alpha)^2}{2\beta^-}\left(\frac{dw}{dt} + [f] - \frac{v}{\beta^-}\left(g - \lambda u^- + \beta_x^-\right)\right)\right\}, \tag{9.31}$$

$k = 1, 2, 3$, $\tilde{S} = \{\tilde{s}_{kj}\} = \tilde{A}^{-1}$, and

$$\tilde{A}^T = \{\tilde{a}_{kj}\}^T = \begin{bmatrix} 1 & 1 & 1 \\ \tilde{a}_{12} & x_{j+1} - \alpha & x_{j+2} - \alpha \\ \dfrac{(x_j - \alpha)^2}{2\rho} & \dfrac{(x_{j+1} - \alpha)^2}{2} & \dfrac{(x_{j+2} - \alpha)^2}{2} \end{bmatrix}.$$

The element $\tilde{a}_{12}$ is given by

$$\tilde{a}_{12} = \frac{(x_j - \alpha)}{\rho} - \frac{(x_j - \alpha)^2}{2\beta^-}\left(g\left(1 - \frac{1}{\rho}\right) + \lambda\left(\frac{u^-}{\rho} - u^+\right) - \frac{\beta_x^-}{\rho} + \beta_x^+\right).$$

9.2.4 Computing interface quantities

As mentioned in §9.2.1, in order to compute (9.11), (9.13), (9.18), and (9.29), we need to compute the interface values such as $u^\pm$, $u_x^\pm$, $[u_t]_\tau$, etc. Again we distinguish two different cases.

Case 1: The solution on the interface $u(\alpha(t), t) = r(t)$ is known. In this case, the solution is continuous, that is, $u^- = u^+ = r(t)$. With the knowledge of the computed solution $\{U_i^n\}$, an estimate of $\{U_i^{n+1}\}$, and the solution on the interface $r(t^n)$, we use the one-sided difference (9.21) to approximate $u_x^{-,n}$ by switching the position between x_j and α^n, $x_j \le \alpha^n < x_{j+1}$. Similarly, we use (9.22) to approximate $u_x^{+,n}$ after switching the position of x_{j+1} and α^n in (9.22). The same approach is used for the next time level t^{n+1}.

If the interface crosses a grid line $x = x_i$ at some time τ, we need to compute $[u_t]_{;\tau}$ in order to get the correction term $Q_i^{n+\frac{1}{2}}$. In this case we simply use

$$[u_t]_{;\tau} \approx \frac{u_i^{n+1} - r(\tau)}{t^{n+1} - \tau} - \frac{r(\tau) - u_i^n}{\tau - t^n}. \tag{9.32}$$

Case 2: The jump conditions are known. In this case, $u^\pm$ and $u_x^\pm$ are computed using (9.27)–(9.28) and (9.30)–(9.31). In order to compute the jump $[u_t]_\tau$, we differentiate the first jump condition

$$u(\alpha^+, t) - u(\alpha^-, t) = w(t)$$

with respect to t to get

$$\big(u_x(\alpha^+, t) - u_x(\alpha^-, t)\big)\frac{d\alpha}{dt} + u_t(\alpha^+, t) - u_t(\alpha^-, t) = w'(t),$$

$$\text{i.e.,} \qquad [u_t] = w'(t) - [u_x]\, g. \tag{9.33}$$

We need to express (9.33) in terms of the quantities at either time level t^n or t^{n+1}. If $\alpha^n \ge \alpha^{n+1}$ (see Figure 9.1(b)), we have, at time $t = \tau$,

$$u^- = u(x_j, t^n) + O(h) \approx U_j^n, \qquad u^+ = u(x_j, t^{n+1}) + O(h) \approx U_j^{n+1},$$
$$u_x^- = u_x(x_j, t^n) + O(h) \approx U_{x,j}^n, \qquad u_x^+ = u_x(x_j, t^{n+1}) + O(h) \approx U_{x,j}^{n+1}.$$

Otherwise, we have

$$u^- = u(x_j, t^{n+1}) + O(h) \approx U_j^{n+1}, \qquad u^+ = u(x_j, t^n) + O(h) \approx U_j^n,$$
$$u_x^- = u_x(x_j, t^{n+1}) + O(h) \approx U_{x,j}^{n+1}, \qquad u_x^+ = u_x(x_j, t^n) + O(h) \approx U_{x,j}^n$$

for $\alpha^n < \alpha^{n+1}$. Thus we use the following scheme to compute $[u_t]_\tau$:

$$[u_t]_{;\tau} \approx \begin{cases} w'(\tau) - g(\tau, U_j^n, U_j^{n+1}, U_{x,j}^n, U_{x,j}^{n+1})(U_{x,j}^{n+1} - U_{x,j}^n) & \text{if } \alpha^n \geq \alpha^{n+1}, \\ w'(\tau) - g(\tau, U_j^{n+1}, U_j^n, U_{x,j}^{n+1}, U_{x,j}^n)(U_{x,j}^{n+1} - U_{x,j}^n) & \text{if } \alpha^n < \alpha^{n+1}. \end{cases}$$

9.2.5 Solving the resulting nonlinear system of equations

From the discussions above, we know that in order to get an approximate solution for $u(x,t)$ at time t^{n+1} using the finite difference method described here, generally we need to solve the following nonlinear system of equations:

$$\frac{U_i^{n+1} - U_i^n}{\Delta t} - Q_i^{n+\frac{1}{2}} + \frac{\lambda}{2}\left(U_i^n U_{x,i}^n + U_i^{n+1} U_{x,i}^{n+1}\right)$$
$$= \frac{1}{2}\left((\beta U_x)_{x,i}^n + (\beta U_x)_{x,i}^{n+1}\right) - \frac{1}{2}\left(f_i^n + f_i^{n+1}\right),$$

$$\frac{\alpha^{n+1} - \alpha^n}{\Delta t} = \frac{1}{2}\left(g^n + g^{n+1}\right),$$

where the quantities $U_{x,i}^n$ and $(\beta U_x)_{x,i}^n$ can be expressed as some linear combinations of U_i^n.

Since a fully implicit discretization is used, the numerical scheme should be stable. The local truncation errors are $O(h^2)$ at most grid points, but they are $O(h)$ at the two grid points that are closest to the interface from the left and the right, and at those grid points where the interface crosses. The global error in the solution is second-order accurate at all grid points as explained in §2.4 for one-dimensional elliptic interface problems.

We need to solve a nonlinear system of equations for $(U_j^{n+1}, \alpha^{n+1})$ at each time step. The difficulty is that some quantities, such as $Q_j^{n+\frac{1}{2}}$ and $C_{j,k}^{n+1}$, are not known until we know the solutions $\{U_i^{n+1}\}$ and α^{n+1}. We use an implicit discretization for the diffusion term $(\beta u_x)_x$ and a *prediction-correction* approach for $\alpha(t)$. An adaptive time step is chosen for (9.5) based on the classic stability theory,

$$\Delta t \leq \min\left\{h, \left|\frac{1}{\partial g/\partial \alpha}\right|\right\}.$$

The constraint $\Delta t \leq h$ is imposed to maintain second-order accuracy in both space and time. From the second equation in (9.5), we can get

$$\frac{d\alpha}{dt} = \frac{\partial \alpha}{\partial g}\frac{\partial g}{\partial t} = g, \qquad \text{i.e.,} \qquad \frac{\partial g}{\partial \alpha} = \frac{\partial g}{\partial t}\Big/ g.$$

That implies

$$\Delta t \le \min\left\{h, \left|\frac{g}{\partial g/\partial t}\right|\right\},$$

or, in the discrete form,

$$\Delta t_{new} = \min\left\{h, \left|\frac{g^n \, \Delta t_{old}}{g^{n+1} - g^n}\right|\right\}. \tag{9.34}$$

Below we give an outline of the iterative process. Suppose we have obtained all quantities at the time level t^n, and the current time step size is Δt (i.e., $t^{n+1} = t^n + \Delta t$). To get all corresponding quantities at the time level t^{n+1}, we follow the procedure outlined below.

- Determine j_0 such that $x_{j_0} \le \alpha^n < x_{j_0+1}$. Approximate u_x^n and $(\beta u_x)_x^n$ at x_{j_0} according to the scheme discussed in §9.2.3.
- Set

$$\alpha_1^{n+1} = \alpha^n + \Delta t \, g(t^n, \alpha^n;\, U^{-,n},\, U^{+,n},\, U_x^{-,n},\, U_x^{+,n}).$$

- Set an initial guess of the solution $\{U_{i,1}^{n+1}\} = \{U_i^n\}$ at the time level t^{n+1}.

  For $m = 1, 2, \ldots,$ do the following:

  - (**) Determine j_m such that $x_{j_m} \le \alpha_1^{n+1} < x_{j_m+1}$. Determine the coefficients and the correction terms for u_x^{n+1} and u_{xx}^{n+1} at x_{j_m}. Substitute $U_{i,m}^{n+1}$ for U_i^{n+1} in the nonlinear term $U_i^{n+1} U_{x,i}^{n+1}$.
  - If $j_0 \ne j_m$, then for $l = j_0 + 1, \ldots, j_m$, when $j_0 < j_m$, or for $l = j_m, j_m + 1, \ldots, j_0$, when $j_0 > j_m$, first get τ_m using (9.18), then determine the correction $Q_{l,m}^{n+\frac{1}{2}}$ to $(U_{l,m}^{n+1} - U_l^n)/\Delta t$ using the approach described in §9.2.2.
  - Solve the tridiagonal system for $U_{i,m+1}^{n+1}$.
  - Interpolate $\{U_{i,m+1}^{n+1}\}$ to get $U_{m+1}^{\pm,n+1}$ and $U_{x,m+1}^{\pm,n+1}$ if necessary (depending on the interface condition).
  - Determine

$$\alpha_{m+1}^{n+1} = \alpha^n + \frac{\Delta t}{2}\left(g^n + g_{m+1}^{n+1}\right),$$

  where

$$g_{m+1}^{n+1} = g\left(t^{n+1}, \alpha_m^{n+1};\, U_{m+1}^{-,n+1}, U_{m+1}^{+,n+1}, U_{x,m+1}^{-,n+1}, U_{x,m+1}^{+,n+1}\right).$$

  - If $|\alpha_m^{n+1} - \alpha_{m+1}^{n+1}| > \epsilon$, a given tolerance, then $m = m + 1$. Go to (**).

- If $|\alpha_m^{n+1} - \alpha_{m+1}^{n+1}| < \epsilon$, then set all quantities $\{\ \}_m^{n+1}$ to $\{\ \}^{n+1}$; in other words, accept the values $U_{i,m+1}^{n+1}$ and α_{m+1}^{n+1} as approximations at time level t^{n+1}. Determine the next time step size Δt,

$$\Delta t = \min\left\{ h, \left| \frac{g^n \, \Delta t}{g^{n+1} - g^n} \right| \right\}. \tag{9.35}$$

Go to the next time step.

9.2.6 Validation of the algorithm for a one-dimensional moving interface problem

We validate the IIM for a one-dimensional moving interface problem through a classical Stefan problem of tracking a freezing front of ice in water. The description of the problem is excerpted from [84], where Furzeland used this example to compare different methods. The thermal properties are the heat conductivity k_i, the specific heat c_i, the density ρ (assumed to be the same in each phase) and the latent heat L. The subscript $i = 1$ denotes the phase 1 (ice) in $0 < x < \alpha(t)$ and $i = 2$ denotes the phase 2 (water) in $\alpha(t) < x < 1$. Define also two constants $C_i = c_i \, \rho$ and $\sigma = L\rho$. The governing equations are

$$C_i \frac{\partial u_i}{\partial t} = k_i \frac{\partial^2 u_i}{\partial x^2}, \qquad i = 1,\ 2, \quad t_0 < t,$$

$$u_1(0, t) = u^* < 0, \qquad t > 0,$$

$$\left.\begin{aligned} & u_1(\alpha, t) = u_2(\alpha, t) = 0 \\ & \sigma \frac{\partial \alpha}{\partial t} = k_1 \frac{\partial u_1}{\partial x} - k_2 \frac{\partial u_2}{\partial x} \end{aligned}\right\} \quad \text{on} \quad x = \alpha(t), \quad t_0 < t,$$

where the solution u represents the temperature. This problem has the exact solution,

$$\alpha(t) = 2\,\phi\,\sqrt{\kappa_1 t}\,,$$

$$u_1(x, t) = u^* \left\{ 1 - \frac{\operatorname{erf}(x/2\sqrt{\kappa_1 t}\,)}{\operatorname{erf}\phi} \right\},$$

$$u_2(x, t) = u_0 \left\{ 1 - \frac{\operatorname{erfc}(x/2\sqrt{\kappa_2 t}\,)}{\operatorname{erfc}(\phi\sqrt{\kappa_1/\kappa_2})} \right\},$$

where $\kappa_i = k_i / C_i$, $\operatorname{erf}(x)$ is the error function, u^* and u_0 are two constants, and ϕ is the root of the transcendental equation

$$\frac{e^{-\phi^2}}{\operatorname{erf}\phi} + \frac{k_2}{k_1}\sqrt{\frac{\kappa_1}{\kappa_2}}\, \frac{u_0\, e^{-\kappa_1\phi^2/\kappa_2}}{u^* \operatorname{erfc}(\phi\sqrt{\kappa_1/\kappa_2})} + \frac{\phi\lambda\sqrt{\pi}}{C_1 u^*} = 0,$$

which can be easily computed, say by using the bisection method. The exact solution is used as the initial condition at time $t_0 = 0.5$, as well as the boundary condition at both ends $x = 0$ and $x = 1$. The following thermal properties are used:

$$k_1 = 2.22, \quad k_2 = 0.556, \quad C_1 = 1.762, \quad C_2 = 4.226, \quad \sigma = 338$$

with $u^* = -20$ and $u_0 = 10$, which gives $\phi = 0.2054269\ldots$.

Table 9.1. *A grid refinement analysis for the Stefan problem at $t = 1.0$.*

M	$\| E_M \|_\infty$	Ratio	$\lvert E_\alpha \rvert$	Ratio
20	4.3067×10^{-3}		1.0941×10^{-4}	
40	9.7147×10^{-4}	4.4333	2.4947×10^{-5}	4.3857
80	2.3713×10^{-4}	4.0967	5.7298×10^{-6}	4.3539
160	5.8160×10^{-5}	4.0772	1.3828×10^{-6}	4.1434
320	1.4213×10^{-5}	4.0920	3.3721×10^{-7}	4.1009

Table 9.1 shows the results of a grid refinement analysis. In the table, $\| E_M \|_\infty$ is the infinity norm of the error at the fixed time $t = 1$. E_α is the difference between the exact $\alpha(t)$ and the computed interface at the final time $t = 1$. We see clearly second-order convergence for both the temperature $u(x, t)$ and the interface location $\alpha(t)$. More examples, including nonlinear moving interface problems, can be found in [162]. In [173], the method is applied to simulate the temperature profile of an ice sheet during the process of glaciation where there are two interfaces: one is fixed, and the other is moving with time.

Multidimensional problems are discussed in Chapter 11. There we use a time splitting method and combine the IIM with evolution schemes such as the front-tracking and level set methods to evolve moving interfaces.

9.3 The modified ADI method for heat equations with discontinuities

In this section, we explain the alternating direction implicit (ADI) method for the heat equation

$$\begin{aligned} &u_t = u_{xx} + u_{yy} - f(x, y, t), \qquad (x, y) \in \Omega \backslash \Gamma, \\ &[u] = w(:, t), \qquad \left[\frac{\partial u}{\partial \mathbf{n}}\right] = v(:, t) \end{aligned} \tag{9.36}$$

with a fixed interface Γ.

For parabolic equations, it is often desirable to use implicit methods because the time step restriction is severe for explicit methods. For a heat equation, the ADI method is often used to solve parabolic PDEs numerically. The ADI method is unconditionally stable with second-order accuracy in both time and space. From one time level to the next, we need only solve a sequence of tridiagonal systems of equations. We refer the reader to [33, 68, 201, 277] for an introduction to the ADI method.

The classical ADI method for the heat equation is

$$\begin{aligned} \frac{U_{ij}^{n+\frac{1}{2}} - U_{ij}^n}{\Delta t/2} &= \delta_{xx} U_{ij}^{n+\frac{1}{2}} + \delta_{yy} U_{ij}^n - f_{ij}^{n+\frac{1}{2}}, \\ \frac{U_{ij}^{n+1} - U_{ij}^{n+\frac{1}{2}}}{\Delta t/2} &= \delta_{xx} U_{ij}^{n+\frac{1}{2}} + \delta_{yy} U_{ij}^{n+1} - f_{ij}^{n+\frac{1}{2}}, \end{aligned} \tag{9.37}$$

where, for example,

$$\delta_{xx} U_{ij} = \frac{U_{i-1,j} - 2U_{ij} + U_{i+1,j}}{h^2}. \tag{9.38}$$

The local truncation error of the above ADI method contains such a term as

$$(\Delta t)^2 \, \delta_{xx} \, \delta_{yy} \, u_t \approx (\Delta t)^2 u_{txxyy}, \tag{9.39}$$

(see (9.52)) which indicates a strong regularity requirement on the solution in order to get second-order accurate results. Note that we again assume that the ratio $h/\Delta t$ is a constant.

Obviously, we cannot apply the ADI method directly to (9.36) since the solution is not even in $C(\Omega)$. It seems that we can use the IIM to add correction terms to the spatial discretization to get an ADI scheme,

$$\begin{aligned} \frac{U_{ij}^{n+\frac{1}{2}} - U_{ij}^n}{\Delta t/2} &= \delta_{xx} U_{ij}^{n+\frac{1}{2}} + \delta_{yy} U_{ij}^n - C_{ij}^{n+\frac{1}{2}} - f_{ij}^{n+\frac{1}{2}}, \\ \frac{U_{ij}^{n+1} - U_{ij}^{n+\frac{1}{2}}}{\Delta t/2} &= \delta_{xx} U_{ij}^{n+\frac{1}{2}} + \delta_{yy} U_{ij}^{n+1} - C_{ij}^{n+\frac{1}{2}} - f_{ij}^{n+\frac{1}{2}}, \end{aligned} \tag{9.40}$$

but this does not lead to a second-order solution. In fact, theoretical analysis and numerical experiments show that scheme (9.40) is only first-order accurate. The failure results from the fact that we have not split the correction term $C_{i,j}^{n+\frac{1}{2}}$ correctly.

9.3.1 The modified ADI scheme

The modified ADI method can be written as

$$\begin{aligned} \frac{U_{ij}^{n+\frac{1}{2}} - U_{ij}^n}{\Delta t/2} &= \delta_{xx} U_{ij}^{n+\frac{1}{2}} - (C_x)_{ij}^{n+\frac{1}{2}} - Q_{ij}^n - R_{ij}^n \\ &\quad + \delta_{yy} U_{ij}^n - (C_y)_{ij}^n - f_{ij}^{n+\frac{1}{2}}, \end{aligned} \tag{9.41}$$

$$\begin{aligned} \frac{U_{ij}^{n+1} - U_{ij}^{n+\frac{1}{2}}}{\Delta t/2} &= \delta_{xx} U_{ij}^{n+\frac{1}{2}} - (C_x)_{ij}^{n+\frac{1}{2}} - Q_{ij}^n - R_{ij}^n \\ &\quad + \delta_{yy} U_{ij}^{n+1} - (C_y)_{ij}^{n+1} - f_{ij}^{n+\frac{1}{2}}, \end{aligned} \tag{9.42}$$

where we add the correction terms Q_{ij}^n, R_{ij}^n, $(C_x)_{ij}^{n+\frac{1}{2}}$, $(C_y)_{ij}^n$, and $(C_y)_{ij}^{n+1}$ to get a second-order scheme. At regular grid points, that is, all the grid points in the standard ADI finite difference stencil that are on the same side of the interface, the standard ADI method is used in which

$$(C_x)_{ij}^{n+\frac{1}{2}} = Q_{ij}^n = R_{ij}^n = (C_y)_{ij}^n = Q_{ij}^n = (C_y)_{ij}^{n+1} = 0.$$

At each irregular grid point we need to determine those correction terms. This will be explained in the following subsections.

9.3.2 Determining the spatial correction terms

The local truncation error at regular grid points is $O(h^2)$. To get second-order accuracy globally, we need to determine the correction terms so that the local truncation error is of order $O(h)$ or smaller in magnitude at irregular grid points. First we explain how to approximate u_{xx} and u_{yy} by choosing the correction terms C_x and C_y. Then we explain how to choose the correction terms R^n_{ij} and Q^n_{ij} so that the local truncation error of the equivalent finite difference scheme at each irregular grid point is $O(h)$.

Let (x_i, y_j) be an irregular grid point. Without loss of generality, we assume that the interface Γ cuts the grid line $y = y_j$ at $x = x^*_{ij}$, where $x_i \le x^*_{ij} \le x_{i+1}$. Using the Taylor expansion in the x-direction at x^*_{ij}, or x^* for short, and substituting U_{ij} with the exact solution $u_{ij} = u(x_i, y_j)$, we have

$$
\begin{aligned}
\frac{u(x_{i-1}, y_j) - 2u(x_i, y_j) + u(x_{i+1}, y_j)}{h^2} &= \frac{u^- - 2u^- + u^+}{h^2} \\
&\quad + u_x^- \left(\frac{x_{i-1} - x^*}{h^2} - 2\frac{x_i - x^*}{h^2} \right) + u_x^+ \frac{x_{i+1} - x^*}{h^2} \\
&\quad + u_{xx}^- \left(\frac{(x_{i-1} - x^*)^2}{2h^2} - 2\frac{(x_i - x^*)^2}{2h^2} \right) + u_{xx}^+ \frac{(x_{i+1} - x^*)^2}{2h^2} + O(h) \\
&= u_{xx}^- + \frac{[u]}{h^2} + [u_x]\frac{(x_{i+1} - x^*)}{h^2} + [u_{xx}]\frac{(x_{i+1} - x^*)^2}{2h^2} + O(h) \\
&= u_{xx}^- + (C_x)_{ij} + O(h),
\end{aligned}
$$

where we have used the Taylor expansion

$$
u(x, y_j) = u^\pm + u_x^\pm \left(x - x^*_{ij}\right) + \frac{1}{2} u_{xx}^\pm \left(x - x^*_{ij}\right)^2 + O(h^3),
$$

and u^-, u_x^-, u_{xx}^-, and the jumps are defined at $(x^*_{ij}, y_j) \in \Gamma$.

Similarly, at the point (x_{i+1}, y_j), we have

$$
\begin{aligned}
&\frac{u(x_i, y_j) - 2u(x_{i+1}, y_j) + u(x_{i+2}, y_j)}{h^2} \\
&\quad = u_{xx}^+ - \frac{[u]}{h^2} - [u_x]\frac{(x_i - x^*)}{h^2} - [u_{xx}]\frac{(x_i - x^*)^2}{2h^2} + O(h) \\
&\quad = u_{xx}^+ + (C_x)_{i+1,j} + O(h).
\end{aligned}
$$

In other words, we can write

$$
\delta_{xx} u_{ij} = u_{xx}^\pm + (C_x)_{ij} + O(h) \quad \text{at irregular points,} \tag{9.43}
$$

where

$$
(C_x)_{ij} = \pm\frac{[u]}{h^2} \pm [u_x]\frac{(x_{i\pm1} - x^*)}{h^2} \pm [u_{xx}]\frac{(x_{i\pm1} - x^*)^2}{2h^2}. \tag{9.44}
$$

If the point (x_i, y_j) is regular, then $(C_x)_{ij} = 0$; otherwise it can be expressed in terms of the jumps of $[u]$, $[u_x]$, and $[u_{xx}]$. The sign is determined by the relative position of the interface Γ and the grid point (x_i, y_j). By the same token we can do the same in the y-direction to get

$$\delta_{yy} u_{ij} = u_{yy}^{\pm} + (C_y)_{ij} + O(h) \quad \text{at irregular points.} \tag{9.45}$$

9.3.3 Decomposing the jump condition in the coordinate directions

To determine the correction terms $(C_x)_{ij}$ and $(C_y)_{ij}$, the values of the jumps $[u_x]$, $[u_y]$, $[u_{xx}]$, $[u_{yy}]$ are needed in terms of the given information $[u]$ and $[u_{\mathbf{n}}]$. These interface relations can be obtained by differentiating the known jump conditions $[u]$ and $[u_{\mathbf{n}}]$ and using the differential equation itself.

Using the local coordinate system (1.34), we obtain $[u_\eta] = [u]_\eta = w_\eta(:, t)$ by differentiating $[u] = w$ along the interface. For the remaining second derivative jumps we can use the interface relations from (3.5), with $\beta = 1$ and f being replaced by $f + u_t$, to get

$$\begin{aligned} [u_{\xi\xi}] &= \chi''[u_\xi] - w_{\eta\eta} + [f] + w_t, \\ [u_{\eta\eta}] &= -\chi''[u_\xi] + w_{\eta\eta}, \\ [u_{\xi\eta}] &= \chi''[u_\eta] + v_\eta, \end{aligned} \tag{9.46}$$

where $[u_\xi] = [u_{\mathbf{n}}]$ is the given jump condition in the flux. Thus, we have expressed all the jumps in the local coordinates in terms of the known quantities $[u]$ and $[u_{\mathbf{n}}]$. The jump relations in the x- and y-directions then are given by the following formulas:

$$\begin{aligned} [u_x] &= [u_\xi]\cos\theta - [u_\eta]\sin\theta, \\ [u_y] &= [u_\xi]\sin\theta + [u_\eta]\cos\theta, \\ [u_{xx}] &= [u_{\xi\xi}]\cos^2\theta - 2[u_{\xi\eta}]\cos\theta\sin\theta + [u_{\eta\eta}]\sin^2\theta, \\ [u_{yy}] &= [u_{\xi\xi}]\sin^2\theta + 2[u_{\xi\eta}]\cos\theta\sin\theta + [u_{\eta\eta}]\cos^2\theta. \end{aligned} \tag{9.47}$$

With these known jumps we can compute the correction terms $(C_x)_{ij}$ and $(C_y)_{ij}$.

9.3.4 The local truncation error analysis for the ADI method

In this subsection we discuss how to determine Q_{ij} and R_{ij} through the local truncation error analysis. Now if we add (9.41) and (9.42) together, we get

$$\begin{aligned} \frac{U_{ij}^{n+1} - U_{ij}^n}{\Delta t} &= \delta_{xx} U_{ij}^{n+\frac{1}{2}} - (C_x)_{ij}^{n+\frac{1}{2}} - Q_{ij}^n - R_{ij}^n \\ &+ \frac{1}{2}\left(\delta_{yy} U_{ij}^n + \delta_{yy} U_{ij}^{n+1}\right) - \frac{1}{2}\left((C_y)_{ij}^n + (C_y)_{ij}^{n+1}\right) - f_{ij}^{n+\frac{1}{2}}. \end{aligned} \tag{9.48}$$

If we subtract (9.42) from (9.41), we have the intermediate result

$$U_{ij}^{n+\frac{1}{2}} = \frac{U_{ij}^n + U_{ij}^{n+1}}{2} + \frac{\Delta t}{4}\left(\delta_{yy}U_{ij}^n - (C_y)_{ij}^n - \delta_{yy}U_{ij}^{n+1} + (C_y)_{ij}^{n+1}\right). \tag{9.49}$$

Plugging this into (9.48), we get

$$\begin{aligned}\frac{U_{ij}^{n+1} - U_{ij}^n}{\Delta t} &= \frac{1}{2}\delta_{xx}\left(U_{ij}^n + U_{ij}^{n+1}\right) - (C_x)_{ij}^{n+\frac{1}{2}} - Q_{ij}^n - R_{ij}^n \\ &+ \frac{1}{2}\left(\delta_{yy}U_{ij}^n - (C_y)_{ij}^n + \delta_{yy}U_{ij}^{n+1} - (C_y)_{ij}^{n+1}\right) \\ &+ \frac{\Delta t}{4}\delta_{xx}\left(\delta_{yy}U_{ij}^n - (C_y)_{ij}^n - \delta_{yy}u_{ij}^{n+1} + (C_y)_{ij}^{n+1}\right).\end{aligned} \tag{9.50}$$

This is the *actual finite difference scheme* for the solution U_{ij}^{n+1} at time t^{n+1}. Note that the interface is fixed and all the quantities are continuous with time. We check the local truncation error of the finite difference scheme by examining each term, with U_{ij}^n being substituted for by the exact solution $u(x_i, y_j, t^n)$, to obtain the local truncation error of order $O(h)$. The left-hand side in (9.50), when substituted with the exact solution, gives

$$\frac{u(x_i, y_j, t^{n+1}) - u(x_i, y_j, t^n)}{\Delta t} = u_t(x_i, y_j, t^{n+1/2}) + O(\Delta t^2). \tag{9.51}$$

If we substitute for U_{ij} with the exact solution, the first few terms on the right-hand side can be written as

$$\begin{aligned}&\frac{1}{2}\delta_{xx}\left(u(x_i, y_j, t^n) + u(x_i, y_j, t^{n+1})\right) - (C_x)_{ij}^{n+\frac{1}{2}} - Q_{ij}^n \\ &= \frac{1}{2}\delta_{xx}u(x_i, y_j, t^{n+\frac{1}{2}}) + (\Delta t)^2\, \delta_{xx}u_{tt}(x_i, y_j, t^{n+\frac{1}{2}}) - (C_x)_{ij}^{n+\frac{1}{2}} - Q_{ij}^n + O(\Delta t).\end{aligned}$$

When u is continuous, that is, $[u] = w = 0$, we conclude that

$$\Delta t^2\, \delta_{xx}u_{tt}(x_i, y_j, t^{n+\frac{1}{2}}) \sim \frac{(\Delta t)^2}{h} \sim h. \tag{9.52}$$

We can simply take $Q_{ij}^n \equiv 0$. If u is not continuous, we rewrite the expression above as

$$\begin{aligned}&\frac{1}{2}\delta_{xx}\left(u(x_i, y_j, t^n) + u(x_i, y_j, t^{n+1})\right) - (C_x)_{ij}^{n+\frac{1}{2}} - Q_{ij}^n \\ &= \frac{1}{2}\left(\delta_{xx}u(x_i, y_j, t^n) - (C_x)_{ij}^n\right) + \frac{1}{2}\left(\delta_{xx}u(x_i, y_j, t^{n+1}) - (C_x)_{ij}^{n+1}\right) \\ &\quad - (C_x)_{ij}^{n+\frac{1}{2}} + \frac{1}{2}\left((C_x)_{ij}^n + (C_x)_{ij}^{n+1}\right) - Q_{ij}^n \\ &= \frac{1}{2}\left((u_{xx})_{ij}^n + (u_{xx})_{ij}^{n+1}\right) - (C_x)_{ij}^{n+\frac{1}{2}} + \frac{1}{2}\left((C_x)_{ij}^n + (C_x)_{ij}^{n+1}\right) - Q_{ij}^n + O(h) \\ &= u_{tt}(x_i, y_j, t^{n+\frac{1}{2}}) - (C_x)_{ij}^{n+\frac{1}{2}} + \frac{1}{2}\left((C_x)_{ij}^n + (C_x)_{ij}^{n+1}\right) - Q_{ij}^n + O(h).\end{aligned}$$

It is clear now that we should take

$$Q_{ij}^n = \frac{1}{2}\left((C_x)_{ij}^n + (C_x)_{ij}^{n+1}\right) - (C_x)_{ij}^{n+\frac{1}{2}}. \tag{9.53}$$

We turn our attention to the terms remaining in (9.50). If $[u] = 0$, we have

$$\begin{aligned}
&\frac{1}{2}\left(\delta_{yy}u(x_i, y_j, t^n) - (C_y)_{ij}^n + \delta_{yy}u(x_i, y_j, t^{n+1}) - (C_y)_{ij}^{n+1}\right) \\
&= \frac{1}{2}\left(u_{yy}(x_i, y_j, t^n) + u_{yy}(x_i, y_j, t^{n+1})\right) + O(h) \\
&= u_{yy}(x_i, y_j, t^{n+\frac{1}{2}}) + O(h).
\end{aligned} \tag{9.54}$$

Thus, there is no need to correct the expression above and we set $R_{ij}^n = 0$. If $[u] \neq 0$, we define

$$S_{ij}^n = \frac{\delta_{yy}u(x_i, y_j, t^n) - (C_y)_{ij}^n}{h}, \tag{9.55}$$

which is an $O(1)$ quantity. Since the interface is fixed and all quantities are continuous with time, we can conclude that $|S_{ij}^n - S_{ij}^{n+1}| = O(h)$. Hence, from (9.45), we know that

$$\begin{aligned}
&\delta_{yy}u(x_i, y_j, t^n) - (C_y)_{ij}^n - \delta_{yy}u(x_i, y_j, t^{n+1}) + (C_y)_{ij}^{n+1} \\
&= u_{yy}(x_i, y_j, t^n) - u_{yy}(x_i, y_j, t^{n+1}) + O(h^2).
\end{aligned}$$

Therefore, we have

$$\begin{aligned}
&\frac{\Delta t}{4}\delta_{xx}\left(\delta_{yy}u(x_i, y_j, t^n) - (C_y)_{ij}^n - \delta_{yy}u(x_i, y_j, t^{n+1}) + (C_y)_{ij}^{n+1}\right) - R_{ij}^n \\
&= \frac{\Delta t}{4}\delta_{xx}\left(u_{yy}(x_i, y_j, t^n) - u_{yy}(x_i, y_j, t^{n+1})\right) + O(h) - R_{ij}^n \\
&= -\frac{\Delta t^2}{4}\delta_{xx}u_{yyt}(x_i, y_j, t^n) - R_{ij}^n + O(h).
\end{aligned}$$

We can see that the term which needs to be offset is $(\Delta t)^2\delta_{xx}u_{yyt}(x_i, y_j, t^n)/4 \sim O(1)$. We need to approximate this term at least to first-order accuracy in order to make the right correction. Note that

$$\begin{aligned}
\frac{(\Delta t)^2}{4}\delta_{xx}u_{yyt}(x_i, y_j, t^n) &= \frac{(\Delta t)^2}{4h^2}\Big(u_{yyt}(x_{i-1}, y_j, t^n) - 2u_{yyt}(x_i, y_j, t^n) \\
&+ u_{yyt}(x_{i+1}, y_j, t^n)\Big) \pm \frac{\Delta t^2}{4h^2}[u_{yyt}](x_{ij}^*, y_j, t^n) + O(h),
\end{aligned} \tag{9.56}$$

where the sign is determined by the relative position of the (x_i, y_j) and the interface. The jump term $[u_{yyt}]$ above can be approximated by

$$[u_{yyt}](x_{ij}^*, y_j, t^n) = \frac{[u_{yyt}](x_{ij}^*, y_j, t^{n+1}) - [u_{yyt}](x_{ij}^*, y_j, t^n)}{\Delta t} + O(h). \qquad (9.57)$$

At last we can determine the correction term R_{ij}^n as

$$R_{ij}^n = \mp \frac{[u_{yy}]^{n+1} - [u_{yy}]^n}{4\,\Delta t}, \qquad (9.58)$$

where the jump is calculated at (x_i, y_{ij}^*). From the analysis above we know that if we take Q_{ij}^n and R_{ij}^n as in (9.53) and (9.58), then we can guarantee that the local truncation errors are $O(h^2)$, at regular grid points, and $O(h)$ at the irregular grid points near the interface. The finite difference scheme will still give a second-order accurate solution globally.

9.3.5 A numerical example of the modified ADI method

We consider an example in which the interface Γ is the circle $x^2 + y^2 = 1/4$. The exact solution is

$$u(x, y, t) = \begin{cases} e^{-t} J_0(r) & \text{if } r \le \dfrac{1}{2}, \\[2ex] \dfrac{e^{-t} J_0(0.5)\, Y_0(r)}{Y_0(0.5)} & \text{if } r > \dfrac{1}{2}, \end{cases} \qquad (9.59)$$

where $J_0(x)$ and $Y_0(x)$ are the Bessel functions of first and second kind of order zero, respectively. In this example the solution $u(x, y, t)$ is continuous across the interface $r = \sqrt{x^2 + y^2} = 1/2$, but it has a jump in the flux which is

$$v(t) = e^{-t}\left(Y_0'(0.5)\, J_0(0.5)/Y_0(0.5) - J_0'(0.5)\right). \qquad (9.60)$$

The heat equation is defined on the square $-1 \le x, y \le 1$. The Dirichlet boundary condition and the initial condition are taken from the exact solution.

Tables 9.2 and 9.3 show the results of grid refinement analysis at $t = 5$ with and without the correction terms $Q_{i,j}^n$ and $R_{i,j}^n$, respectively. In the tables, $\|E_N\|$ is the maximum norm of the computed solution over all the grid points, while $\|T_N\|$ is the maximum norm of the local truncation error. For a second-order method, the ratio of the errors approaches 4. We see that the method with correction terms $Q_{i,j}^n$ and $R_{i,j}^n$ behaves better. But the method without correction terms Q_{ij}^n and R_{ij}^n seems also to approach second-order accuracy. This is likely due to fortunate cancellations of errors for this particular example.

Table 9.2. *A grid refinement analysis of the example with continuous solution using the modified ADI method with correction terms $Q^n_{i,j}$ and $R^n_{i,j}$.*

N	$\| E_N \|_\infty$, $t=5$	Ratio 1	$\| T_N \|_\infty$, $t=0$	Ratio 2
20	5.39851×10^{-5}		4.04973×10^{-1}	
40	1.01368×10^{-5}	4.6345	1.885347×10^{-1}	2.1480
80	2.30004×10^{-6}	4.4072	8.94152×10^{-2}	2.1085
160	5.56747×10^{-7}	4.1312	4.33057×10^{-2}	2.0647

Table 9.3. *A grid refinement analysis of the example with continuous solution using the modified ADI method without correction terms Q^n_{ij} and R^n_{ij}.*

N	$\| E_N \|_\infty$, $t=5$	Ratio 1	$\| T_N \|_\infty$, $t=0$
20	7.70093×10^{-5}		2.32077
40	2.07288×10^{-5}	3.7151	2.60534
80	5.44382×10^{-6}	3.8078	2.73938
160	1.42010×10^{-6}	3.8334	2.80317

9.4 The IIM for diffusion and advection equations

The maximum principle preserving IIM for elliptic interface problems can be readily applied to diffusion and advection equations of the following form:

$$u_t + \mathbf{a}(\mathbf{x},t)\cdot\nabla u = \nabla\cdot(\beta\nabla u) + f, \quad \mathbf{x}\in\Omega = \Omega^+\cup\Omega^-, \quad x\notin\Gamma, \tag{9.61a}$$

$$[u]|_\Gamma = w, \qquad [\beta u_{\mathbf{n}}]|_\Gamma = v, \tag{9.61b}$$

$$u(\mathbf{x},0) = u_0(\mathbf{x}), \qquad u(\mathbf{x},t)|_{\partial\Omega} = g(t), \tag{9.61c}$$

where $\beta(\mathbf{x},t) \ge \beta_{min} > 0$, $\mathbf{a}(\mathbf{x},t) = (\alpha_1,\alpha_2)$, and $f(\mathbf{x},t)$ are piecewise continuous but may have a finite jump across a fixed interface Γ. The time discretization is based on the prediction-correction Crank–Nicolson discretization,

$$\begin{aligned}\frac{U^{n+1}-U^n}{\Delta t} &+ (\mathbf{a}\cdot\nabla_h U)^{n+\frac12} \\ &= \frac12\left((\nabla_h\cdot\beta\nabla_h U)^n + (\nabla_h\cdot\beta\nabla_h U)^{n+1}\right) + f^{n+\frac12},\end{aligned} \tag{9.62}$$

where

$$(\mathbf{a}\cdot\nabla_h U)^{n+\frac12} = \frac32(\mathbf{a}\cdot\nabla_h U)^n - \frac12(\mathbf{a}\cdot\nabla_h U)^{n-1}, \tag{9.63}$$

and ∇_h is the discrete gradient operator. This discretization is second-order accurate in time. The diffusion term is discretized implicitly so that we can take large time steps, while the first-order derivative term is discretized explicitly so that second-order accuracy can be achieved without affecting the stability of the discretization of the diffusion part.

At a regular grid point (x_i, y_j), where all the points in the centered 5-point stencil are on the same side of the interface, the discretization is the standard one,

$$(\nabla_h U)_{ij} = \left(\frac{U_{i+1,j} - U_{i-1,j}}{2h}, \; \frac{U_{i,j+1} - U_{i,j-1}}{2h} \right), \tag{9.64}$$

$$\begin{aligned}(\nabla_h \cdot (\beta \nabla_h) U)_{ij} &= \frac{1}{h} \left(\beta_{i+\frac{1}{2},j} \frac{(U_{i+1,j} - U_{ij})}{h} - \beta_{i-\frac{1}{2},j} \frac{(U_{i,j} - U_{i-1,j})}{h} \right) \\ &+ \frac{1}{h} \left(\beta_{i,j+\frac{1}{2}} \frac{(U_{i,j+1} - U_{i,j})}{h} - \beta_{i,j-\frac{1}{2}} \frac{(U_{i,j} - U_{i,j-1})}{h} \right),\end{aligned} \tag{9.65}$$

where we have omitted the time index for simplicity.

At an irregular grid point (x_i, y_j), where the central 5-point stencil consists of grid points from different sides of the interface Γ, the discretization is done using a method of undetermined coefficients. We let

$$(\mathbf{a} \cdot \nabla_h U)_{ij} = \sum_{k=1}^{L_2} \bar{\gamma}_k U_{i+i_k, j+j_k} - Q_{ij}, \tag{9.66}$$

$$(\nabla_h \cdot \beta \nabla_h)\, U_{ij} = \sum_{k=1}^{L_1} \gamma_k U_{i+i_k, j+j_k} - C_{ij}, \tag{9.67}$$

where L_1 and L_2 are the numbers of grid points involved in the finite difference stencil. The sum over k involves a finite number of points neighboring (x_i, y_j). So each i_k and j_k will take values in the set $\{0, \pm 1, \pm 2, \ldots\}$. The coefficients $\{\gamma_k\}$ and the indexes i_k, j_k depend on (i, j), but for simplicity of notation, the dependency has been dropped. In order to determine the coefficients, we choose a point (x_i^*, y_j^*) on the interface. Below we discuss how to find the coefficients for the diffusion and advection terms, respectively.

9.4.1 Determining the finite difference coefficients for the diffusion term

The coefficients for the approximation of the diffusion term $\nabla \cdot (\beta \nabla u)$ are almost the same as in the maximum preserving principle IIM discussed in §3.5. The modification is needed because the last interface condition at the local coordinates (1.34) centered at (x_i^*, y_j^*) in (3.5) now is

$$\begin{aligned}u^+_{\xi\xi} &= \rho u^-_{\xi\xi} + (\rho - 1)\, u^-_{\eta\eta} + \left((\rho - 1)\chi'' + \frac{\beta^-_\xi - \rho\beta^+_\xi + \rho\tilde{\alpha}^+_1 - \tilde{\alpha}^-_1}{\beta^+} \right) u^-_\xi \\ &+ \frac{[\tilde{\alpha}_2 - \beta_\eta]}{\beta^+} u^-_\eta + \frac{\tilde{\alpha}^+_2 - \beta^+_\eta}{\beta^+} w_\eta - w_{\eta\eta} \\ &+ \frac{1}{\beta^+} w_t + \left(\frac{\chi''}{\beta^+} + \frac{\tilde{\alpha}^+_1 - {\beta_\xi}^+}{(\beta^+)^2} \right) v - \frac{[f]}{\beta^+},\end{aligned} \tag{9.68}$$

due to the time derivative, where

$$\tilde{\alpha}_1 = \alpha_1 \cos\theta + \alpha_2 \sin\theta, \qquad \tilde{\alpha}_2 = -\alpha_1 \sin\theta + \alpha_2 \cos\theta \tag{9.69}$$

in the local coordinates (1.34). The equality constraints now are

$$
\begin{aligned}
a_1 + a_2 &= 0, \\
a_3 + \rho a_4 + a_8 \left((\rho - 1)\chi'' + \frac{\beta_\xi^- - \rho\beta_\xi^+ + \rho\tilde{\alpha}_1^+ - \tilde{\alpha}_1^-}{\beta^+} \right) & \\
+ a_{10}\,(1-\rho)\chi'' + a_{12}\,\frac{\beta_\eta{}^- - \beta_\eta{}^+\rho}{\beta^+} &= \beta_\xi{}^-, \\
a_5 + a_6 + a_8\frac{[\tilde{\alpha}_2 - \beta_\eta]}{\beta^+} + a_{12}(1-\rho)\,\chi'' &= \beta_\eta{}^-, \\
a_7 + a_8\,\rho &= \beta^-, \\
a_9 + a_{10} + a_8\,(\rho - 1) &= \beta^-, \\
a_{11} + a_{12}\,\rho &= 0.
\end{aligned}
\tag{9.70}
$$

The sign constraints are exactly the same as in §3.5. The correction term is

$$
\begin{aligned}
C_{ij} &= a_2\, w + \left(a_6 + a_8 \frac{\tilde{\alpha}_2^+ - \beta_\xi{}^+}{\beta^+} \right) w_\eta + (a_{10} - a_8) w_{\eta\eta} + \frac{a_8}{\beta^+}\, w_t \\
&+ \frac{1}{\beta^+} \left\{ a_4 + a_8 \left(\chi'' + \frac{\tilde{\alpha}_1^+ - \beta_\xi{}^+}{\beta^+} \right) - a_{10}\chi'' - a_{12}\frac{\beta_\eta{}^+}{\beta^+} \right\} v \\
&+ a_{12}\,\frac{1}{\beta^+}\, v_\eta - a_8 \frac{[f]}{\beta^+},
\end{aligned}
\tag{9.71}
$$

where the $\{a_i\}$'s are defined as in (3.17).

9.4.2 Determining the finite difference coefficients for the advection term

Borrowing an idea from the projection method for the Navier–Stokes equations (see [168] for example), we use an explicit discretization for the first-order derivative term since it involves only first-order derivatives of the solution. The truncation error of the finite difference approximation (9.66) to the first-order derivative term at (x_i^*, y_j^*) is

$$
T_{ij}^r = \sum_{k=1}^{L_2} \bar{\gamma}_k\, u(x_{i+i_k}, y_{j+j_k}, t) - Q_{ij} - (\mathbf{a} \cdot \nabla u)^- . \tag{9.72}
$$

As mentioned in [154, 166], we can require the truncation error to be $O(h)$ at irregular grid points without affecting the second-order accuracy as h approaches zero. Therefore, we use the standard centered 5-point stencil ($L_2 = 5$), and the linear system of equations is

$$
\begin{aligned}
\bar{a}_1 + \bar{a}_2 &= 0, \\
\bar{a}_3 + \rho\bar{a}_4 &= \tilde{\alpha}_1^-, \\
\bar{a}_5 + \bar{a}_6 &= \tilde{\alpha}_2^-,
\end{aligned}
\tag{9.73}
$$

where we have neglected higher-order terms of h. The $\{\bar{a}_k\}$'s, for example, are defined as

$$\bar{a}_1 = \sum_{k \in K^-} \bar{\gamma}_k, \qquad \bar{a}_2 = \sum_{k \in K^+} \bar{\gamma}_k, \qquad \bar{a}_3 = \sum_{k \in K^-} \xi_k \bar{\gamma}_k.$$

They are literally the same as those defined in (3.17), with $\{\gamma_k\}$ being replaced by $\bar{\gamma}_k$. The solution to the linear system of equations is also different. The system is an underdetermined system and the solution is defined as the least squares solution with the least 2-norm. The correction term then is

$$Q_{ij} = \bar{a}_2\, w + \bar{a}_6\, w_\eta + \frac{\bar{a}_4}{\beta^+}\, v. \tag{9.74}$$

Note that the algorithm described in this section is second-order accurate, that is, the global error is proportional to h^2 as h approaches zero. In practice, however, h is fixed. If the advection term is very strong ($\|\mathbf{a}\| > 1/h$), the advection may carry larger truncation errors near the interface to other parts of the domain. The global error then may be affected by such error propagation.

From the analysis of the projection method for Navier–Stokes equations (the problem here can be regarded as a special case) (see for example, [16]), we know that the time step size should satisfy

$$\Delta t \le \frac{h}{\sqrt{2}\,\|\mathbf{a}\|_2}$$

if there is no interface. For the interface problem, we suggest taking a conservative time step size,

$$\Delta t \le \frac{h}{2\,\|\mathbf{a}\|_2}. \tag{9.75}$$

That is, we reduce the time step size by a factor of $\sqrt{2}$.

Recently, Adams and Li developed a new multigrid method for solving the linear system of equations using a 9-point stencil. See [6] for the method and some numerical examples.

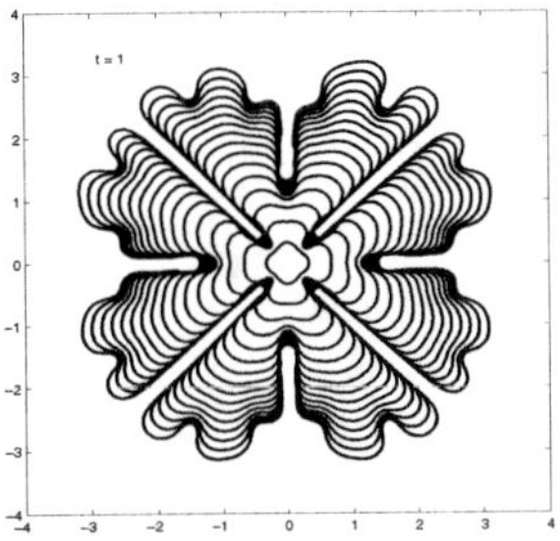

Chapter 10

The IIM for Stokes and Navier–Stokes Equations

Incompressible Stokes and Navier–Stokes equations are fundamental equations for multiphase flows with moving interfaces or free boundary problems in computational fluid dynamics (CFD). The nature of multiphase flows is reflected in the coefficients and the source terms of the governing equations. The viscosity and density in a multiphase flow may have a finite jump discontinuity across the interface separating two fluids. The source terms may have a source/dipole distribution in terms of the Dirac delta function.

One well-known example of such a multiphase flow is Peskin's immersed boundary (IB) model [210, 209, 212, 211, 215] for the study of the blood flow in a human heart. In Peskin's IB model, the complicated boundary is embedded in a rectangular box with a periodic boundary condition. The boundary condition at the heart valve is modeled as forces exerted by fibers on the fluid in the domain. Mathematically, it is a force distribution along the immersed interface after the embedding. Therefore the Navier–Stokes equations with a complicated boundary condition are converted into a problem defined on a rectangular region with singular forces. The IB method has been applied to a variety of problems, particularly in biomechanics (see for example [22, 77, 79, 263]) since these problems often involve complex geometries where standard "body-fitted" or unstructured grid approaches can run into difficulties. On the other hand, the IB method introduces a thin layer around an interface and the solution will be smeared in the layer. This leads to a first-order accurate solution in general.

One of the motivations of the IIM is to improve the accuracy of the IB method by using a sharp interface method that accurately captures discontinuities of the solution and its derivatives without smearing. In this chapter, we present the IIM for incompressible Stokes and Navier–Stokes equations with interfaces. Related work can also be found in [156, 155, 160, 168, 167, 177, 182].

10.1 The derivation of the jump conditions for Stokes and Navier–Stokes equations

In order to apply the IIM for Stokes or Navier–Stokes equations, we need to derive the jump conditions for the pressure and the velocity. These jump conditions are not obvious for

Stokes and Navier–Stokes equations and will be derived in this section. Since the inertial term $\rho\frac{d\mathbf{u}}{dt}$ is continuous, the jump conditions for the Navier–Stokes equations are the same as those for the Stokes equations. Some classical jump conditions involving balancing the force in the normal and tangential directions can be found in the literature; see [129, 216]. In order to get second-order accurate methods using the IIM, an additional jump condition for the normal derivative in the pressure is also needed. For Stokes equations with continuous viscosity, the jump conditions are derived in [156, 144, 160]. The jump conditions for Stokes equations with discontinuous viscosity were derived recently.

Consider the incompressible Stokes equations

$$\begin{aligned} &\operatorname{div}\mathbf{S} = \mathbf{g} + \mathbf{F}(\mathbf{x}), \quad \operatorname{div}\mathbf{u} = 0, \quad \text{in} \quad \Omega \in R^d, \; d = 2, 3, \\ &\mathbf{F} = \int_\Gamma \mathbf{f}(s)\,\delta(\mathbf{x} - \mathbf{X}(s))ds, \end{aligned} \tag{10.1}$$

where $\mathbf{g}$ is the body force and $\mathbf{f}$ is the density function of the surface force along the interface Γ. The strain tensor S is given by

$$\mathbf{S} = -\mu\,\mathbf{T} + p\,\mathbf{I}, \tag{10.2}$$

and the tensor $\mathbf{T}$ is defined by

$$T_{i,j} = D_{x_i}u_j + D_{x_j}u_i, \tag{10.3}$$

where $\mathbf{u}$ is the divergence-free velocity field, p is the pressure, Γ is an arbitrary interface, $\mathbf{f}$ is the density of the force strength along Γ,[14] $\mathbf{g}(\mathbf{x})$ is a bounded forcing term, e.g., the gravitational force, which can be discontinuous across Γ as well, and Ω is a bounded domain. We refer the readers to [15, 53, 141, 147, 220, 253] for general discussions of Stokes flows. An equivalent form of the Stokes equations is

$$\nabla p = \nabla\cdot\mu\left(\nabla\mathbf{u} + (\nabla\mathbf{u})^T\right) + \mathbf{g} + \mathbf{F}, \quad \mathbf{x} \in \Omega, \tag{10.4a}$$

$$\nabla\cdot\mathbf{u} = \mathbf{0}, \quad \mathbf{x} \in \Omega, \tag{10.4b}$$

$$\mathbf{F} = \int_\Gamma \mathbf{f}(s)\,\delta(\mathbf{x} - \mathbf{X}(s))ds. \tag{10.4c}$$

The domain Ω is divided into two subdomains, Ω^+ and Ω^-, by Γ. We assume that the viscosity μ is piecewise constant,

$$\mu(\mathbf{x}) = \begin{cases} \mu^+, & \mathbf{x} \in \Omega^+, \\ \mu^-, & \mathbf{x} \in \Omega^-. \end{cases} \tag{10.5}$$

For symmetric matrices $\mathbf{S}$, $\mathbf{T} \in R^{d\times d}$, we define $\mathbf{S} : \mathbf{T} \in R$ by

$$\mathbf{S} : \mathbf{T} = \sum_i S_i \cdot T_i.$$

[14]The singular source term $\int_\Gamma \mathbf{f}(s)\,\delta(\mathbf{x} - \mathbf{X}(s))ds$ can also be written as $((\mathbf{f}\cdot\mathbf{n})\mathbf{n} + (\mathbf{f}\cdot\boldsymbol{\tau})\boldsymbol{\tau})\,\delta(\Gamma)$, or in the form of $((\mathbf{f}\cdot\mathbf{n})\mathbf{n} + (\mathbf{f}\cdot\boldsymbol{\tau})\boldsymbol{\tau})|\nabla\varphi|\,\delta(\varphi)$, where φ is a level set function whose zero level set represents the interface Γ.

For the existence and uniqueness of the weak solution to (10.1), we refer the reader to [254]. The pair $(\mathbf{u}, p)$ is the saddle point of

$$L(\mathbf{u}, p) = \frac{1}{2}\int_\Omega \mu \mathbf{T} : \mathbf{T}\, dx - \int_\Gamma \mathbf{f} \cdot \mathbf{u}\, ds - \int_\Omega \mathbf{g} \cdot \mathbf{u}\, dx - \int_\Omega (\operatorname{div} \mathbf{u})\, p\, dx; \quad (10.6)$$

see, for example, [92, 29] for the definition of the saddle point problem. Thus, (10.1) is equivalent to

$$-\int_\Omega \mathbf{S} : \nabla\boldsymbol{\phi}\, dx = \int_\Omega \mathbf{g} \cdot \boldsymbol{\phi}\, dx + \int_\Gamma \mathbf{f} \cdot \boldsymbol{\phi} ds \quad (10.7)$$

for $\boldsymbol{\phi} \in H_0^1(\Omega)^d$.

For the two-phase Stokes flow across the interface Γ, the pressure is discontinuous and the velocity has discontinuous derivatives. We assume that $(\mathbf{u}, p)$ is piecewise smooth. Applying Green's formula in each domain $\Omega^\pm$, we get

$$\int_\Omega \boldsymbol{\psi} \cdot \nabla\chi\, dx = \int_\Gamma [\mathbf{n} \cdot \boldsymbol{\psi}]\chi\, ds - \int_{\Omega^+} \operatorname{div} \boldsymbol{\psi}\, \chi\, dx - \int_{\Omega^-} \operatorname{div} \boldsymbol{\psi}\, \chi\, dx, \quad (10.8)$$

for $\boldsymbol{\psi} \in H^1(\Omega^\pm)$ and $\chi \in H_0^1(\Omega)$. Thus, we apply Green's formula to (10.7) to get

$$-\int_\Gamma ([\mathbf{n} \cdot \mathbf{S}] + \mathbf{f}) \cdot \boldsymbol{\phi}\, ds + \int_{\Omega^+} \operatorname{div} \mathbf{S} \cdot \boldsymbol{\phi}\, dx + \int_{\Omega^-} \operatorname{div} \mathbf{S} \cdot \boldsymbol{\phi}\, dx = \int_\Omega \mathbf{g} \cdot \boldsymbol{\phi}\, dx,$$

where

$$(\operatorname{div} \mathbf{S})_i = \operatorname{div} \mathbf{S}_i.$$

Hence, we obtain

$$\operatorname{div} \mathbf{S} = \mathbf{g} \quad \text{a.e. in } \Omega^\pm \quad (10.9)$$

and

$$[\mathbf{n} \cdot \mathbf{S}] + \mathbf{f} = \mathbf{0} \quad \text{a.e. in } \Gamma. \quad (10.10)$$

Note that the equation above is the result of balancing force along the normal direction at the interface Γ. This jump condition is also derived in [129, 216] and elsewhere in the literature.

Next, from (10.1), we have

$$\operatorname{div}(\operatorname{div} \mathbf{S} - \mathbf{g} - \mathbf{F}) = 0. \quad (10.11)$$

Therefore, we get

$$\int_\Omega \mathbf{S} : \nabla(\nabla\chi)\, dx + \int_\Omega \mathbf{g} \cdot \nabla\chi\, dx + \int_\Gamma \mathbf{f} \cdot \nabla\chi\, ds = 0, \quad (10.12)$$

for all $\chi \in C_0^2(\Omega)$.[15] By applying Green's formula in each domain and then combining them, we also get

$$\int_\Omega \mathbf{S} : \nabla(\nabla\chi)\, dx = \int_\Gamma [\mathbf{n}\cdot\mathbf{S}]\cdot\nabla\chi\, ds - \int_\Omega \operatorname{div}\mathbf{S}\cdot\nabla\chi\, dx. \tag{10.13}$$

From (10.12) and (10.13), we obtain

$$\int_\Gamma ([\mathbf{n}\cdot\mathbf{S}] + \mathbf{f})\cdot\nabla\chi\, ds - \int_\Omega (\operatorname{div}\mathbf{S} - \mathbf{g})\cdot\nabla\chi\, dx = 0. \tag{10.14}$$

Applying Green's formula in each domain to the volume integration in the expression above, we get

$$\int_\Gamma ([\mathbf{n}\cdot\mathbf{S}] + \mathbf{f})\cdot\nabla\chi\, ds - \int_\Gamma [\mathbf{n}\cdot(\operatorname{div}\mathbf{S} - \mathbf{g})]\,\chi\, ds$$
$$+ \int_{\Omega^+} \operatorname{div}(\operatorname{div}\mathbf{S} - \mathbf{g})\,\chi\, dx + \int_{\Omega^-} \operatorname{div}(\operatorname{div}\mathbf{S} - \mathbf{g})\,\chi\, dx = 0.$$

We use (10.10) to simplify the expression above to get

$$\int_\Gamma ([\mathbf{n}\cdot\mathbf{S}] + \mathbf{f})\cdot\nabla\chi\, ds = \int_\Gamma \{([\mathbf{n}\cdot\mathbf{S}] + \mathbf{f})\cdot\boldsymbol{\tau}\}\cdot\nabla_\tau\,\chi\, ds = -\int_\Gamma \nabla_\tau\cdot\{([\mathbf{n}\cdot\mathbf{S}] + \mathbf{f})\cdot\boldsymbol{\tau}\}\chi\, ds,$$

where $\boldsymbol{\tau}$ is the unit tangent direction. Thus we obtain

$$\operatorname{div}(\operatorname{div}\mathbf{S} - \mathbf{g}) = 0 \quad \text{a.e. in } \Omega^\pm \tag{10.15}$$

and

$$[\mathbf{n}\cdot(\operatorname{div}\mathbf{S} - \mathbf{g})] + \nabla_{\boldsymbol{\tau}}\cdot\{([\mathbf{n}\cdot\mathbf{S}] + \mathbf{f})\cdot\boldsymbol{\tau}\} = 0 \quad \text{a.e. in } \Gamma, \tag{10.16}$$

or equivalently

$$[\mathbf{n}\cdot(\operatorname{div}\mathbf{S} - \mathbf{g})] = 0 \quad \text{a.e. in } \Gamma. \tag{10.17}$$

We summarize the results so far in the following theorem.

Theorem 10.1. *Let* $(\mathbf{u}, p)$ *be the solution of the incompressible Stokes equations* (10.1)–(10.3) *and* (10.5). *Then the following equations (excluding the interface),*

$$-\mu\,\Delta\mathbf{u} + \nabla p = \mathbf{g} \quad \textit{in } \Omega^\pm, \tag{10.18}$$

$$\Delta p = \operatorname{div}\mathbf{g} \quad \textit{in } \Omega^\pm \tag{10.19}$$

hold along with the interface conditions (10.10), (10.16), *and*

$$[\mathbf{n}\cdot\mathbf{S}] + \mathbf{f} = 0, \quad [\mathbf{n}\cdot(\operatorname{div}\mathbf{S} - \mathbf{g})] + \nabla_\tau\cdot\{([\mathbf{n}\cdot\mathbf{S}] + \mathbf{f})\cdot\boldsymbol{\tau}\} = 0, \tag{10.20}$$

[15] This can also be obtained by setting $\boldsymbol{\phi} = \nabla\chi$ in (10.7).

as well as the continuity conditions

$$[\mathbf{u}] = 0, \quad [\operatorname{div} \mathbf{u}] = 0$$

a.e. in Γ.

Proof: The expression (10.9) is equivalent to (10.18), and (10.15) is equivalent to (10.19).

Let us use $\hat{f}_1 = \mathbf{f} \cdot \mathbf{n}$ and $\hat{f}_2 = \mathbf{f} \cdot \boldsymbol{\tau}$ to represent the force densities in the normal and tangential directions, respectively. Then the interface conditions can be summarized in the following theorem. □

Theorem 10.2. *Assume* $\Gamma \in C^2$, $\hat{f}_1 \in C^1$, *and* $\hat{f}_2 \in C^1$. *Let* κ *be the curvature of* Γ, *and let* $(p, \mathbf{u})$ *be the solution to the Stokes equations* (10.1)–(10.3) *and* (10.5). *We have the following jump conditions across the interface* Γ:

$$[p] = 2\left[\mu \frac{\partial \mathbf{u}}{\partial \mathbf{n}} \cdot \mathbf{n}\right] + \hat{f}_1, \tag{10.21}$$

$$\left[\frac{\partial p}{\partial n}\right] = [\mathbf{g} \cdot \mathbf{n}] + \frac{\partial}{\partial \boldsymbol{\tau}} \hat{f}_2 + 2\left[\mu \frac{\partial^2 \hat{\mathbf{u}}}{\partial \eta^2}\right] + 2\kappa \left(\left[\mu \frac{\partial (\mathbf{u} \cdot \mathbf{n})}{\partial \mathbf{n}}\right] - \left[\mu \frac{\partial (\mathbf{u} \cdot \boldsymbol{\tau})}{\partial \boldsymbol{\tau}}\right]\right) \tag{10.22}$$

$$\left[\mu \frac{\partial \mathbf{u}}{\partial \mathbf{n}} \cdot \boldsymbol{\tau}\right] + \left[\mu \frac{\partial \mathbf{u}}{\partial \boldsymbol{\tau}} \cdot \mathbf{n}\right] + \hat{f}_2 = 0, \tag{10.23}$$

$$[\mu \nabla \cdot \mathbf{u}] = 0. \tag{10.24}$$

Again, the first jump condition above is the result of balancing the force in the normal direction at the interface, while the third is in the tangential condition. The method used in the proof is the same as that in §3.1 to derive the interface relations for the elliptic interface problems. Where (ξ, η) is the local coordinates at the particular point where the jumps are defined. We did not use the local coordinates for the first order derivatives since they are invariant under orthogonal coordinates transforms.

Sketch of the proof: The first and third jump conditions are obtained by applying the vector product of $\mathbf{n}$ and $\boldsymbol{\tau}$ and the jump condition (10.10), respectively. To get the second jump condition $[p_{\mathbf{n}}]$, we define the normal and tangential velocity components as

$$\hat{u} = \mathbf{u} \cdot \mathbf{n}, \quad \hat{v} = \mathbf{u} \cdot \boldsymbol{\tau}. \tag{10.25}$$

Given a fixed point (X, Y) on the interface, we use the local coordinates centered at (X, Y). The interface Γ can be written as $\xi = \chi(\eta)$ with $\chi(0) = 0$ and $\chi'(0) = 0$. The normal direction of the interface in the neighborhood of (X, Y) is $\mathbf{n} = (1, \ \chi'(\eta))/\sqrt{1 + \chi'^2}$. Note that the expression (10.20) is invariant under an orthogonal coordinate transformation. Using the local coordinates and expressing all the quantities in terms of $\hat{u}$ and $\hat{v}$, we obtain the jump condition for $[p_{\mathbf{n}}]$ after some careful manipulations.

Remark 10.1. *For the incompressible Navier–Stokes equations, since the material derivative is continuous, the interface conditions are exactly the same as those for incompressible Stokes equations.*

Remark 10.2. *Let the unit normal direction be* $\mathbf{n} = (\cos\theta, \sin\theta)$ *as before. Then the jump conditions* (10.23) *and* (10.24) *are equivalent to*

$$[\mu u_{\mathbf{n}}] - [\mu v_{\boldsymbol{\tau}}] = \hat{f}_2(:,t)\sin\theta + 2\left[\mu\frac{\partial \mathbf{u}}{\partial \mathbf{n}}\cdot\mathbf{n}\right]\cos\theta, \tag{10.26}$$

$$[\mu v_{\mathbf{n}}] + [\mu u_{\boldsymbol{\tau}}] = -\hat{f}_2(:,t)\cos\theta + 2\left[\mu\frac{\partial \mathbf{u}}{\partial \mathbf{n}}\cdot\mathbf{n}\right]\sin\theta. \tag{10.27}$$

Proof: If $\sin\theta = 0$, then $[\mu u_{\mathbf{n}}] = [\mu u_x]$, $[\mu v_{\boldsymbol{\tau}}] = [\mu v_y]$, and (10.26) is

$$[\mu u_n] - [\mu v_{\tau}] = 2[\mu u_n] \quad \text{which is} \quad [\mu(u_x + v_y)] = 0.$$

The expression (10.27) becomes

$$[\mu v_x] + [\mu u_y] = -\hat{f}_2,$$

which is equivalent to (10.23).

Now we assume that $\sin\theta \neq 0$ and we multiply (10.27) by $\sin\theta$ to get

$$\sin\theta\,\left([\mu v_{\mathbf{n}}] + [\mu u_{\boldsymbol{\tau}}]\right) = -\hat{f}_2(s,t)\cos\theta\sin\theta + 2\left[\mu\frac{\partial \mathbf{u}}{\partial \mathbf{n}}\cdot\mathbf{n}\right]\sin^2\theta.$$

We multiply (10.26) by $\cos\theta$ and add it to the expression above to get

$$([\mu u_{\mathbf{n}}] - [\mu v_{\boldsymbol{\tau}}])\cos\theta + ([\mu v_{\mathbf{n}}] + [\mu u_{\boldsymbol{\tau}}])\sin\theta = 2[\mu(u_{\mathbf{n}}\cos\theta + v_{\mathbf{n}}\sin\theta)].$$

From the relation between x-y and $\mathbf{n}$-$\boldsymbol{\tau}$ coordinates, the expression above is simplified to

$$[\mu(u_x + v_y)] = 0.$$

If we multiply (10.26) by $\sin\theta$, and multiply (10.27) by $-\cos\theta$, and add them together, then we get

$$[\mu u_{\mathbf{n}}]\sin\theta - [\mu v_{\boldsymbol{\tau}}]\sin\theta - ([\mu v_{\mathbf{n}}] + [\mu u_{\boldsymbol{\tau}}])\cos\theta = \hat{f}_2,$$

from which we can get (10.23) easily since $\mathbf{n} = (\cos\theta, \sin\theta)$ and $\boldsymbol{\tau} = (-\sin\theta, \cos\theta)$. □

10.2 The IIM for Stokes equations with singular sources: The membrane model

In this section, we explain the IIM for incompressible Stokes equations with singular sources but constant viscosity. The method was first introduced in [156, 160] and was further discussed in [155, 185, 174]. One of the motivations is to improve the accuracy of the IB method for this type of problem, particularly, the benchmark problem described in [262].

We assume a periodic boundary condition for the velocity and the pressure. For a nonslip-type velocity boundary condition, an augmented approach for dealing with the pressure boundary condition is discussed in §10.5. We also assume that the interface $\Gamma \in C^2$ is a closed interface within the solution domain Ω. When μ is continuous, the jump conditions (10.21)–(10.22) and (10.26)–(10.27) are simplified to the following decoupled relations:

$$[p] = \hat{f}_1(:,t), \tag{10.28a}$$

$$[p_n] = \frac{\partial}{\partial \boldsymbol{\tau}} \hat{f}_2(:,t), \tag{10.28b}$$

$$[\mu u_n] = \hat{f}_2(:,t) \sin\theta, \tag{10.28c}$$

$$[\mu v_n] = -\hat{f}_2(:,t) \cos\theta. \tag{10.28d}$$

The system of equations (10.4) can be solved as a coupled system (as is done in [262]) or alternatively reduced to a sequence of three Poisson problems, one for each variable, as explained below. If we apply the divergence operator to (10.4a) and use the fact that $\nabla \cdot \Delta \mathbf{u} = 0$ due to the incompressibility condition, we get

$$\Delta p = \nabla \cdot (\mathbf{F} + \mathbf{g}) \tag{10.29}$$

in the entire domain excluding the interface. Since the right-hand side is known, this is a Poisson equation for the pressure with known jump conditions, (10.28a) and (10.28b), in the solution and the normal derivative. Once p is known, (10.4a) consists of two independent Poisson equations for u and v. Since we know all the jump conditions for the pressure and the velocity, the IIM for the elliptic interface problems described in Chapter 3 and Chapter 5 can be applied to solve all three Poisson equations if we know the source strength $\mathbf{f}$. The discretization is the standard discrete Laplacian plus correction terms at irregular grid points. The system of linear finite difference equations can be solved by a fast Poisson solver, for example, Fishpack [2].

10.2.1 The force density of the elastic membrane model

The elastic band problem has been studied intensively in [156, 160, 262]. For this application, $\mathbf{g} = 0$ and the force $\mathbf{F}$ is the elastic force that is defined only along the membrane.

Assume that an elastic band is parameterized by $\mathbf{X}(s,t) = (X(s,t), Y(s,t))$ at any time t, where s is the arc length along the unstretched band, $0 \le s \le L_0$, measured from some arbitrary but fixed origin. According to [156, 262], the force exerted by the band at $\mathbf{X}(s,t)$ is given by (10.4c) with strength

$$\mathbf{f}(s,t) = \frac{\partial}{\partial s} \left(T(s,t) \boldsymbol{\tau}(s,t) \right), \tag{10.30}$$

where $T(s,t)$ is the tension given by

$$T(s,t) = T_0 \left(\left| \frac{\partial \mathbf{X}}{\partial s} \right| - 1 \right) \tag{10.31}$$

and $\boldsymbol{\tau}(s,t)$ is the tangent vector to the band,

$$\boldsymbol{\tau}(s,t) = \frac{\partial \mathbf{X}}{\partial s} \Big/ \left| \frac{\partial \mathbf{X}}{\partial s} \right|. \tag{10.32}$$

Note that in the relaxed state, we have $|\partial \mathbf{x}/\partial s| \equiv 1$ since s is the arc length. The scalar T_0 describes the elastic properties of the band, which are assumed to be uniform along its length. The larger T_0 is, the stiffer the elastic band becomes and the greater the force induced by a stretching of the band. The tension given by (10.31) is a linear Hooke's law relation, which could easily be replaced with a more general nonlinear relation. The velocity $\mathbf{u}$ is continuous across the interface and satisfies the differential equation

$$\frac{\partial}{\partial t}\mathbf{X}(s,t) = \mathbf{u}(\mathbf{X}(s,t),\ t). \tag{10.33}$$

Note that the time evolution of the flow is governed entirely by the time dependence of the force $\mathbf{F}$ determined by the force density $\mathbf{f}$. If $\mathbf{F}$ is known at a given instant in time, then system (10.4) is elliptic and the solution $(\mathbf{u}, p)$ is determined independently of the past history of the flow. This is a reflection of the fact that there is no inertia in the system, i.e., the convective and time-derivative acceleration terms have been dropped from the momentum equations.

The jumps in the solution result from the fact that the force $\mathbf{F}$ is singular and is supported only along the interface (resulting from the elastic force or the surface tension). The singular force in (10.4a) leads to jumps in the first derivatives of u and v across the interface. Since the Poisson problem (10.29) for p involves derivatives of $\mathbf{F}$, and hence a dipole source, the pressure will be discontinuous, along with its derivatives.

While both the particle (Lagrangian) approach and a level set (Eulerian) method (for example, the level set method in [72]) can be used to represent the moving interface, for this particular application, however, the particle approach is more appropriate since we need to know how far the interface is stretched in reference to the resting configuration.

We consider the Lagrangian approach to track the motion of the membrane. The location of the interface at time t^n is represented by a finite set of control points (X_k^n, Y_k^n) for $k = 0, 1, \ldots, N_b$. Since the boundary is a closed curve in the model problem, we have $(X_0^n, Y_0^n) = (X_{N_b}^n, Y_{N_b}^n)$. The kth control point gives an approximation to $(X(s_k, t^n), Y(s_k, t^n))$, where $s_k = kL_0/N_b$. Based on these control points, the cubic spline package [160, 165] is used to represent the membrane. Given a function,[16] defined along the interface, the cubic spline package can evaluate the function value and its first- and second-order derivatives at any point on the interface if the function values are known at those control points $\{(X_k^n, Y_k^n)\}$. From the cubic spline $(X^n(s,:), Y^n(s,:))$ we can compute $\partial \mathbf{X}/\partial s$ and hence the tension $T(s, t^n)$. Multiplying by the tangent vector and differentiating again give an approximation to (10.30). We evaluate this function at each of the control points to obtain the values of $\mathbf{f}_k^n$ and then interpolate these values by a new cubic spline $\mathbf{f}^n(s)$ to obtain the representation of the force along the interface.

Other representations of the interface are possible. One approach is to use a Fourier series representation, which is quite convenient for the smooth closed curves since $X(s)$ and $Y(s)$ are both periodic in s; see [160, 262]. With this approach, it is easy to compute the necessary derivatives and also apply filtering to remove high-frequency oscillations of the interface. In some problems, this has been found to improve stability properties; see [160, 262] for more details.

[16] The function can be $X(s,:)$ or $Y(s,:)$. Using the cubic spline package, we can compute the normal and tangential directions, the curvature, and other information at any point on the interface.

10.2.2 Solving the Poisson equation for the pressure

Given the location of the interface at time t^n and the jump conditions for p across it, from (10.28a) and (10.28b), we can solve for the pressure equation (10.29) with these specified jump conditions. The finite difference equations have the following simple form from (3.22)–(3.23) (or from (5.20) and (5.23)):

$$\frac{1}{h^2}(P_{i+1,j} + P_{i-1,j} + P_{i,j+1} + P_{i,j-1} - 4P_{ij}) = C_{ij} \tag{10.34}$$

on a uniform Cartesian grid, where the correction term C_{ij} is determined from either (3.23) in Chapter 3 or (5.23) in Chapter 5. The value C_{ij} will be nonzero only at *irregular* grid points, those for which the 5-point stencil includes grid points from both sides of the interface.

Away from the interface, we simply have $\Delta p = 0$. For a periodic boundary condition, the discrete compatibility condition is

$$\sum_{i,j} C_{ij} = 0.$$

Unfortunately, this condition may not exactly hold. If we set

$$\bar{C} = \sum_{i,j} C_{ij}, \tag{10.35}$$

one practice is to perturb each C_{ij} according to

$$\hat{C}_{ij} = C_{ij} - \bar{C}/N^2$$

to obtain a solvable perturbed system,

$$\frac{1}{h^2}(P_{i+1,j} + P_{i-1,j} + P_{i,j+1} + P_{i,j-1} - 4P_{ij}) = \hat{C}_{ij} \tag{10.36}$$

whose solution is the least squares solution to the unperturbed equation (see [252]). Notice that $\bar{C}/N^2$ is of order h^2. This means that the order of the local truncation errors, which are $O(h^2)$ away from the interface and $O(h)$ near the interface, have not been changed. The solution to the perturbed equation (10.36) approximates the true solution of the Poisson equation to second-order accuracy. A fast Poisson solver can be used to solve (10.36). Often, we set $p_{00} = 0$ to get a particular solution.

10.2.3 Solving the Poisson equations for the velocity (u,v)

Next we need to solve the Poisson equation (10.4a) for u and v. These are essentially identical and we need only explain the procedure to obtain u. The component of the forcing term in the x-direction is singular along the interface, but it is a delta function singularity rather than a dipole, leading to a jump in the normal derivative of u but not in u itself. Away from the interface, we have

$$\Delta u = \frac{1}{\mu} p_x,$$

where p_x is smooth, coupled with the jump conditions $[u] = 0$ and $[u_n] = \hat{f}_2 \sin\theta/\mu$ across the interface. Using the IIM described in Chapter 3 or Chapter 5, the finite difference equation at a grid point (x_i, y_j) is

$$\frac{1}{h^2}(U_{i+1,j} + U_{i-1,j} + U_{i,j+1} + U_{i,j-1} - 4U_{ij}) = \frac{1}{\mu}(P_x)_{ij} + \tilde{C}_{ij}, \tag{10.37}$$

where $(P_x)_{ij} \approx p_x(x_i, y_j)$ and the correction terms $\bar{C}_{ij}$, resulting from the jump conditions, are nonzero only at irregular grid points.

At regular grid points we can approximate p_x by the standard central finite difference,

$$(P_x)_{ij} = \frac{1}{2h}(P_{i+1,j} - P_{i-1,j}).$$

At irregular grid points we normally use one-sided finite differences to obtain first-order accurate approximations. Alternatively, one could use the known jump conditions for p and p_x to correct the centered formula, but this may not be necessary because the first-order local accuracy for the right-hand side of (10.37) at these points is sufficient to maintain the second-order accuracy globally; see §3.6 and [23, 110, 154].

There is one situation where a one-sided approximation cannot be used, and that is if both points (x_{i-1}, y_j) and (x_{i+1}, y_j) lie on the opposite side of the interface from (x_i, y_j). This could happen, for example, at the top or the bottom of a circular interface. In this case we interpolate $p_{i-1,j}$, p_{ij}, and $p_{i+1,j}$ to get $(p_x)_{ij}$ to first order by using the known jump conditions $[p]$, $[p_x]$, and $[p_y]$ from (10.28a) and the relations

$$[p_x] = [p_n]\cos\theta - [p_\tau]\sin\theta, \tag{10.38}$$
$$[p_y] = [p_n]\sin\theta + [p_\tau]\cos\theta. \tag{10.39}$$

We know the jump $[p_n]$ already from (10.28b) and

$$[p_\tau] = \frac{\partial}{\partial s}[p] = \frac{\partial \hat{f}_1}{\partial s}. \tag{10.40}$$

Thus we find

$$[p_x] = \frac{\partial \hat{f}_2}{\partial s}\cos\theta - \frac{\partial \hat{f}_1}{\partial s}\sin\theta,$$
$$[p_y] = \frac{\partial \hat{f}_2}{\partial s}\sin\theta + \frac{\partial \hat{f}_1}{\partial s}\cos\theta.$$

Let (X_k, Y_k) be the control point closest to (x_i, y_j) on the interface Γ, and let x_l $(l = i+1$ or $i-1)$ be the point of x_i and x_{i+1} that is closer to X_k. Then we can use the following interpolation formula:

$$(P_x)^{\pm}{}_{ij} = \frac{P_{ij} - P_{lj} \mp [p] \mp [p_x]\,(x_l - X_k) \mp [p_y]\,\left(y_j - Y_k\right)}{(x_i - x_l)}, \tag{10.41}$$

where the sign in the expression depends on which side of the interface the point (i, j) is on, and $[p]$, $[p_x]$, and $[p_y]$ are evaluated at (X_k, Y_k). It is shown in [160] that (10.41) gives a first-order accurate approximation to p_x.

Solving the system (10.37) gives the velocities $\{U_{ij}\}$ that are second-order accurate at all grid points. An analogous procedure on (10.4b) gives the component of the velocity $\{V_{ij}\}$ in the y-direction.

10.2.4 Evolving the interface using an explicit method

Using the Lagrangian approach, the interface evolves at the local fluid velocity, which can be accomplished by moving each control point (X_k^n, Y_k^n) according to the velocity (U_k^n, V_k^n), determined by interpolating the velocity fields $\{U_{ij}^n\}$ and $\{V_{ij}^n\}$ to the kth control point.

Dropping superscripts and subscripts for simplicity, let (X, Y) be an arbitrary point on the interface and consider the problem of interpolating from the $\{U_{ij}\}$ to obtain the x-component U of the velocity at (X, Y). Recall Peskin's IB interpolation formula (1.25),

$$U(X, Y) = \sum_{ij} U_{ij}\delta_h(x_i - X)\delta_h(y_j - Y)h_x h_y$$

from the continuous analogue

$$U(X, Y) = \iint_\Omega u(x, y)\delta\left((x, y) - (X, Y)\right) dxdy, \tag{10.42}$$

where δ_h is a discrete delta function. Such an interpolation is first-order accurate and smears the velocity because u is not smooth across the interface. To get second-order interpolation, the known jump in the normal derivatives of the velocity should be taken into account in the interpolation formulas. There are various ways to do this. One approach is based on using three nearby points to perform linear interpolation, modified to incorporate the jump conditions at the interface.

First, we choose the first three grid points (x_{i1}, y_{j1}), (x_{i2}, y_{j2}), and (x_{i3}, y_{j3}) closest to (X, Y). Then, we form a linear combination of the grid values at these points plus a correction term to approximate U,

$$U = \gamma_1 U_{i1,j1} + \gamma_2 U_{i2,j2} + \gamma_3 U_{i3,j3} - C. \tag{10.43}$$

We use the Taylor expansion about (X, Y) to get the equations for the coefficients $\{\gamma_k\}$ so that we have a second-order approximation,

$$\begin{aligned}
U &= \gamma_1 u(x_{i1}, y_{j1}) + \gamma_2 u(x_{i2}, y_{j2}) + \gamma_3 u(x_{i3}, y_{j3}) - C \\
&= a_1 u^- + a_2 u^+ + a_3 u_x^- + a_4 u_x^+ + a_5 u_y^- + a_6 u_y^+ - C + O(h^2) \\
&= (a_1 + a_2)u^- + (a_3 + a_4)u_x^- + (a_5 + a_6)u_y^- + a_2[u] \\
&\quad + a_4[u_x] + a_6[u_y] - C + O(h^2),
\end{aligned}$$

where the $\{a_i\}$'s are linear combinations of the $\{\gamma_k\}$'s as given in (3.17) in Chapter 3, obtained from the Taylor series expansion of each (x_i, y_j) about (X, Y). We then should set

$$\begin{aligned}
a_1 + a_2 &= 1, \\
a_3 + a_4 &= 0, \\
a_5 + a_6 &= 0.
\end{aligned}$$

This gives a linear system of equations for the $\{\gamma_k\}$'s which can be solved analytically,

$$\begin{aligned}\gamma_2 &= \frac{\left(y_{j1}-Y\right)\left(x_{i3}-x_{i1}\right)-\left(x_{i1}-X\right)\left(y_{j3}-y_{j1}\right)}{\left(x_{i2}-x_{i1}\right)\left(y_{j3}-y_{j1}\right)-\left(x_{i3}-x_{i2}\right)\left(y_{j2}-y_{j1}\right)},\\ \gamma_3 &= \frac{\left(y_{j1}-y_{j1}\right)\left(x_{i1}-X\right)-\left(x_{i2}-x_{i1}\right)\left(y_{j1}-Y\right)}{\left(x_{i2}-x_{i1}\right)\left(y_{j3}-y_{j1}\right)-\left(x_{i3}-x_{i2}\right)\left(y_{j2}-y_{j1}\right)},\\ \gamma_1 &= -\left(\gamma_2+\gamma_3\right).\end{aligned} \tag{10.44}$$

Once we get the coefficients γ_1, γ_2, and γ_3, we also obtain the correction term

$$C = a_2[u] + a_4[u_x] + a_6[u_y]. \tag{10.45}$$

We can use the same coefficients $\{\gamma_i\}$ and a new correction term based on the jumps in v to find the y-component V of the velocity at (X, Y). These velocities (U, V) can then be used to evolve the interface at the control point (X, Y). Notice that the coefficients $\gamma_k \sim O(1)$ and are independent of (X, Y). Thus the interpolation scheme is stable and second-order accurate. Another efficient approach is to use the least squares interpolation described in §6.1.3 with the interpolation stencil being four corner grid points of the rectangle that contains (X, Y). This approach saves the cost of finding the closest three grid points, but needs to solve an underdetermined system of equations using a singular value decomposition (SVD) approach.

Applying this procedure at each control point (X_k^n, Y_k^n) gives the velocities (U_k^n, V_k^n). The simplest explicit method is the forward Euler method, in which we move the interface by shifting each control point according to

$$\begin{aligned}X_k^{n+1} &= X_k^n + \Delta t U_k^n,\\ Y_k^{n+1} &= Y_k^n + \Delta t V_k^n.\end{aligned}$$

In the next time step the whole process is repeated. To summarize, the process consists of the following.

1. Use the location of the interface, as determined by the control points, to determine the forces and jump conditions.
2. Solve three Poisson equations using these jump conditions to determine $\{U_{ij}^n\}$ and $\{V_{ij}^n\}$ on the uniform grid.
3. Interpolate $\{U_{ij}^n\}$ and $\{V_{ij}^n\}$ to determine $\{U_k^n\}$ and $\{V_k^n\}$ at the control points.
4. Evolve the control points at these velocities for time Δt.

There are two difficulties with the explicit method. One is that Euler's method is only first-order accurate in time. A more serious difficulty is that the system is very stiff (for reasonable values of T_0), and very small time steps must be taken to maintain stability. This difficulty is discussed in detail in [108, 196, 262]. In order to take reasonable time steps, an implicit method, as described below, is preferred.

10.2.5 Evolving the interface using an implicit method

Steps 1 through 3 of the procedure described in the previous subsection can be used to define an operator $\mathcal{U}$ that maps a set of control points $\mathbf{X} = (\mathbf{X}_1, \mathbf{X}_2, \ldots, \mathbf{X}_{N_b})$ to the resulting velocities $\mathbf{U} = (\mathbf{U}_1, \mathbf{U}_2, \ldots, \mathbf{U}_{N_b})$ at the control points, which can be written as

$$\mathbf{U} = \mathcal{U}(\mathbf{X}).$$

Applying $\mathcal{U}$ to $\mathbf{X}$ requires computing forces and jump conditions along the interface, solving three Poisson equations, and interpolating the resulting velocities back to the control points. The forward Euler method can now be written succinctly as

$$\mathbf{U}^{n+1} = \mathbf{U}^n + \Delta t\, \mathcal{U}(\mathbf{X}^n).$$

To enable large time steps and to have better accuracy, a Crank–Nicolson-type method,

$$\mathbf{U}^{n+1} = \mathbf{U}^n + \frac{1}{2}\Delta t \left(\mathcal{U}(\mathbf{X}^n) + \mathcal{U}(\mathbf{X}^{n+1}) \right),$$

has been proposed in [156, 160]. This implicit method is second-order accurate and also eliminates most stability problems, but of course it is more difficult to solve the nonlinear system. At time t^n, $\mathbf{X}^n$ is known and so $\mathbf{U}^n = \mathcal{U}(\mathbf{X}^n)$ can be computed as before. But then $\mathbf{X}^{n+1}$ must be determined from the implicit system $g(\mathbf{X}^{n+1}) = 0$, where

$$g(\mathbf{X}) = \mathbf{X} - \frac{1}{2}\Delta t\, \mathcal{U}(\mathbf{X}) - \left(\mathbf{X}^n + \frac{1}{2}\Delta t\, \mathbf{U}^n \right).$$

Normally, a Newton-like method would be used in order to obtain quadratic convergence. Newton's method requires the Jacobian matrix

$$J(\mathbf{X}) \equiv Dg(\mathbf{X}) = I - \frac{1}{2}\Delta t\, D\mathcal{U}(\mathbf{X}). \tag{10.46}$$

Unfortunately, the matrix $D\mathcal{U}(\mathbf{X})$ is almost impossible to calculate exactly, and even obtaining a finite difference approximation would be prohibitively expensive.

Instead, a quasi-Newton method was proposed in [156, 160] using the following BFGS (Broyden–Fletcher–Goldfarb–Shanno) method (see for example [122]):

For $m = 0, 1, 2, \ldots, m^*$,

(m^* is the step when the iterative method converges)

$$\mathbf{X}_0^{n+1} = \mathbf{X}_{m^*}^n, \quad B_0^n = B_{m^*}^{n-1},$$

$$\mathbf{X}_{m+1}^{n+1} = \mathbf{X}_m^{n+1} - B_m^n\, g\left(\mathbf{X}_m^{n+1}\right), \tag{10.47}$$

$$B_{m+1}^n = B_m^n + \frac{\mu_m s_m s_m^T - s_m y_m^T B_m^n - B_m^n y_m s_m^T}{s_m^T y_m}, \tag{10.48}$$

where

$$
\begin{aligned}
s_m &= -B_m^n\, g\left(\mathbf{X}_{\mathbf{m}}^{\mathbf{n+1}}\right), \\
y_m &= g\left(\mathbf{X}_{m+1}^{n+1}\right) - g\left(\mathbf{X}_m^{n+1}\right), \\
\mu_m &= 1 + \frac{y_m^T B_m^n y_m}{s_m^T y_m}.
\end{aligned}
\tag{10.49}
$$

Note that we have omitted the time index n for s_m, μ_m, and y_m. At the initial time step $t = 0$, we take $B_0^0 = I$. This is reasonable since $J = I + O(\Delta t)$. Note that in the BFGS method, the matrix B is symmetric, and thus it will not converge to the Jacobian matrix of $D\mathcal{U}(\mathbf{X})$ unless it is symmetric. Nevertheless, the quasi-Newton BFGS method has been found to be a good method for the nonlinear system of equations from the IIM applied to the membrane model.

The efficiency of this method is enhanced by the fact that we have a very similar system to solve at each time step. At each time step we begin with the current approximation from the previous time step for both $\mathbf{X}^{n+1}$ and B^{n+1}. The above comments are valid once the process has begun. At the first step we initialize B to the identity matrix, which is reasonable since, from (10.46), we see that $J = I + O(\Delta t)$. In almost all the numerical tests, it took about $3 \sim 5$ iterations for the BFGS method.

10.2.6 The validation of the IIM for moving elastic membranes

We use the example from Tu and Peskin [262] to validate the IIM for the elastic membrane problem. The initial interface (the solid line in Figure 10.1) is an ellipse with major and minor axes $a = 0.75$, $b = 0.5$, respectively. The unstretched interface (the dashed-dotted line in Figure 10.1) is a circle with radius $r_0 = 0.5$. Due to the restoring force, the ellipse will

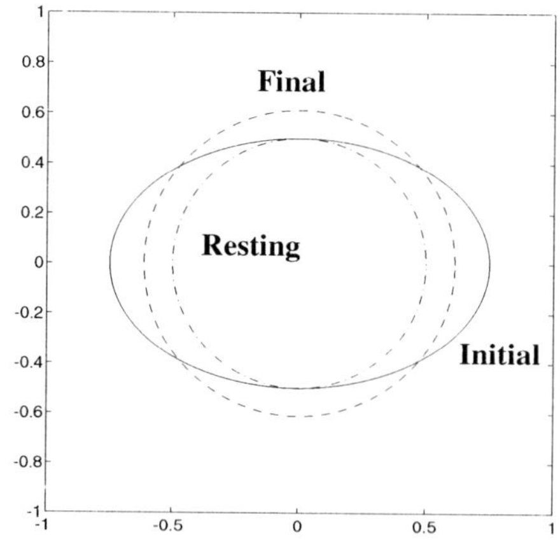

Figure 10.1. *The interface at different states. The initial interface is the solid ellipse with $a = 0.75$ and $b = 0.5$. The equilibrium position is the dashed circle with $r_e = \sqrt{ab} \approx 0.6123\ldots$. The resting circle, shown as a dashed-dotted line, has radius $r_0 = 0.5$.*

converge to an equilibrium circle (the dashed line in Figure 10.1) with radius $r_e = \sqrt{ab} \approx 0.61237$; this is larger than the unstretched interface because of the incompressibility of the enclosed fluid. Thus, the interface is still stretched at the equilibrium state, and the nonzero boundary force is balanced by a nonzero jump in the pressure (see Figure 10.2(b)).

We begin by showing the velocity and pressure at time $t = 0$ in Figure 10.2 based on the initial elliptical interface, before the interface has moved at all. As expected, p is discontinuous across the interface while u is continuous but not smooth. Figure 10.3 shows this more clearly, displaying cross sections of u and p along the line $y = -0.4$. In Figure 10.3(b) we see that the discontinuity in pressure is captured sharply by the immersed interface approach.

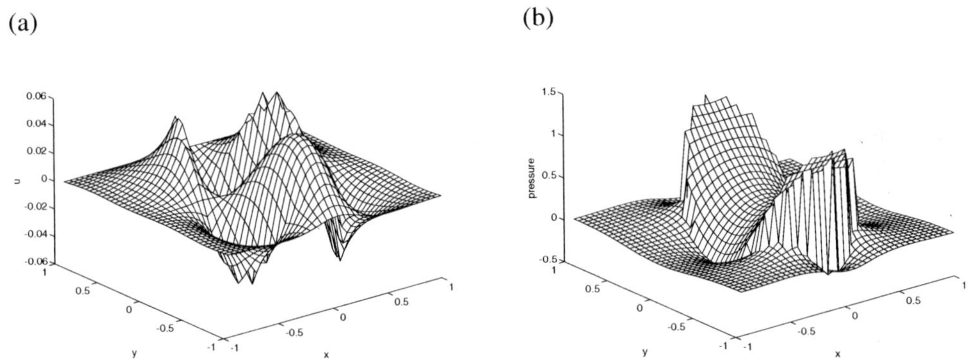

Figure 10.2. (a) *The x-component of the velocity u in the Stokes flow at* $t = 0$. *It is continuous but not smooth across the interface.* (b) *The computed pressure distribution of the Stokes flow at* $t = 0$. *The pressure is discontinuous.*

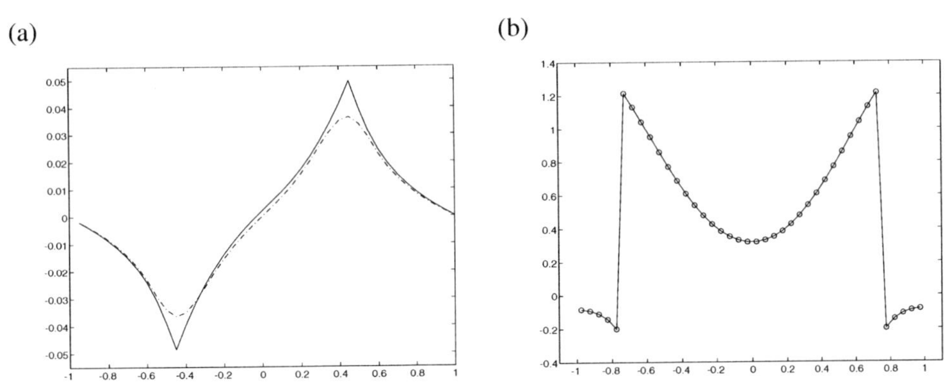

Figure 10.3. (a) *A slice of the computed velocity u at* $t = 0$ *and* $y = -0.4$. *It is continuous but not smooth. The solid line is the result from the IIM, while the dotted-dashed line is from the IB method.* (b) *A slice of computed pressure at* $t = 0$ *and* $y = 0$. *The points and solid line both show the computed results with the IIM at the grid points. Note that the large jump in pressure across the interface is captured without smearing.*

Table 10.1. (a) *The errors in the computed p, u, and v at $t = 0$ using the IIM via three Poisson equations. Second-order accuracy can be observed.* (b) *The errors in the computed u and v at $t = 0$ using the IB method. First-order accuracy can be observed.*

(a)

N	$\|p_N - p_{320}\|_\infty$	r_1	$\|u_N - u_{320}\|_\infty$	r_2	$\|v_N - v_{320}\|_\infty$	r_3
40	1.973×10^{-2}		2.674×10^{-3}		5.041×10^{-3}	
80	1.542×10^{-3}	12.80	6.361×10^{-4}	4.204	5.542×10^{-4}	9.097
160	2.609×10^{-4}	5.909	1.116×10^{-4}	5.700	1.071×10^{-4}	5.173

(b)

N	$\|u_N - u_{320}\|_\infty$	Ratio	$\|v_N - v_{320}\|_\infty$	Ratio
40	1.0170×10^{-2}		5.0540×10^{-3}	
80	4.4694×10^{-3}	2.2755	2.0512×10^{-3}	2.4639
160	1.5012×10^{-3}	2.9773	7.4032×10^{-4}	2.7707

Figure 10.3(a) also shows the plot of the cross section of the velocity u that was computed using our implementation of the IB method on the same grid. This gives a similar result except near the interface, where it is smeared with the sharp peak in the velocity being lost. Since it is the velocity right at the interface that is used to evolve the interface, this can be expected to have a substantial impact on the overall performance of the algorithm.

Table 10.1(a) shows the results of a grid refinement study on the IIM where the values on three different $N \times N$ grids with $N = 40,\ 80,\ 160$ are compared with a fine grid solution with $N = 320$. The errors in p, u, and v are measured in the maximum norm over all N^2 grid points and displayed along with the ratios of successive errors. Since we are comparing this with a computed solution on a grid that is not much finer, we do not expect a standard error ratio of 2 for a first-order method, or 4 for a second-order method. In particular, when going from $N = 80$ to $N = 160$ we expect a qth-order accurate method to produce a ratio

$$\frac{\|p_{80} - p_{320}\|_\infty}{\|p_{160} - p_{320}\|_\infty} \approx \frac{C(1/80)^q - C(1/320)^q}{C(1/160)^q - C(1/320)^q} = \frac{4^q - 1}{2^q - 1},$$

rather than the ratio 2^q; see [160] for details about the grid refinement analysis. For $q = 1$ this ratio is 3 while for $q = 2$ it is 5. Table 10.1(a) shows the final ratio to be between 5 and 6 for all three variables, indicating second-order accuracy.

Table 10.1(b) shows results for the IB method, for the velocity components only. Now the final ratios are all roughly 3, indicating the expected first-order accuracy.

Comparing the solution at all the uniform grid points at later times is difficult, since the interface may lie on one side of a given point in one calculation, but slightly on the other side in a different calculation. Instead, we focus on the error of the interface location, which

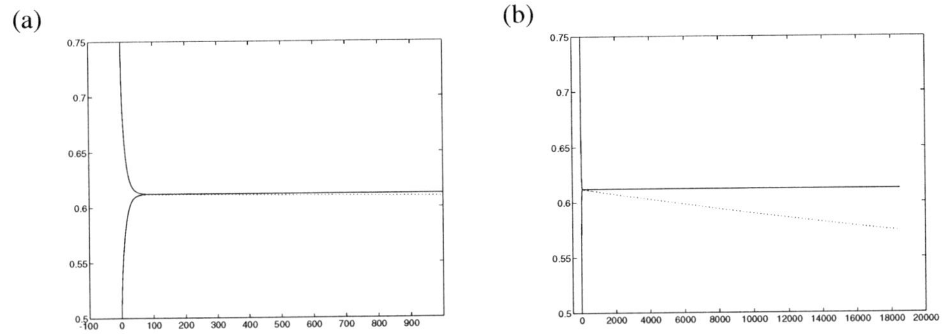

Figure 10.4. *A plot of* r_{max} *(upper curve) and* r_{min} *(lower curve) on a* 160×160 *grid with* $N_b = 160$ *on the boundary. Solid line: IIM results. Dotted line: IB method results.* (a) $0 \leq t \leq 150$*; the convergence to a near-circle is apparent.* (b) *Over a longer time scale,* $0 \leq t \leq 2 \times 10^4$.

is appropriate since this is often what we are most interested in. One simple and effective measure is to study the values r_{min} and r_{max}, defined as

$$r^n_{min} = \min_{1 \leq k \leq N_b} \sqrt{(X^n_k)^2 + (Y^n_k)^2}, \qquad r^n_{max} = \max_{1 \leq k \leq N_b} \sqrt{(X^n_k)^2 + (Y^n_k)^2}.$$

They measure the smallest and greatest distance from the origin to the interface. Note that, since we expect the interface to converge to a circle centered at the origin, we expect that $r^n_{min} \to r_e$ and $r^n_{max} \to r_e$ as $n \to \infty$.

Figure 10.4(a) shows how r_{min} and r_{max} behave computationally over a short time scale. The solid line represents the results of the IIM with $N = N_b = 160$ and the dotted line represents the result using the IB method on the same grid. Figure 10.4(b) shows what happens over a longer time scale of the two methods near the true equilibrium position $r_e \approx 0.61237$. With the IIM, a numerical equilibrium is reached that agrees well with the true equilibrium, and this equilibrium is then maintained. At $t = 2000$ we have $r_{min} \approx 0.61232$ and $r_{max} \approx 0.61248$, and this is maintained at later times. With the IB method, some leaking is apparent which causes the circle to shrink. This problem is also mentioned by Tu and Peskin [262]. Various research (see, for example, [216]), has been conducted to fix the leaking problem.

To check the error along the entire interface in the 2-norm, we take $N^* = N^*_b$ as the finest grid. For the coarser grid with $N \times N$, we take $N_b = N^*/l$, where $l = int(N^*/N)$. In this way we are guaranteed that each control point $(x_i^{(N)}, y_i^{(N)})$, $i = 1, 2, \ldots, N_b$, on the interface is also a control point for the finest grid $N^* \times N^*$ and N^*_b. Then it is possible to compute the error

$$e_N = \frac{1}{N_b} \sqrt{\sum_{i=1}^{N_b} \left(x_i^{(N)} - x_{i*l}^{(N^*)}\right)^2 + \left(y_i^{(N)} - y_{i*l}^{(N^*)}\right)^2}. \tag{10.50}$$

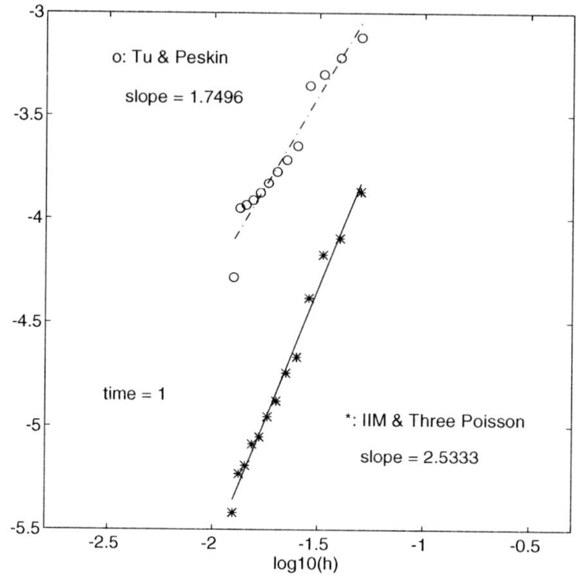

Figure 10.5. *The error in the interface location at $t = 1$ as measured in the norm (10.50). Solid line and the star "$*$": IIM results. Dashed-dotted line and the "o": IB method results.*

In Figure 10.5, we plot the global error at $t = 1$ with the finest grid being $N^* = 320$, and N and N_b being the pairs of $(40 + 10k, 40)$, $k = 0, 1, \ldots, 6$; $(80 + 10k, 80)$, $k = 0, \ldots, 7$; and $(160, 160)$. The choice of N_b allows direct comparison at control points with the fine grid solution, as required in (10.50).

Figure 10.5 shows that the IIM converges with a smaller error than that of the IB method. Again the slope is greater than we would expect if we had the exact solution to compare with, rather than a fine grid solution.

Generally, we need to increase the number of control points when we refine the Cartesian grid so that no error is dominated. If the cubic spline package [160, 165] is used, we can take fewer control points on the interface with little effect on the accuracy with the IIM, as we can see from Figure 10.5, where we have the same sequence of grids. The jumps in the error when N_b is increased are much less obvious than that in the IB method.

Figure 10.6 shows the simulation of a more complicated interface. The initial interface is $\rho = 0.6 + 0.3 \sin 8\theta$ in polar coordinates. The unstretched interface is the circle centered at the origin with radius $r_0 = 0.3$. Figure 10.6 shows the interface at different times computed using a 160×160 grid and $N_b = 160$. The problem is very stiff and we need to take a fairly small time step even with the implicit method at a first few steps because of the fast restoring process. We start with $\Delta t = O(h^2)$, but we increase the time step at later times. A comparison with the IB method reveals behavior similar to that in the previous example.

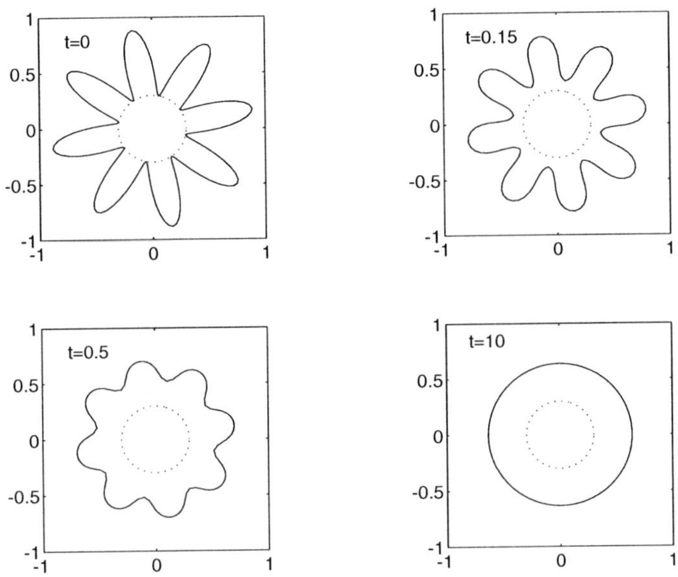

Figure 10.6. *The interface at different times with a* 160×160 *grid. The dotted circle is the unstretched interface with* $r = 0.3$ *(IIM results only).*

10.3 The IIM for Stokes equations with singular sources: The surface tension model

The method described in the previous sections can be easily adapted for interfaces between two different fluids, with the surface tension providing the singular force rather than an elastic membrane. The force strength $\mathbf{f}(s,t)$ is now given by

$$\mathbf{f}(s,t) = \gamma \frac{\partial^2}{\partial s^2} \mathbf{X}(s,t), \tag{10.51}$$

where γ is the coefficient of the surface tension between the two fluids. We still assume that the viscosity is a constant. The vector $\partial^2 \mathbf{X}/\partial s^2$ is normal to the interface with magnitude equal to the curvature. In this case, the motion of the interface does not depend on the stretch of the interface. Therefore, both the particle approach and the level set method can be used to track the motion. If the interface develops topological changes such as merging and splitting, the level set method tracks those changes more easily. On the other hand, the particle method can preserve the area better if enough particles are used. The level set method has better stability than the particle method.

The new feature needed to be included is the effect of gravity, which is important in most applications since the two fluids may have different densities. If gravity is directed in the negative y-direction, we need only modify (10.4b) to

$$p_y = \mu(v_{xx} + v_{yy}) + F_2 - g\rho, \tag{10.52}$$

where g is the gravitational constant and ρ is the density, which are assumed to have the constant value ρ_1 in one fluid and ρ_2 in the other. The Poisson problem for p then becomes

$$p_{xx} + p_{yy} = \nabla \cdot \mathbf{F} - g\rho_y.$$

Since ρ is a piecewise constant, the term $g\rho_y$ gives only an additional delta function source along the interface, which contributes to the jump in the normal derivative p_n across the interface,

$$[p_n](s) = g[\rho] \sin\theta.$$

Note the fact that the force (10.51) is normal to the interface and so $\hat{f}_2(s,t) = 0$ in (10.28). The $\sin\theta$ term arises from the fact that the delta function source is directed vertically, and hence at angle θ to the interface. The jump conditions for p, $u_{\mathbf{n}}$, and $v_{\mathbf{n}}$ are still given by (10.28) with $\hat{f}_1 = \gamma\kappa$ and $\hat{f}_2 = 0$. Note that the velocity is now continuously differentiable across the interface, simplifying the procedure for interpolation to the interface that was presented in §10.2.4.

The addition of gravity will induce a hydrostatic pressure gradient that is linear in y. This means that periodic boundary conditions are no longer reasonable. However, if we are computing on the rectangle $\Omega = [a,b] \times [c,d]$ and we set

$$\rho_0 = \frac{1}{(b-a)(d-c)} \iint_\Omega \rho(x,y)\,dx\,dy,$$

then we can write p as

$$p(x,y,t) = g\rho_0(d-y) + \tilde{p}(x,y,t),$$

where $\tilde{p}$ is the deviation from the linear profile obtained from the average density ρ_0. If the boundaries are well away from the interface, then we expect $\tilde{p}$ to be roughly constant along the entire boundary $\partial\Omega$, so the periodic boundary conditions are physically reasonable. In terms of the pressure deviation $\tilde{p}$, (10.52) becomes

$$\tilde{p}_y = \mu(v_{xx} + v_{yy}) + F_2 - g(\rho - \rho_0).$$

Now we can solve for $\mathbf{u}$ and p in exactly the same way as for the elastic membrane case.

As an example, consider a rising bubble of fluid computed using the IIM described above with density $\rho_1 = 1$ inside the bubble and density $\rho_2 = 2$ outside. Figure 10.7 shows experiments with four different values of the surface tension coefficient $\gamma = 10,\ 1,\ 0.5$, and 0. In each case the bubble was initialized to an elliptical shape, $X = 0.5\cos(\theta)$, $Y = 0.3\sin(\theta) - 2.2$ with $0 \le \theta \le 2\pi$.

When γ is large, the surface tension is sufficiently strong to bring the bubble back to a nearly circular shape even as it rises. For smaller values of γ, the bubble is distorted. For sufficiently small values, the bubble eventually breaks up; see Figure 10.8. The computation is done by combining the IIM with the level set method [174]. This behavior agrees qualitatively with the known behavior of axisymmetric three-dimensional bubbles, the case most frequently treated in the literature; see, for example, [15, 53, 244]. Note also that if we

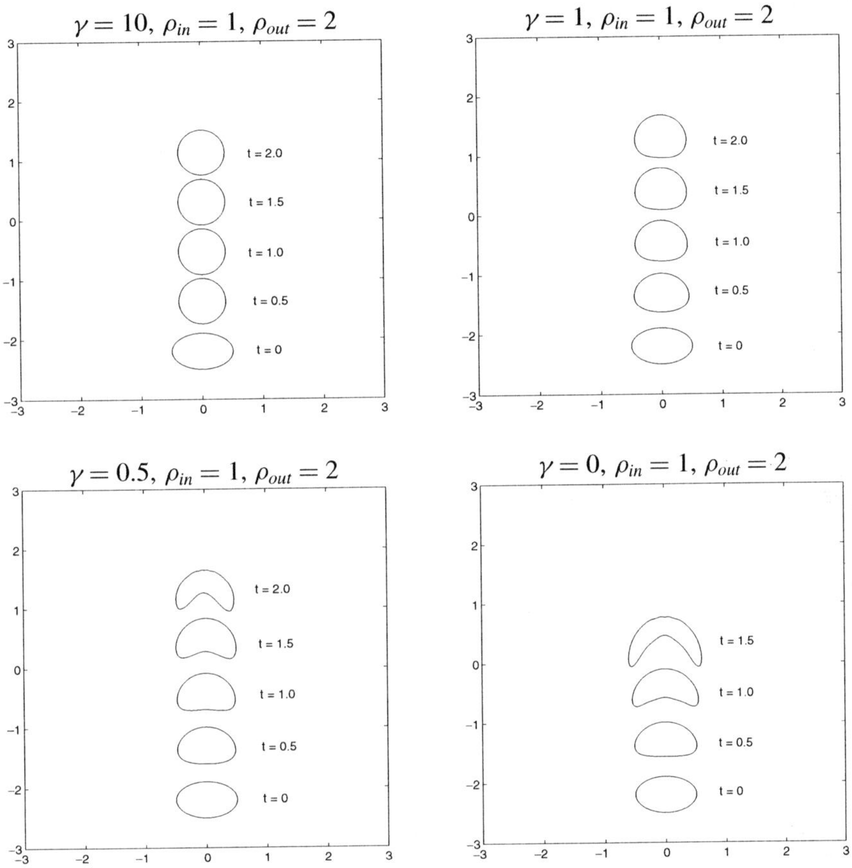

Figure 10.7. *Bubble computations with various surface tensions $\gamma = 10, 1, 0.5, 10^{-6}$. In these computations $\rho = 2$ outside and $\rho = 1$ inside the bubble and $\mu = 1$ everywhere. The computations are done on a 160×160 grid with $N_b = 80$ points on the boundary. The computation with very small surface tension ($\gamma = 10^{-6}$) breaks down before $t = 2$ when the bubble breaks up.*

start with a circular bubble, rather than an ellipse, the bubble remains circular (to reasonable accuracy) for all values of γ. This also agrees with the expected behavior [53].

We should point out that if a coarse grid is used, the level set method does not preserve the area well compared with a front-tracking method with particles. The main reason is that the computed velocity is not totally divergence free. It has $O(h^2)$ error near the interface. How to make the computed velocity divergence free near the interface with the level set method is still a challenging problem for the Stokes equations without the $\mathbf{u}_t$ term. In §10.6, we use the projection method to solve full Navier–Stokes equations with interfaces in which the incompressibility condition is enforced by the projection method. A possible solution is the combination of the level set method with the particle approach; see, for example, [71].

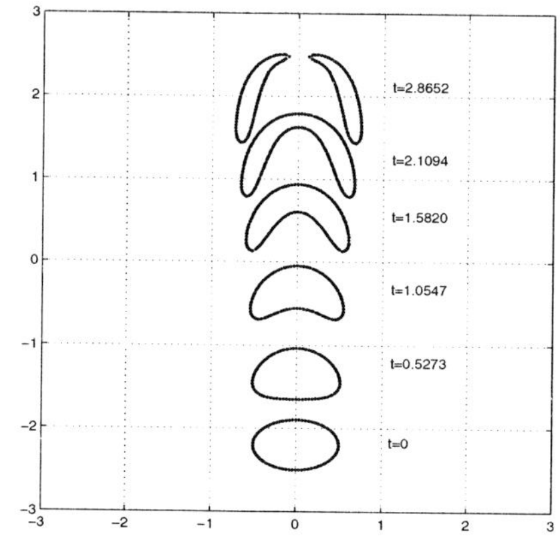

Figure 10.8. *A rising bubble breaks into two pieces when the surface tension is small ($\gamma = 10^{-6}$).*

10.4 An augmented approach for Stokes equations with discontinuous viscosity

For Stokes equations with discontinuous viscosity, the jump conditions for the pressure and the velocity are coupled together; see (10.21)–(10.24) in Theorem 10.2. This makes it difficult to discretize the system accurately.

In this section, we explain the augmented IIM for two-phase Stokes equations with discontinuous viscosity; see [167] for the reference of the method. The idea, in the spirit of Chapter 6, is to introduce two augmented variables that are defined only along the interface so that the jump conditions can be decoupled. The GMRES iterative method is then applied to the Schur complement system for the discrete augmented variables. Furthermore, the augmented approach rescales the original problem and enables us to use a fast Poisson solver in the iterative process. Each GMRES iteration requires solving the rescaled Stokes equations with decoupled jump conditions, which can be done by calling a fast Poisson solver three times and an interpolation scheme to evaluate the residual of the Schur complement.

There is more than one way to introduce augmented variables so that the jump conditions can be decoupled. Different augmented variables and equations will lead to different algorithms. It is reasonable to assume that the viscosity is a piecewise constant. In choosing the augmented variable, we take into account two different scales corresponding to the viscosity in each phase. We also wish to use a fast Poisson solver to solve the Stokes equations, as discussed in previous sections, once the augmented variables are known. Based on these two considerations, we introduce the jumps $[\mu u]$ and $[\mu v]$ along the interface as two augmented variables. The advantages and details can be seen in the rest of this section.

Using the local coordinate system (1.34), we can rewrite the two jump conditions (10.26)–(10.27) in terms of the augmented variables $[\mu u]$ and $[\mu v]$ as follows.

Lemma 10.3. *Let p, u, and v be the solution to the Stokes equations* (10.4a)–(10.4b). *We define*

$$\tilde{u} = \mu u, \qquad \tilde{v} = \mu v, \qquad \tilde{\mathbf{u}} = (\tilde{u}, \tilde{v}). \tag{10.53}$$

Then the following jump relations hold for $\tilde{u}$ and $\tilde{v}$:

$$\left[\frac{\partial \tilde{u}}{\partial \mathbf{n}}\right] = \left(\hat{f}_2 + \left[\frac{\partial \tilde{\mathbf{u}}}{\partial \boldsymbol{\tau}} \cdot \mathbf{n}\right]\right) \sin\theta - \left[\frac{\partial \tilde{\mathbf{u}}}{\partial \boldsymbol{\tau}} \cdot \boldsymbol{\tau}\right] \cos\theta, \tag{10.54}$$

$$\left[\frac{\partial \tilde{v}}{\partial \mathbf{n}}\right] = -\left(\hat{f}_2 + \left[\frac{\partial \tilde{\mathbf{u}}}{\partial \boldsymbol{\tau}} \cdot \mathbf{n}\right]\right) \cos\theta - \left[\frac{\partial \tilde{\mathbf{u}}}{\partial \boldsymbol{\tau}} \cdot \boldsymbol{\tau}\right] \sin\theta, \tag{10.55}$$

$$\left[\frac{\partial \tilde{\mathbf{u}}}{\partial \mathbf{n}} \cdot \mathbf{n}\right] = -\left[\frac{\partial \tilde{\mathbf{u}}}{\partial \boldsymbol{\tau}} \cdot \boldsymbol{\tau}\right]. \tag{10.56}$$

Proof: Note that $\mathbf{n} = (\cos\theta, \sin\theta)$ and $\boldsymbol{\tau} = (-\sin\theta, \cos\theta)$. Rewriting the incompressibility condition $[\mu \nabla \cdot \mathbf{u}] = 0$ in the local coordinates, we have

$$\left[\frac{\partial \tilde{u}}{\partial \mathbf{n}}\right] \cos\theta - \left[\frac{\partial \tilde{u}}{\partial \boldsymbol{\tau}}\right] \sin\theta + \left[\frac{\partial \tilde{v}}{\partial \mathbf{n}}\right] \sin\theta + \left[\frac{\partial \tilde{v}}{\partial \boldsymbol{\tau}}\right] \cos\theta = 0, \tag{10.57}$$

which is

$$\left[\frac{\partial \tilde{u}}{\partial \mathbf{n}}\right] \cos\theta + \left[\frac{\partial \tilde{v}}{\partial \mathbf{n}}\right] \sin\theta = -\left[\frac{\partial \tilde{\mathbf{u}}}{\partial \boldsymbol{\tau}} \cdot \boldsymbol{\tau}\right]. \tag{10.58}$$

Rewriting the interface relation (10.27) in the local coordinates, we have

$$-\left[\frac{\partial \tilde{u}}{\partial \mathbf{n}}\right] \sin\theta + \left[\frac{\partial \tilde{v}}{\partial \mathbf{n}}\right] \cos\theta = -\hat{f}_2 - \left[\frac{\partial \tilde{\mathbf{u}}}{\partial \boldsymbol{\tau}} \cdot \mathbf{n}\right]. \tag{10.59}$$

From the two equalities (10.58) and (10.59) above, we solve $[\frac{\partial \tilde{u}}{\partial \mathbf{n}}]$ and $[\frac{\partial \tilde{v}}{\partial \mathbf{n}}]$ to get (10.54) and (10.55). The last equality is verified by substituting $[\frac{\partial \tilde{u}}{\partial \mathbf{n}}]$ and $[\frac{\partial \tilde{v}}{\partial \mathbf{n}}]$ in

$$\left[\frac{\partial \tilde{\mathbf{u}}}{\partial \mathbf{n}} \cdot \mathbf{n}\right] = \left[\frac{\partial \tilde{u}}{\partial \mathbf{n}}\right] \cos\theta + \left[\frac{\partial \tilde{v}}{\partial \mathbf{n}}\right] \sin\theta$$

to get (10.56) after some manipulations. □

10.4.1 The augmented algorithm for Stokes equations

The augmented algorithm for Stokes equations with discontinuous viscosity is based on the following theorem.

Theorem 10.4. *Let p, u, and v be the solution to the Stokes equations* (10.4a)–(10.4b). *Let $q_1 = [\tilde{u}] = [\mu u]$, $q_2 = [\tilde{v}] = [\mu v]$, and $\mathbf{q} = (q_1, q_2)$. Then $\tilde{u}$, $\tilde{v}$, p, q_1, q_2 are the* *solution of the following augmented system of PDEs:*

$$\begin{cases} \Delta p = \nabla \cdot \mathbf{g}, \\ [p] = \hat{f}_1 - 2\dfrac{\partial \mathbf{q}}{\partial \boldsymbol{\tau}} \cdot \boldsymbol{\tau}, \\ \left[\dfrac{\partial p}{\partial \boldsymbol{n}}\right] = \dfrac{\partial \hat{f}_2}{\partial \boldsymbol{\tau}} + 2\dfrac{\partial^2}{\partial \eta^2}(\mathbf{q}\cdot\mathbf{n}) + 2\kappa\left(\left[\dfrac{\partial(\tilde{u}\cdot\mathbf{n})}{\partial \mathbf{n}}\right] - \left[\dfrac{\partial(\mathbf{q}\cdot\boldsymbol{\tau})}{\partial \boldsymbol{\tau}}\right]\right), \end{cases} \tag{10.60}$$

$$\begin{cases} \Delta \tilde{u} = p_x - g_1, \\ [\tilde{u}] = q_1, \quad \left[\dfrac{\partial \tilde{u}}{\partial \mathbf{n}}\right] = \left(\hat{f}_2 + \dfrac{\partial \mathbf{q}}{\partial \boldsymbol{\tau}}\cdot\mathbf{n}\right)\sin\theta - \left(\dfrac{\partial \mathbf{q}}{\partial \boldsymbol{\tau}}\cdot\boldsymbol{\tau}\right)\cos\theta, \end{cases} \tag{10.61}$$

$$\begin{cases} \Delta \tilde{v} = p_y - g_2, \\ [\tilde{v}] = q_2, \quad \left[\dfrac{\partial \tilde{v}}{\partial \mathbf{n}}\right] = -\left(\hat{f}_2 + \dfrac{\partial \mathbf{q}}{\partial \boldsymbol{\tau}}\cdot\mathbf{n}\right)\cos\theta - \left(\dfrac{\partial \mathbf{q}}{\partial \boldsymbol{\tau}}\cdot\boldsymbol{\tau}\right)\sin\theta, \end{cases} \tag{10.62}$$

$$\left[\frac{\tilde{u}}{\mu}\right] = 0, \qquad \left[\frac{\tilde{v}}{\mu}\right] = 0. \tag{10.63}$$

Where (ξ, η) is the local coordinates at the particular point where the jumps are defined. We did not use the local coordinates for the first order derivatives since they are invariant under orthogonal coordinates transforms. The proof of the theorem is straightforward from the Stokes equations (10.4a)–(10.4b) and the jump conditions in Theorem 10.2 and Lemma 10.3. Note that we have the following equality:

$$\left[\frac{\partial(\tilde{\mathbf{u}}\cdot\mathbf{n})}{\partial \mathbf{n}}\right] = \left[\frac{\partial \tilde{\mathbf{u}}}{\partial \mathbf{n}}\cdot\mathbf{n}\right] + \left[\tilde{\mathbf{u}}\cdot\nabla\mathbf{n}\right], \tag{10.64}$$

which is computable once we know $\mathbf{q} = [\tilde{\mathbf{u}}]$.

The existence and uniqueness of the solution to the system above is the same as the original incompressible Stokes equations (10.4a)–(10.4b). This is because if (u, v) and p are the solution to the original Stokes equations, then they are also the solution to the system above according to the definition of (u, v), Theorem 10.2, and Lemma 10.3. On the other hand, if (u, v) and p are the solution to the system (10.60)–(10.63) above in addition to the periodic boundary condition, then they satisfy all the equations in (10.4a)–(10.4b) and the incompressibility condition. So they are also the solution to the original problem.

Notice that if we know $\mathbf{q}$, then the jump conditions for the pressure ($[p]$ and $[p_n]$) are known and we can solve for the pressure independently of the velocity. After the pressure is solved, we can solve for the velocity from (10.61) and (10.62). The three Poisson equations with the given jump conditions can be solved using the IIM described in previous sections, in which a fast solver can be called with modified right-hand sides at grid points near or on the interface. This observation is the basis of the augmented method. The compatibility condition for the two augmented variables are the two equations in (10.63) (i.e., the velocity is continuous across the interface). It is also important to mention that the incompressibility condition is used to obtain the pressure Poisson equation of (10.60).

Once the augmented variables ($[\mu u]$ and $[\mu v]$) and the augmented equations (the two equations in (10.63)) are chosen, the success of the numerical algorithm depends on how efficiently we can solve for the augmented variables. While the augmented approach has been explained for various problems in Chapter 6 as an alternative algorithm, it may be the only way to get a second-order finite difference method to solve the Stokes flow with discontinuous viscosity. The key to the success of an augmented approach is the choice of the augmented variable(s) and equations, which is a research process, and it is problem dependent.

Let $\{\mathbf{X}_k\} = \{(X_k, Y_k)\}$, $k = 1, 2, \ldots, N_b$, be a set of selected *control points* on the interface, either from a number of given particles, or from the orthogonal projections of irregular grid points from a selected side of the interface implicitly defined by a level set function. The auxiliary variable $\mathbf{q} = (q_1, q_2)$ is defined, and the augmented equations (10.63) are discretized at $\{\mathbf{X}_k\}$.

Given an initial guess of $\mathbf{Q}$ at the control points, we can approximate its first- and second-order tangential derivatives $\frac{\partial \mathbf{q}}{\partial \boldsymbol{\tau}}$ and $\frac{\partial^2 \mathbf{q}}{\partial \boldsymbol{\tau}^2}$. Thus all the jump conditions in the system (10.60)–(10.62) are known and we can solve the system (10.60)–(10.62) sequentially using the IIM discussed in §3.4 or Chapter 5. Note that the jump conditions for the pressure and the velocity now are decoupled if we know $\mathbf{Q}$. Since p and (u, v) depend on $\mathbf{q}$, we can write $\mathbf{P}(\mathbf{Q})$, $\tilde{\mathbf{U}}(\mathbf{Q})$, and $\tilde{\mathbf{V}}(\mathbf{Q})$ in discretization.

If the computed $\tilde{\mathbf{U}}(\mathbf{Q})$ and $\tilde{\mathbf{V}}(\mathbf{Q})$ satisfy the two equations in (10.63), then $\mathbf{P}(\mathbf{Q})$, $\tilde{\mathbf{U}}(\mathbf{Q})/\mu$, and $\tilde{\mathbf{V}}(\mathbf{Q})/\mu$ are approximated solutions to the original system (10.4a)–(10.4b). The core of the augmented method is to use the GMRES iterative method [231] to solve the augmented variables $\tilde{\mathbf{U}}$ and $\tilde{\mathbf{V}}$ that are one-dimensional vectors defined along the interface Γ, or more precisely, at the control points $\{\mathbf{X}_k\}$. It is not necessary to form the coefficient matrix explicitly, as we will explain below. Each GMRES iteration requires solving three Poisson equations with given jump conditions in the solution and its normal derivative, and a least squares interpolation scheme to evaluate the residual. It is worth pointing out that the most expensive part of the algorithm, which is solving the three Poisson equations, can be done by calling a fast Poisson solver three times.

The discrete system of equations in the matrix-vector form

Given a discrete approximation of (q_1, q_2) at $\{\mathbf{X}_k\}$, we can solve the first three equations (10.60)–(10.62) to get an approximate solution: the pressure $\mathbf{P}(\mathbf{Q})$, the scaled velocity $\tilde{\mathbf{U}}(\mathbf{Q})$, and $\tilde{\mathbf{V}}(\mathbf{Q})$. Generally the computed velocity $(\tilde{\mathbf{U}}(\mathbf{Q}), \tilde{\mathbf{V}}(\mathbf{Q}))$ does not satisfy the two augmented equations in (10.63), that is, $(\mathbf{U}, \mathbf{V}) = (\tilde{\mathbf{U}}/\mu, \tilde{\mathbf{V}}/\mu)$ may not be continuous across the interface.

Let us put the discrete solutions $\{P_{ij}\}$, $\{U_{ij}\}$, and $\{V_{ij}\}$ together as a big vector $\tilde{\mathcal{U}}$ whose dimension is $O(3MN)$, where M and N are the number of grid lines in the x- and y-direction, respectively. We denote also the vector of the discrete values of (q_1, q_2) at the control points $\{\mathbf{X}_k\}$ by $\mathbf{Q}$ whose dimension is $O(2N_b)$. Then the discrete solution of (10.60)–(10.62) given $\mathbf{Q}$ can be written as

$$A\tilde{\mathcal{U}} + B\mathbf{Q} = \mathbf{F}_1 \tag{10.65}$$

for some vector $\mathbf{F}_1$ and sparse matrices A and B. It requires solving three Poisson equations with different source terms and jump conditions to get $\tilde{\mathcal{U}}$.

Once we know the solution $\tilde{\mathcal{U}}$ given $\mathbf{Q}$, we can use $(\tilde{\mathbf{U}}, \tilde{\mathbf{V}})$ and the jump conditions $[\tilde{\mathbf{U}}_n]$ and $[\tilde{\mathbf{V}}_n]$, which also depend on $\mathbf{Q}$, to get $[\mathbf{U}(\mathbf{Q})] = [\tilde{\mathbf{U}}(\mathbf{Q})/\mu]$ and $[\mathbf{V}(\mathbf{Q})] = [\tilde{\mathbf{V}}(\mathbf{Q})/\mu]$ at those control points $\{\mathbf{X}_k\}$, $1 \le k \le N_b$. If both $\| [\mathbf{U}(\mathbf{Q})] \|$ and $\| [\mathbf{V}(\mathbf{Q})] \|$ are smaller than a given tolerance, then the method has already converged and the set of $\mathbf{Q}$, $\tilde{\mathbf{U}}/\mu$, $\tilde{\mathbf{V}}/\mu$ is an approximate solution. The interpolation scheme to get $[\tilde{\mathbf{U}}(\mathbf{Q})/\mu]$ and $[\tilde{\mathbf{V}}(\mathbf{Q})/\mu]$, which will be explained in the next subsection, depends on $\tilde{\mathcal{U}}$ and $\mathbf{Q}$ *linearly*. Therefore, we can write

$$[\mathbf{U}(\mathbf{Q})]_\Gamma = \left([\tilde{\mathbf{U}}/\mu],\ [\tilde{\mathbf{U}}/\mu]\right)^T = S\tilde{\mathcal{U}} + E\mathbf{Q} - \mathbf{F}_2, \tag{10.66}$$

where S and E are two sparse matrices, and $\mathbf{F}_2$ is a vector. The matrices depend on the interpolation scheme but need not be actually constructed in the algorithm. We need to choose a vector $\mathbf{Q}$ such that the continuity condition for the velocity is satisfied along the interface Γ. If we put the two matrix-vector equations (10.65) and (10.66) together, we get

$$\begin{bmatrix} A & B \\ S & E \end{bmatrix} \begin{bmatrix} \tilde{\mathcal{U}} \\ \mathbf{Q} \end{bmatrix} = \begin{bmatrix} \mathbf{F}_1 \\ \mathbf{F}_2 \end{bmatrix}. \tag{10.67}$$

Note that $\mathbf{Q}$ is defined only on a set of points $\{\mathbf{X}_k\}$ on the interface while $\tilde{\mathcal{U}}$ is defined at all grid points. The Schur complement for $\mathbf{Q}$ is

$$(E - SA^{-1}B)\mathbf{Q} = \mathbf{F}_2 - SA^{-1}\mathbf{F}_1 = \bar{\mathbf{F}}. \tag{10.68}$$

If we can solve the system above to get $\mathbf{Q}$, then we can get $\tilde{\mathcal{U}}$ easily. Because the dimension of $\mathbf{Q}$ is much smaller than that of $\tilde{\mathcal{U}}$, we expect to get a reasonably fast algorithm for the two-phase Stokes equations if we can solve (10.68) efficiently.

In implementation, the GMRES iterative method [231] is used to solve (10.68). The GMRES method requires only the matrix vector multiplication. We explain below how to evaluate the right-hand side $\bar{\mathbf{F}}$ of the Schur complement and how to evaluate the matrix-vector multiplication needed by the GMRES iteration. We can see why we do not need to form the coefficient matrix $E - SA^{-1}B$ explicitly.

Evaluation of the right-hand side of the Schur complement

First, we set $\mathbf{Q} = \mathbf{0}$ and solve the decoupled system (10.60)–(10.62), or (10.65) in the discrete form, to get $\tilde{\mathcal{U}}(\mathbf{0})$ which is $A^{-1}\mathbf{F}_1$ from (10.65). From the interpolation (10.66), we also have

$$\left[\mathbf{U}(\mathbf{0})\right]_\Gamma = S\tilde{\mathcal{U}}(\mathbf{0}) + E\,\mathbf{0} - \mathbf{F}_2 = S\tilde{\mathcal{U}}\,(\mathbf{0}) - \mathbf{F}_2. \tag{10.69}$$

Note that the residual of the Schur complement for $\mathbf{Q} = \mathbf{0}$ is

$$\begin{aligned} R(\mathbf{0}) &= (E - SA^{-1}B)\,\mathbf{0} - \bar{\mathbf{F}} = -\bar{\mathbf{F}} \\ &= -\left(\mathbf{F}_2 - SA^{-1}\mathbf{F}_1\right) = -\mathbf{F}_2 + S\tilde{\mathcal{U}}\,(\mathbf{0}) \\ &= \left[\mathbf{U}(\mathbf{0})\right]_\Gamma, \end{aligned} \tag{10.70}$$

which gives the right-hand side of the Schur complement system with an opposite sign.

Evaluation of the matrix-vector multiplication

The matrix-vector multiplication of the Schur complement system given $\mathbf{Q}$ is obtained from the following two steps.

Step 1: Solve the coupled system (10.60)–(10.62), or (10.65) in the discrete form, to get $\tilde{\mathcal{U}}(\mathbf{Q})$.

Step 2: Interpolate $\tilde{\mathcal{U}}(\mathbf{Q})$ using (10.66) to get $[\mathbf{U}(\mathbf{Q})]|_\Gamma$. Then the matrix-vector multiplication is

$$\left(E - SA^{-1}B\right)\mathbf{Q} = \left[\mathbf{U}(\mathbf{Q})\right]_\Gamma - \left[\mathbf{U}(\mathbf{0})\right]_\Gamma. \tag{10.71}$$

This is because

$$\begin{aligned}\left(E - SA^{-1}B\right)\mathbf{Q} &= E\mathbf{Q} - SA^{-1}B\mathbf{Q} \\ &= E\mathbf{Q} - S\left(A^{-1}\mathbf{F}_1 - \tilde{\mathcal{U}}(\mathbf{Q})\right) \qquad \text{(from (10.65))} \\ &= E\mathbf{Q} + S\tilde{\mathcal{U}}(\mathbf{Q}) - \mathbf{F}_2 + \mathbf{F}_2 - SA^{-1}\mathbf{F}_1 \\ &= \left[\mathbf{U}(\mathbf{Q})\right]_\Gamma + \bar{\mathbf{F}} \qquad \text{(from (10.66))} \\ &= \left[\mathbf{U}(\mathbf{Q})\right]_\Gamma - \left[\mathbf{U}(\mathbf{0})\right]_\Gamma \qquad \text{(from (10.70)).}\end{aligned}$$

Now we can see that a matrix-vector multiplication is equivalent to solving the coupled system (10.60)–(10.62), or (10.65) in the discrete form, to get $\tilde{\mathcal{U}}$, and using an interpolation scheme (10.66) to get $[\mathbf{U}(\mathbf{Q})]_\Gamma$ at the control points.

Since we know the right-hand side of the linear system of equations and the matrix-vector multiplication of the coefficient matrix, it is straightforward to use the GMRES or other iterative methods.

The least squares interpolation scheme to compute the residual

The interpolation scheme to evaluate $[\tilde{\mathbf{U}}/\mu]$ and $[\tilde{\mathbf{V}}/\mu]$ needed in (10.66) is crucial to the efficiency (accuracy and the number of iterations of the GMRES iteration) of the augmented method. To reduce the number of iterations, it is important to couple the solutions on both sides of the interface using the jump conditions. Although the least squares interpolation scheme is not a new idea, the details vary with the problem. We explain the least squares interpolation scheme for computing (10.66) to see why we have the second matrix-vector equation in expression (10.67).

Given an approximation of the augmented variable $\mathbf{Q}$, we can solve the pressure, and then the velocity $(\mathbf{U}, \mathbf{V}) = (\tilde{\mathbf{U}}/\mu,\ \tilde{\mathbf{V}}/\mu)$ from (10.60)–(10.62). Since

$$\left[\frac{\tilde{\mathbf{U}}}{\mu}\right] = \frac{\tilde{\mathbf{U}}^+}{\mu^+} - \frac{\tilde{\mathbf{U}}^-}{\mu^-},$$

we need to evaluate $\{\tilde{U}^+\}$ and $\{\tilde{U}^-\}$ at all control points $\mathbf{X}_k$ to get the vector $[\tilde{\mathbf{U}}/\mu]$ to see whether its norm is small enough. To explain the idea, we just need to explain the interpolation scheme for $\tilde{U}^-(\mathbf{X})$ at a point $\mathbf{X}$ on the interface. The interpolation scheme can be written as

$$\tilde{U}^-(\mathbf{X}) = \sum_{k=0}^{k_s-1} \gamma_k \, \tilde{U}_{i^*+k,\, j^*+k} - C, \tag{10.72}$$

where k_s is the number of grid points involved in the interpolation scheme, (x_{i^*}, y_{j^*}) is the closest grid point to $\mathbf{X}$, and C is a correction term.

The least squares interpolation scheme discussed in §6.1.3 can be used to find the coefficients $\{\gamma_k\}$ with minor modifications. The main change is the right-hand side of the system of equations (6.20). The first two equations now become

$$a_1 + a_2 = 0, \qquad a_3 + a_4 = 1. \tag{10.73}$$

The correction term is

$$\begin{aligned} C = {} & a_2[\tilde{u}] + a_4[\tilde{u}_\xi] + a_6[\tilde{u}_\eta] + a_8[\tilde{u}_{\xi\xi}] + a_{10}[\tilde{u}_{\xi\eta}] + a_{12}[\tilde{u}_{\xi\,\eta}] \\ = {} & a_2\, q_1 + a_4 \left[\frac{\partial \tilde{u}}{\partial \mathbf{n}}\right] + a_6 \frac{dq_1}{d\eta} + a_8 \left(\left[\frac{\partial \tilde{u}}{\partial \mathbf{n}}\right] \kappa - \frac{d^2[\tilde{u}]}{d\eta^2} + [\, p_x - g_1\,] \right) \\ & + a_{10} \left(\frac{d^2 q_1}{d\eta^2} - \left[\frac{\partial \tilde{u}}{\partial \mathbf{n}}\right] \kappa \right) + a_{12} \left(\frac{dq_1}{d\eta}\kappa + \frac{d}{d\eta}\left[\frac{\partial \tilde{u}}{\partial \mathbf{n}}\right] \right), \end{aligned} \tag{10.74}$$

where $[\tilde{u}_n]$ is determined from (10.61) in terms of $\hat{f}_2$, $\mathbf{q}$, and its first surface derivative.

We should point out that a one-sided interpolation scheme works poorly in the sense that the convergence speed is slow for the GMRES iteration. Note that $\{\gamma_k\}$ and C depend on $\mathbf{X}$. But for simplicity of notation, we have omitted the dependency.

We choose a neighborhood of $\{\mathbf{X}\}$ that contains more than six different grid points so that we have an underdetermined system. In the numerical tests, we chose $k_s = 9 \sim 12$; that is, we selected $9 \sim 12$ closest grid points to $\mathbf{X} = (X, Y)$ as the interpolation stencil. We use the SVD to find the least squares solution which also has the least L^2 norm among all the solutions. In this way, the magnitude of the coefficients γ_k is controlled and balanced.

The only trade-off of the least squares interpolation is that we have to solve an underdetermined system of equations. However, the size of the linear system is small and the coefficients can be predetermined before the GMRES iteration. The extra time needed in dealing with the interface is usually less than 5% of the total CPU time and the percentage decreases as the mesh size (M and N) increases.

10.4.2 The validation of the augmented method for Stokes equations

We show several numerical experiments for the augmented method for the two-phase Stokes equations. More examples can be found in [167]. It is quite challenging to construct examples with known exact solutions for incompressible Stokes flow with an interface. Throughout this section, the computational domain is taken as $\Omega = [-2,\ 2] \times [-2,\ 2]$.

Example 10.1. Consider the example with the following exact solution:

$$u = \begin{cases} y\left(x^2+y^2-1\right) & \text{if } x^2+y^2 \le 1, \\ 0 & \text{if } x^2+y^2 > 1, \end{cases} \tag{10.75}$$

$$v = \begin{cases} -x\left(x^2+y^2-1\right) & \text{if } x^2+y^2 \le 1, \\ 0 & \text{if } x^2+y^2 > 1, \end{cases} \tag{10.76}$$

$$p = \begin{cases} 1 & \text{if } x^2+y^2 \le 1, \\ 0 & \text{if } x^2+y^2 > 1. \end{cases} \tag{10.77}$$

The interface is the unit circle. The viscosity is

$$\mu = \begin{cases} 1 & \text{if } x^2+y^2 \le 1, \\ \dfrac{1}{2} & \text{if } x^2+y^2 > 1. \end{cases} \tag{10.78}$$

The bounded external forcing term $\mathbf{g}$ is given by

$$g_1 = \begin{cases} -8y & \text{if } x^2+y^2 \le 1, \\ 0 & \text{if } x^2+y^2 > 1, \end{cases} \tag{10.79}$$

$$g_2 = \begin{cases} 8x & \text{if } x^2+y^2 \le 1, \\ 0 & \text{if } x^2+y^2 > 1, \end{cases} \tag{10.80}$$

which has a finite jump across the interface. The normal and tangential force densities are

$$\hat{f}_1 = [p] - 2\left[\mu\frac{\partial \mathbf{u}}{\partial \mathbf{n}}\cdot\mathbf{n}\right] = -1, \tag{10.81}$$

$$\hat{f}_2 = -\left[\mu\frac{\partial \mathbf{u}}{\partial \mathbf{n}}\cdot\boldsymbol{\tau}\right] - \left[\mu\frac{\partial \mathbf{u}}{\partial \boldsymbol{\tau}}\cdot\mathbf{n}\right] = -2, \tag{10.82}$$

obtained from (10.21) and (10.23), respectively. In this example, the periodic boundary conditions are used for p and $\mathbf{u}$. The velocity is nonsmooth and the jump in the normal derivative of the velocity is not a constant along the interface. There is a finite jump in $\mathbf{g}$ as well.

In Table 10.2, we show the result of a grid refinement analysis. The tolerance for the GMRES iteration is taken as 10^{-6}. The errors in Table 10.2 are measured in the maximum norm at all grid points, for example,

$$E_{\mathbf{u}}(N) = \frac{1}{2}\left(\max_{0\le i,j\le N}\left|U_{ij}-u(x_i,y_j)\right| + \max_{0\le i,j\le N}\left|V_{ij}-v(x_i,y_j)\right|\right). \tag{10.83}$$

Table 10.2. *A grid refinement analysis for Example 10.1.*

N	E_p	p-order	$E_{\mathbf{u}}$	$\mathbf{u}$-order	No.
32	8.4430×10^{-3}		3.4549×10^{-3}		9
64	2.8405×10^{-3}	1.5716	8.8800×10^{-4}	1.9600	10
128	8.0952×10^{-4}	1.8110	2.2666×10^{-4}	1.9700	11
256	2.5417×10^{-4}	1.6713	4.7693×10^{-5}	2.2487	12
512	1.4086×10^{-5}	2.1296	1.4086×10^{-5}	1.7595	13

In all the tables in this section, N is the number of grid lines in both the x- and the y-direction. The ratio

$$p\text{-order} = \frac{\log\left(E_p(N)/E_p(2N)\right)}{\log 2}, \tag{10.84}$$

for example, is an indication of the order of accuracy for the pressure. We can see roughly second-order convergence for all the quantities. The last column is the number of iterations (No.) for the GMRES method. We can see that the number of iterations remains roughly the same as we double the grid lines in each direction.

To be more precise, we also use a linear regression analysis to find an approximate order of accuracy. In Figure 10.9, we show the error plot in log-log scale for the pressure and the velocity versus the grid spacing h. From the slopes, we can observe that the average convergence order of the pressure and the velocity are 1.9168 and 2.1519, respectively. The mesh size varies from $N = 100$ to $N = 320$ according to $N = 100 + 5k$, $k = 0, 1, \ldots, 44$. The number of iterations is small and does not change much with the mesh size.

Example 10.2. In the previous example, the force density is a constant. In this example, we construct the exact solutions in such a way that all the jumps and their derivatives along the interface are nonconstant functions. The exact velocity and the pressure are given by

$$u = \begin{cases} \dfrac{y}{4} & \text{if } x^2 + y^2 \le 1, \\[2mm] \dfrac{y}{4}\left(x^2 + y^2\right) & \text{if } x^2 + y^2 > 1, \end{cases} \tag{10.85}$$

$$v = \begin{cases} -\dfrac{x}{4}\left(1 - x^2\right) & \text{if } x^2 + y^2 \le 1, \\[2mm] -\dfrac{xy^2}{4} & \text{if } x^2 + y^2 > 1, \end{cases} \tag{10.86}$$

$$p = \begin{cases} \left(-\dfrac{3}{4}x^3 + \dfrac{3}{8}x\right) y & \text{if } x^2 + y^2 \le 1, \\[2mm] 0 & \text{if } x^2 + y^2 > 1. \end{cases} \tag{10.87}$$

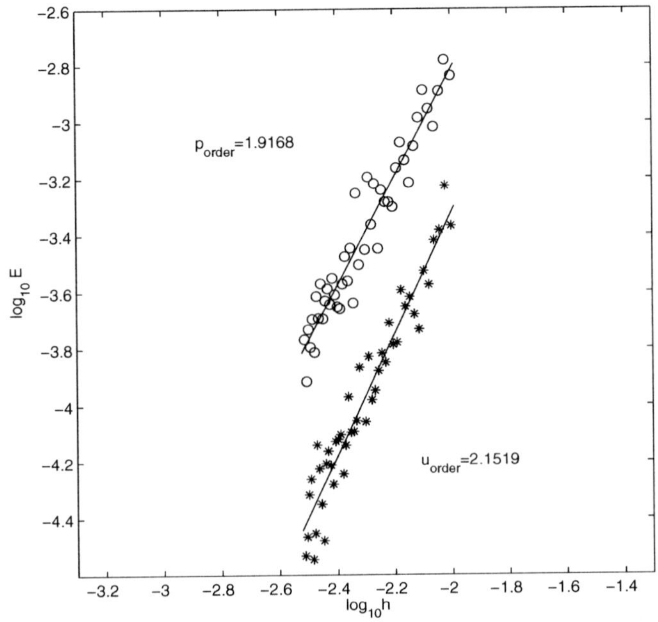

Figure 10.9. *Linear regression analysis of the convergence order in* log-log *scale for Example* 10.1. *The average convergence order for the pressure and the velocity are* $p_{order} = 1.9168$ *and* $u_{order} = 2.1519$*, respectively.*

The bounded external forcing term **g** is

$$g_1 = \begin{cases} \left(-\frac{9}{4}x^2 + \frac{3}{8}\right) y & \text{if } x^2 + y^2 \le 1, \\ -2\mu^+ y & \text{if } x^2 + y^2 > 1, \end{cases} \tag{10.88}$$

$$g_2 = \begin{cases} -\frac{3}{4}x^3 + \frac{3}{8}x - \frac{3\mu^-}{2}x & \text{if } x^2 + y^2 \le 1, \\ \frac{\mu^+}{2}x & \text{if } x^2 + y^2 > 1, \end{cases} \tag{10.89}$$

which is discontinuous across the interface. The force densities corresponding to the singular Dirac delta function in the normal and tangential directions are

$$\hat{f}_1 = \left(\frac{3}{4}\cos^3\theta - \frac{3}{8}\cos\theta\right)\sin\theta - \frac{3}{2}[\mu]\cos^3\theta\sin\theta, \tag{10.90}$$

$$\hat{f}_2 = \frac{1}{2}\mu^+ + \frac{3}{4}[\mu]\cos^2\theta\left(1 - 2\cos^2\theta\right), \tag{10.91}$$

respectively. All the jump conditions (10.21)–(10.24) are satisfied except for (10.22). Since the exact solution is not periodic, we use the Dirichlet boundary condition when we solve

Table 10.3. *A grid refinement analysis for Example* 10.2.

(a) $\mu^- = 1$, $\mu^+ = 0.1$.

N	E_p	p-order	$E_{\mathbf{u}}$	$\mathbf{u}$-order	No.
32	1.1252×10^{-2}		1.0331×10^{-2}		10
64	3.1248×10^{-3}	1.8484	2.2832×10^{-3}	2.1779	11
128	8.9338×10^{-4}	1.8064	5.5784×10^{-4}	2.0331	10
256	2.4296×10^{-4}	1.8786	1.1291×10^{-4}	2.3047	9
512	5.5515×10^{-5}	2.1297	2.8135×10^{-5}	2.0047	8

(b) $\mu^- = 0.001$, $\mu^+ = 1$.

N	E_p	p-order	$E_{\mathbf{u}}$	$\mathbf{u}$-order	No.
32	1.4537×10^{-2}		9.9572×10^{-1}		12
64	3.8694×10^{-3}	1.9095	2.2471×10^{-1}	2.1477	12
128	1.0974×10^{-3}	1.8180	5.7087×10^{-2}	1.9768	11
256	3.4110×10^{-4}	1.6858	1.1255×10^{-2}	2.3426	11
512	6.6231×10^{-5}	2.3646	2.8277×10^{-3}	1.9929	9

(c) $\mu^- = 1$, $\mu^+ = 0.001$.

N	E_p	p-order	$E_{\mathbf{u}}$	$\mathbf{u}$-order	No.
32	2.8939×10^{-2}		2.2153		27
64	5.4693×10^{-3}	2.4036	3.6865×10^{-1}	2.5872	25
128	1.6101×10^{-3}	1.7642	9.1288×10^{-2}	2.0138	25
256	4.1020×10^{-4}	1.9728	2.4121×10^{-2}	1.9201	31
512	1.3470×10^{-4}	1.6066	6.7498×10^{-3}	1.8374	30

the three Poisson equations (10.60)–(10.62) so that we can check the accuracy of the computed solution. This did not cause any inconsistency since we know that the solution exists. Numerical results using Neumann boundary conditions for the pressure can be found in [167].

Table 10.3 shows a grid refinement analysis for different jumps in μ. The problem is scaled such that $\max\{\mu^-, \mu^+\} = 1$. The results for $\mu^-/\mu^+ = 10,\ 10^{-3},\ 10^3$ are tested in Table 10.3. While the accuracy does depend on $[\mu]$, the average convergence rates are about the same (second-order accurate). Note that there are two very different scales for the problems in Table 10.3(b), (c). The number of iterations seems to be dependent on the jump in μ but not on the mesh size h.

Example 10.3. We show an example of a moving interface problem by tracking the motion and deformation of a liquid drop in a simple shear flow in two space dimensions. We refer the readers to [137] and the references therein for a detailed description of the problem and physical explanations.

The setup of the problem is as follows. A circular viscous drop with viscosity μ_2 is placed in a shear flow with viscosity μ_1. The velocity of the shear flow without the drop is $\mathbf{u} = (Gy, 0)$, where G is the shear rate. The computational domain is $\Omega = [-2,\ 2] \times [-2,\ 2]$. The Dirichlet boundary condition $\mathbf{u}|_{\partial\Omega} = (Gy, 0)$ is used along $\partial\Omega$; and the Neumann boundary condition $p_n = 0$ is used for the pressure along $\partial\Omega$. The capillary number is defined as $Ca = Gr_0\mu_1/\gamma$, where r_0 is the radius of the circular drop and γ is the surface tension. The source strength is $\hat{f}_1 = \mathbf{f} \cdot \mathbf{n} = \gamma\kappa$, where κ is the curvature. The ratio of the viscosity is $\lambda = \mu_2/\mu_1$. A front-tracking method is used to evolve the interface. The crucial parameters are the capillary number Ca and the viscosity ratio λ. Depending on the values of Ca and λ, the drop can develop internal circulations and can have an asymptotic equilibrium state and other configurations.

In Figure 10.10 we present a sequence of contour plots of the stream function of the velocity for the two-phase Stokes flow in comparison with the results in [137]. In our simulation, the initial drop is a circle with radius $r_0 = 0.75$ and the shear rate is $G = 0.9143$. The parameters are chosen to mimic the setup in [137]. We try to reproduce the results there except now it is in two space dimensions. Our results agree qualitatively with those presented in [137]. The stream function is obtained by solving the Poisson equation $\Delta\psi = v_x - u_y$ with the boundary condition determined from the far field boundary condition of the shear flow. When λ is small, we observe two regions of recirculating fluid within the drop. Generally, for smaller λ, we observe more deformation but larger angle (less tilting) between the longer axis of the drop and the x-axis. As λ gets larger, we see less deformation but smaller angle (more tilting) between the longer axis of the drop and the x-axis. All these agree with the results and analysis discussed in [137]. Note that for various parameters, internal circulations do not always occur. For instance, for circular drop at $t = 0$ with $\lambda = 1$, the solution of the pressure is piecewise constant and the solution of the velocity is simply the shear flow itself: $\mathbf{u} = (Gy, 0)$. While we do observe internal circulation for $Ca = 0.4695$ and $\gamma = 1.4605$ as presented in Figure 10.10(d), this is not true for $Ca = 1.3898$ and $\gamma = 0.4605$. Also at $t = 0.75$, for all the simulations in Figure 10.10 we have observed that the drop has reached its steady state.

10.5 An augmented approach for pressure boundary conditions

If a nonslip boundary condition is prescribed for the velocity, then it is well known that one cannot specify the boundary condition for the pressure along the boundary. Otherwise the problem is overdetermined. This brings some difficulties to numerical schemes using uniform Cartesian grids. For time-dependent Stokes or Navier–Stokes equations, often a staggered grid is used to avoid the boundary condition for the pressure; see, for example, [217]. For numerical methods that need an artificial pressure boundary condition, some approaches are discussed in [99, 128] for time-dependent problems. Often a Neumann boundary condition coupled with the velocity is used. However, for *stationary* Stokes flow,

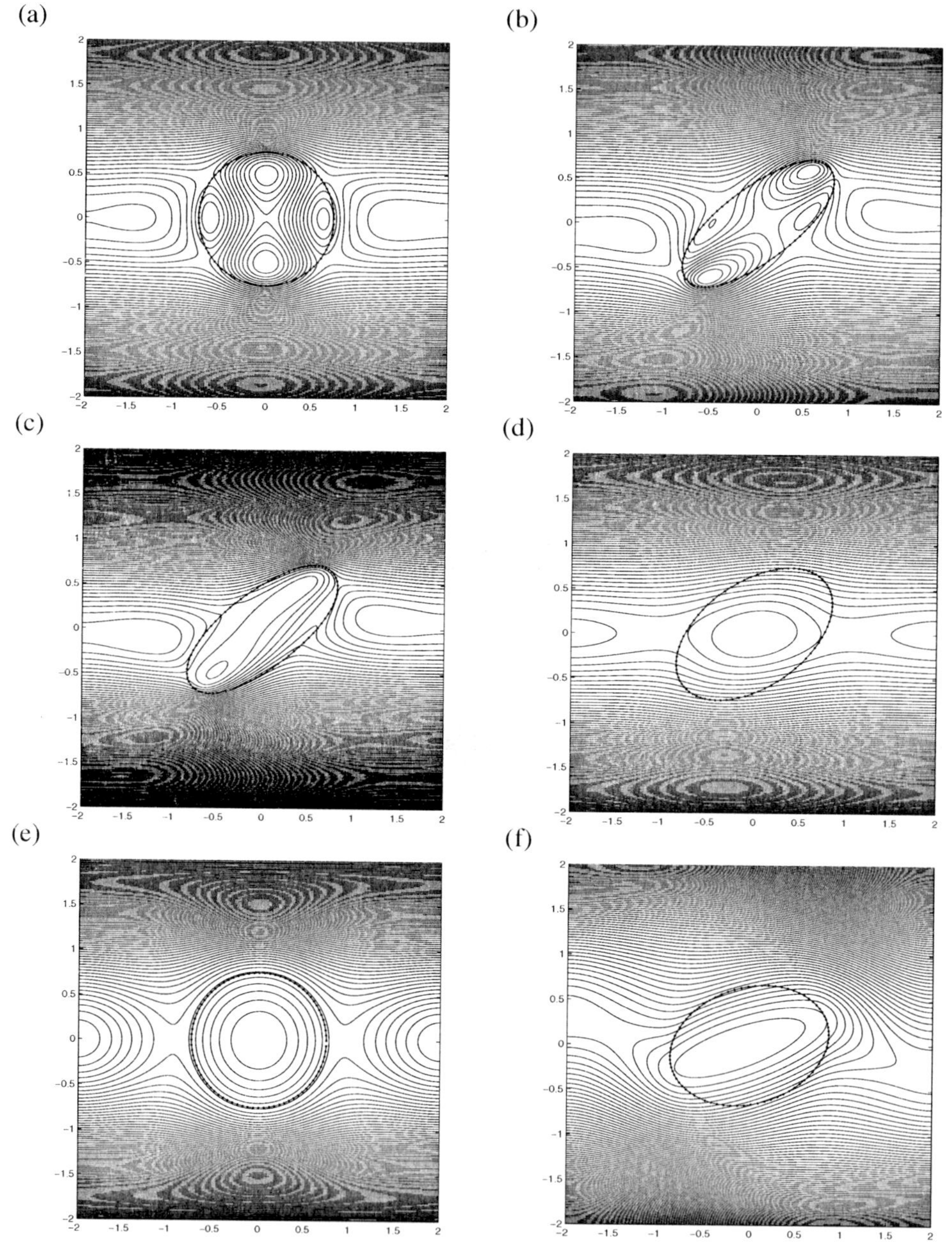

Figure 10.10. *Contour plots of the stream function of the shear flow with various viscosity ratios* λ *and* Ca. *In all cases* $Ca = 1.3898$, $\gamma = 0.4605$ *except for* (d) *in which* $Ca = 0.4695$, $\gamma = 1.4605$. (a) $\lambda = 0.04$ *and* $t = 0$. (b) $\lambda = 0.04$ *and* $t = 0.75$. (c) $\lambda = 0.08$ *and* $t = 0.75$. (d) $\lambda = 1$ *and* $t = 0.75$. (e) $\lambda = 6.4$ *and* $t = 0$. (f) $\lambda = 6.4$ *and* $t = 0.75$. *The drop is the solid-dotted line in the plots.*

one cannot use fast Poisson solvers if the pressure boundary condition is coupled with the velocity.

Similarly to the augmented approach that is used to decouple the jump condition for the pressure and the velocity, we can use an augmented approach to decouple the pressure boundary from the velocity as well. We use the boundary $y = d$, along which we assume that the velocity is zero, to illustrate this idea.

We set the Neumann boundary condition for the pressure along $y = d$ as the augmented variable

$$\left.\frac{\partial p}{\partial \mathbf{n}}\right|_{y=d} = \mathbf{q}_p. \tag{10.92}$$

The equation above, together with the Stokes equations and the following equation (from the momentum equation),

$$\mathbf{q}_p = (\mu\Delta v + g_2)|_{y=d}\,, \tag{10.93}$$

form a closed system. In discretization, we can set

$$\mathbf{Q}_p = \begin{bmatrix} q_{p1} & q_{p2} & \cdots & q_{p,M-1} \end{bmatrix}^T \tag{10.94}$$

to define q_p at (x_1, d), (x_2, d), $\ldots$, (x_{M-1}, d). Again GMRES can be used to solve the augmented variable q_p. Given q_p, we can solve the three Poisson equations for the pressure $p(q_p)$, the velocity $u(q_p)$, and $v(q_p)$ using a fast Poisson solver three times. The residual equation now is

$$R(q_p) = \mathbf{q}_p - \left(\mu\Delta v(q_p) + g_2\right)\Big|_{y=d}\,. \tag{10.95}$$

The rest of the discussion is similar to that in §10.4. Our numerical tests show that only 2–5 iterations are needed for the GMRES method.

10.5.1 Computing the Laplacian of the velocity along a boundary for a nonslip boundary condition

To compute the residual (10.95) with the assumption that $u = v = 0$ along the boundary $y = d$, we need to approximate $\Delta v = v_{xx} + v_{yy} = v_{yy}$. Note that $v_{xx} = 0$ along $y = d$ since $v = 0$ there. To compute v_{yy} along $y = d$, we make use of the incompressibility condition $u_x + v_y = 0$. From the Taylor expansion, we have

$$\begin{aligned} v(x, d - h_y) &= v(x, d) - v_y(x, d)h_y + \frac{1}{2}v_{yy}(x, d)h_y^2 + O(h_y^3) \\ &= v(x, d) + u_x(x, d)h_y + \frac{1}{2}v_{yy}(x, d)h_y^2 + O(h_y^3). \end{aligned} \tag{10.96}$$

Thus we get an approximation for v_{yy},

$$v_{yy}(x, d) \approx \frac{2\left(v(x, d - h_y) - v(x, d) - u_x(x, d)h_y\right)}{h_y^2}. \tag{10.97}$$

Note that $u_x(x,d)$ is computable since $u(x,t)$ is given along $y=d$, and thus its first-order derivative, $v(x, d-h_y)$, has been obtained after we have solved the three Poisson equations for $p(q_p)$, $u(q_p)$, $v(q_p)$ given q_p. The augmented method described above for dealing with pressure boundary conditions has been tested in [176] for problems with an interface that touches the two sides $x = a$ and $x = b$. Once again, second-order accuracy has been verified numerically.

10.6 The IIM for Navier–Stokes equations with singular sources

In this section, we discuss the IIM for incompressible Navier–Stokes equations,

$$\rho\left(\frac{\partial \mathbf{u}}{\partial t} + (\mathbf{u}\cdot\nabla)\mathbf{u}\right) + \nabla p = \mu\Delta\mathbf{u} + \mathbf{g} + \mathbf{F}, \quad \mathbf{x}\in\Omega, \tag{10.98a}$$

$$\nabla\cdot\mathbf{u} = 0, \tag{10.98b}$$

$$\mathbf{u}|_{\partial\Omega} = \mathbf{u}_b, \quad \text{BC}, \tag{10.98c}$$

$$\mathbf{u}(\mathbf{x},0) = \mathbf{u}_0, \quad \text{IC}, \tag{10.98d}$$

in a bounded domain $\Omega \subset \mathcal{R}^2$. The singular force $\mathbf{F}$ has support only on an arbitrary interface $\Gamma(t)$ and has the form

$$\mathbf{F}(\mathbf{x},t) = \int_{\Gamma(t)} \mathbf{f}(s,t)\,\delta(\mathbf{x}-\mathbf{X}(s,t))ds. \tag{10.99}$$

The external force $\mathbf{g}$ may have a finite jump across the interface, but is bounded and piecewise continuous in the entire domain. Throughout this section, we simply assume that the density $\rho \equiv 1$ and that the viscosity μ is a constant.

The IB method was originally developed for the incompressible Navier–Stokes model (10.98a)–(10.98d). It is well known that, generally, the IB method is first-order accurate for nonsmooth but continuous quantities. Different variations of the IB method have been proposed to improve accuracy of the method. For example, Lai and Peskin [142] proposed a new, formally second-order, IB method with reduced numerical viscosity and applied the new method to simulate the flow around a circular cylinder. By formal second-order accuracy, the authors meant that the scheme is second-order accurate if the delta function is replaced by a fixed smooth function, independent of the mesh. While the numerical result is in good agreement with the experimental data, the fully second-order accuracy in the maximum norm has not been achieved. Cortez and Minion [57] proposed another high-order scheme using the smoothed blob projection to compute the force. Their method shows better accuracy in L_1 and L_2 norms and preserves the volume better than the original IB method. Nevertheless, both approaches described in [57],[142] use a discrete delta function, which usually smears the solution across the interface. The ghost fluid method (GFM), a first-order sharp interface method, has also been applied to multiphase incompressible flow [136] and two-phase incompressible flame simulations [203]. Another sharp interface method is the hybrid finite volume–level set method [102, 103] in which the jumps in the coefficients were taken into account in constructing the algorithm.

The IIM has been applied to the Navier–Stokes equations on irregular regions using the stream-vorticity formulation in [36, 177]. In [150], a method is proposed for the full Navier–Stokes equations with a moving interface by combining the IIM with LeVeque's Clawpack [153].

In this section, we explain the IIM for solving the full Navier–Stokes equations (10.98a)–(10.99) developed in [168] and compare the method with different versions of the IB method. The method is a sharp interface method, which means that the jump conditions in the solution will be enforced. Besides the differences in the methodology, the IIM discussed here is distinguished from [136, 183, 203] by its accuracy and is distinguished from [102, 103] by the way it deals with different problems. There exist some other models and methods for interface problems; notably, the phase field model and the finite volume method, for example, [222]. Each model has advantages and disadvantages and may have specific applications. It is easy to compute surface tension and curvature with sharp interface models. It is also advantageous to use the level set method with sharp interface models.

Since the inertia term, or material derivative, is continuous, the jump conditions for the Navier–Stokes equations are the same as those for the Stokes equations given in (10.21)–(10.24) and (10.26)–(10.27) or, if the viscosity is continuous for those given in (10.28a)–(10.28d).

10.6.1 Additional interface relations

Additional interface relations can be derived by differentiating the known jump conditions (10.28a)–(10.28d) along the interface using the momentum equation (10.98a). These interface relations are summarized in Theorem 10.5 and are used in the numerical scheme. For interface problems, most physical quantities along/across the interface are associated with the normal and tangential directions.

Theorem 10.5. *Let* $\mathbf{u}$ *and* p *be the solution to* (10.98a)–(10.98d) *with a continuous viscosity, that is,* $[\mu] = 0$. *The following interface relations are true at a point* (X, Y) *on the interface:*

$$
\begin{aligned}
&[p] = \hat{f}_1, \qquad [p_{\mathbf{n}}] = \frac{\partial \hat{f}_2}{\partial \boldsymbol{\tau}}, \\
&[\mathbf{u}] = \mathbf{0}, \qquad [\mu \mathbf{u}_{\mathbf{n}}] = -\hat{f}_2\, \boldsymbol{\tau}, \qquad [\mathbf{u}_{\tau}] = \mathbf{0}, \\
&[\mu \mathbf{u}_{\tau\tau}] = \kappa \hat{f}_2\, \boldsymbol{\tau}, \qquad [\mu \mathbf{u}_{\mathbf{n}\tau}] = -\frac{\partial \hat{f}_2}{\partial \boldsymbol{\tau}}\, \boldsymbol{\tau} - \kappa \hat{f}_2\, \mathbf{n}, \\
&[\mu \mathbf{u}_{\mathbf{nn}}] = -[\mu \mathbf{u}_{\tau\tau}] + [p_{\mathbf{n}}]\, \mathbf{n} + [p_{\tau}]\, \boldsymbol{\tau} + [\mathbf{u}_{\mathbf{n}}]\, (\mathbf{u} \cdot \mathbf{n}) - [\mathbf{g}].
\end{aligned}
\tag{10.100}
$$

Sketch of the proof: The first four equalities are directly copied from the jump conditions in (10.28a)–(10.28d) and the continuity condition of the velocity. Using the local coordinates (1.34) centered at (X, Y), the interface Γ can be written as $\xi = \chi(\eta)$ in a neighborhood of (X, Y), or $(\xi, \eta) = (0, 0)$ in the local coordinates, with $\chi(0) = 0$, $\chi'(0) = 0$, and $\chi''(0) = \kappa$. Differentiating $[\mathbf{u}] = \mathbf{0}$ along the interface and using $\chi'(0) = 0$, we get the fifth equality. Differentiating $[\mathbf{u}] = \mathbf{0}$ along the interface twice and using $\chi'(0) = 0$ and $\chi''(0) = \kappa$, we get the sixth equality. Differentiating the fourth equality along the interface, we get the seventh equality. Finally, expressing the momentum equation (10.98a) in the local coordinate (1.34), we get the last equality.

In order to be numerically useful, those interface relations in (10.100) in the local coordinate (1.34) are transformed into the jump relations in the Cartesian coordinates as expressed in (9.47).

Using the identities (9.47) and the interface relations in (10.100), we can easily write $[\mathbf{u}]$, $[\mathbf{u}_x]$, $[\mathbf{u}_y]$, $[\mathbf{u}_{xx}]$, $[\mathbf{u}_{yy}]$, $[p]$, $[p_x]$, and $[p_y]$ at any point on the interface in terms of the force strength, its surface derivatives, and the geometric information of the interface.

10.6.2 The modified finite difference method for Navier–Stokes equations with interfaces

The finite difference method is based on the projection method for solving incompressible Navier–Stokes equations with some modifications at grid points that are near or on the interface. There are several versions of the projection methods for solving the incompressible Navier–Stokes equations; see, for example, [16, 134, 139] and many others. We use the projection method described in [34] that is based on the pressure increment formulation of [16, 134] with an additional accurate pressure correction (in [34], the method is denoted by Pm II). The key modification to Pm II is adding some correction terms at grid points near or on the interface.

The IIM method from time t^n to t^{n+1} can be written as

$$\begin{aligned} &\frac{\mathbf{u}^* - \mathbf{u}^n}{\Delta t} + (\mathbf{u} \cdot \nabla_h \mathbf{u})^{n+\frac{1}{2}} \\ &\qquad = \nabla_h p^{n-\frac{1}{2}} + \frac{\mu}{2}\left(\Delta_h \mathbf{u}^* + \Delta_h \mathbf{u}^n\right) + \mathbf{g}^{n+\frac{1}{2}} + \mathbf{C}_1^n, \\ &\mathbf{u}^*|_{\partial\Omega} = \mathbf{u}_b^{n+1}, \end{aligned} \tag{10.101}$$

where $(\mathbf{u} \cdot \nabla_h \mathbf{u})^{n+\frac{1}{2}}$ is approximated using

$$(\mathbf{u} \cdot \nabla_h \mathbf{u})^{n+\frac{1}{2}} = \frac{3}{2}\left(\mathbf{u}^n \cdot \nabla_h\right)\mathbf{u}^n - \frac{1}{2}\left(\mathbf{u}^{n-1} \cdot \nabla_h\right)\mathbf{u}^{n-1} + \mathbf{C}_2^n. \tag{10.102}$$

The projection step is the following:

$$\Delta_h \phi^{n+1} = \frac{\nabla_h \cdot \mathbf{u}^*}{\Delta t} + C_3^n, \tag{10.103}$$

$$\left.\frac{\partial \phi^{n+1}}{\partial \mathbf{n}}\right|_{\partial\Omega} = 0, \tag{10.104}$$

$$\mathbf{u}^{n+1} = \mathbf{u}^* - \Delta t\, \nabla_h \phi^{n+1} + \mathbf{C}_4^n, \tag{10.105}$$

$$\nabla_h p^{n+1/2} = \nabla_h p^{n-1/2} + \nabla_h \phi^{n+1} + \mathbf{C}_5^n. \tag{10.106}$$

In the expressions above, ∇_h and Δ_h are the standard central finite difference operators defined in (9.64) and (5.18), respectively. To compute the velocity, only $\nabla_h p^{n-1/2}$ is needed, but not the pressure p itself. The pressure can also be obtained using

$$p^{n+1/2} = p^{n-1/2} + \phi^{n+1} + \mathbf{C}_6^n. \tag{10.107}$$

It is clear that the modifications we made to the projection method are those correction terms C_k^n, $k = 1, 2, \ldots, 5$, at grid points near or on the interface.

10.6.3 Determining the correction terms

At a regular grid point (x_i, y_j), we use the standard finite difference scheme to discretize (10.98a)–(10.98d) without any correction, that is, $C_k^n \equiv 0$, $k = 1, 2, \ldots, 5$. At an irregular grid point (x_i, y_j), we need to determine the correction terms so that the finite difference scheme can be as accurate as possible. The correction terms can be determined in the x- and y-direction, respectively. The following lemma is the basis for determining the correction terms.

Lemma 10.6. *Let $u(x)$ be a* piecewise *twice differentiable function. Assume that $u(x)$ and its derivatives have finite jumps $[u]$, $[u_x]$, and $[u_{xx}]$ at $x^* = x + \alpha h$, $-1 \le \alpha \le 1$; then the following relations hold:*

$$\frac{u(x+h) - u(x-h)}{2h} = \begin{cases} u'(x) + \dfrac{C(x,\alpha)}{2h} + O(h^2) & \textit{if}\ 0 \le \alpha \le 1, \\ u'(x) - \dfrac{C(x,\alpha)}{2h} + O(h^2) & \textit{if}\ -1 \le \alpha < 0, \end{cases} \tag{10.108}$$

$$\frac{u(x+h) - 2u(x) + u(x-h)}{h^2} = u''(x) + \frac{C(x,\alpha)}{h^2} + O(h), \tag{10.109}$$

where

$$C(x,\alpha) = [u] + [u_x]\,(1 - |\alpha|)\,h + [u_{xx}]\,\frac{(1-|\alpha|)^2\,h^2}{2}. \tag{10.110}$$

Note that (10.109) is a special case in the expressions (2.15) and (2.17), and (10.108) is a special case in the expressions (9.28) and (9.30). Therefore we omit the proof. Nevertheless, the lemma gives a "uniform" correction formula for both the first- and second-order derivatives approximation.

Remark 10.3.

- Lemma 10.6 *tells us that the correction term is simply the product of the finite difference coefficient, corresponding to the grid point from different sides of the interface in reference to the master grid point, and $C(x,\alpha)$ which corresponds to the singular source term. For example, if $-1 < \alpha < 0$ in* Lemma 10.6, *which means that $x - h$ and x are on different sides of the interface, then the correction term for u_x is the product of the coefficient $-1/(2h)$ and $C(x,\alpha)$.*

- *If there are two interfaces between $(x-h, x+h)$, say, $x_1^* = x + \alpha_1 h$ and $x_2^* = x + \alpha_2 h$, with $\alpha_1 < 0$ and $\alpha_2 \ge 0$, then we just need to add another correction term, for example,*

$$\frac{u(x+h) - 2u(x) + u(x-h)}{h^2} = u''(x) + \frac{C(x,\alpha_1)}{h^2} + \frac{C(x,\alpha_2)}{h^2} + O(h),$$

$$\frac{u(x+h) - u(x-h)}{2h} = u'(x) + \frac{C(x,\alpha_1)}{2h} - \frac{C(x,\alpha_2)}{2h} + O(h^2).$$

10.6.4 Correction terms to the projection method

In the numerical scheme (10.101)–(10.105), there are several correction terms to be determined. For the sake of simplicity, we only explain how to evaluate the correction terms $\mathbf{C}_1^n$. The other correction terms can be treated in the same way. Assume that x^*, where $x_i \le x^* \le x_{i+1}$, is the intersection of the interface and the x-axis. The contribution to $\mathbf{C}_1^n$ in the x-direction from this interface point includes the following:

- the correction term for $u^n(D_{x,h})u^n$ from the nonlinear term $(\mathbf{u}^n \cdot \nabla_h)\,\mathbf{u}^n$,
$$-\frac{3}{4h}\left([u_x^n]\,(x_{i+1}-x^*)+[u_{xx}^n]\,\frac{(x_{i+1}-x^*)^2}{2}\right)\,u^n(x^*,y_j);$$
- the correction term for $u^{n-1}(D_{x,h}u^{n-1})$ from the nonlinear term $(\mathbf{u}^{n-1}\cdot\nabla_h)\mathbf{u}^{n-1}$,
$$\frac{1}{4h}\left([u_x^{n-1}]\,(x_{i+1}-x^*)+[u_{xx}^{n-1}]\,\frac{(x_{i+1}-x^*)^2}{2}\right)\,u^{n-1}(x^*,y_j);$$
- the correction term for $D_{x,h}\,p^{n-\frac{1}{2}}$ from $\nabla_h\,p^{n-\frac{1}{2}}$,
$$-\frac{1}{2h}\left([p^{n-\frac{1}{2}}]+[p_x^{n-\frac{1}{2}}]\,(x_{i+1}-x^*)\right);$$
- the correction term for $\mu(u_{i-1,j}^*-2u_{i,j}^*+u_{i+1,j}^*)/(2h^2)$ from $\mu\Delta_h(u^*/2)$,
$$-\frac{\mu}{2h^2}\left([u_x^{n+1}]\,(x_{i+1}-x^*)+[u_{xx}^{n+1}]\,\frac{(x_{i+1}-x^*)^2}{2}\right);$$
- the correction term for $\mu(u_{i-1,j}^n-2u_{i,j}^n+u_{i+1,j}^n)/(2h^2)$ from $\mu\Delta_h\,(u^n/2)$,
$$-\frac{\mu}{2h^2}\left([u_x^n]\,(x_{i+1}-x^*)+[u_{xx}^n]\,\frac{(x_{i+1}-x^*)^2}{2}\right).$$

The *bilinear interpolation scheme* (see (11.7) in §11.1.3) is used to approximate the values of $u^n(x^*,y_j)$ and $v^n(x^*,y_j)$ at (x^*,y_j) from the values of $\{U_{ij}^n\}$ and $\{V_{ij}^n\}$ at the neighboring four grid points. Note that the jump conditions at t^{n+1} have been used to approximate the jump conditions at t^* corresponding to the same treatment for the velocity boundary condition in the projection method.

Similarly, if the interface cuts between x_{i-1} and x_i, we need to add the corresponding correction terms as well. In this way, we can determine the correction terms in the x-direction for $\mathbf{C}_1^n$. Similarly, we can determine the correction terms in the y-direction.

It is worth pointing out that the modifications made to the projection method affect only the right-hand sides of the original projection method. Therefore the method does not change the stability nature of the original projection method, which is stable if the CFL condition is satisfied.

Once those correction terms are computed, two generalized Helmholtz equations need to be solved to get $\mathbf{u}^*$ in (10.101), plus one Poisson equation for ϕ^{n+1} in (10.103). The discrete systems for the generalized Helmholtz equation can be solved using a fast solver.

10.6.5 Further corrections near the boundary and the interface

Using the projection method with the correction terms added at irregular grid points, it has been observed that the velocity is second-order accurate but the pressure is only first-order accurate. The largest error in the pressure appears near the boundary or the interface. This is due to the inherent numerical boundary layer in the projection method as analyzed in [34, 69, 239, 254]. In [34] (see also [139]) a correction scheme is proposed,

$$p^{n+\frac{1}{2}} = p^{n-\frac{1}{2}} + \phi^{n+1} - \frac{\mu}{2}\left(\nabla_h \cdot \mathbf{u}^* + C_7^n\right), \tag{10.111}$$

where ϕ^{n+1} is the solution of the Poisson equation of (10.103)–(10.104) from t^n to t^{n+1} and where C_7^n is a correction term only needed at irregular grid points. This correction does dramatically improve the accuracy for the pressure; see §10.6.6.

In implementation, the centered finite difference scheme is used to compute the correction term $\nabla_h \cdot \mathbf{u}^*$ at interior grid points. An extrapolation scheme is used to extend the quantity to the boundary of the rectangle. A second-order extrapolation scheme at the boundary $x = a$, for example, is

$$\nabla_h \cdot \mathbf{u}^*(0, j) = 3\nabla_h \cdot \mathbf{u}^*(1, j) - 3\nabla_h \cdot \mathbf{u}^*(2, j) + \nabla_h \cdot \mathbf{u}^*(3, j). \tag{10.112}$$

Near the interface, even though the velocity is second-order accurate, the term $\nabla_h \cdot \mathbf{u}^*/\Delta t$ may have only zeroth-order accuracy since the error may not be smooth near the interface and $\Delta t \sim h$. This will reduce the accuracy of the pressure to first order. One of the solutions to this problem is approximating $\nabla_h \cdot \mathbf{u}^*/\Delta t$ at an irregular grid point using the $\nabla_h \cdot \mathbf{u}^*/\Delta t$ computed at the nearest regular grid point.

The numerical method described in this section is called the projection IIM.

10.6.6 Comparisons and validation of the IIM for Navier–Stokes equations with interfaces

In this section, we show some numerical experiments for solving the velocity and the pressure in (10.98a)–(10.98d) at some fixed time t. The velocity and the pressure, up to a constant, are uniquely determined by the initial conditions $\mathbf{u}(\mathbf{x}, 0)$, $p(\mathbf{x}, 0)$, the initial interface $\Gamma(:, 0)$, the boundary condition $\mathbf{u}(\mathbf{x}, t)|_{\partial\Omega}$, and the force strength $(\hat{f}_1, \hat{f}_2)$ in the normal and tangential directions along the interface. More examples can be found in [168].

Example 10.4. *An interface problem with a constant jump in the pressure.* Let the interface be a fixed ellipse,

$$\frac{(x - 0.5)^2}{a^2} + \frac{(y - 0.5)^2}{b^2} = 1$$

in the domain $[0, 1] \times [0, 1]$. The normal and tangential force strengths are $\hat{f}_1 = 10$ and

$\hat{f}_2 = 0$, respectively. The external force is $\mathbf{g} = \mathbf{0}$. The exact solution is given by

$$u = 0, \quad v = 0,$$
$$p = \begin{cases} C & \text{if } \dfrac{(x-0.5)^2}{a^2} + \dfrac{(y-0.5)^2}{b^2} - 1 > 0, \\ -10 + C & \text{if } \dfrac{(x-0.5)^2}{a^2} + \dfrac{(y-0.5)^2}{b^2} - 1 \le 0, \end{cases} \tag{10.113}$$

where C is an arbitrary constant. The domain is the unit square $\Omega = [0, 1] \times [0, 1]$. We compare the projection IIM with Peskin's IB method using the discrete cosine delta function. Depending on how the interface is represented, there are two different versions of the IB methods.

The Lagrangian particle approach

In this traditional approach, the interface is represented by a set of Lagrangian points $\mathbf{X}_k$, $k = 1, 2, \dots, N_b$, where N_b is the total number of particle points and is usually chosen in such a way that

$$\max \Delta s_k = \max \|\mathbf{X}_{k+1} - \mathbf{X}_k\| \sim h.$$

The forces defined on those particle points are then distributed to the nearby grid points (x, y) by the formula

$$\mathbf{F}_{ij} = \sum_{k=1}^{N_b} \mathbf{f}(s_k)\, \delta_w(x_i - X_k)\, \delta_w(y_j - Y_k)\, \Delta s_k,$$

where δ_w is the one-dimensional discrete delta function given in (1.20). Although there are other discrete delta functions in the literature, the computational results are not much different for two- and three-dimensional problems. The most common choice of w is h, the spatial mesh size. As we can see, when w gets larger, the cost of spreading the singular force to the grid points increases significantly. The numerical tests for this and other examples show that the accuracy of the computed velocity remains pretty much the same for different choices of w. However, the pressure, if it is discontinuous, smears more widely as w increases. Therefore the best choice of w is h. The sixth column in Table 10.4 is a grid refinement analysis for the velocity which is clearly first-order accurate.

The level set representation

If the interface is represented by the zero level set of a two-dimensional function $\varphi(x, y)$, it is not straightforward how to use the IB method directly. However, if the tangential component of the singular force is zero, then we can write

$$\int_{\Gamma(t)} \mathbf{f}(s)\, \delta(\mathbf{x} - \mathbf{X}(s))ds = (\mathbf{f} \cdot \mathbf{n})\, \delta(\varphi)\, \nabla\varphi \tag{10.114}$$

in the distribution sense; see [40], for example. It is easy to apply the discrete delta function to the above form. Unfortunately, the choice of the width of the discrete delta function is

Table 10.4. *A grid refinement analysis and comparison of the three projection IB methods for Example* 10.4 *with* $\mu = 0.1$, $a = 0.35$, $b = 0.25$, *and* $\Delta t = h$. *The pressure, which is not listed in the table, has zeroth-order accuracy since it is smeared.*

	Level set IB width $w = h$		Level set IB width $w = \sqrt{h}$		Lagrangian IB width $w = h$	
N	$\|E(\mathbf{u})\|_\infty$	Order	$\|E(\mathbf{u})\|_\infty$	Order	$\|E(\mathbf{u})\|_\infty$	Order
32	1.2434×10^{-1}		4.9254×10^{-3}		5.1204×10^{-2}	
64	6.3619×10^{-2}	0.96673	2.9594×10^{-3}	0.7349	2.5839×10^{-2}	0.9867
128	4.0730×10^{-2}	0.64339	1.1062×10^{-3}	1.4197	1.2968×10^{-2}	0.9929
256	4.3059×10^{-2}	0.09459	4.1558×10^{-4}	1.4124	6.5055×10^{-3}	0.9952

crucial for the IB method applied to (10.114). If $w = h$, the IB method barely converges; see the second column of Table 10.4. T. Hou [106] may be the first one to note this problem and suggest that the best choice of the width is $w = \sqrt{h}$. This has also been confirmed in the numerical tests; see the fourth column of Table 10.4. However, $w = \sqrt{h}$ is much larger than h, if h is small. The IB method smears the pressure and shows zeroth-order convergence for the pressure. Since the width has to be taken as $w = \sqrt{h}$, we cannot guarantee the right pressure even at those grid points that are away from the interface.

A remedy to the problem is to use the IB method in which the control points are selected orthogonal projections of irregular grid points of the level set function. Then the Peskin's original IB method can be applied with those control points. Using this approach, we still can use a thin layer ($w = h$) of the interface in the IB method, so that the solution will be smeared only in the thin layer.

In Table 10.5, we show a grid refinement analysis of the projection IIM discussed in this section. Without the modification of the pressure equation (10.111), the velocity in the second column is second-order accurate, while the pressure in the fourth column is first-order accurate with the largest errors occurring near the boundary or the interface. With the correction term in (10.111) added, we see second-order accuracy for both the velocity

Table 10.5. *A grid refinement analysis of the projection IIM for Example* 10.4. *The parameters are* $\mu = 0.1$, $a = 0.35$, $b = 0.25$, *and* $\Delta t = h$.

	Projection IIM without (10.111)				With (10.111)	
N	$\|E(\mathbf{u})\|_\infty$	Order	$\|E(p)\|_\infty$	Order	$\|E(p)\|_\infty$	Order
32	1.6412×10^{-4}		1.5785×10^{-2}		9.9058×10^{-3}	
64	3.9857×10^{-5}	2.0418	7.9281×10^{-3}	0.9935	1.1533×10^{-3}	3.1026
128	9.8987×10^{-6}	2.0095	3.9676×10^{-3}	0.9987	1.5946×10^{-4}	2.8544
256	1.2367×10^{-6}	2.0008	1.9842×10^{-3}	0.9997	2.4832×10^{-5}	2.6830

and the pressure. Since the pressure is piecewise constant in this case, the accuracy for the pressure is much better than second order. However, this is not true for general problems.

In a further numerical experiment, we add the following particular solution taken from [264] to (10.113):

$$u(x, y, t) = -\sin^2(\pi x)\sin(2\pi y)\cos t,$$
$$v(x, y, t) = \sin^2(\pi y)\sin(2\pi x)\cos t. \tag{10.115}$$

The pressure is

$$p(x, y, t) = -\frac{\partial \psi}{\partial t} + \nu \Delta \psi + \begin{cases} C & \text{if } \dfrac{(x-0.5)^2}{a^2} + \dfrac{(y-0.5)^2}{b^2} - 1 > 0, \\ -10 + C & \text{if } \dfrac{(x-0.5)^2}{a^2} + \dfrac{(y-0.5)^2}{b^2} - 1 \le 0, \end{cases}$$

where

$$\psi = \left[(x - x^2)\sin(\pi x)(y - y^2)\sin(\pi y) - \frac{16}{\pi^6} \right] \cos t.$$

The force term $\mathbf{g}$ is determined from the exact solution. In this example, we have $\frac{\partial p}{\partial \mathbf{n}} \neq 0$ on the boundary. Thus, the numerical boundary layer might deteriorate the accuracy of the pressure. Figure 10.11(a) shows the pressure computed using the projection IIM. The computed pressure has the right jump up to second-order accuracy. Figures 10.11(b), (c) show the pressure computed using the IB method with the level set representation of the interface ($w = \sqrt{h}$) and the original IB method with particle representation of the interface ($w = h$), respectively. In Figure 10.11(b), we see that the pressure is smeared by the width $w = \sqrt{h}$ while in Figure 10.11(c), we see larger errors but a thinner layer ($w = h$) near the interface.

For this test problem, the projection IIM is faster compared with the IB method because the triple loops of the IB method spread the forces to the nearby grid points.

Example 10.5. *An example with nonconstant jump in the pressure.* In this example, the interface Γ and the domain Ω are the same as in Example 10.4. The source term $\mathbf{g}$ is the gradient of a function $q(x, y)$, which is the solution of the following:

$$\Delta q = 0, \quad (x, y) \in \Omega, \qquad \left.\frac{\partial q}{\partial \mathbf{n}}\right|_{\partial\Omega} = 0, \tag{10.116}$$

$$[q]_\Gamma = x + y, \qquad \left[\frac{\partial q}{\partial \mathbf{n}}\right]_\Gamma = 0, \quad (x, y) \in \Gamma. \tag{10.117}$$

The function $q(x, y)$ and $\mathbf{g} = \nabla q$ are computed using the IIM [163] with a 512×512 fine grid. Note that from $[\frac{\partial q}{\partial \mathbf{n}}] = 0$ and $[q] = x + y$, it is easy to derive that

$$[g_1] = [q_x] = \sin^2\theta - \sin\theta\cos\theta, \qquad [g_2] = [q_y] = \cos^2\theta - \sin\theta\cos\theta,$$

where $\mathbf{n} = (\cos\theta, \sin\theta)$ is the unit normal direction of Γ pointing outward.

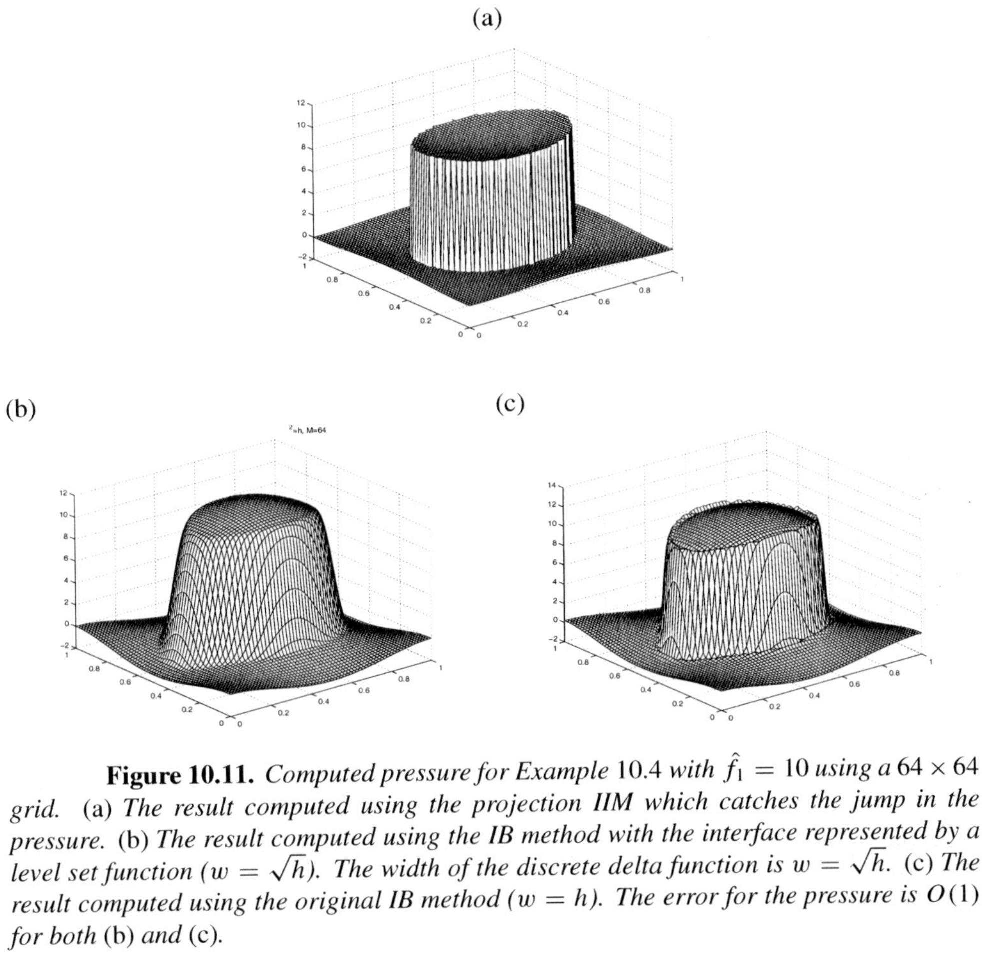

Figure 10.11. *Computed pressure for Example* 10.4 *with* $\hat{f}_1 = 10$ *using a* 64×64 *grid.* (a) *The result computed using the projection IIM which catches the jump in the pressure.* (b) *The result computed using the IB method with the interface represented by a level set function* ($w = \sqrt{h}$). *The width of the discrete delta function is* $w = \sqrt{h}$. (c) *The result computed using the original IB method* ($w = h$). *The error for the pressure is* $O(1)$ *for both* (b) *and* (c).

The following parameters and functions are used in conjunction with the full Navier–Stokes equations (10.98a)–(10.99):

$$\mathbf{u}|_{\partial\Omega} = \mathbf{0}, \qquad \mu = 1, \qquad \mathbf{g} = \nabla q(x, y),$$

$$\hat{f}_2 = 0, \qquad \hat{f}_1 = x + y, \quad (x, y) \in \Gamma.$$

For this example, the exact solution is not available. However, the steady state solution is

$$\lim_{t \to \infty} \mathbf{u}(\mathbf{x}, t) = \mathbf{0}, \qquad \lim_{t \to \infty} p(\mathbf{x}, t) = q(\mathbf{x}).$$

In Table 10.6, we show a grid refinement analysis for the *steady state solution* obtained using the projection IIM at $t = 10$. The initial velocity is taken from (10.115), while the pressure is

$$p_0(x, y) = \cos \pi x \, \cos \pi y + q(x, y).$$

Table 10.6. *A grid refinement analysis of the projection IIM applied to Example* 10.5 *with* $\mu = 1$ *and* $\Delta t = h$ *at* $t = 10$. *The initial velocity is taken as* (10.115). *The initial pressure is taken as* $\cos \pi x \, \cos \pi y + q(x, y)$.

N	$\|E_{IIM}(\mathbf{u})\|_\infty$	Order	$\|E_{IIM}(p)\|_\infty$	Order
32	2.4930×10^{-5}		5.1614×10^{-4}	
64	3.5197×10^{-6}	2.8244	8.4365×10^{-5}	2.6131
128	4.6660×10^{-7}	2.9152	1.6308×10^{-5}	2.3710
256	3.5073×10^{-8}	2.9332	5.7023×10^{-6}	1.5160

In this way, the initial pressure is guaranteed to have the right jump, $[p] = \hat{f}_1$. Table 10.6 shows that the projection IIM gives a second-order accurate solution for both the velocity and the pressure. The results are qualitatively the same if different initial conditions are used as long as the jump condition in p is initially satisfied. Note that $\cos \pi x \, \cos \pi y$ satisfies the homogeneous Neumann boundary condition $\nabla p \cdot \mathbf{n} = 0$ on the boundary. If we take the initial pressure as a smooth function, then the computed pressure is only first-order accurate since $[p] \neq \hat{f}_1$ initially while the velocity is still second-order accurate.

Example 10.6. *The circular flow with a line force.* The velocity is nonsmooth in this example. The exact solution is the following:

$$u(x, y, t) = \begin{cases} h(t)\left(\dfrac{y}{r} - 2y\right) & \text{if } r > \dfrac{1}{2}, \\ 0 & \text{if } r \le \dfrac{1}{2}, \end{cases} \tag{10.118}$$

$$v(x, y, t) = \begin{cases} h(t)\left(-\dfrac{x}{r} + 2x\right) & \text{if } r > \dfrac{1}{2}, \\ 0 & \text{if } r \le \dfrac{1}{2}, \end{cases} \tag{10.119}$$

$$p(x, y, t) = 0, \tag{10.120}$$

where $r = \sqrt{x^2 + y^2}$. The interface is the circle $r = 1/2$ and the solution domain is $[-1, 1] \times [-1, 1]$. It is easy to verify that the velocity satisfies the incompressibility constraint, and it is continuous but has a finite jump in the normal direction,

$$[u_{\mathbf{n}}] = [u_r] = -2h(t)\sin\theta, \qquad [v_{\mathbf{n}}] = [v_r] = 2h(t)\cos\theta,$$

across the interface. The normal and tangential force strengths are

$$\hat{f}_1 = 0, \qquad \hat{f}_2 = -2h(t).$$

Note that, $\mathbf{u}_t \cdot \mathbf{n}$ is not zero on the boundary if $\frac{dh}{dt} \neq 0$, which may affect the performance of the projection method. Outside of the interface $r = 0.5$, the nonzero and bounded source

Table 10.7. *A grid refinement analysis of the projection IIM applied to Example 10.6 with $\mu = 0.02$ and $\Delta t = h$ at $t = 10$. Second-order convergence is observed.*

N	$\|E_{IIM}(\mathbf{u})\|_\infty$	Order	$\|E_{IIM}(p)\|_\infty$	Order
32	2.4215×10^{-3}		1.1513×10^{-2}	
64	5.3547×10^{-4}	2.1771	3.2255×10^{-3}	1.8357
128	1.4970×10^{-4}	1.8388	9.1307×10^{-4}	1.8357
256	3.6173×10^{-5}	2.0491	1.9727×10^{-4}	2.2106

term $\mathbf{g}$ is derived directly from the exact solution using the Maple software package. Note that there is also a finite jump in $\mathbf{g}$.

With $h(t) = 1$, the method converges to the steady state solution. The grid refinement analysis reveals results similar to Example 10.5. Table 10.7 shows a grid refinement analysis for $h(t) = 1 - e^{-t}$ at $t = 10$. One can easily see that the velocity is second-order accurate and the pressure is nearly second-order accurate in Table 10.7.

We also tested a slightly different example by setting

$$\mathbf{g} = \mathbf{0}, \qquad \hat{f}_1 = 0, \qquad \hat{f}_2 = 10, \qquad \mathbf{u}|_{\partial\Omega} = \mathbf{0}.$$

The domain and the interface are the same. The initial velocity and the pressure are taken to be zero on the rectangular domain. Since the force is only along the tangential direction, the pressure is continuous, but the normal derivative of the velocity has a nonconstant jump across the interface. While the exact solution is difficult to find, the motion of the steady state is nothing but a simple rotation along the interface. In Figure 10.12(a), we plot the x-component of the velocity $-u(x, y)$ at $t = 10$ computed using a 64×64 grid. We can clearly see the nonconstant jump in the normal derivative. In Figure 10.12(b), we plot the velocity field at $t = 10$. The flow approaches a steady rotation along the circular interface.

Example 10.7. *An example of moving interface with the full Navier–Stokes equations.* Consider a similar moving interface problem as in the example of the Stokes flow on page 233 except that now the governing equations are full Navier–Stokes equations with a uniform density and viscosity. The initial interface is given by

$$r(\theta) = r_0 + \epsilon \sin(k\theta), \quad 0 \le \theta \le 2\pi. \tag{10.121}$$

The initial velocity and the pressure are all set to zero. The only driving force of the fluid motion is the surface tension which is proportional to the curvature κ, that is,

$$\hat{f}_1(\Gamma, t) = \epsilon\kappa, \quad \hat{f}_2(\Gamma, t) = 0. \tag{10.122}$$

The velocity is smooth but the pressure has a nonconstant jump,

$$[p]_{\Gamma(t)} = \epsilon\kappa, \quad [p_{\mathbf{n}}]_{\Gamma(t)} = 0.$$

In the test, the parameters are taken as $r_0 = 0.5$, $k = 5$, $\epsilon = 0.05$, and $\mu = 0.1$.

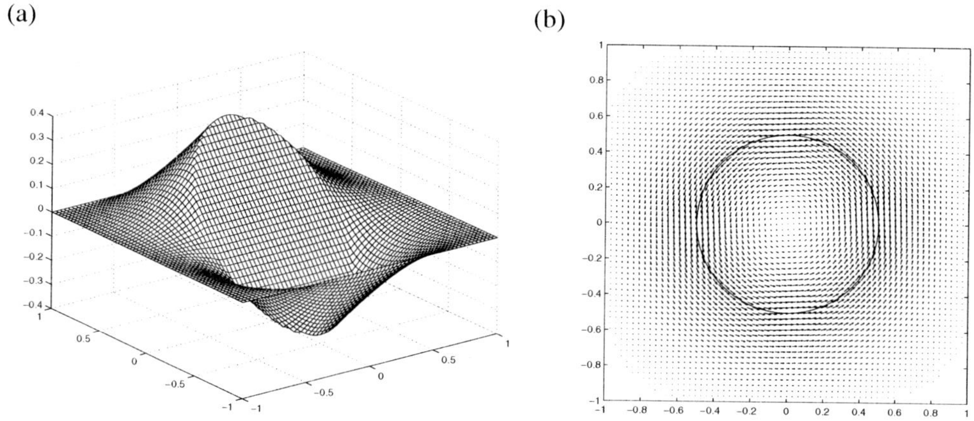

Figure 10.12. *The computed steady state velocity for the modified Example* 10.6 *with a circular interface $r = 0.5$ with a 64×64 grid. Other parameters are $\mathbf{g} = \mathbf{0}$, $\hat{f}_1 = 0$, $\hat{f}_2 = 10$, and $\mu = 0.02$. The time step size is $\Delta t = h$.* (a) *A plot of $-u(x, y)$ at $t = 10$ which is not smooth.* (b) *A plot of the velocity field at $t = 10$; the flow approaches a steady rotation along the interface.*

The level set method is used to evolve the interface. Readers are referred to Chapter 11 and [107, 175, 180] for details on how to combine the IIM with the level set method. At each time step, the projection IIM is applied to compute the velocity. Then the velocity is used to evolve the level set function whose zero level set represents the new location of the interface. Since we use the level set method to update the interface, the time step is chosen as

$$\Delta t = \min\left\{ \frac{h}{2\|\mathbf{u}\|_\infty},\ h \right\} \tag{10.123}$$

to maintain the stability for the level set method and the accuracy for the projection IIM.

Figure 10.13(a) is the plot of the computed interface at $t = 0$, $t = 7.0172$, $t = 21.0867$, and $t = 35.1563$ with a 160×160 grid. The relative error in the area change is 0.25% at $t = 7.0172$, 0.45% at $t = 21.0867$, and 0.47% at $t = 35.1563$. Figure 10.13(b) is the plot of the interface at $t = 0$ and $t = 100$ which is, almost in the equilibrium state, a circle. The relative area change (loss) at $t = 100$ is about 1.61%. Note that, in the equilibrium state, the velocity approaches zero everywhere and the pressure approaches two different constants inside and outside the interface.

In Table 10.8, a grid refinement analysis is given at time $t = 15$. Since the exact solution is not available, the error is calculated between the computed solution and the one obtained by the finest grid, 512×512. We denote such an error as $\tilde{E}_N$, for example, $\tilde{E}_N(p) = \|p_N - p_{512}\|_\infty$. For a second-order method, the error ratio $\tilde{E}_N/\tilde{E}_{2N}$ is close to 4 if N is much smaller than 512, the finest number of grid lines. The ratio approaches 5 as $2N$ gets closer to the finest number of the grid lines; see [160] for the methodology. From Table 10.8, we can see that the IIM for Navier–Stokes equations behaves like a second-order method and agrees with the results of other examples in this section.

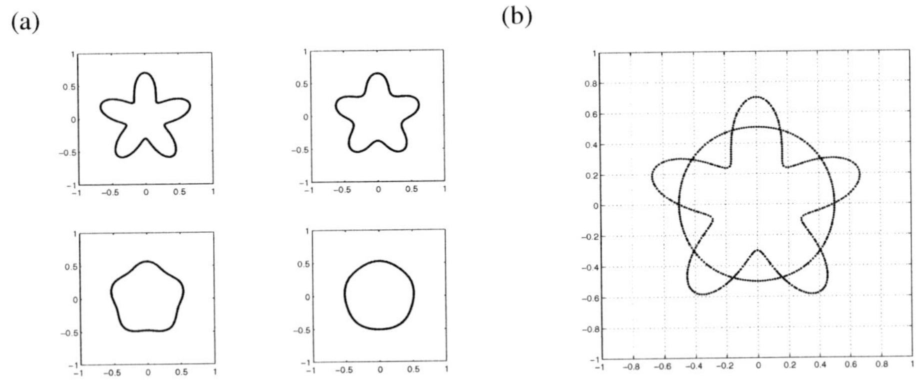

Figure 10.13. *The numerical evolution of the moving interface for Example* 10.7 *using the projection IIM coupled with the level set formulation. The parameters are* $\mu = 0.1$, $\epsilon = 0.05$, *and* $M = N = 160$. *Due to the surface tension, the interface is relaxing to a circle.* (a) *The computed interface at* $t = 0$, $t = 7.0172$, $t = 21.0867$, *and* $t = 35.1563$. (b) *The initial interface at* $t = 0$ *and the computed interface at* $t = 100$.

Table 10.8. *A grid refinement analysis of the projection IIM applied to Example* 10.7 *at* $t = 15$. *The error is computed against the result obtained using the finest grid* (512×512).

N	$\|\tilde{E}_N(\mathbf{u})\|_\infty$	Ratio($\mathbf{u}$)	$\|\tilde{E}_N(p)\|_\infty$	Ratio(p)
32	1.8080×10^{-3}		0.2427	
64	4.5062×10^{-4}	4.0123	0.0717	3.3849
128	1.1920×10^{-4}	3.7804	0.0181	3.9613
256	2.4400×10^{-5}	4.8852	4.2500×10^{-3}	4.2588

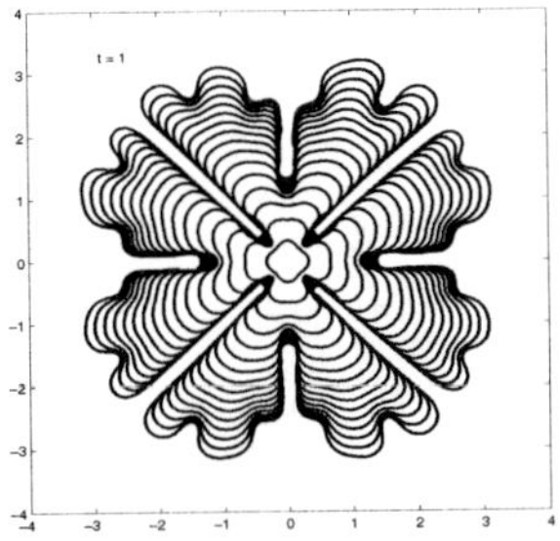

Chapter 11

Some Applications of the IIM

The immersed interface method (IIM) has been successfully applied to a variety of linear and nonlinear problems. It has been applied to a mixed hyperbolic elliptic system in [181]; hyperbolic systems in [158]; elasticity systems in [179, 278, 230]; glacier prediction in [173]; the Hele–Shaw flow in [107]; electromigration of voids in [180]; the traffic flow in [269]; drop spreading in [113]; Stokes and Navier–Stokes equations with interfaces and singular sources in [36, 150, 156, 168, 177]; the Stefan problems and crystal growth in [175]; nonlinear interface problems in magnetorheological fluids containing iron particles in [119]; weighted minimal surface computations in [171]; simulations of porosity evolution in chemical vapor infiltration in [125]; and shape identification of an inverse problem in [116].

In this chapter, we explain how the IIM can be coupled with evolution schemes for free boundary and moving interface problems through a few particular applications.

11.1 The framework coupling the IIM with evolution schemes

To simulate moving interface/free boundary problems, it is crucial to have (or compute) the velocity of the interface or boundary. In pure geometrical motion such as the mean curvature flow, see [207] for several examples there, the velocity is given and the location of the boundary at different times can be computed from an evolution scheme. However, for many other applications, the velocity field has to be computed from the governing differential equations as in the examples of multiphase flows governed by the Stokes or Navier–Stokes equations. In these applications, the governing PDEs for the velocity are coupled with differential equations for the motion of the interface or the boundary. A commonly used approach is a splitting method in which we fix (or freeze) the interface or boundary temporarily and solve the governing equations to get the velocity. The computed velocity then is used to evolve the interface or the boundary. Such a process can be done once or iteratively until a convergence criterion is satisfied. The framework using the IIM for moving interface/free boundary problems is to use the IIM to solve the governing equations to obtain an accurate velocity, and then evolve the interface or the boundary with the computed velocity by an evolution scheme. There are several commonly used evolution schemes in the literature, which we explain briefly below.

11.1.1 The front-tracking method

One of the evolution methods is the front-tracking method based on a Lagrangian formulation; see, for example, [87, 91, 88, 89, 90, 132, 156, 157, 259, 260, 262, 263]. In this approach, a set of ordered marked particles $\{\mathbf{X}_k(t)\}$, $k = 1, 2, \ldots, N_b$, also called the *control points*, are tracked at different time levels according to $\frac{d\mathbf{X}}{dt} = \mathbf{u}$, where $\mathbf{u}$ is the velocity. If the governing PDEs are solved on a uniform Cartesian grid (Eulerian coordinates) to get the velocity, then we have two different coordinate systems. The key to combining the IIM and an evolution scheme is exchanging information between the two grids with adequate accuracy, or precisely, how to interpolate the velocity $\{\mathbf{U}_{ij}\}$ from the grid points in the Eulerian coordinates to the control points in the Lagrangian coordinates. Recall Peskin's interpolation formula,

$$\mathbf{U}(X, Y) \approx \sum_{ij} \mathbf{u}_{ij}\, \delta_h(X - x_i)\delta_h(Y - y_j) h_x h_y \tag{11.1}$$

from

$$\mathbf{U}(X, Y) = \iint_{\Omega} \mathbf{u}(x, y)\, \delta\left((x, y) - (X, Y)\right) dxdy \tag{11.2}$$

in two space dimensions. While this approach is simple and generally first-order accurate, it is widely used along with the immersed boundary (IB) method since it is also first-order accurate.

In the IIM, the velocity at a point on the interface is obtained either from the 3-point interpolation formula (10.43)–(10.45) or from the least squares interpolation

$$\mathbf{U}(X, Y) \approx \sum_{ij} \gamma_{ij}\mathbf{u}_{ij} - \mathbf{C}, \tag{11.3}$$

described in §10.4.1, where C is a correction term determined from the interpolation coefficients and the jump conditions. The coefficients $\{\gamma_{ij}\}$ are chosen such that the interpolation scheme is second-order accurate. The interpolation points are chosen in a robust way, for example, in a disk or in some special set. In the least squares interpolation, we take a few more points than needed for the accuracy to get an underdetermined system of equations. The solution is then the one with the least 2-norm among all feasible solution. With the least squares interpolation, the magnitude of the coefficients of the interpolation are well balanced, which also helps the stability of the entire evolving process. The only trade-off may be the computational cost used in the SVD decomposition.

For many application problems, the motion of the interface depends on the interface quantities, particularly, the first-order derivatives, such as the normal and tangential directions, and the second-order derivatives such as the curvature. Typically, this information can be obtained from interpolation of the control points $\{\mathbf{X}_k(t)\}$. For example, one approximation for the unit tangential vector is

$$\boldsymbol{\tau}_k = \frac{\mathbf{X}_{k+1} - \mathbf{X}_k}{\|\mathbf{X}_{k+1} - \mathbf{X}_k\|}. \tag{11.4}$$

A better approach is to use a cubic spline interpolation, as used in [160, 165], in terms of the arc length. Once we have expressed the interface in terms of the arc length $(X(s), Y(s))$, we can get the first and second surface derivatives easily.

One difficulty associated with the conventional front-capturing approach is the possible loss of high-order accuracy at the moving front. This is especially the case for incompressible fluid interfaces with surface tension [28, 40, 251], especially if surface tension is modeled by a singular delta function source.

One of the advantages of a front-tracking method is that the Lagrangian coordinates are independent of the underlying Cartesian grid. One can control the accuracy of the interface by the number of control points. Reparameterization (or re-griding) and filtering techniques may be needed for front-tracking methods.

11.1.2 Coupling the level set method with the IIM

A basic front-tracking method may break down when an interface develops topological changes such as breaking and merging. The implementation of the front-tracking method for three-dimensional problems can also be rather complicated. An alternative is the level set method first proposed by Osher and Sethian in [207]; see §1.6.4. The level set method is a front-capturing method that avoids the explicit tracking of the moving interface. The moving interface is implicitly captured on the same Eulerian grid by the zero level set of a Lipschitz continuous function $\varphi(\mathbf{x}, t)$,

$$\Gamma(\mathbf{X}(t), t) = \{\mathbf{x},\ \varphi(\mathbf{x}, t) = 0\}. \tag{11.5}$$

By differentiating $\varphi(\mathbf{x}, t) = 0$ with time t, we get the evolution equations for the level set function:

$$\varphi_t + \nabla\varphi \cdot \mathbf{u} = 0,$$

or

$$\varphi_t + |\nabla\varphi| V_n = 0, \tag{11.6}$$

where $V_n = \mathbf{u} \cdot \mathbf{n}$ is the component of the velocity in the normal direction given by $\mathbf{n} = \nabla\varphi/|\nabla\varphi|$. The level set equation (11.6) is a Hamilton–Jacobi-type equation that is often solved using numerical method for conservation laws, particularly the ENO (essential nonoscillating) or WENO (weighted ENO) schemes; see, for example, [123, 124] and the references therein.

The level set method deals relatively easily with complex interface structures, topological changes, and problems in three or higher dimensions. We refer readers to the two books [206, 238] on the level set method. The level set method for two-phase flows can be found, for example, in [40, 136, 168, 250, 251] and many others. The variational level set method has been developed with applications in [282, 283]. There are intensive studies in the literature to improve the level set method to reduce its computational cost, due to embedding the interface into a higher-dimensional space, to better preserve the volume for incompressible flows. These efforts lead to a fast level set method using a computational tube [180, 208, 107], reinitialization techniques [1, 40, 47, 123, 238, 250, 251], hybrid methods with particle method [71], finite volume-level set method [247, 248], adaptive level set method [249], and finite element-level set methods [14, 256].

We can combine the advantages of the IIM and the level set approach to evolve moving fronts. The information regarding the location, the normal and tangential directions, of the interface can be extracted from the level set function. When we incorporate this information into the immersed interface discretization, we obtain a uniformly high-order discretization. Clearly, we get the advantages of both methods and avoid the shortcomings of these two methods. This gives rise to a robust and accurate Eulerian discretization for interface problems. The key is how to accurately exchange the information between the interface and the solution (or a grid function) on the grid, and how to extend a surface quantity to nearby grid points if necessary.

11.1.3 Orthogonal projections and the bilinear interpolation

Using the level set method, the interface is not explicitly given. Instead we only have information $\varphi(x_i, y_j)$ at each grid point in two dimensions. The interface can be reconstructed from the orthogonal projections of some grid points on the interface or the intersections of the interface and the grid lines. The procedure to find the orthogonal projections has been already described in §1.6.4 in two dimensions and §4.3.2 in three dimensions.

It is reasonable to assume that the level set function has continuous first- and second-order derivatives in a neighborhood of the interface where $\varphi(x, y) = 0$. This is true if the level set function is chosen as the signed distance function and the mesh is fine enough. Therefore, we can use the standard central finite difference formulas to get $\varphi_x, \varphi_y, \ldots, \varphi_{xy}$, and thus the tangential and normal directions and the curvature of the level curve/surface at each grid point. The bilinear interpolation formula described below then can be used to get this interface information at a nongrid point, say at an orthogonal projection.

In two space dimensions, given any point (x, y), let the vertices of the square that contains (x, y) be (x_i, y_j), (x_{i+1}, y_j), (x_i, y_{j+1}), and (x_{i+1}, y_{j+1}). Let G_{ij} be a grid function.

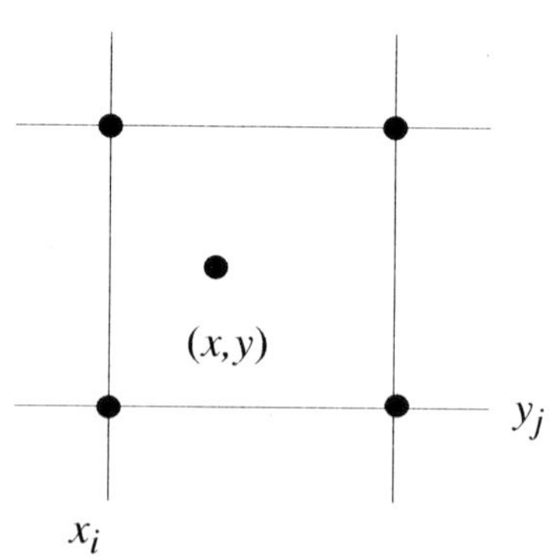

Figure 11.1. *A diagram for the bilinear interpolation.*

The bilinear interpolation formula, in two dimensions, to get $G(x, y)$ in terms of G_{ij} is the following:

$$G(x, y) = \frac{1}{4} \sum_{k=0,l=0}^{1} G_{i+k,j+l}\, \bar{x}_k\, \bar{y}_l, \tag{11.7}$$

where

$$\bar{x}_k = 1 + (2k - 1)\left(\frac{2(x - x_i)}{h_x} - 1\right),$$

$$\bar{y}_l = 1 + (2l - 1)\left(\frac{2(y - y_j)}{h_y} - 1\right);$$

see Figure 11.1 for an illustration. The bilinear interpolation is second-order accurate if $G(x, y) \in C^1$ in the neighborhood of interpolation. The bilinear interpolation in three space dimensions is given in §4.3.2.

11.1.4 Velocity extension along normal directions

In the level set method, the level set function (11.6) is solved at grid points if a finite difference method is used. The velocity field is often computed before we can solve the level set function. For some problems, the velocity is only defined at the moving front; in this case, the velocity has to be extended to the grid points near a computational tube $|\varphi| \le \delta$, where $\delta \sim Ch$ is a given parameter. There are several ways of extending the velocity off the front. One is the fast marching method, see [47, 238]; the other one is the PDE approach using the following evolution equation:

$$\frac{\partial V_n}{\partial t} \pm \nabla V_n \cdot \frac{\nabla \varphi}{|\nabla \varphi|} = 0, \tag{11.8}$$

where t is an artificial variable; see, for example, [107, 113, 175, 180]. The sign is determined from the normal direction of the level set function. Figure 11.2(a) plots the normal velocity at irregular grid points for a test of the Stefan problem (see §11.3), while Figure 11.2(b) is the normal velocity after the extension.

For some other problems, for example, the Stokes and Navier–Stokes equations, the velocity field is defined in the entire domain; we can use the velocity field at grid points directly. However, for the fast level set method in which the level set function is only evolved in a small tube, it is believed and somewhat confirmed that the extension gives a better result for the location of the moving front. It is also crucial to reinitialize the level set function to get an approximate signed distance function $|\nabla \varphi| \sim 1$ in the neighborhood of the interface to avoid the level set function being too flat or steep. The reinitialization can be done using the fast marching method [1, 47, 238], the fast weeping method [261], or solving the following PDE:

$$\varphi_t^{new} + sign(\varphi^{old})\,(1 - |\nabla \varphi|) = 0 \tag{11.9}$$

to steady state, where φ^{old} is the level set function to be reinitialized, and t is an artificial time variable. In implementation, only a few iterations ($1 \sim 5$) are needed at every or every other few time steps; see [40, 123, 250, 251].

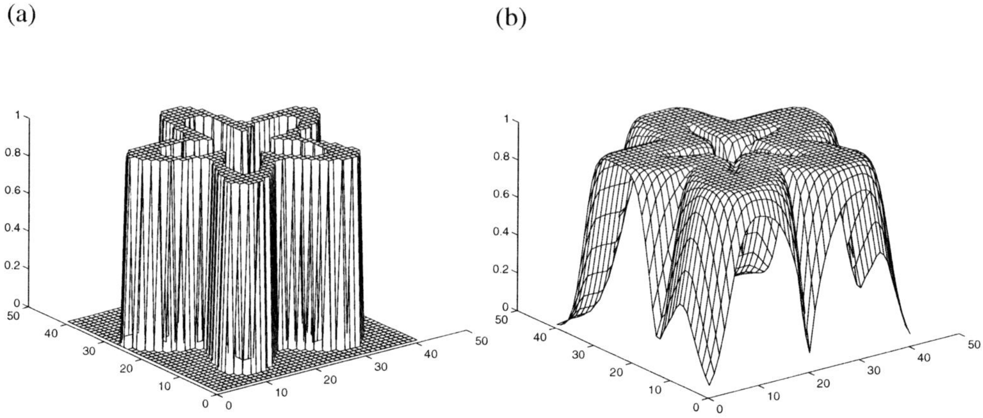

Figure 11.2. *A diagram of velocity extension.* (a) *A computed normal velocity at irregular grid points.* (b) *The normal velocity field in a tube after the extension.*

11.1.5 Reconstructing the interface locally from a level set function

For many applications, it is necessary to find the surface derivatives of a surface quantity. This may be trivial if the interface is expressed in terms of a cubic spline interpolation, but not so if the interface is implicitly given by a level set function. Some of the approaches can be found in [180, 274]. Here, we explain a direct approach by reconstructing the interface locally, so that an interpolation scheme can be applied to obtain the surface derivatives.

Let (x_i, y_j) be an irregular grid point in reference to the standard central 5-point stencil. Let the orthogonal projection of (x_i, y_j) on the interface $\varphi(x, y) = 0$ be $\mathbf{X}^* = (X^*, Y^*)$. We describe how to reconstruct the interface locally in terms of the orthogonal projections of irregular grid points in a neighborhood of (x_i, y_j).

Under the local coordinates system (1.34) centered at (X^*, Y^*), the interface has the form

$$\xi(\eta) = C\eta^2 + D\eta^3 + O(\eta^4). \tag{11.10}$$

Once we know the coefficients C and D, we have an analytic expression of the interface in a neighborhood of (X^*, Y^*). Let $\mathbf{X}_1^* = (x_1^*, y_1^*)$ be the orthogonal projection of a different irregular grid point. We can determine C and D using the interface information. Denoting $(\boldsymbol{n}_\xi, \boldsymbol{n}_\eta)$ as the unit normal direction of the interface at $\mathbf{X}_1^*$, we have

$$\begin{aligned} \frac{2C\eta_1 + 3D\eta_1^2}{\sqrt{1 + \left(2C\eta_1 + 3D\eta_1^2\right)^2}} &= -\boldsymbol{n}_\eta, \\ \frac{1}{\sqrt{1 + \left(2C\eta_1 + 3D\eta_1^2\right)^2}} &= \boldsymbol{n}_\xi, \end{aligned} \tag{11.11}$$

where (ξ_1, η_1) is the local coordinate of $\mathbf{X}_1^*$; see (1.34). We arrive at the following linear system of equations for C and D:

$$\begin{aligned} C\eta_1^2 + D\eta_1^3 &= \xi_1, \\ 2C\eta_1 + 3D\eta_1^2 &= -\frac{\boldsymbol{n}_\eta}{\boldsymbol{n}_\xi}. \end{aligned} \tag{11.12}$$

In other words, the curve is approximated by a Hermite spline interpolation between $\mathbf{X}^*$ and $\mathbf{X}_1^*$. Once we have solved for C and D, we have an analytic expression to approximate the interface. The arc length between $\mathbf{X}^*$ and $\mathbf{X}_1^*$ is determined from

$$|s_1| = \int_0^{\eta_1} \sqrt{1 + \left(2C\eta + 3D\eta^2\right)^2} d\eta. \tag{11.13}$$

This definite integral can be approximated by the Simpson rule or a Gaussian quadrature formula using the approximate analytic expression of the interface. In this way, the arc length is evaluated to third-order accuracy. Finally, we need to determine the sign of the arc length according to the relative position between $\mathbf{X}^*$ and $\mathbf{X}_1^*$,

$$s_1 = \begin{cases} |s_1| & \text{if } \left(\mathbf{X}_1^* - \mathbf{X}^*\right) \cdot \boldsymbol{\tau}^* > 0, \\ -|s_1| & \text{otherwise,} \end{cases} \tag{11.14}$$

where $\boldsymbol{\tau}^*$ is the tangential vector at $\mathbf{X}^*$.

Once we have the signed arc length between $\mathbf{X}^*$ and other projections, it is easy to interpolate a surface function $g(s)$ to obtain $g'(s^*)$ and $g''(s^*)$ at the projection on the interface.

11.2 The hybrid IIM-level set method for the Hele–Shaw flow

In 1958, Saffman and Taylor [233] performed experiments replacing a viscous fluid from between two closely spaced, parallel plates with a less viscous fluid. The shape of the interface is well known to exhibit a fingering phenomenon. The velocity field $\mathbf{u} = (u, v)$ of the flow is proportional to the gradient of the pressure p. The governing equations are

$$\mathbf{u} = -\beta \nabla p, \tag{11.15a}$$

$$\nabla \cdot \mathbf{u} = \phi \tag{11.15b}$$

with $\beta = b^2/(12\mu)$, where b is the gap width and μ is the viscosity, which is very different inside and outside the interface separating the two fluids. The source term ϕ is the result of the injection of the less viscous fluid into the Hele–Shaw cell. It is reasonable to take

$$\phi = \begin{cases} \phi_0(r_0)\left(1 + \cos(r\pi/r_0)\right) & \text{if } r \le r_0, \\ 0 & \text{if } r > r_0, \end{cases} \tag{11.16}$$

where $r = \sqrt{x^2 + y^2}$. The total injection rate is

$$\bar{\phi} = \iint \phi(x, y) dx dy = \phi_0(r_0)\left(\pi - \frac{4}{\pi}\right) r_0^2. \tag{11.17}$$

The jump conditions across the interface are

$$[p] = \gamma\kappa, \qquad \text{the Laplace–Young condition,} \tag{11.18}$$

$$[\beta p_{\mathbf{n}}] = 0, \qquad \text{the kinematic interface condition,} \tag{11.19}$$

where γ is the surface tension and κ is the curvature of the interface.

11.2.1 Dynamic stability of the Hele–Shaw flow

It is known that the Hele–Shaw flow is intrinsically unstable. To see this, we follow the discussion in [61] and assume that the less viscous fluid inside the Hele–Shaw cell has a negligible viscosity, while the more viscous fluid has a finite viscosity β^+ and is incompressible. The amalgamated surface tension parameter with the dimension of length is defined as[17]

$$d_0 = \frac{2\gamma\pi\beta^+}{\bar{\phi}}, \tag{11.20}$$

where $\bar{\phi}$ is the injection rate.

Linearization about a Hele–Shaw cell with radius $R(t)$ and injection rate $\bar{\phi}$ gives the instantaneous growth rate [61, 108]

$$\lambda_k(t) = \frac{1}{R^2(t)}\left((k-1)\frac{\bar{\phi}}{2\pi} - \frac{d_0}{R(t)}(k-1)k(k+1)\right), \tag{11.21}$$

where k represents the kth Fourier mode. For a constant injection rate, this implies a Mullins–Sekerka-type instability, which shows the competition between the destabilizing effect due to the injection and the stabilizing effect due to the surface tension.

Note that, for high-frequency modes, $\lambda_k(t)$ is negative, indicating that the Hele–Shaw flow is stable for these frequencies. For lower frequencies, depending on the parameters, $\lambda_k(t)$ can be positive, indicating unstable growth. Usually, it is relatively easy to control high-frequency noise. But it is more difficult to control the round-off errors of low to intermediate frequencies.

11.2.2 The IIM for the Hele–Shaw flow

The Hele–Shaw flow has served as a benchmark problem for numerical algorithms to compute unstable fronts and has attracted much attention in the literature; see, for example, [60, 61, 63, 108, 200, 223, 259, 267]. This is an ideal test model for the IIM with a moving interface since (11.15a) can be written as

$$\nabla \cdot (\beta \nabla p) = -\phi, \tag{11.22}$$

together with the jump conditions (11.18)–(11.19). To determine the outer boundary condition of the pressure, we assume that the interface is far away from the boundary so that

[17] This is similar to the Atwood ratio described in [259].

the flow at the boundary agrees with the radial outflow which arises from the source term in a uniform fluid, i.e.,

$$p(x, y) = p_0 - \frac{\phi_0}{2\pi\beta} \log r \qquad (11.23)$$

is specified on the boundary, where p_0 is some arbitrary constant. Numerical simulations have shown that the boundary condition has little effect on the motion of the interface until the interface gets very close to the boundary.

We employ a splitting method to solve the moving interface problem. That is, at a time step, we solve (11.22) to get the pressure, and then find the velocity through (11.15a). The computed velocity is then used to evolve the interface by solving the level set function through (11.6) since $\frac{d\mathbf{X}}{dt} = \mathbf{u}$ at the interface $\Gamma(t)$.

The interface problem (11.22)–(11.23) with piecewise constant β can be solved readily using the augmented IIM described in §6.1 and [163]. We use the orthogonal projections of irregular grid points on the interface $\Gamma(t)$ from one side, say $\varphi \geq 0$, as the control points; see §6.1.2.

After we have solved the Poisson equation (11.22)–(11.23) to get the pressure, we use two approaches to get the velocity. The first approach is to use (11.15a) to get the velocity at all the grid points in the computational tube ($|\varphi_{ij}| \leq \delta \sim h$). At a regular grid point, this can be done using the central finite difference scheme. At an irregular grid point, the interpolation is explained below. The second approach is to evaluate the normal velocity only at irregular grid points and then extend the normal velocity to the computational domain.

In both approaches mentioned above, we need to compute the normal velocity at irregular grid points. Note that a simple one-sided interpolation scheme often leads to loss of accuracy, say from second order to first order. One of the important properties of the augmented IIM for Poisson interface problems is that we can obtain a second-order accurate gradient of the solution at those orthogonal projections of irregular grid points from one particular side of the interface, say the $\varphi \geq 0$ side; see Table 6.1 in §6.1. Since we have $V_n = \mathbf{u} \cdot \mathbf{n} = \beta \nabla p \cdot \mathbf{n}$, the normal velocity is directly extended to the irregular grid points from the $\varphi \geq 0$ side.

To get the normal velocity at an irregular grid point (x_i, y_j) from the other side, say $\varphi_{ij} < 0$, we take advantage of the fact that the normal velocity $V_{\boldsymbol{n}}$ is continuous. The following approach is used to find the normal velocity of the level set function at the grid point; see Figure 11.3 for an illustration.

1. Find the two closest orthogonal projections $\mathbf{X}_k$ and $\mathbf{X}_{k+1}$. We know the normal velocity at these two control points from the Poisson solver using the augmented approach.

2. Find a regular grid point, say (x_{i_0}, y_{j_0}), $\varphi_{i_0,j_0} \geq 0$, which is closest to (x_i, y_j), $\mathbf{X}_k$, and $\mathbf{X}_{k+1}$. The normal velocity at this point can be calculated by the standard central finite difference scheme from $\mathbf{u} \cdot \nabla\varphi / |\nabla\varphi|$.

3. Interpolate the normal velocity at the three points to get an approximation to the normal velocity at the irregular grid point (x_i, y_j),

$$V_n(x_i, y_j) = \alpha_1 V_n(x_{i_0}, y_{j_0}) + \alpha_2 V_n(X_k, Y_k) + \alpha_3 V_n(X_{k+1}, Y_{k+1}),$$

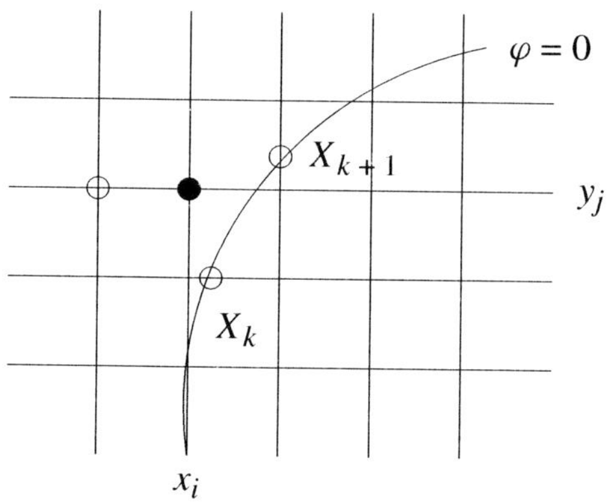

Figure 11.3. *A diagram of an irregular grid point and the choice of three points for interpolating the normal velocity.*

where α_1, α_2, and α_3 are the solution of the following linear system:

$$\begin{aligned} \alpha_1 + \alpha_2 + \alpha_3 &= 1, \\ \alpha_1 \left(x_i - x_{i_0}\right) + \alpha_2 \left(x_i - X_k\right) + \alpha_3 \left(x_i - X_{k+1}\right) &= 0, \\ \alpha_1 \left(y_j - y_{j_0}\right) + \alpha_2 \left(y_j - Y_k\right) + \alpha_3 \left(y_j - Y_{k+1}\right) &= 0. \end{aligned}$$

The system of linear equations is obtained using Taylor expansions of $V_n(x_i, y_j)$, $V_n(X_k, Y_k)$, and $V_n(X_{k+1}, Y_{k+1})$ at (x_i, y_j). The interpolation scheme is second-order accurate.

Once we have the normal velocity at all the irregular grid points in reference to the standard central 5-point stencil, then we can use either the fast marching method or the PDE approach (11.8) to extend the normal velocity to the computational tube and solve the level set equation (11.6) to evolve the moving front.

11.2.3 Numerical experiments of the Hele–Shaw flow

The results of extensive numerical experiments of the IIM, with different initial interfaces, viscosities, and surface tensions, agree with the theoretical analysis and numerical results in the literature. Since the Hele–Shaw flow is unstable, the results may not converge to a unique solution due to round-off and the discretization errors. This is consistent with the experiments in which different shapes are observed with the same settings. However, this should not invalidate numerical simulations because they can still be used to predict roughly the shape and the location of the expanding front as time evolves. Moreover, for a short period of time, the computed solution is consistent with grid refinement convergence tests. The crucial parameter which affects the stability is the amalgamated surface tension d_0 defined in (11.20). The smaller d_0 is, the more unstable is the Hele–Shaw flow. We show a few examples below and refer the reader to [107] for more examples and analysis. More detailed discussion of the physical mechanism can also be found, e.g., in [61, 108, 200, 259].

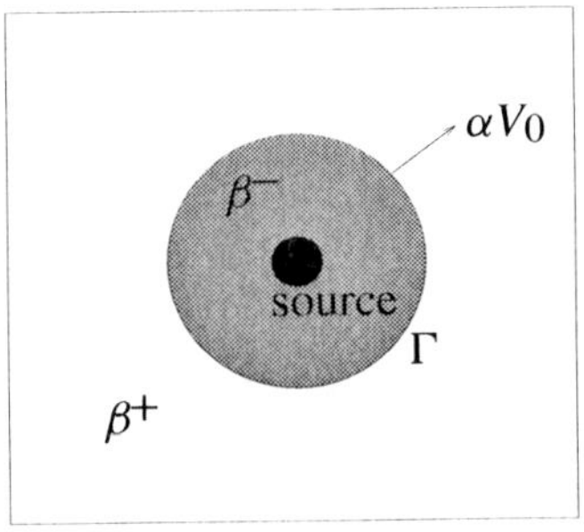

Figure 11.4. *An expanding interface with a constant speed αV_0.*

Example 11.1. *A benchmark problem.* If the initial interface is a circle, there is only a single mode, $k = 1$, in the linear stability analysis (11.21). An exact solution for this axisymmetric flow corresponding to a known source term can be constructed. The exact pressure is

$$p(r) = \begin{cases} \dfrac{V_0}{\beta^-}\left(\dfrac{2r^3}{3\alpha^2} - \dfrac{3r^2}{2\alpha}\right) + C_1 & \text{if } 0 \le r \le \alpha, \\[2ex] -\dfrac{V_0\,\alpha}{\beta^-}\log(r) + C_0 & \text{if } \alpha < r \le r_\Gamma, \\[2ex] -\dfrac{V_0\,\alpha}{\beta^+}\log(r) & \text{if } r_\Gamma \le r, \end{cases} \tag{11.24}$$

where $r_\Gamma = \sqrt{2\alpha V_0 t + r_1^2}$, r_1 is the radius of the initial interface; see Figure 11.4 for an illustration. Note that the pressure is continuous across the circle $r = \alpha$. C_0 is chosen as

$$C_0 = \frac{\gamma}{r_\Gamma} + V_0\alpha\,\log(r_\Gamma)\left(\frac{1}{\beta^-} - \frac{1}{\beta^+}\right), \tag{11.25}$$

so that $[p]_{r_\Gamma} = \gamma\kappa = -\gamma/r_\Gamma$. The pressure is continuous across the inner boundary $r = \alpha$ surrounding the source. The constant C_1,

$$C_1 = C_0 - \frac{V_0\alpha\,\log(\alpha)}{\beta^-} + \frac{5V_0\alpha}{6\beta^-}, \tag{11.26}$$

is chosen such that $[\beta p_n]_{r=\alpha} = 0$. The velocity is determined from (11.15a). In polar coordinates, we have

$$u_r = -\beta\frac{\partial p}{\partial r} = \begin{cases} -V_0\left(\dfrac{2r^2}{\alpha^2} - \dfrac{3r}{\alpha}\right) & \text{if } 0 \le r \le \alpha, \\[2ex] \dfrac{\alpha V_0}{r} & \text{if } \alpha \le r, \end{cases} \tag{11.27}$$

$$u_\theta = \frac{1}{r}\frac{\partial p}{\partial \theta} = 0. \tag{11.28}$$

Table 11.1. *A grid refinement analysis in the infinity norm, $t_0 = 0$, $t_{out} = 0.1$. The parameters are $V_0 = 0.25$, $\beta^+ = 1$, $\beta^- = 100$, $\alpha = 0.1$, and $d_0 \approx 2.5 \times 10^{-3}$.*

N	$E_p(t = 0.1)$	Rate
40	2.4230×10^{-3}	
80	7.8189×10^{-4}	1.6318
160	1.4923×10^{-4}	2.3894
320	3.7564×10^{-5}	1.9901

The pressure p satisfies the following Poisson equation:

$$\begin{aligned}\nabla \cdot (\beta \nabla p) &= \frac{1}{r}\frac{\partial}{\partial r}\left(r\frac{\partial p}{\partial r}\right) + \frac{1}{r^2}\frac{\partial^2 p}{\partial \theta^2} \\ &= \begin{cases} V_0\left(\dfrac{6r}{\alpha^2} - \dfrac{6r}{\alpha}\right) & \text{if } 0 \le r \le \alpha, \\ 0 & \text{if } \alpha < r, \end{cases} \\ &= -\nabla \cdot \mathbf{u}. \end{aligned} \tag{11.29}$$

Note that the benchmark problem is not a trivial one. It is indeed the solution of the Hele–Shaw flow with a specific source and an initial interface. The solutions of the pressure and the velocity depend on the surface tension, the radius of the initial circle, and the injection rate.

The numerical computations show that, for a short period of time and modest surface tension, second-order accuracy for the pressure p, the interface location $\Gamma(t) = \{r = r_\Gamma\}$, and the velocity vector (u, v) is achieved. Table 11.1 shows the result of a grid refinement analysis for the pressure at the fixed time $t_{out} = 0.1$ in the infinity norm. Second-order accuracy is observed. The initial interface is $r = 0.5$. The other parameters are

$$V_0 = 0.25, \quad \beta^+ = 1, \quad \beta^- = 100, \quad \alpha = 0.1.$$

The amalgamated surface tension corresponding to the above parameters is about 2.5×10^{-3}.

For a longer time computation, there will be some low to intermediate frequency unstable modes if the surface tension is small enough to produce finger splitting. In this case, the round-off errors associated with these modes become significant. The reinitialization process controls only the round-off errors of high frequencies, but it does not effectively remove round-off errors for low to intermediate modes. Thus the round-off error perturbations in the low to intermediate modes can trigger instability in this problem.

Example 11.2. *Effect of the surface tension.* Let the initial interface in polar coordinates be

$$\rho = 0.5 + 0.1\,(\sin(2\theta) + \cos(3\theta)), \qquad 0 \le \theta \le 2\pi.$$

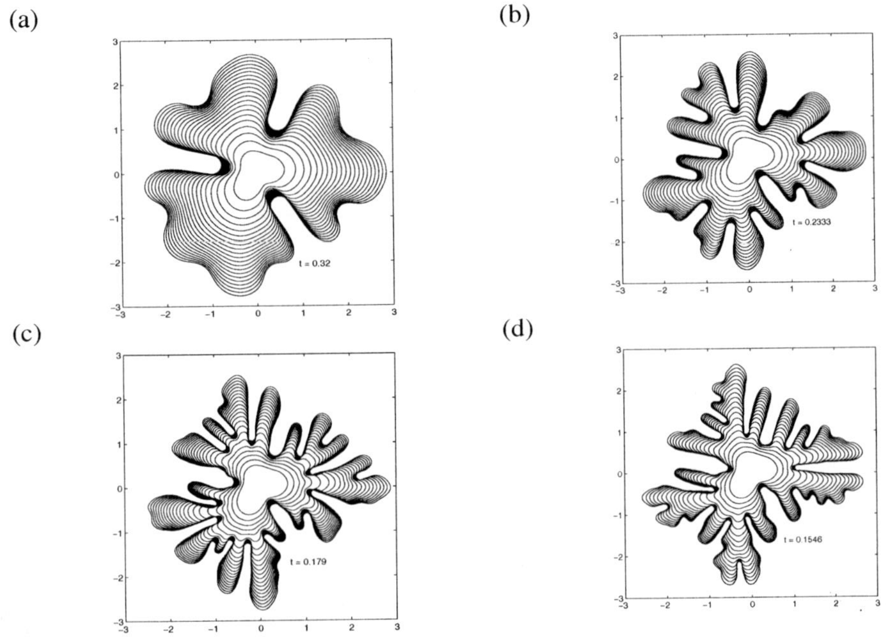

Figure 11.5. *An expanding Hele–Shaw bubble with different surface tensions computed from a* 320×320 *grid.* (a) $d_0 = 6.3 \times 10^{-3}$; (b) $d_0 = 2.513 \times 10^{-3}$; (c) $d_0 = 1.257 \times 10^{-3}$; (d) $d_0 = 7.5 \times 10^{-4}$.

A similar example has been used in [108] with different scaling. There is no particular symmetry in the motion. Because of the surface tension, the interface remains smooth all the time. It is referred to as q-pole data in [61]. Figure 11.5 shows the expansion of a Hele–Shaw bubble with different surface tensions at almost equally spaced time increments $\Delta t \sim O(h^2)$. The stiffness in time integration is less severe compared to a Lagrangian boundary integral method because of a fixed underlying grid. Moreover, the reinitialization process plays the role of geometric regularization and stabilizing high-frequency components of the solution.

The simulations display much of the behavior that has become known to the numerical analysts working on this subject. At an early stage, three main "fjords" were developed on the interface corresponding to the three Fourier modes in the initial interface. Then the three fjords separated into three expanding fronts. The expanding fronts developed more fingers and petals depending on the surface tension. For large surface tension, only a few low frequencies, k between 1 and 4, are unstable for each "fjord"; see Figure 11.5(a). As we decrease the surface tension, more Fourier modes become unstable and we see more fingers and petals; see Figures 11.5(b), (c). The petals expand outward and eventually tip-split into two petals while fingers have either stopped growing or receded and have been absorbed back into the main bulk of the bubble.

For a short time, the shape of the interface varies little for different values of surface tension. The smaller the surface tension is, the quicker the secondary structure (or finger

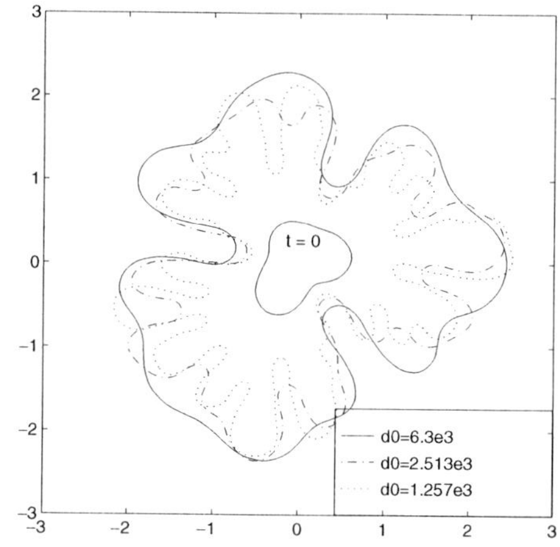

Figure 11.6. *Three expanding Hele–Shaw bubbles with different surface tensions.*

tip-splitting) develops. As we decrease the physical surface tension further, the numerical surface tension or dissipation becomes more apparent, indicating the limitation on the real surface tension we can resolve. But without numerical dissipation, the interface will develop unphysical *cusps* and it will take more time to solve the Poisson equation because of the corners developed on the interface if the surface tension is very small.

The envelopes of the interface in a fjord are almost the same regardless of the surface tension; see Figure 11.6. This agrees with the result in [61]. Figure 11.7 shows a fully developed Hele–Shaw bubble corresponding to the initial interface,

$$\rho = 0.2 + 0.05\sin(3\theta), \qquad 0 \le \theta \le 2\pi. \tag{11.30}$$

We can see that many fingers and petals are developed with time.

11.3 Simulations of Stefan problems and crystal growth

Solving Stefan problems and unstable crystal growth is one of many important free boundary and moving interface problems. In [146], Langer has given an easily understood, but insightful, description of the problem which we briefly explain below. Consider two typical cases involving the same substance but at different states, say liquid and solid, as shown in Figure 11.8. In both cases, a pure fluid is contained in a vessel whose walls are held at some temperature T_W, which is less than the melting temperature T_M. In the first case (a), the liquid is initially at a temperature $T_0 \ge T_M$ and the solid is allowed to start forming on the walls. In this case, the process is completely stable. The solidification front Γ moves smoothly and uniformly toward the center. Such a process is often referred to as a stable Stefan problem. The second case (b) is more interesting. Here the liquid is initially undercooled to a temperature $T_0 < T_M$ and the solidification is initiated at a seed crystal at the center. The situation is intrinsically unstable. The moving front Γ develops an unstable dendrite. Such

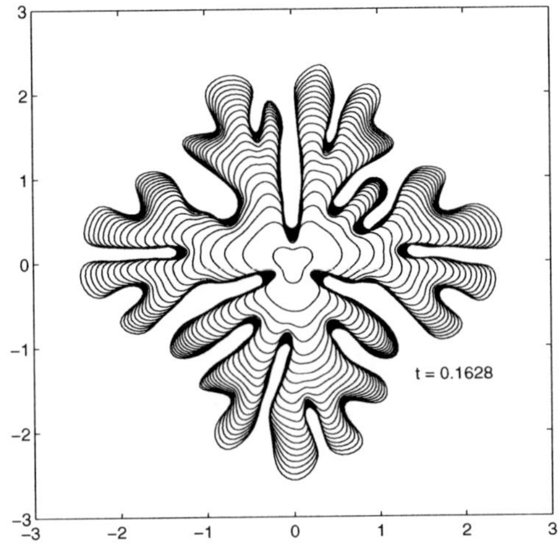

Figure 11.7. *An expanding Hele–Shaw bubble computed on a* 320×320 *grid with initial interface:* $\rho = 0.2 + 0.05\sin(3\theta)$, $0 \leq \theta \leq 2\pi$, *and* $d_0 = 1.257 \times 10^{-3}$.

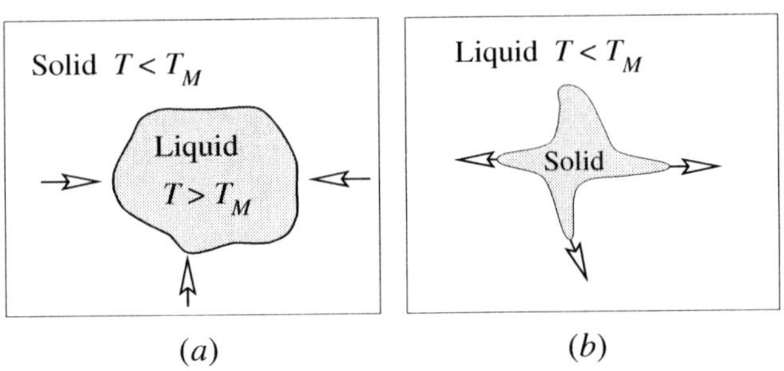

Figure 11.8. *A schematic illustration of solidification occurring in* (a) *stable and* (b) *unstable configurations; see* [146].

processes have wide applications in industry and science including the aerospace, materials science, semiconductor, and even medical, industries.

A simple but widely accepted mathematical model consists of the heat equation for the temperature,

$$\rho c \frac{\partial T}{\partial t} = \nabla \cdot (\sigma \nabla T), \tag{11.31}$$

where ρ is the density, c is the heat capacity, and σ is the thermal conductivity, respectively. Generally, these quantities are discontinuous across the interface. The normal velocity $V_{\mathbf{n}}$ of the moving front is coupled by the change in the heat flux,

$$\rho L V_n = -\left[\sigma \frac{\partial T}{\partial \mathbf{n}}\right]. \tag{11.32}$$

In the equation above, L is the latent heat of solidification and $\mathbf{n}$ is the outward normal direction. For the classical Stefan problems, the temperature at the front is a constant,

$$T(\mathbf{x}, t) = T_M, \tag{11.33}$$

which is the melting temperature. For problems involving crystal growth, we should consider the Gibbs–Thomson relation

$$T(\mathbf{x}, t) = -\epsilon_C \kappa - \epsilon_V V_{\mathbf{n}}, \tag{11.34}$$

where κ is the curvature, ϵ_C is the surface tension coefficient, and ϵ_V is the molecular kinetic coefficient. In the isotropic case, both ϵ_C and ϵ_V are constants. For the anisotropic case, they have the following form:

$$\epsilon_C = \bar{\epsilon}_C \left(1 - A\cos(k_A\theta + \theta_o)\right), \tag{11.35}$$

$$\epsilon_V = \bar{\epsilon}_V \left(1 - A\cos(k_A\theta + \theta_o)\right), \tag{11.36}$$

where θ is the angle between the x-axis and $\mathbf{n}$, and θ_o controls the angle of the symmetry axis upon which the crystal grows. The constants A, k_A, θ_o, $\bar{\epsilon}_C$, and $\bar{\epsilon}_V$ depend on the material.

There are a variety of methods in the literature [9, 31, 42, 132, 232, 235, 236] for Stefan problems and crystal growth. Among finite difference methods, the front-tracking method proposed in [132] presented some amazing results and analysis of the physical phenomena. It can also handle certain topological changes. However, it employs an explicit method to solve the heat equation, which has a severe time step limitation. It also used Peskin's discrete delta function which limits the accuracy near or on the front to first order. As an alternative, the level set method developed in [207] is easier to implement and can handle topological changes. With the level set method used in [207], one can expect that the solution near or on the front is only first-order accurate. Hence a first-order implicit method such as that used in [42, 236] is appropriate.

The IIM coupled with a modified ADI method based on the Crank–Nicolson discretization and the level set method is a much more efficient method for simulations of Stefan problems and crystal growth. The method was originally developed in [175]. Here we explain the main ideas and analysis.

We use the nondimensional equations

$$\frac{\partial T}{\partial t} = \nabla \cdot (\sigma \nabla T), \tag{11.37}$$

$$V_{\mathbf{n}} = -S\left[\beta \frac{\partial T}{\partial \mathbf{n}}\right] \tag{11.38}$$

in the discussion, along with the interface relations (11.33) or (11.34)–(11.36). In the expressions above, S is the Stefan number and β, σ, t, and T are all nondimensional variables.

11.3.1 A modified Crank–Nicolson discretization

One of the primary costs of the simulation comes from updating the temperature field from time t^n to t^{n+1}. Due to the time step limitation, an implicit second-order discretization is preferred. For Stefan problems with known temperature $T(\mathbf{x},t)_\Gamma = T_M$, we can use the Crank–Nicolson scheme at regular grid points, but the backward Euler's method at irregular grid points,

$$T_{ij}^{n+1} = \begin{cases} T_M & \text{if } \varphi_{ij}^{n+1} = 0, \\ T_{ij}^n + \Delta t \left(\delta_{xx} T_{ij}^{n+1} + \delta_{yy} T_{ij}^{n+1}\right) + Q_{ij}^n & \text{otherwise,} \end{cases} \tag{11.39}$$

where $\delta_{xx} T_{ij}^{n+1}$ and $\delta_{yy} T_{ij}^{n+1}$ are discrete versions of $(\sigma T_x)_x$ and $(\sigma T_y)_y$ at time level t^{n+1}. These two operators will, in general, depend on (i, j), but we drop the indices for simplicity, Moreover, Q_{ij} is a correction term that will be needed if the interface crosses the grid line between t^n and t^{n+1}.

At an irregular grid point (see Figure 11.9), we use a one-sided finite difference scheme for $\delta_{xx} T_{ij}^{n+1}$ and $\delta_{yy} T_{ij}^{n+1}$. For example, if the interface cuts through the grid line $y = y_j$ at (x^*, y_j), $x_i < x^* < x_{i+1}$, then

$$\delta_{xx} T_{ij}^{n+1} = \gamma_1^{n+1} T_{i-1,j}^{n+1} + \gamma_2^{n+1} T_{i,j}^{n+1} + \gamma_3^{n+1} T_M, \tag{11.40}$$

where

$$\gamma_1^{n+1} = \frac{2\sigma}{h^2 + h(x^* - x_i)}, \qquad \gamma_3^{n+1} = \frac{2\sigma}{(x^* - x_i)^2 + h(x^* - x_i)},$$
$$\gamma_2^{n+1} = -\left(\gamma_{n+1}^1 + \gamma_{n+1}^3\right).$$

The local truncation errors are $O(h^2)$ at regular grid points and $O(h)$ at irregular grid points.

Grid crossing

If the interface cuts though the grid line, i.e., when $\varphi_{ij}^n \varphi_{ij}^{n+1} < 0$, then the truncation error for $T_t^{n+1/2}$ will be $O(1)$. Thus a correction term has to be added to ensure second-order accuracy. This can be done according to the following theorem about the jump $[T_t]$ in time t.

Theorem 11.1. *If* $\varphi(:, t^n)\,\varphi(:, t^{n+1}) < 0$ *and* $t^* \in (t^n, t^{n+1})$ *is the unique solution of* $\varphi(:, t) = 0$ *in the interval* (t^n, t^{n+1})*, then*

$$[T_t] = -V_n \left[\frac{\partial T}{\partial \mathbf{n}}\right] \quad at \quad t = t^*, \tag{11.41}$$

$$\begin{aligned} \frac{T(:, t^{n+1}) - T(:, t^n)}{\Delta t} &= \frac{1}{2}\left(T_t(:, t^n) + T_t(:, t^{n+1})\right) + [T_t]^{n+1/2} \frac{(t - t^*)}{\Delta t} + O(\Delta t) \\ &= T_t(:, t^{n+1/2}) + [T_t]^{n+1/2} \frac{(t^{n+1/2} - t^*)}{\Delta t} + O(\Delta t), \end{aligned} \tag{11.42}$$

where $(:)$ *denotes* (x_i, y_j).

Proof: We know that the temperature is continuous, that is, $[T] = 0$ across the interface. Differentiating this with respect to time t along the interface gives

$$[T]_t + [\nabla T] \cdot \mathbf{u} = 0.$$

The velocity field $\mathbf{u}$ can be decomposed into the tangential and the normal components. All the jumps are defined as the difference between the limiting values from above and below the time t^*. Using the fact that both the temperature T and its tangential derivative are continuous, we have the expression

$$\begin{aligned}[T]_t + [\nabla T] \cdot (V_n \mathbf{n} + u_\tau \boldsymbol{\tau}) &= [T]_t + [\nabla T] \cdot V_{\mathbf{n}} \mathbf{n} \\ &= [T]_t + \left[\frac{\partial T}{\partial \mathbf{n}}\right] V_{\mathbf{n}} = 0,\end{aligned}$$

where $\boldsymbol{\tau}$ is the unit tangential vector of the interface. This proves the first part of the theorem. The second part has been proved in Theorem 9.1; see also [23, 162]. □

This theorem tells us what the correction term should be when the interface crosses the grid line. If β in (11.38) is continuous across the interface, then the jump $[T_t]$ is given by

$$[T_t] = -\frac{L\, V_n^2}{S\, \beta} \tag{11.43}$$

from (11.38). In this case we can find the correction term exactly. If β is discontinuous, then the correction terms can be determined iteratively using the approximation of T^{n+1} and T^n or from an interpolation scheme from previous time levels.

11.3.2 The modified ADI method for Stefan problems

While the Crank–Nicolson method is simple and second-order accurate, the disadvantage is that we need to solve a linear system of equations, which may take significant computational time. To speed up the process, the following ADI method, which requires only solving a sequence of tridiagonal systems, has shown to be very efficient. This method is unconditionally stable in the asymptotic sense, as discussed later in this section. Since T_M is known at the interface, the ADI method for Stefan problems is simpler than for the interface problem described in §9.3. The ADI method can be written as

$$\frac{T_{ij}^{n+1/2} - T_{ij}^n}{\Delta t/2} = \delta_{xx} T_{ij}^{n+1/2} + \delta_{yy} T_{ij}^n + Q_{ij}^n, \tag{11.44}$$

$$\frac{T_{ij}^{n+1} - T_{ij}^{n+1/2}}{\Delta t/2} = \delta_{xx} T_{ij}^{n+1/2} + \delta_{yy} T_{ij}^{n+1} + Q_{ij}^{n+1}, \tag{11.45}$$

where $\delta_{xx} T_{ij}$ and $\delta_{yy} T_{ij}$ are again discrete analogues of $(\sigma T_x)_x$ and $(\sigma T_y)_y$, respectively, which should depend on the indices i and j. We drop the dependencies i and j for simplicity. At regular grid points, these finite difference operators are standard 3-point centered finite difference formulas. At irregular grid points, they are determined from the one-sided difference formula. In other words, during a sweep, a large tridiagonal system can break up into several subtridiagonal systems. Q_{ij}^n is the correction term when the interface crosses the grid line from time t^n to $t^{n+1/2}$ as we discussed in the previous section.

The error analysis of the ADI method

We have the following theorem about the convergence of this ADI method.

Theorem 11.2. *The truncation errors of the ADI method described above are $O(h^2)$ at regular grid points, and at least of order $O(1)$ at irregular grid points. Therefore the global convergence is at least first order in the infinity norm.*

Proof: By adding (11.45) to (11.44) and subtracting (11.45) from (11.44), after some manipulations, we have

$$\begin{aligned}\frac{T_{ij}^{n+1} - T_{ij}^{n}}{\Delta t} - Q_{ij}^{n} - Q_{ij}^{n+1} &= \frac{1}{2}\delta_{xx}\left(T_{ij}^{n} + T_{ij}^{n+1}\right) \\ &+ \frac{1}{2}\left(\delta_{yy}T_{ij}^{n} + \delta_{yy}T_{ij}^{n+1}\right) + \frac{\Delta t}{4}\delta_{xx}\left(\delta_{yy}^{n}T_{ij}^{n} - \delta_{yy}T_{ij}^{n+1}\right).\end{aligned} \tag{11.46}$$

The left-hand side approximates T_t to first-order accuracy. The one-sided finite difference operator is continuous, so $\frac{1}{2}\delta_{xx}(T_{ij}^{n}+T_{ij}^{n+1})$ approximates $(\sigma T_x)_x$ to first order also. Similarly $\frac{1}{2}(\delta_{yy}T_{ij}^{n} + \delta_{yy}T_{ij}^{n+1})$ approximates $(\sigma T_y)_y$ to first order. Unfortunately, the last term in the equation above is $O(1)$. Thus at the worst, the local truncation error is $O(1)$ at irregular grid points. Since the interface is one dimension lower than the computational domain, the global error is still at least first-order accurate. This completes the proof. □

Remark 11.1. *As we did in* §9.3.4 *and as in* [172], *we are able to add another correction term so that the truncation error in the ADI method is $O(h)$. However, as observed in* §9.3.4 *and* [172], *even without correction terms, the numerical results exhibit second-order convergence globally in the infinity norm, while the local truncation error is $O(1)$. This has also been verified numerically. Hence, it is not necessary to spend the effort to force the truncation error to be of order $O(h)$ at irregular grid points. The reason, we believe, is due to cancellations of errors.*

The stability of the ADI method

To show the stability of the ADI method, it is enough to assume homogeneous jump conditions, which means that the correction terms are zero. Under such an assumption, we have the following theorem.

Theorem 11.3. *Assuming that the interface cuts through the grid lines $x = x_i$ and $y = y_j$, no more than once in the central 5-point stencil between time levels t^n, $t^{n+1/2}$, $t^{n+1/2}$, and t^{n+1}; then the ADI method is at least asymptotically stable in terms of the von Neumann stability as h approaches zero.*

Proof: Without correction terms, the ADI method can be written in the following operator form:

$$\left(I - \frac{1}{2}\Delta t\,\delta_{xx}\right)T^{n+1/2} = \left(I + \frac{1}{2}\Delta t\,\delta_{yy}\right)T^{n}, \tag{11.47}$$

$$\left(I - \frac{1}{2}\Delta t\,\delta_{yy}\right)T^{n+1} = \left(I + \frac{1}{2}\Delta t\,\delta_{xx}\right)T^{n+1/2}. \tag{11.48}$$

Using the standard stability analysis technique [202, 245] we set

$$T^n \sim \lambda^n e^{ik_x x + k_y y}. \tag{11.49}$$

Substituting this into the first equation in (11.47), we have

$$\sqrt{\lambda} = \frac{1 + \frac{\Delta t}{2}\left(\gamma_1^n e^{-ik_y h} - \gamma_2^n + \gamma_3^n e^{ik_y h}\right)}{1 - \frac{\Delta t}{2}\left(\gamma_1^{n+1/2} e^{-ik_x h} - \gamma_2^{n+1/2} + \gamma_3^{n+1/2} e^{ik_x h}\right)}. \tag{11.50}$$

Similarly, we have the following equation from (11.48):

$$\sqrt{\lambda} = \frac{1 + \frac{\Delta t}{2}\left(\gamma_1^{n+1/2} e^{-ik_x h} - \gamma_2^{n+1/2} + \gamma_3^{n+1/2} e^{ik_x h}\right)}{1 - \frac{\Delta t}{2}\left(\gamma_1^{n+1} e^{-ik_y h} - \gamma_2^{n+1} + \gamma_3^{n+1} e^{ik_y h}\right)}, \tag{11.51}$$

where the $\{\gamma_k\}$'s are the coefficients of the finite difference scheme and are nonnegative. At regular grid points, they are simply k/h^2 for γ_1 and γ_3, and are $2k/h^2$ for γ_2 if the heat conductivity is a constant. If the interface cuts through the grid point, then γ_1 or γ_3 is zero. The amplification factor is

$$\begin{aligned}\lambda &= \frac{1 + \frac{\Delta t}{2}\left(\gamma_1^{n+1/2} e^{-ik_x h} - \gamma_2^{n+1/2} + \gamma_3^{n+1/2} e^{ik_x h}\right)}{1 - \frac{\Delta t}{2}\left(\gamma_1^{n+1/2} e^{-ik_x h} - \gamma_2^{n+1/2} + \gamma_3^{n+1/2} e^{ik_x h}\right)} \\ &\times \frac{1 + \frac{\Delta t}{2}\left(\gamma_1^n e^{-ik_y h} - \gamma_2^n + \gamma_3^n e^{ik_y h}\right)}{1 - \frac{\Delta t}{2}\left(\gamma_1^{n+1} e^{-ik_y h} - \gamma_2^{n+1} + \gamma_3^{n+1} e^{ik_y h}\right)} = \lambda_1 \lambda_2.\end{aligned} \tag{11.52}$$

At a regular grid point, we have

$$\lambda_1 = \frac{1 - 2\frac{\Delta t}{h^2}\sin^2\frac{1}{2}k_x h}{1 + 2\frac{\Delta t}{h^2}\sin^2\frac{1}{2}k_x h},$$

which is always between zero and one. At an irregular grid point, the difference scheme is the one-sided difference as described earlier. In the case depicted in Figure 11.9, we have $\gamma_3^{n+1/2} = 0$, and λ_1 becomes

$$\lambda_1 = \frac{1 - \frac{\Delta t}{2}\left(\gamma_2^{n+1/2} - \gamma_1^{n+1/2}\cos(k_x h) - i\,\gamma_1^{n+1/2}\sin(k_x h)\right)}{1 + \frac{\Delta t}{2}\left(\gamma_2^{n+1/2} - \gamma_1^{n+1/2}\cos(k_x h) - i,\gamma_1^{n+1/2}\sin(k_x h)\right)}.$$

Since $\gamma_2^{n+1/2} > \gamma_1^{n+1/2}$ from the one-sided interpolation scheme, we conclude that

$$\frac{\Delta t}{2}\left(\gamma_2^{n+1/2} - \gamma_1^{n+1/2}\cos(k_x h)\right) > 0.$$

Thus the modulus of λ_1 is always less than one.

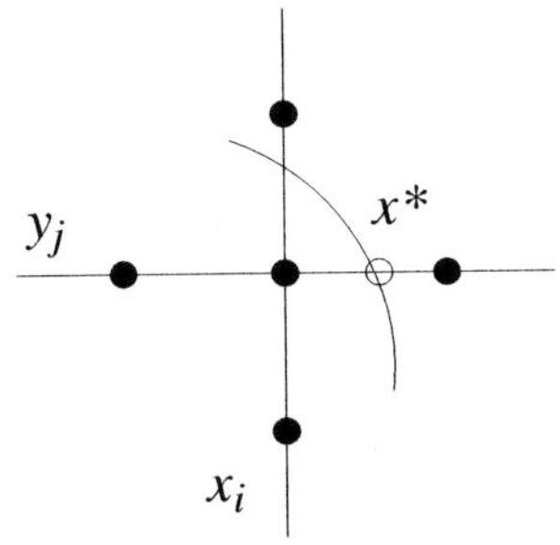

Figure 11.9. *A sketch of the geometry at an irregular grid point. The interface cuts through the grid line y_j at (x^*, y_j).*

The discussion of λ_2 is complicated by the fact that the coefficients are taken from different time levels. It is obvious that the real part of the denominator is always bigger than one, while the real part of the numerator is always smaller than one. We do not have clear knowledge about the imaginary parts. However, the coefficients of the finite difference scheme, either the standard scheme at regular grid points or the one-sided interpolation at irregular grid points, is a continuous function of time t. Therefore, if Δt is sufficiently small, then the coefficients $\{\gamma_l^n\}$ and $\{\gamma_l^{n+1}\}$ are sufficiently close, and we will still have $|\lambda_2| < 1$. That is why we can conclude only the asymptotic stability of the ADI method. □

Remark 11.2. *In practice, the ADI method starts the sweep in the x-direction, then in the y-direction. In the next time level, the method should switch first to the y-direction, and then to the x-direction. The advantage of this switching is that it keeps the symmetry of the algorithm, which should also have better stability according to the analysis above.*

For the modified ADI method, we can obtain a second-order accurate temperature and normal velocity. We use a second-order method to evolve the interface. For the Hamilton–Jacobi equation, we know that the CFL restriction is $\Delta t \sim h/|V_{\mathbf{n}}|$; therefore we can use a multistep method. The simplest one is the Adams–Bashforth scheme,

$$\varphi_{ij}^{n+1} = \varphi_{ij}^n + \frac{\Delta t}{2}\left(3R_{ij}^n - R_{ij}^{n-1}\right), \qquad (11.53)$$

where $R_{ij}^n = -|\nabla\varphi_{ij}^n|(V_n)_{ij}^n$. Often a third-order WENO scheme (see [124] for example) is used to compute $|\nabla\varphi_{ij}^n|$.

11.3.3 Numerical simulations of the Stefan problem

To verify and study the convergence rate of the hybrid IIM, we tested the method against an axisymmetric case, where the exact solution is available. The example is called a growing Frank sphere; see [9, 42]. The Stefan number is $S = 1$ in all the simulations. The symbol S is used differently in the following example.

Example 11.3. *The growing Frank sphere with condition* (11.33). This example often serves as a benchmark problem for moving interface and free boundary problems. In this example,

$\sigma = 1$, $L = 1$, the solid region is a cylinder of radius $R = S\sqrt{t}$, and the temperature is given by

$$T(r,t) = \begin{cases} T_\infty \left(1 - \dfrac{F(s)}{F(S)}\right) & \text{if } s > S, \\ 0 & \text{otherwise,} \end{cases} \tag{11.54}$$

where $r = \sqrt{x^2 + y^2}$, $s = r/\sqrt{t}$, and T_∞ is a given undercooling temperature. The function $F(s)$ is defined as

$$F(s) = E_1(s^2/4), \tag{11.55}$$

where E_1 is the special function

$$E_1(z) = \int_z^\infty \frac{e^{-t}}{t} dt. \tag{11.56}$$

The undercooling constant $T_\infty < 0$ and the parameter S are related by the jump condition $V_n = -[\frac{\partial T}{\partial \mathbf{n}}]$. After simple derivation, we can get

$$u_\infty = \frac{S}{2} \frac{F(S)}{F'(S)}. \tag{11.57}$$

If we take $S = 1.56$ as chosen in [9, 42], then

$$u_\infty = -0.499469490768 \cdots, \tag{11.58}$$

instead of -0.5 as used in [9, 42], to ensure that we have the exact solution for comparison. We set $t_0 = 1$ and $t = 1.5$. The computational domain is $-2 \le x, y \le 2$ and the radius range is between 1.5 and 1.9.

First, we want to know the convergence rate for the ADI method itself. Therefore, we use the *exact* velocity at irregular grid points and then extend the velocity to a tube surrounding the interface. Table 11.2 shows a grid refinement analysis with a reinitialization process. In the table, E_r is the infinity norm of the errors of the location of the interface, defined as

$$E_r = \max_{proj} \left\{ \left| \ |\mathbf{X}_{proj}| - r_{exact} \right| \right\} \tag{11.59}$$

at $t = 1.5$, where $\mathbf{X}_{proj}$ are the orthogonal projections of irregular grid points and $r_{exact} = S\sqrt{1.5}$. The second term E_T in the table is the error of the computed temperature at $t = 1.5$ in the infinity norm. Nearly second-order accuracy can be seen for both the location and the temperature. The results imply that the modified ADI method is indeed second-order accurate without the correction terms for the cross derivatives in (11.46). Also the extension algorithm for the velocity works well and the reinitialization algorithm has little effect on the accuracy for this problem.

Example 11.4. *Unstable crystal growth with condition* (11.34). We also tested the method for unstable crystal growth. The computation domain is a rectangular box, $-4 \le x, y \le 4$,

Table 11.2. *A grid refinement analysis using the modified ADI method in the infinity norm,* $t_0 = 1$, $t = 1.5$.

N	E_r	Ratio	E_T	Ratio
40	5.2846×10^{-2}		2.4511×10^{-2}	
80	1.6037×10^{-2}	3.2952	5.5328×10^{-3}	4.4301
160	4.970×10^{-3}	3.2266	1.3900×10^{-3}	5.6210
320	1.032×10^{-3}	4.8161	3.8394×10^{-4}	5.6947

with the insulated boundary condition for the temperature field, that is, $\partial T/\partial \mathbf{n} = 0$. The initial front between the liquid and the solid is given in polar coordinates as

$$r = r_0 + \epsilon \cos(l\theta), \qquad 0 \le \theta \le 2\pi. \tag{11.60}$$

The enclosed solid seed has initial temperature higher than the liquid outside the interface. The temperature field for the liquid is given initially as

$$T_{ij} = \begin{cases} T_\infty & \text{if } \varphi_{ij} < -l_w, \\ -T_\infty \left(2 + \sin(\varphi_{ij}\pi/2l_w)\right) & \text{if } |\varphi_{ij}| \le l_w, \\ -T_\infty & \text{otherwise,} \end{cases} \tag{11.61}$$

where T_∞ and l_w are two positive parameters. Thus the initial temperature field is continuous. In the tests, the time step is taken as $\Delta t = h/8$ initially and is adjusted according to the magnitude of the normal velocity. Latent heat is set to $L = 0.8$ and the transition width is set to $l_w = 5h$. The Stefan number and other parameters are $S = 1$, $T_\infty = 2.0$, and $l = 4$ for a symmetric interface, and $l = 5$ for general cases. The tube width used for the fast level set method is $5h$. Figure 11.10 shows the simulations with different surface tensions. As we decrease the surface tension, we see more structure. For Figures 11.10(a)–(c), the initial front is symmetric with respect to the axes, and the symmetry is very well preserved. In Figure 11.10(d), we see the tip-splitting forming from each branch for a general initial front. We refer readers to [175] for more numerical examples.

11.4 An application to an inverse problem of shape identification

The IIM has been successfully applied to a number of free boundary problems. One example is the simulation of electromigration of voids in integrated circuits [180], in which the motion depends on the surface Laplacian of electrical field and chemical potentials. In another application, we use a lubrication theory to formulate a model for the reactive spreading of drops that deposit an autophobic monolayer of surfactant on a surface in [113]. The success of these simulations is based on the fast IIM for Poisson/generalized Helmholtz equations on irregular domains described in §6.2.

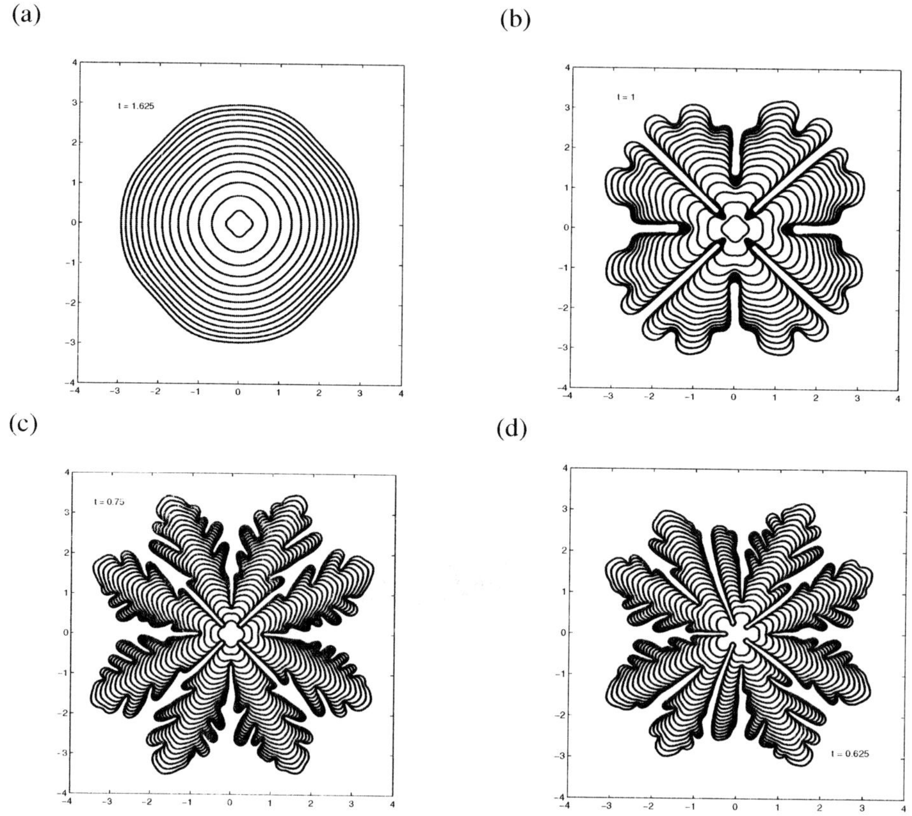

Figure 11.10. *Simulations of the crystal growth with different surface tensions for Example* 11.4. *The latent heat is* $L = 1/0.8 = 1.25$. (a) $\epsilon_C = 0.05$, $\epsilon_v = 10^{-5}$, $M = 160$, $l = 4$; (b) $\epsilon_c = 0.01$, $\epsilon_v = 0$, $M = 320$, $l = 4$; (c) $\epsilon_C = 0.001$, $\epsilon_v = 0.001$, $M = 320$, $l = 4$; (d) $\epsilon_C = 10^{-4}$, $\epsilon_v = 0$, $M = 320$, $l = 5$. *As the surface tension* ϵ_C *decreases, we see that more dendrites are formed.*

In this section, we use an example of an inverse problem in shape identification to explain the hybrid method of the fast IIM for problems on irregular domains and how to combine it with the level set method. In this application, the level set method is used as a tool since the intermediate zero level sets do not have physical meanings, but rather approximations to the boundary that is to be identified. The mathematical model and a numerical method are studied in [116].

Let Ω be a given domain that can be decomposed into two disjoint subdomains, Ω^+ and Ω^-, and where the value of the conductivity β is known on each of them. Let u be the potential which is the solution of

$$\begin{aligned} &-\nabla \cdot (\beta(\mathbf{x})\, \nabla u) = 0, \\ &\frac{\partial u}{\partial \mathbf{n}} = g \quad \text{on } \partial\Omega. \end{aligned} \tag{11.62}$$

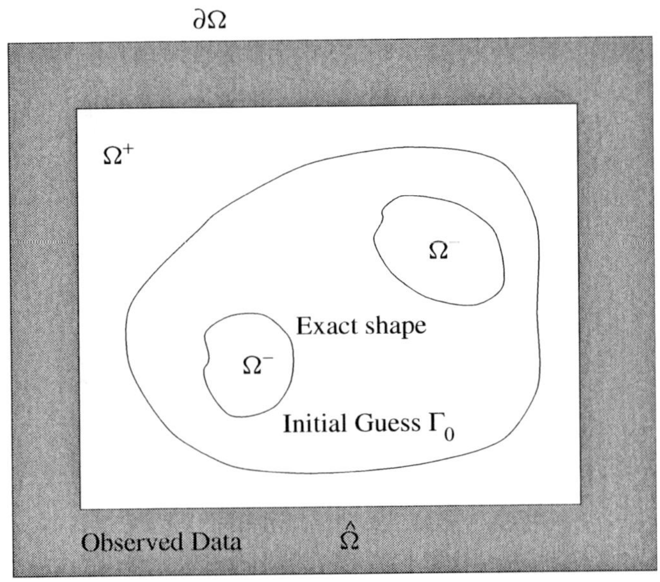

Figure 11.11. *A diagram of the inverse problem of identifying an unknown shape from observed data $\hat{\Omega}$ (shaded area).*

The conductivity β is assumed to be piecewise constant and is defined by

$$\beta(x) = \begin{cases} \beta^+, & x \in \Omega^+, \\ \beta^-, & x \in \Omega^-. \end{cases}$$

We assume that Ω^- is the finite union of simply connected open sets in Ω. Their boundary Γ represents the interface between the two open domains Ω^+ and Ω^-. It is assumed to be the union of closed C^2 curves. Let $\hat{\Omega}$ be the region of observation defined by

$$\hat{\Omega} = \{x \in \Omega : \text{ dist } (x, \partial\Omega) \le \alpha\}, \tag{11.63}$$

$\alpha > 0$, on which data of the potential function u are assumed to be available. We further assume that the interface Γ is strictly contained in Ω; see Figure 11.11 for an illustration.

We consider the inverse problem of identifying an unknown interface Γ from the observation z of u on $\hat{\Omega}$. Given the interface Γ, let $u(\Gamma)$ denote the solution to the boundary value problem (11.62). We formulate the least squares problem

$$\min J(\Gamma) = \int_{\hat{\Omega}} \frac{1}{2} |u(\Gamma) - z|^2 \, dx + \epsilon \int_{\Gamma} 1 \, ds \quad \text{over } \Gamma \in Q_{ad}, \tag{11.64}$$

where $\epsilon > 0$ and Q_{ad} is the admissible class of interfaces defined uniformly bounded away from $\hat{\Omega}$. The second term in (11.64) is referred to as the regularization term, with ϵ the regularization parameter. The quantity used for regularization here is the arc length.

We consider the case when $\beta^- = \infty$ and thus, in the case when Ω^- consists of only one connected component, the boundary value problem (11.62) reduces to

$$-\Delta u = 0 \quad \text{in } x \in \Omega^+, \tag{11.65}$$

with boundary conditions

$$u = 0 \quad \text{on } \Gamma \qquad \text{and} \qquad \frac{\partial u}{\partial \mathbf{n}} = g \quad \text{on } \partial\Omega, \tag{11.66}$$

where Γ describes the shape that we are looking for.

In order to determine the steepest descent direction for the functional, we consider the shape derivative of $u(\Gamma)$ and $J(\Gamma)$ with respect to Γ for the problem (11.65)–(11.66). Let $\Gamma \in Q_{ad}$ be fixed and, for $|t|$ sufficiently small, let $\Omega_t = F_t(\Omega^+)$ be the image of Ω^+ obtained by the mapping $F_t : R^2 \to R^2$ defined as

$$F_t(x_1, x_2) = (x_1, x_2) + t\, h(x_1, x_2).$$

For functions $u \in H^1(\Omega^+)$ and $u_t \in H^1(\Omega_t)$, the material derivative of u for the field $h \in H^1(\Omega^+)^2$ is given by

$$\dot{u}(\mathbf{x}) = \lim_{t\to 0} \frac{u_t(\mathbf{x} + t\,\mathbf{h}(\mathbf{x})) - u(\mathbf{x})}{t} \quad \text{for } \mathbf{x} \in \Omega^+. \tag{11.67}$$

If $u_t(\mathbf{x})$ has a regular extension to a neighborhood of $\bar{\Omega}_t$, then

$$u'(\mathbf{x}) = \lim_{t\to 0} \frac{u_t(\mathbf{x}) - u(\mathbf{x})}{t} = \dot{u}(\mathbf{x}) - \mathbf{h}(\mathbf{x}) \cdot \nabla u(\mathbf{x}), \quad \mathbf{x} \in \Omega^+, \tag{11.68}$$

is called the shape derivative of u. These notions are standard in the theory of shape optimization; see [101, 116, 241], for example.

It can be shown that the shape derivative u' of the solution u to (11.65)–(11.66) is given by

$$-\Delta u' = 0 \tag{11.69}$$

with boundary conditions

$$u' + \frac{\partial u}{\partial \mathbf{n}} (\mathbf{n} \cdot h) = 0 \quad \text{on } \Gamma \qquad \text{and} \qquad \frac{\partial u'}{\partial \mathbf{n}} = 0 \quad \text{on } \partial\Omega. \tag{11.70}$$

For the proof we refer to the above mentioned references.

Let $p \in H^1(\Omega^+)$ be the adjoint function satisfying

$$-\Delta p = (u - z)\chi_{\hat{\Omega}} \tag{11.71}$$

with boundary conditions

$$p = 0 \quad \text{on } \Gamma \qquad \text{and} \qquad \frac{\partial p}{\partial \mathbf{n}} = 0 \quad \text{on } \partial\Omega, \tag{11.72}$$

where $\chi_{\hat{\Omega}}$ is the indicator function of the domain $\hat{\Omega}$. We denote $\Gamma_t = \{\mathbf{x} : \varphi(\mathbf{x}+t\,\mathbf{h}(\mathbf{x})) = 0\}$ and set $\epsilon = 0$ for simplicity of the analysis. Since Γ is uniformly bounded away from $\hat{\Omega}$, for $\Gamma \in Q_{ad}$, we have

$$\lim_{t\to 0} \frac{J(\Gamma_t) - J(\Gamma)}{t} = \int_{\hat{\Omega}} (u(\mathbf{x}) - z(\mathbf{x}), u'(\mathbf{x}))\,dx$$
$$= \langle -\Delta p, u' \rangle = \int_\Gamma (\nabla u \cdot \nabla p)\,(\mathbf{n}\cdot\mathbf{h})\,ds,$$

where we used the fact that the tangential derivatives $u_\tau = p_\tau = 0$. Thus the steepest descent direction $\mathbf{V}$ of J is given by

$$\mathbf{V}(\mathbf{x}) = -(\nabla u \cdot \nabla p)\,\mathbf{n} \quad \text{on } \Gamma. \tag{11.73}$$

The level set method allows the interface Γ to move along the direction $\mathbf{V}$ within a neighborhood of Γ. To derive the equation for the level set function, let $\varphi(t,\cdot)$ denote a family of functions from R^2 to R. If $\mathbf{x} \to \varphi(t,\mathbf{x})$ is perturbed to $\mathbf{x} \to \varphi(t, \mathbf{x} + t\,\mathbf{h}(\mathbf{x}))$, by differentiating the level contours $\{\mathbf{x} : \varphi(t, \mathbf{x}+t\,\mathbf{h}(\mathbf{x})) = \text{const}\}$ with respect to t, we obtain

$$\varphi_t + \mathbf{h}\,\nabla\varphi = 0.$$

To realize a gradient technique we choose $\mathbf{h}(\mathbf{x}) = -\mathbf{V}(\mathbf{x})$. Note that $\mathbf{n} = \nabla\varphi/|\nabla\varphi|$ at Γ. This leads to the Hamilton–Jacobi equation for the level set function $\varphi = \varphi(t,\mathbf{x})$,

$$\varphi_t - V_{\mathbf{n}}(t,\mathbf{x})\,|\nabla\varphi| = 0, \tag{11.74}$$

where the normal velocity $V_{\mathbf{n}}(t,\mathbf{x})$ is defined by

$$V_{\mathbf{n}}(t,\mathbf{x}) = \nabla u \cdot \nabla p \quad \text{on } \Gamma, \tag{11.75}$$

and $(u, p) \in H_0^1(\Omega_t^+) \times H^1(\Omega_t^+)$ satisfies

$$-\Delta u = 0 \qquad \text{and} \qquad -\Delta p = (u - z)\chi_{\hat{\Omega}} \tag{11.76}$$

with boundary conditions

$$u = p = 0 \quad \text{on } \Gamma_t \qquad \text{and} \qquad \frac{\partial u}{\partial \mathbf{n}} = g, \quad \frac{\partial p}{\partial \mathbf{n}} = 0 \ \text{ on } \partial\Omega. \tag{11.77}$$

The Hamilton–Jacobi equation is used within an iterative scheme. Given a current zero level set Γ^k, the initial condition for (11.74) is chosen such that $\varphi = 0$ on Γ^k. Equation (11.74) is solved in a computational tube surrounding Γ^k with no inflow boundary condition.

The sensitivity of the cost J at $\epsilon = 0$ can be obtained from $\epsilon > 0$,

$$\lim_{t\to 0} \frac{J(\Gamma_t) - J(\Gamma)}{t} = \int_\Gamma (\nabla u \cdot \nabla p + \epsilon\,\kappa)\,(n \cdot \mathbf{h})\,ds,$$

where κ denotes the curvature along Γ. Here we used the fact that

$$\frac{d}{dt}\int_{\Gamma_t} = \int_\Gamma \kappa\,(\mathbf{n}\cdot\mathbf{h})\,ds.$$

This in turn results in $V_{\mathbf{n}}(t,\mathbf{x}) = \nabla u \cdot \nabla p + \epsilon\,\kappa$ on Γ. From the literature on mean curvature flow, it is known that the effect of the curvature term results in a tangential smoothing of the interface Γ; see, e.g., [94].

11.4.1 An outline of the algorithm for the inverse problem

Given data on the domain $\hat{\Omega}$, and g along $\partial\Omega$, the algorithm for identifying the unknown Γ inside the domain Ω is outlined below.

- Set an initial level set function $\varphi(x, y)$ as an initial guess of the envelope of the unknown shape $\Gamma^0 = \{\varphi(x, y) = 0\}$.

For $k = 0, k = 1, \ldots$, do the following until the algorithm converges.

- Solve the Laplace equation (11.65)–(11.66) in $(\Omega^k)^+$ for $u(\Gamma^k)$, where we use k to indicate the kth step.
- Compute the difference of the computed solution with the observed data, that is, $(u(\Gamma^k) - z)\chi_{\hat{\Omega}}$.
- Solve the Poisson equation (11.71)–(11.72) in $(\Omega^k)^+$.
- Evaluate the normal velocity Γ_k using the weighted least squares interpolation to get

$$V_{\mathbf{n}}^k = \nabla p^k \cdot \nabla u^k + \epsilon\kappa(\Gamma^k), \tag{11.78}$$

where ϵ is a regularization parameter and $\kappa(\Gamma^k)$ is the curvature of the boundary of the shape Γ^k.

- Extend the velocity $V_{\mathbf{n}}^k$ to a computational tube $|\varphi^k| \le \delta$, where δ is the width of the tube; see Figure 11.2.
- Update the level set function by solving the Hamilton–Jacobi equation $\varphi_t^k - V_n^k|\nabla\varphi^k| = 0$ for φ^k using

$$\frac{\varphi^{k+1} - \varphi^k}{\Delta t} - V_{\mathbf{n}}^k|\nabla\varphi^k| = 0, \tag{11.79}$$

where Δt is chosen as

$$\Delta t = \min\left\{ \frac{h}{2\|V_n^k\|_\infty}, \frac{h}{2} \right\}. \tag{11.80}$$

- Check convergence. Options include repeating the process, stopping if convergence criteria are satisfied, and starting from another initial level curve. Update $\Gamma^k = \{\varphi^k(x, y) = 0\}$ and Ω^k.

11.4.2 Identifying several minima

For the minimization problem (11.64), there can be several local and global minima. Starting with an initial guess that contains all possible expected shapes, the algorithm typically finds quickly an *envelope* of all shapes. The following algorithm is successful in some situations for determining possible further minima inside this envelope.

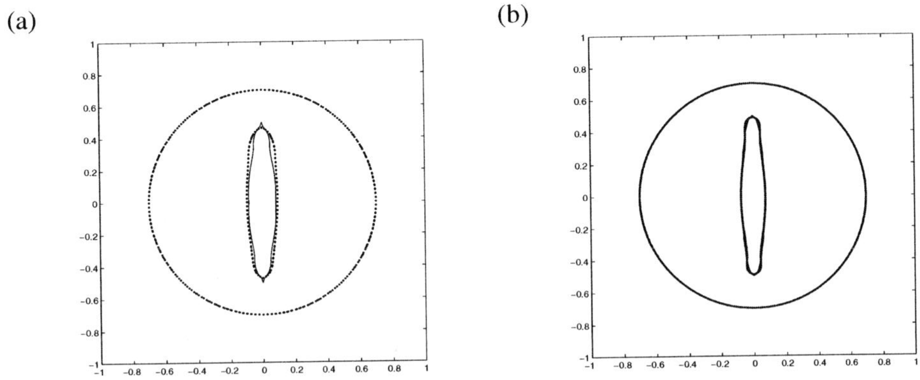

Figure 11.12. *Computed shape boundaries with* $\epsilon = 10^{-6}$. *The initial guess is a big circle of radius* 0.6 *centered at the origin.* (a) *The result obtained using an* 80×80 *grid.* (b) *The result obtained using a* 160×160 *grid.*

We denote the level set function of the first equilibrium, the envelope, by $\varphi^e(x, y)$ and set $\varphi_{min} = \min\{\varphi(x, y)\}$. Then the algorithm is repeated with the initial level set function chosen by shifting $\varphi^e(x, y)$. That is, given a step size $\Delta\varphi = |\varphi_{min}|/L$, we use the level set functions

$$\varphi^0(x, y) = \varphi^e(x, y) + k\Delta\varphi, \quad k = 1, 2, L-1, \tag{11.81}$$

as initial guesses to repeat the algorithm $L - 1$ times. In this way, we can search for possible multiple local or global minima.

11.4.3 Numerical examples of shape identification

We provide some numerical examples here of the inverse problem of shape identification. The domain is $[-1, 1] \times [-1, 1]$ unless specified otherwise. The boundary condition is taken as $g = 1$ along $x = -1$ and $y = -1$, and as $g = -1$ along $x = 1$ and $y = 1$. Unless otherwise specified, the width of the observed data is $\alpha = 0.2$. More examples can be found in [116].

Example 11.5. *A grid refinement analysis for a known shape.* First we test the hybrid method for a single object in the domain. The exact shape is a skinny ellipse,

$$\frac{x^2}{0.07^2} + \frac{y^2}{0.5^2} = 1, \tag{11.82}$$

which is the solid line in Figures 11.12(a), (b). Figure 11.12(a) gives the result computed on an 80×80 grid and Figure 11.12(b) depicts the result computed using a 160×160 grid.

The initial guess is a circle that surrounds the exact shape. The circle shrinks quickly in the radial direction to a small circle. Then it gradually extends in the direction of the major axis and slowly expands in the direction of the minor axis. The average number of GMRES iterations is between 4 and 10. Each iteration requires two calls to the Poisson

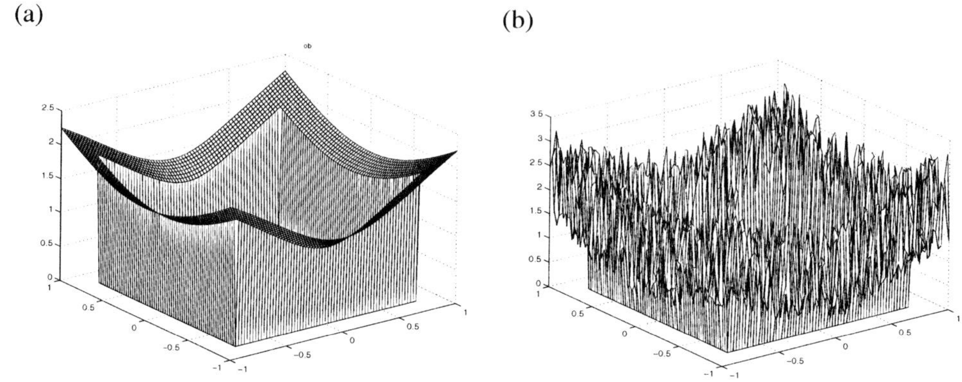

Figure 11.13. (a) *Exact "observed data" $-u$ without noises used in Figure* 11.12(a). (b) *The observed data $-u$ with 68% relative uniform noise level.*

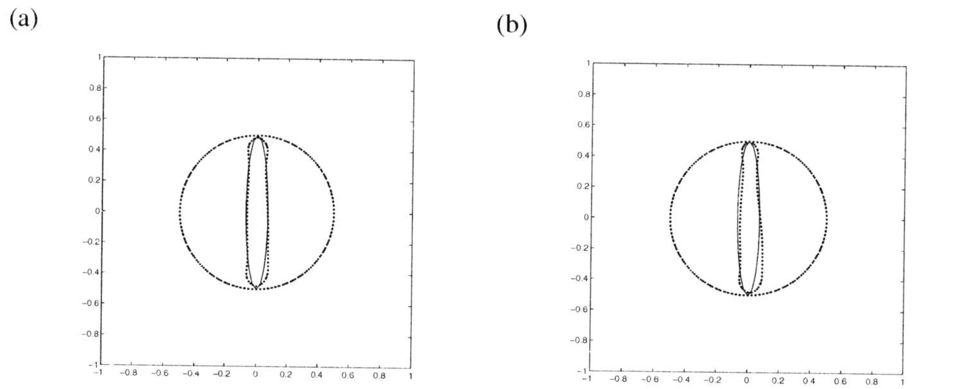

Figure 11.14. *Computed shape boundaries with noisy data.* (a) *With* 17% *uniformly distributed relative noise level.* (b) *With* 68% *uniformly distributed relative noise level.*

solver, one for the primal variable u and one for the adjoint variable p. It is observed that the algorithm is very efficient for single objects in the domain.

Example 11.6. *Noise analysis.* In practice, the observed data are likely corrupted by noise. We, therefore, also tested the algorithm with data given by $z_{i,j} = u(\Gamma^*)_{i,j} + \delta_{i,j}$, where Γ^* denotes the "true" interface and $\delta_{i,j}$ is chosen as a uniformly distributed random noise.

In the example, the infinity norm of the perturbation is $||\delta||_\infty = 48.8332$ and the relative perturbation is $||\delta||_\infty/||u_{\Gamma^*}||_\infty = 0.6768 \approx 68\%$. In Figure 11.13(a) we show the unperturbed data, while Figure 11.13(b) gives the observed data with a 68% relative noise level. Figure 11.14(b) is a computed result using the noise data with a 68% relative noise level compared with Figure 11.14(a) at 17%. While ϵ was chosen to be 10^{-6} in the noise-free examples, we chose it to be 10^{-4} when noise is present. We can see that the hybrid method also works well in the presence of noise.

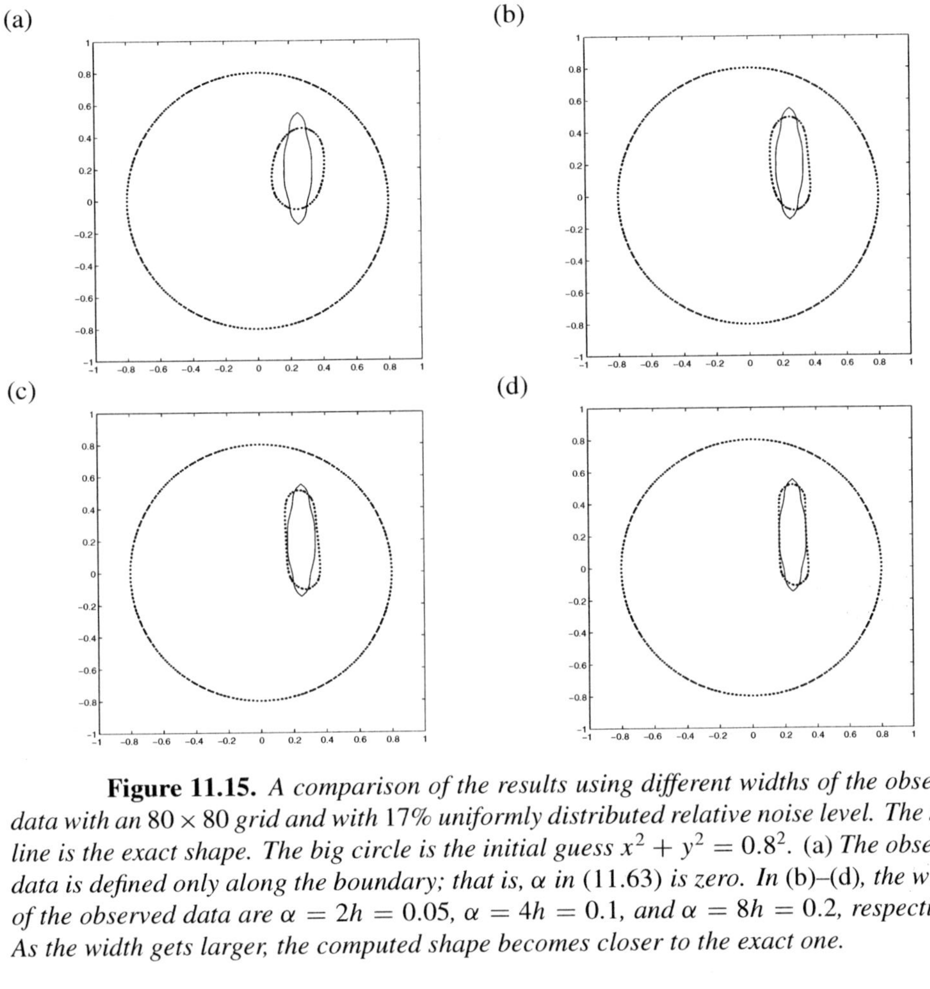

Figure 11.15. *A comparison of the results using different widths of the observed data with an* 80×80 *grid and with 17% uniformly distributed relative noise level. The solid line is the exact shape. The big circle is the initial guess* $x^2 + y^2 = 0.8^2$. (a) *The observed data is defined only along the boundary; that is,* α *in* (11.63) *is zero. In* (b)–(d), *the widths of the observed data are* $\alpha = 2h = 0.05$, $\alpha = 4h = 0.1$, *and* $\alpha = 8h = 0.2$, *respectively. As the width gets larger, the computed shape becomes closer to the exact one.*

Example 11.7. *The effect of the width of the observed data on the solution.* Numerical tests confirm the conjecture that the wider the observation region is, the better the reconstruction of the shape becomes. In fact, if the observed data are noise free and the region of observation is the entire domain, we get the exact solution in one step.

In Figure 11.15, we show a comparison of the computed solution with different widths of the observed data. The exact shape boundary is the zero level set implicitly defined by φ_{ij} obtained after the reinitialization process applied to the original level set function $\varphi(x, y) = \frac{(x-0.25)^2}{0.07^2} + \frac{(y-0.2)^2}{0.35^2} - 1$. The regularization parameter is $\epsilon = 10^{-3}$. In Figure 11.15(a), the observed data are defined *only* on the boundary; the computed shape roughly agrees with the exact shape but not as well as that in Figure 11.15(b), where the width of the observed data is $8h = 0.2$ along the boundary. The result is further improved in Figure 11.15(c) for which the width is $\alpha = 4h = 0.1$. Increasing the width further to $\alpha = 8h = 0.2$ results in only marginal improvement in Figure 11.15(d). We also tested the case in which the exact

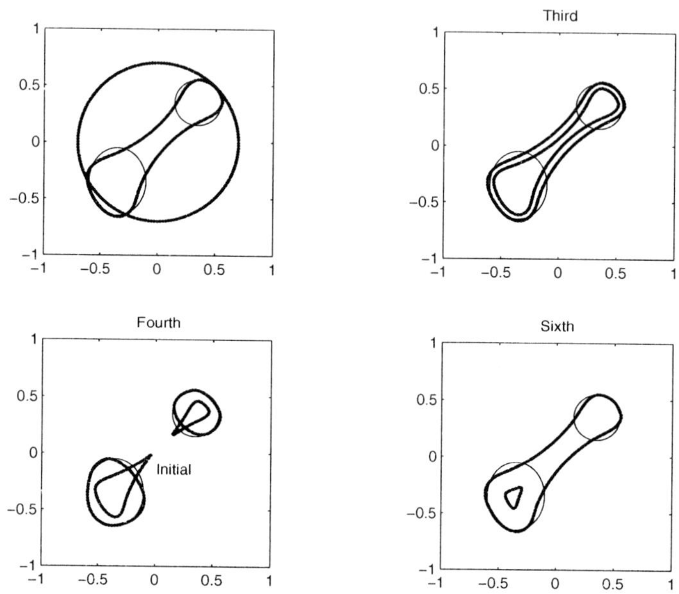

Figure 11.16. *Computed results using automatic adjusting of the initial zero level set function with an* 80×80 *grid. The initial guess is the circle* $r = 0.7$ *in the top left graph. The initial level set function is* $\varphi = \varphi_{env} + 0.06$ *in the top right graph, where* φ^{env} *is the level set function corresponding to the envelope. The initial level set function is* $\varphi = \varphi^{env} + 0.09$ *in the bottom left graph. The initial level set function is* $\varphi = \varphi^{env} + 0.18$ *in the bottom right graph.*

shape is a circle or fat ellipse. In these cases, the effect of the width is not as obvious as for the skinny ellipse in Figure 11.15.

Example 11.8. *An example with two objects.* Now we consider an example that has two objects in the domain. The exact shape is composed of the following two closed curves:

$$\frac{(x+0.35)^2}{0.25^2} + \frac{(y+0.35)^2}{0.3^2} = 1, \tag{11.83}$$

$$\frac{(x-0.35)^2}{0.2^2} + \frac{(y-0.35)^2}{0.2^2} = 1. \tag{11.84}$$

The numerical algorithm soon produces an envelope of the two objects and it can be checked numerically that the envelope is indeed a local minimum of the variational form (11.64). Now we proceed with the idea explained in §11.4.2 and denote the level set function corresponding to the envelope by φ^{env}. Then we use

$$\varphi^{env} + k\Delta\varphi, \quad k = 1, 2, \ldots, \tag{11.85}$$

as new initial guesses to see if we can find further local extrema. In the example, $\Delta\varphi = 0.03$. Figure 11.16 shows plots of the computed objects with different values of k. Figure 11.16(a) is the result of φ^{env} from an initial big circle of radius 0.7 centered at the origin. In Figure 11.16(b), the result is obtained from the initial level set function that was perturbed by a quantity 0.06 from φ^{env}. The approximate shape boundary is attracted back to the envelope. It appears that the envelope is a stable local extremum. In Figure 11.16(c), the initial zero level set function was perturbed further by a quantity 0.09 from φ^{env}. The initial boundary is the union of two closed curves that contain the exact boundary. The zero level set function converges to the desired shape with reasonable accuracy. In Figure 11.16(d), the initial zero level set function was perturbed by a quantity 0.18 from φ^{env} and it contains only one closed boundary. The zero level set function produced by the algorithm again approaches the envelope.

We can check the residual of the variational form to see which equilibrium is closer to the global minimum. For the results corresponding to the envelopes depicted in Figures 11.16(a), (b), (d), the residual is about 1.3363×10^{-2}, whereas it is 5.80216×10^{-3} for the result in Figure 11.16(c). Since there are only two equilibrium states, we can conclude that the result in Figure 11.16(c) is a good approximation to the solution of the problem.

11.5 Applications to nonlinear interface problems

The IIM has been successfully applied to nonlinear elliptic interface problems in which the discontinuous coefficient depends on the solution and its gradient. Applications include the simulation of magnetorheological (MR) fluid containing iron particles [119] and the weighted minimal surface problem through a heterogeneous medium [171]. We discuss the MR fluid application below as a demonstration of the IIM.

In the application of the MR fluid, the potential is the solution to the nonlinear elliptic PDE,

$$-\nabla \cdot (\beta(\mathbf{x}, |\nabla u|)\nabla u) = f. \tag{11.86}$$

The coefficient $\beta(\mathbf{x}, |\nabla u|)$ usually represents a nonlinear constitutive law and depends on the local material property and composition. In the model for the MR fluid, u is the magnetic potential and (11.87) is the empirical constitutive law (Frohlich–Kennelly relation); see [187] and the references therein. The magnetic permeability of the inclusion, i.e., the iron particles, is given by (in nondimensionlized form)

$$\beta(\mathbf{x}, H) = \begin{cases} 1 + \dfrac{(\mu - 1)\, B^{sat}}{(\mu - 1)\, |\nabla u| + B^{sat}} & \text{if } \mathbf{x} \text{ is within the iron,} \\ \beta^{+} & \text{if } \mathbf{x} \text{ is within the fluid,} \end{cases} \tag{11.87}$$

where B^{sat} is the maximum attainable magnetization of the inclusion, β^{+} is a constant, and μ is the permeability of the iron. Across the interface between the MR fluid and the iron particles, the natural jump conditions

$$[u] = 0, \qquad [\beta u_{\mathbf{n}}] = 0 \tag{11.88}$$

hold if there are no sources or sinks present.

For the linear model in which β in (11.87) is approximated by a piecewise constant (see [186]), finite element methods based on body-fitted grids, or boundary integral methods, may be quite effective. For the nonlinear model, however, it is difficult to use a finite element or boundary integral method because the coefficient of (11.89) depends on the solution and a Green's function may not be available.

11.5.1 The substitution method

Given an initial guess u^0, the substitution method at the kth step is to solve the elliptic interface problem,

$$\begin{aligned} &-\nabla \cdot \left(\beta(\mathbf{x}, |\nabla u^k|)\, \nabla u^{k+1}\right) = f, \\ &\left[u^{k+1}\right] = 0, \qquad \left[\beta\left(\mathbf{x}, |\nabla u^k|\right)\, u_n^{k+1}\right] = 0 \end{aligned} \tag{11.89}$$

for u^{k+1} using the information u^k. The substitution method has shown to be globally convergent in [115, 119].

Without loss of generality, we set $u|_{\partial\Omega} = 0$. The weak form of (11.89) is given by

$$\left(\beta(x, |\nabla u|)\, \nabla u, \nabla\psi\right) - (f, \psi) = 0 \tag{11.90}$$

for all $\psi \in H_0^1(\Omega)$, where $(\cdot, \cdot)$ denotes the $L^2(\Omega)$ inner product. Let us define

$$\Phi(x, s) = \int_0^s \beta(x, \sqrt{\sigma})\, d\sigma, \qquad s \geq 0, \quad x \in \Omega.$$

Then for the β given in (11.88), $s \to \Phi(x, s^2)$ is convex and there are two constants c_1 and c_2, $c_1 \leq c_2 < \infty$, such that $c_1 \leq \beta(x, s) \leq c_2$, for all $s \in R^+$ and $x \in \Omega$. It can be shown using the methodology from [13], for example, that the solution u to (11.90) in $H_0^1(\Omega)$ is the unique minimizer of

$$J(u) = \int_\Omega \left(\frac{1}{2}\,\Phi\left(x, |\nabla u(x)|^2\right) - f(x)u(x)\right) d\mathbf{x} \tag{11.91}$$

over $H_0^1(\Omega)$. Since $s \to \Phi(x, s^2)$ is convex, we have the monotonicity

$$\left(\beta(x, |\nabla u|)\, \nabla u - \beta(x, |\nabla v|)\, \nabla v,\ \nabla u - \nabla v\right) \geq 0 \tag{11.92}$$

for all $u,\ v \in H_0^1(\Omega)$. With the above preparations, we have the following theorem, which is proved in [119].

Theorem 11.4. *Assume that*

$$s \in R^+ \to \beta(x, s) \in R^+ \tag{11.93}$$

is monotonically nonincreasing for all $x \in \Omega$. Let $\{u^k\} \in H_0^1(\Omega)$ be a sequence generated by

$$\left(\beta(x, |\nabla u^k|)\, \nabla u^{k+1},\ \nabla\psi\right) - (f, \psi) = 0, \qquad \psi \in H_0^1(\Omega). \tag{11.94}$$

Then $J(u_k)$ is monotonically decreasing and u^k converges strongly to u in $H_0^1(\Omega)$.

Proof: First we note that $c_1 \le \beta(x, |\nabla u^k|) \le c_2$. Thus (11.94) has a unique solution $u^{k+1} \in H_0^1(\Omega)$. Setting $\psi = u^{k+1} - u^k$ in (11.94), we can get

$$\left(\beta^k \nabla u^{k+1},\ \nabla u^{k+1} - \nabla u^k\right) - (f, u^{k+1} - u^k) = 0,$$

where $\beta^k(x) = \beta(x, |\nabla u^k|)$. Note that

$$\begin{aligned}&(\beta^k \nabla u^{k+1},\ \nabla u^{k+1} - \nabla u^k))\\ &\qquad = \frac{1}{2}\int_\Omega \Big((\beta^k, |\nabla u^{k+1} - \nabla u^k)|^2 + (\beta^k, |\nabla u^{k+1}| - \nabla u^k)|^2) \Big)\, d\mathbf{x}.\end{aligned}$$

Since

$$\Phi(x, s^2) - \Phi(x, t^2) - \beta(x, t)(s^2 - t^2) \le 0$$

for all $s, t \in R$ and $x \in \Omega$, which can be seen from (11.93), we have

$$\int_\Omega \beta^k\, |\nabla u^{k+1} - \nabla u^k|^2\, d\mathbf{x} \ge \int_\Omega \left(\Phi\left(x, |\nabla u^{k+1}|^2\right) - \Phi\left(x, |\nabla u^k|^2\right)\right) d\mathbf{x}.$$

Hence, we obtain

$$J(u^{k+1}) - J(u^k) + \frac{1}{2}\int_\Omega \beta^k\, |\nabla u^{k+1} - \nabla u^k)|^2\, d\mathbf{x} \le 0.$$

Summing this over k, we obtain

$$J(u^m) + \frac{1}{2}\sum_{k=1}^{m} c_1\, |u^k - u^{k-1}|_{H_0^1}^2 \le J(u_0)$$

for all $m \ge 0$. This implies that $u^{k+1} - u^k \to 0$ in $H_0^1(\Omega)$. Since

$$\left|\beta^k \nabla\left(u^{k+1} - u^k\right)\right|_{L^2} \le c_2\, \left|u^{k+1} - u^k\right|_{H_0^1},$$

we have

$$\Big(\beta(x, |\nabla u^k|)\, \nabla u^k,\ \nabla u^k\Big) - (f, u^k) \to 0 \tag{11.95}$$

as $k \to \infty$.

It follows from (11.94) that $|u^k|_{H_0^1} \le |f|_{H^{-1}}/c_1$, and thus there exists a weakly convergent subsequence (denoted by the same k) that converges to $u \in H_0^1(\Omega)$. Define the nonlinear map $A : H_0^1(\Omega) \to H^{-1}(\Omega)$ by

$$Au = -\nabla \cdot \Big(\beta(x, |\nabla u|)\, \nabla u\Big).$$

Then, it follows from (11.92) that

$$\langle Au^k - Av, u^k - v\rangle_{H^{-1}\times H_0^1} \ge 0$$

for all $v \in H_0^1(\Omega)$, where $\langle\cdot,\cdot\rangle$ denotes the dual product.

From (11.95), we have

$$\langle f - Av, u - v\rangle \geq 0 \quad \forall v \in H_0^1(\Omega).$$

Since A is monotone, semicontinuous, and coercive, and A satisfies the maximum principle, we conclude that $Au + f = 0$; see [12, 13]. The strong convergence follows from the fact that

$$\langle Au - Av, u - v\rangle_{H^{-1}\times H_0^1} \geq c_1 \, |u - v|^2_{H_0^1}. \tag{11.96}$$

□

11.5.2 Computing β and its derivatives

In each iteration, the maximum principle preserving scheme (see §3.5 and [166]) is used to solve the linearized elliptic interface equation (11.89). The key is to compute the variable coefficient accurately since it depends on the gradient of the approximate solution from the previous iteration. The coefficient β is needed at $(x_{i\pm\frac{1}{2}}, y_j)$ and $(x_i, y_{j\pm\frac{1}{2}})$ if (x_i, y_j) is a regular grid point in reference to the standard 5-point stencil, and at orthogonal projections of irregular grid points. This is the key step in the algorithm, and its implementation sometimes has a significant effect on the accuracy of the computed solution.

Computing β_{ij} at a regular staggered grid point

We discretize (11.91) by

$$\sum_{i,j} \left\{ \frac{1}{2}\,\Phi\left(x, |\nabla u|^2_{ij}\right) - f_{i,j}\, u_{i,j} \right\} h_x h_y, \tag{11.97}$$

where

$$2|\nabla u|^2_{ij} = \left(\frac{u_{i+1,j} - u_{ij}}{h_x}\right)^2 + \left(\frac{u_{ij} - u_{i-1,j}}{h_x}\right)^2 + \left(\frac{u_{i,j+1} - u_{ij}}{h_y}\right)^2 + \left(\frac{u_{ij} - u_{i,j-1}}{h_y}\right)^2 .$$

The reason for using the variational form instead of the PDE (Euler–Lagrange equation) is that the solution to the minimization problem is guaranteed to exist for the discrete system. The discretization actually is equivalent to the discrete Euler–Lagrange equation, which can be seen below.

At a regular grid point (x_i, y_j), we can get the discrete system of equations for u_{ij},

$$-\frac{1}{h_x}\left(\frac{\beta_{i+1/2,j}}{h_x}(u_{i+1,j} - u_{i,j}) - \frac{\beta_{i-1/2,j}}{h_x}(u_{i,j} - u_{i-1,j})\right)$$
$$-\frac{1}{h_y}\left(\frac{\beta_{i,j+1/2}}{h_y}(u_{i,j+1} - u_{i,j}) - \frac{\beta_{i,j-1/2}}{h_y}(u_{i,j} - u_{i,j-1})\right) + f_{i,j} = 0,$$

by differentiating the discrete variational form (11.97) with respect to u_{ij}, where

$$\beta_{i\pm 1/2,j} = \frac{1}{2}\left(\beta_{i\pm 1,j} + \beta_{i,j}\right) \quad \beta_{i,j\pm 1/2} = \frac{1}{2}\left(\beta_{i,j\pm 1} + \beta_{i,j}\right)$$

with

$$\beta_{ij} = \beta\left(\mathbf{x}_{ij}, |\nabla u|_{ij}\right),$$

and $\beta(x, |\nabla u|)$ is defined as in (11.87). It is not surprising that the discrete system of equations derived from the variational form is the same as the discrete system using the central finite difference equations for the Euler–Lagrange equation.

Computing β at irregular grid points

Given a point $\mathbf{X}^* = (x^*, y^*) \in \Omega^-$, which can be an irregular grid point (x_i, y_j) or the orthogonal projection of an irregular grid point, we need to interpolate U_{ij} to get a second-order approximation of u_x and u_y in computing $|\nabla u|$. The information then is used to evaluate β at the grid point from the definition of β in (11.87). The least squares interpolation scheme is used for this purpose, for example,

$$U_x^-(\mathbf{X}^*) = \sum_{\mathbf{x}_{ij}\in\Omega^-,\, |\mathbf{x}_{ij}-\mathbf{X}^*|\le R_\epsilon} \gamma_{ij}\, U_{ij}, \tag{11.98}$$

where $\mathbf{x}_{ij} = (x_i, y_j)$ and R_ϵ is a prechosen parameter. The summation is taken over the neighboring grid points of (x^*, y^*). To get a second-order interpolation, R_ϵ is chosen in such a way that at least six grid points from the Ω^- side are enclosed. If there are more than six grid points enclosed, we take the first n_s points closest to $\mathbf{X}^*$ from the set of $|\mathbf{x}_{ij} - \mathbf{X}^*| \le R_\epsilon$. Note that the coefficients $\{\gamma_{ij}\}$ depend on $\mathbf{X}^*$ and the iteration index k, but for simplicity, the dependency has been omitted. The system of equations for the coefficients $\{\gamma_{ij}\}$ is determined from the Taylor expansion and is solved by an SVD decomposition; see §6.1.3 for details about the least squares interpolation. If the interface is fixed, the coefficients of the interpolation scheme can be computed and stored outside of the iteration.

Computing β at the interface using the flux jump condition

Let $\mathbf{X}^* = (x^*, y^*)$ be the orthogonal projection of an irregular grid point on the interface. Let N_ϵ^+ and N_ϵ^- be the number of the grid points from the Ω^+ and Ω^- sides in the circle $|\mathbf{X}^* - \mathbf{x}_{ij}| \le R_\epsilon$, respectively. To compute β^-, we distinguish two different cases.

Case I: $N_\epsilon^+ \le N_\epsilon^-$. In this case, there are more grid points from Ω^- than from Ω^+. Thus we use the least squares interpolation scheme (11.98) to compute u_x^- and u_y^-, and determine β^- from $\beta^- = \beta^-(X^*, |\nabla u^-(X^*)|)$ defined in (11.87).

Case II: $N_\epsilon^+ > N_\epsilon^-$. In this case, we use the least squares interpolation scheme to compute u_x^+ and u_y^+. Using the local coordinates (1.34), the formulas

$$u_\xi^+ = u_x^+ \cos\theta + u_y^+ \sin\theta,$$

$$u_\eta^+ = -u_x^+ \sin\theta + u_y^+ \cos\theta$$

are used to get ∇u. Since u is continuous, we have $u_\eta^- = u_\eta^+$. To compute u_ξ^-, we use the flux jump condition

$$\beta^- u_\xi^- = \beta^+ u_\xi^+.$$

Using the constitutive law (11.87), we have a nonlinear equation of single variable for u_ξ^-, which is

$$\left(1 + \frac{(\mu - 1)\, B^{sat}}{(\mu - 1)\sqrt{(u_\eta^+)^2 + (u_\xi^-)^2} + B^{sat}} \right) u_\xi^- = \beta^+ u_\xi^+. \tag{11.99}$$

Since the left-hand side is monotone with respect to u_ξ^- and its range is R, there exists a unique solution in the interval $I(0, \beta^+ u_\xi^+)$. A few steps of the bisection method can be used to get a good approximation to the solution of the above nonlinear equation. Then Newton's method can be used to obtain a more accurate solution since it is easy to find the derivative of (11.99) with respect to u_ξ^-. Once we have solved u_ξ^-, we have $|\nabla u|$ as well since $u_\eta^- = u_\eta^+$. Thus we get $\beta^- = \beta^-(X^*, |\nabla u^-|)$.

Remark 11.3. *A one-sided interpolation scheme usually has a wider stencil. This may cause inaccuracy if the geometry of the interface is complicated (i.e., skinny ellipses). This difficulty is avoided when using the alternative approach above.*

Computing the derivative of β along the interface

To use the maximum principle preserving scheme (see §3.5 and [166]) as a black-box solver, we also need to compute the tangential and normal derivatives β_ξ^- and β_η^-. From the discussion of the maximum principle preserving scheme, we know that we need only compute these derivatives to first-order accuracy. Therefore, to compute β_x^- at a projection (x_i^*, y_j^*) corresponding to an irregular grid point (x_i, y_j), we use

$$\beta_x^- = \begin{cases} \dfrac{\beta_{i,j} - \beta_{i-1,j}}{h_x} & \text{if } \varphi_{i-1,j} \le 0 \text{ and } \varphi_{ij} \le 0, \\[2ex] \dfrac{\beta_{i+1,j} - \beta_{i,j}}{h_x} & \text{if } \varphi_{i+1,j} \le 0 \text{ and } \varphi_{ij} \le 0, \\[2ex] \dfrac{\beta_{i-1,j} - \beta_{i-2,j}}{h_x} & \text{if } \varphi_{ij} > 0 \text{ but } \varphi_{i-1,j} \le 0 \text{ and } \varphi_{i-2,j} \le 0, \\[2ex] 0 & \text{if } \varphi_{ij} > 0 \text{ and } \varphi_{i-1,j} \le 0 \text{ and } \varphi_{i+1,j} \le 0. \end{cases} \tag{11.100}$$

Note that the last situation can happen only when a relatively coarse grid is used and the shape of the particle either has large curvature or is skinny, Even if the last situation happens, it can occur only at a few grid points. For an elliptic equation, a point source of $O(1)$ usually affects the solution by a magnitude of $h^2 \log h$. Therefore, this will not affect global second-order accuracy.

11.5.3 Numerical experiments of MR fluids with particles

Example 11.9. *Validation against an exact solution.* Let the level set function be $\varphi(x, y) = \sqrt{x^2 + y^2} - \frac{1}{2}$; that is, the interface is the circle centered at the origin with radius 0.5. The

Table 11.3. *A grid refinement analysis for Example* 11.9. *The stopping criteria for the nonlinear solver and for the multigrid linear solver are both* 10^{-8}.

	Using flux jump			One-sided	
$M \times N$	$\Vert E_N \Vert_\infty$	Order	No. of Iter.	$\Vert E_N \Vert_\infty$	Ratio
80×40	2.6588×10^{-2}		13	3.0267×10^{-2}	
160×80	5.9796×10^{-3}	2.1526	12	7.5448×10^{-3}	2.0042
320×160	1.4047×10^{-3}	2.0898	12	2.5609×10^{-3}	1.5588
640×320	3.5189×10^{-4}	1.9971	12	9.6665×10^{-4}	1.4056

computational domain is a rectangle $-2 \le x \le 2$ and $-1 \le y \le 1$. The source term $f(x, y)$ is

$$-f(x,y) = \begin{cases} \dfrac{2}{r} & \text{if } \varphi(x,y) > 0, \\ 4\left(1 + \dfrac{B^{sat}(\mu - 1)}{2(\mu - 1)r + B^{sat}}\right) - \dfrac{4((\mu-1)^2 B^{sat} r}{(2(\mu - 1)r + B^{sat})^2} & \text{if } \varphi(x,y) \le 0, \end{cases}$$

where $r = \sqrt{x^2 + y^2}$. Note that $f(x, y)$ is discontinuous across the interface. The material parameters are

$$\beta^+ = 1, \qquad B^{sat} = \frac{9}{4}, \qquad \mu = 1 + \frac{5B^{sat}}{3}. \tag{11.101}$$

It is easy to check that the function

$$u(x,y) = \begin{cases} 2\sqrt{x^2 + y^2} - \dfrac{7}{8} & \text{if } x^2 + y^2 > \dfrac{1}{4}, \\ \left(x^2 + y^2\right)^{3/2} & \text{if } x^2 + y^2 \le \dfrac{1}{4} \end{cases} \tag{11.102}$$

is a solution to the elliptic interface problem (11.89)–(11.88) with the source term $f(x, y)$.

The solution (11.102) is used to provide the boundary condition on the boundary of $\partial\Omega$. Dirichlet, Neumann, and periodic boundary conditions along parts of or the entire boundary have all been tested.

In Table 11.3, we show a grid refinement analysis with the Dirichlet boundary condition at $x = -2$ and $x = 2$, and the Neumann boundary condition at $y = -1$ and $y = 1$. The second column shows the errors of the computed solution in the infinity norm, and the third column shows the approximate order of convergence. The results are computed with the flux jump condition (11.88) being used to compute the derivatives of β_ξ^- and β_η^- at projections of irregular grid points. Second-order convergence is achieved. In the fifth and

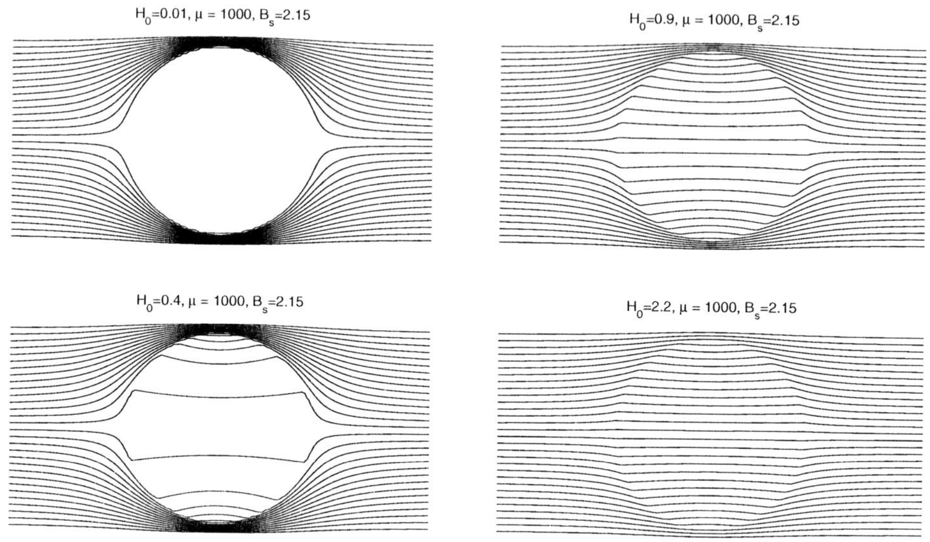

Figure 11.17. *Equipotential curves for the Frohlich–Kennelly model with applied fields, H_0 being* 0.01, 0.4, 0.9, *and* 2.2 *and* $B^{sat} = 2.25$ *with a* 256×128 *grid. The results in* [187] *have been reproduced.*

sixth columns, we show the results using the one-sided, second-order interpolation scheme to compute β_ξ^- and β_η^-. These results are not as good as those of the approach using the flux jump condition.

Example 11.10. *Numerical experiments with the results from the literature.* Below we repeat the experiments presented in [187] with $\beta^+ = 1$. The IIM discussed here is more flexible and more accurate, and allows for general linear boundary conditions and the presence of multiparticles.

In Figure 11.17, the results in [187] have been reproduced. In the example, the interface is a circle with radius $r = 0.45$ centered at the origin. The computational domains are $-1 \le x \le 1$ and $0 \le y \le 1$. A Dirichlet boundary condition is used as follows:

$$u(x,y) = \begin{cases} H_0 & \text{if } y = 0, \\ 0 & \text{if } y = 1, \\ H_0(1-y) & \text{if } x = -1 \text{ or } x = 1. \end{cases} \tag{11.103}$$

In Figure 11.18, we show a simulation of multiparticle inclusions. We use a homogeneous Neumann or a periodic boundary condition, which are both more realistic than a Dirichlet boundary condition. In Table 11.4, we show the CPU time (in seconds) of simulations with different mesh sizes and different numbers of particles in the domain. The number of iterations is around 10 for most of the simulations.

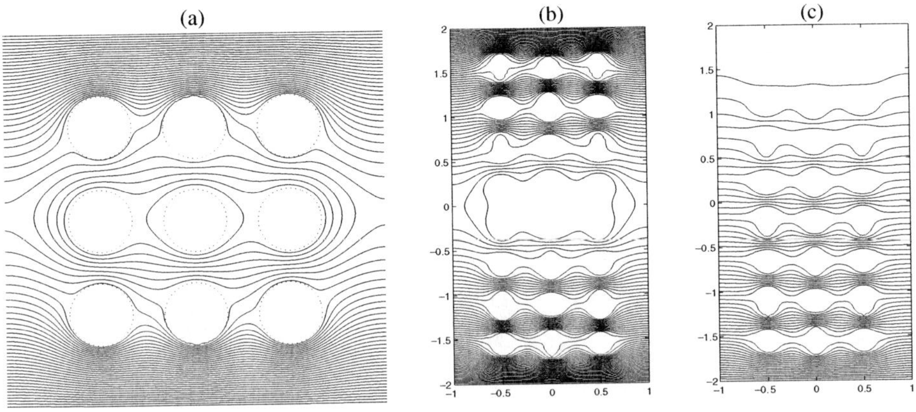

Figure 11.18. *Equipotential curves for the Frohlich–Kennelly model with different applied fields H_0.* (a) *Multiconnected domain with homogeneous Neumann boundary condition along $x = -1$ and $x = 1$ computed with a 320×320 grid. The parameters are* $H_0 = 0.01$, $\mu = 1000$, $B^{sat} = 2.15$. (b) *Multiconnected domain with a periodic boundary condition along $x = -2$ and $x = 2$ computed with a 320×640 grid. The parameters are* $H_0 = 0.01$ *for* (b) *and* $H_0 = 1$ *for* (c); $\mu = 1000$, $B^{sat} = 2.15$.

Table 11.4. *The CPU time (in seconds) of the simulations with different mesh sizes and numbers of particles. In the table, (M, N) is the mesh size in the x- and y-directions; (N_r, N_c) is the number of rows and columns of the particles. Therefore $N_r \times N_c$ is the total number of particles. The simulations are done on a PC workstation (2.2 GHz).*

(M, N)	(N_r, N_c)	CPU (seconds)
(80, 160)	(2, 5)	2.4800
(320, 640)	(1, 1)	7.23000
(320, 640)	(7, 15)	178.950
(480, 960)	(2, 5)	156.34
(480, 960)	(7, 15)	613.46

Remark 11.4.

- *Initially, we can set $\beta^- = \beta^+$ to get an initial approximation for the nonlinear iteration.*

- *The main computational cost is the elliptic solver if the number of particles is relatively small. If there are so many particles that the number of irregular grid points is large, then the computational cost in dealing with the interfaces may be significant.*

11.6 Other methods related to the IIM

The IIM has been well recognized because of its many merits. Several new methods based on the IIM have been developed. In this section, we give an incomplete review of those related methods.

In [279], an immersed interface method for boundary value problems (IIMB) on irregular domains is developed, particularly for Dirichlet boundary conditions. It was shown in [279] that the IIMB is second-order accurate in the maximum norm and that the Schur complement system is well-conditioned. The IIMB was applied to underground water simulations using the stream-vorticity function in [279].

Fogelson and Keener [81] have developed an embedding method for Laplacian equations on irregular domains with a Neumann boundary condition in two and three dimensions. With careful selection of the stencils, the method is second-order accurate and produces a stable matrix (diagonally semidominant). Dumett and Keener [67] have also extended the embedding method to anisotropic elliptic boundary value problems on irregular domains in two space dimensions when $\beta(\mathbf{x})$ in (6.1) is an anisotropic matrix.

In [20], an IIM is developed by approximating the correction terms as part of the iterative procedure. In [182], the IIM is applied to unsteady incompressible flows on irregular domains by constructing high-order interpolation schemes. In [148, 149], the authors developed an IIM that is capable of handling rigid boundaries. The idea is to set the source strength (or force density) as an unknown that is then determined by solving a small system of equations.

Xu and Wang [276] have developed a systematic way of deriving the jump conditions for three-dimensional flow simulations. First, principal jump conditions of the velocity, pressure, and their normal derivatives are derived, which are functions of parameterized coordinates of the interface. Then, by differentiating the principal jump conditions along the interface, they derived a set of linear algebraic equations for the jump conditions of the spatial derivatives of the velocity and pressure. The system can be solved to give analytical expressions of those spatial jump conditions. Last, temporal jump conditions can be derived from their relationship with the spatial jump conditions. Xu and Wang [275] implemented the IIM with the incorporation of their derived jump conditions. They also discussed the effect of the temporal jump conditions on the accuracy of the IIM. They found that in their toy problem [276], it is crucial to include the temporal jump conditions in the numerical algorithm.

11.6.1 The IIM for hyperbolic systems of PDEs

LeVeque and Zhang [158, 281] extended the IIM for hyperbolic equations including one- and two-dimensional acoustic wave equations and two-dimensional elasticity equations. Interface relations have been derived for those systems and used in constructing the finite difference schemes. Nonsmooth interfaces are allowed in their methods. Although the techniques they used are based on finite difference discretizations, some of the techniques have been successfully merged with the finite volume implementation of Clawpack [153]. Such a combination is also used to solve the incompressible Navier–Stokes equations for two-phase flow [150], and is used on irregular domains using the vorticity stream-function formulation [35, 36].

Piraux and Lombard [219] have proposed an explicit simplified interface method (ESIM) for hyperbolic interface problems. The finite difference equations at irregular grid points are constructed implicitly through an explicit modification of numerical values used for time stepping. The modification is based on the spring-mass conditions satisfied at the interface.

Jovanović and Vulkov [130, 131] have studied parabolic and hyperbolic equations with unbounded coefficients of the form

$$\left[c(x) + K\,\delta(x-\xi)\right]\frac{\partial^2 u}{\partial t^2} - \frac{\partial}{\partial x}\left(a(x)\frac{\partial u}{\partial x}\right) = f(x,t), \quad (x,t) \in (0,\,1) \times (0,\,T),$$

where $0 < c_1 \le a(x) \le c_2$, $0 < c_3 \le c(x) \le c_4$, and c_i and $K > 0$ are constants. An abstract operator method is proposed for studying these equations. Estimates for the rate of convergence of the averaging operator difference schemes on special energetic Sobolev norms, compatible with the smoothness of the solutions, are obtained; see [130, 131] and the reference therein. The IIM is also applied and analyzed for reaction-diffusion problems with singular sources in [135].

11.6.2 The explicit jump immersed interface method (EJIIM)

Motivated by the augmented approach (fast IIM) for elliptic interface problems with piecewise constant coefficients (see §6.1), Wiegmann and Bube [268, 270] developed the EJIIM for elliptic interface problems.

The EJIIM works by focusing on the jumps in the solution and its derivatives, rather than on finding coefficients of a new finite difference scheme. That explains the name explicit jump IIM. As we mentioned earlier in §10.6.3, if we know the jumps in $[u]$, $[u_x]$, $[u_y]$, $[u_{xx}]$, $[u_{yy}]$, etc., then we can use a standard finite difference method at all grid points plus some correction terms at irregular grid points. The EJIIM takes advantage of this property and expresses all jumps in terms of constants and the limiting values from a particular side, say, u^-, u_x^-, u_y^-, u_{xx}^-, etc. These quantities are used as augmented unknowns. The additional jump variables are specifically introduced at the intersections of the grid lines with the interface (similar to [188]). This allows treating higher-dimensional corrections as a cross product of one-dimensional corrections. What is also different from the IIM is the way in which the one-sided limits appearing in the jump conditions are approximated. In the EJIIM approach it is done by solving an overdetermined system for the coefficients of a second-order polynomial that approximates the original variables at grid points in a neighborhood of the intersection point.

The augmented system of equations involving the solution of the PDE at grid points and the jumps at intersection points along the interface is solved. Similarly as in the fast IIM, for Poisson and irregular domain elasticity the solution is found in two steps. At first, the Schur complement system for the jump variables is solved iteratively by application of direct solvers. From these jumps, the primary variables are found with one more applications of the direct solver. For the piecewise constant coefficient case in linear elasticity, the jump variables are eliminated and an implicitly derived finite difference formulation is solved.

The EJIIM has been applied to nonlinear one-dimensional interface problems in [269] and two-dimensional elasticity equations in shape design in conjunction with the level set method in [237]. For linear elasticity equations with discontinuous coefficients, the EJIIM was applied in [228, 230]. The EJIIM and its new version, called reduced EJIIM, have been developed in [229] for three-dimensional elliptic irregular domain problems with applications.

11.6.3 The high-order matched interface and boundary method

Recently, a high-order Cartesian grid scheme, the matched interface and boundary (MIB) method, has been proposed by Wei and his coworkers for solving various PDEs with discontinuous coefficients and singular sources [286, 287, 290]. The MIB method has also been applied as a general scheme for accommodating some complex boundary conditions in high-order spatial discretization of PDEs [289, 288]. For elliptic equations, the MIB method can be regarded as a generalization of the IIM [290].

It is well known that the presence of interfaces leads to the loss of the regularity of the solution to the PDE, and thus degrades a conventional high-order finite difference method to a lower order or, even worse, does not converge. Similar to the IIM, the MIB method locally modifies the finite difference stencils near the interface to enforce physical jump conditions. However, unlike the IIM, the MIB approach disassociates the enforcement of physical jump conditions from the discretization of the PDE under study. Such a disassociation is made possible by appropriate use of auxiliary line and/or fictitious points. Consequently, standard central difference schemes are always used in the MIB method for the whole computational domain. When the interface is absent, the MIB method automatically reduces to the standard central difference schemes. To construct higher-order interface schemes, the MIB approach bypasses the major challenge of implementing high-order jump conditions by repeatedly enforcing the lowest-order jump conditions. The MIB approach is of arbitrarily high order in principle. In treating straight, regular interfaces, up to 16th order MIB schemes have been constructed in two and three dimensions. For practical curve interfaces, second-, fourth-, and sixth-order MIB schemes have been demonstrated [290].

The MIB method was originally constructed for electromagnetic wave propagation and scattering in inhomogeneous media [286, 287] in the framework of high-order finite difference time domain (FDTD) schemes. Recently, the MIB method has been applied as a boundary scheme for the treatment of standard boundary conditions, such as Dirichlet, Neumann, and Robin, and nonconventional or complex boundary conditions in high-order spatial discretization of PDEs [289, 288]. As a fictitious domain boundary scheme, a small fictitious domain outside the boundary is usually assumed in the MIB method, and the highly accurate estimates of fictitious function values are essential. The central challenge in the fictitious domain boundary treatments is that the number of available boundary conditions is far smaller than the required ones to determine fictitious points outside each boundary. The MIB method overcomes this difficulty through repeatedly utilizing the given boundary conditions to generate enough algebraic equations. The MIB boundary closure can also be made to arbitrarily high order, in principle. The MIB method was validated for general applications, including multidimensional elliptic boundary value problems with arbitrary combinations of Dirichlet, Neumann, and Robin conditions, elliptic eigenvalue problems, high-order differential equations with multiple boundary conditions, time-dependent boundary conditions, free-edged supports in beam analysis, etc.

11.7 Future directions

In summary, the IIM is an efficient sharp interface method for PDEs involving interfaces or irregular domains; discontinuities in the coefficients, the solution and its derivatives; and Dirac delta function singularities in the source term. The IIM is based on Cartesian grids with second- or fourth-order accuracy. It can be, and has been, coupled with evolution schemes, such as particle methods and the level set method, for free boundary and moving interface problems in two and three dimensions.

There are still some challenging problems for the IIM. (1) It is desirable to develop a second-order *conservative* finite difference scheme for self-adjoint elliptic and parabolic interface problems based on Cartesian grids. (2) It is desirable to have the system of finite difference equations be *symmetric* positive definite. (3) It is desirable to simplify the IIM further and/or develop a reliable IIM software package. (4) It is desirable to have a simple and accurate (second-order accurate) finite element method for interface problems. (5) It is not clear how to apply the IIM directly to problems in which the jump conditions are not explicitly given, or coupled together through some nonlinear relations; the augmented approach provides an alternative. (6) Theoretically, the convergence properties of the IIM need further investigation for moving interface and free boundary problems.

Bibliography

[1] D. Adalsteinsson and J. Sethian. A fast level set method for propagating interfaces. Lawrence Berkeley Lab Report LBL-34893, 1993.

[2] J. Adams, P. Swarztrauber, and R. Sweet. Fishpack: Efficient Fortran subprograms for the solution of separable elliptic partial differential equations. http://www.cisl.ucer.edu/css/software/fishpack; http://www.netlib.org/fishpack/.

[3] L. Adams and T. P. Chartier. New geometric immersed interface multigrid solvers. *SIAM J. Sci. Comput.*, 25:1516–1533, 2004.

[4] L. Adams and T. P. Chartier. A comparison of algebraic multigrid and geometric immersed interface multigrid methods for interface problems. *SIAM J. Sci. Comput.*, 26:762–784, 2005.

[5] L. M. Adams. A multigrid algorithm for immersed interface problems. In *Proceedings of Copper Mountain Multigrid Conference*, NASA Conference Publication 3339, pages 1–14, 1995.

[6] L. Adams and Z. Li. The immersed interface/multigrid methods for interface problems. *SIAM J. Sci. Comput.*, 24:463–479, 2002.

[7] A. Almgren, J. Bell, P. Collella, L. Howell, and M. Welcome. A conservative adaptive projection method for the variable density incompressible Navier-Stokes equations. *J. Comput. Phys.*, 142:1–46, 1998.

[8] A. S. Almgren, J. B. Bell, P. Collella, and T. Marthaler. A Cartesian grid projection method for the incompressible Euler equations in complex geometries. *SIAM J. Sci. Comput.*, 18:1289–1309, 1997.

[9] R. Almgren. Variational algorithms and pattern formulation in dendritic solidification. *J. Comput. Phys.*, 106:337–354, 1993.

[10] K. Arthurs, L. Moore, C. Peskin, E. Pitman, and H. Layton. Modeling arteriolar flow and mass transport using the immersed boundary method. *J. Comput. Phys.*, 147:402–440, 1998.

[11] I. Babuška. The finite element method for elliptic equations with discontinuous coefficients. *Computing*, 5:207–213, 1970.

[12] V. Barbu. *Equations of Evolution*. Pitman, London, 1979.

[13] V. Barbu. *Analysis and Control of Nonlinear Infinite Dimensional Systems*. Academic Press, Boston, 1993.

[14] T. Barth and J. A. Sethian. Numerical schemes for the Hamilton-Jacobi and level set equations on triangulated domains. *J. Comput. Phys.*, 145:1–40, 1998.

[15] G. K. Batchelor. *An Introduction to Fluid Dynamics*. Cambridge University Press, Cambridge, UK, 1967.

[16] J. B. Bell, P. Colella, and H. M. Glaz. A second-order projection method for the incompressible Navier-Stokes equations. *J. Comput. Phys.*, 85:257–283, 1989.

[17] J. B. Bell, C. N. Dawson, and G. R. Shubin. An unsplit, higher order Godunov method for scalar conservation laws in multiple dimensions. *J. Comput. Phys.*, 74:1–24, 1988.

[18] M. J. Berger and R. J. LeVeque. Adaptive mesh refinement using wave-propagation algorithms for hyperbolic systems. *SIAM J. Numer. Anal.*, 35:2298–2316, 1998.

[19] A. Berman and R. J. Plemmons. *Nonegative Matrices in the Mathematical Sciences*. Academic Press, New York, San Francisco, London, 1979.

[20] P. A. Berthelsen. A decomposed immersed interface method for variable coefficient elliptic equations with non-smooth and discontinuous solutions. *J. Comput. Phys.*, 197:364–386, 2004.

[21] A. Bespalov, Y. Kuznetsov, O. Pironneau, and M.-G. Vallet. Fictitious domains with separable preconditioners versus unstructured adapted meshes. *Impact Comput. Sci. Engrg.*, 4:217–249, 1992.

[22] R. P. Beyer. A computational model of the cochlea using the immersed boundary method. *J. Comput. Phys.*, 98:145–162, 1992.

[23] R. P. Beyer and R. J. LeVeque. Analysis of a one-dimensional model for the immersed boundary method. *SIAM J. Numer. Anal.*, 29:332–364, 1992.

[24] G. Biros, L. Ying, and Z. Denis. An embedded boundary integral solver for the Stokes equations. *J. Comput. Phys.*, 196:591–626, 2004.

[25] C. Börgers. A triangulation algorithm for fast elliptic solvers based on domain imbedding. *SIAM J. Numer. Anal.*, 27:1187–1196, 1990.

[26] C. Börgers and O. B. Widlund. On finite element domain imbedding methods. *SIAM J. Numer. Anal.*, 27:963–978, 1990.

[27] D. Bottino. Modeling viscoelastic networks and cell deformation in the context of the immersed boundary method. *J. Comput. Phys.*, 147:86–113, 1998.

[28] J. Brackbill, D. Kothe, and C. Zemach. A continuum method for modeling surface tension. *J. Comput. Phys.*, 100:335–353, 1992.

[29] D. Braess. *Finite Elements.* Cambridge University Press, Cambridge, UK, 1997.

[30] J. Bramble and J. King. A finite element method for interface problems in domains with smooth boundaries and interfaces. *Adv. Comput. Math.*, 6:109–138, 1996.

[31] K. Brattkus and D. I. Meiron. Numerical simulations of unsteady crystal growth. *SIAM J. Appl. Math.*, 52:1303–1320, 1992.

[32] W. L. Briggs, V. E. Henson, and S. F. McCormick. *A Multigrid Tutorial.* 2nd ed., SIAM, Philadelphia, 2000.

[33] W. R. Briley and H. McDonald. On the structure and use of linearized block implicit schemes. *J. Comput. Phys.*, 34:54–73, 1980.

[34] D. L. Brown, R. Cortez, and M. L. Minion. Accurate projection methods for the incompressible Navier-Stokes equations. *J. Comput. Phys.*, 168:464, 2001.

[35] D. Calhoun. *A Cartesian Grid Method for Solving the Streamfunction-Vorticity Equations in Irregular Geometries.* Ph.D. thesis, University of Washington, 1999.

[36] D. Calhoun. A Cartesian grid method for solving the streamfunction-vorticity equation in irregular regions. *J. Comput. Phys.*, 176:231–275, 2002.

[37] B. Camp, T. Lin, Y. Lin, and W. Sun. Quadratic immersed finite element spaces and their approximation capabilities. *Adv. Comput. Math.*, 24:81–112, 2006.

[38] J. M. Chadam and H. Rasmussen. *Free Boundary Problems in Fluid Flow with Applications.* Priman Advanced Publishing Program, 1993.

[39] K. Chan, K. Zhang, X. Liao, J. Zou, and G. Schubert. A three-dimensional spherical nonlinear interface dynamo. *Astrophys. J.*, 596:663–679, 2003.

[40] Y. C. Chang, T. Y. Hou, B. Merriman, and S. Osher. A level set formulation of Eulerian interface capturing method for incompressible fluid flows. *J. Comput. Phys.*, 124:449–464, 1996.

[41] G. Chen, Z. Li, and P. Lin. A fast finite difference method for biharmonic equations on irregular domains. Report CRSC-TR04-09, North Carolina State University, 2004. http://www.ncsu.edu/crsc/reports. Also in *Adv. Comput. Math.*, in press.

[42] S. Chen, B. Merriman, P. Smereka, and S. Osher. A fast level set based algorithm for Stefan problems. *J. Comput. Phys.*, 135:8–29, 1997.

[43] Z. Chen, Q. Du, and J. Zou. Finite element methods with matching and nonmatching meshes for Maxwell equations with discontinuous coefficients. *SIAM J. Numer. Anal.*, 37:1542–1570, 2000.

[44] Z. Chen and J. Zou. Finite element methods and their convergence for elliptic and parabolic interface problems. *Numer. Math.*, 79:175–202, 1998.

[45] Z. Chen and J. Zou. An augmented Lagrangian method for identifying discontinuous parameters in elliptic systems. *SIAM J. Control Optim.*, 37:892–910, 1999.

[46] L.-T. Cheng, R. Fedkiw, F. Gibou, and M. Kang. A symmetric method for implicit time discretization of the Stefan problem. UCLA CAM report #00-37, 2000. http://www.math.ucla.edu/applied/cam/index.html.

[47] D. L. Chopp. Some improvements of the fast marching method. *SIAM J. Sci. Comput.*, 23:230–244, 2001.

[48] A. J. Chorin. Numerical solution of the Navier-Stokes equations. *Math. Comp.*, 22:745–762, 1968.

[49] A. J. Chorin. On the convergence of discrete approximations of the Navier-Stokes equations. *Math. Comp.*, 23:341–353, 1969.

[50] E. T. Chung, Q. Du, and J. Zou. Convergence analysis of a finite volume method for Maxwell's equations in nonhomogeneous media. *SIAM J. Numer. Anal.*, 41:37–63, 2003.

[51] E. T. Chung and J. Zou. A finite volume method for Maxwell's equations with discontinuous physical coefficients. *Int. J. Appl. Math.*, 7:201–223, 2001.

[52] P. G. Ciarlet. *The Finite Element Method for Elliptic Problems.* North–Holland, Austerdam, 1978.

[53] R. Clift, J. R. Grace, and M. E. Weber. *Bubbles, Drops and Particles.* Academic Press, New York, 1978.

[54] R. Cortez. A vortex/impulse method for immersed boundary motion in high Reynolds number flows. *J. Comput. Phys.*, 160:385–400, 2000.

[55] R. Cortez, N. Cowen, R. Dillon, and L. Fauci. Simulation of swimming organisms: Coupling internal mechanics with external fluid dynamics. *Comput. Sci. Engrg.*, 6:38–45, 2004.

[56] R. Cortez, L. Fauci, and A. Medovikov. The method of regularized Stokeslets in three dimensions: Analysis, validation, and application to helical swimming. *Phys. Fluids*, 17:031504, 2005.

[57] R. Cortez and M. L. Minion. The blob projection method for immersed boundary problems. *J. Comput. Phys.*, 161:428–453, 2000.

[58] R. Cortez and D. Varela. Accurate projection methods for the incompressible Navier-Stokes equations. *J. Comput. Phys.*, 138:224–247, 1997.

[59] J. Crank. *Free and Moving Boundary Problems.* Oxford University Press, New York, 1984.

[60] W. Dai, L. Kadanoff, and S. Zhou. Interface dynamics and the motion of complex singularities. *Phys. Rev. A*, 43:6672, 1991.

[61] W. Dai and M. J. Shelley. A numerical study of the effect of surface tension and noise on an expanding Hele-Shaw bubble. *Phys. Fluids* A, 5:2131–2146, 1993.

[62] D. De Zeeuw. Matrix-dependent prolongations and restrictions in a blackbox multigrid solver. *J. Comput. Appl. Math.*, 33:1–27, 1990.

[63] A. J. Degregoria and L. W. Schwartz. A boundary-integral method for two-phase displacement in Hele-Shaw cells. *J. Fluid Mech.*, 164:383–400, 1986.

[64] S. Deng. *Immersed Interface Method for Three Dimensional Interface Problems and Applications*, Ph.D. thesis. North Carolina State University, 2000.

[65] S. Deng, K. Ito, and Z. Li. Three dimensional elliptic solvers for interface problems and applications. *J. Comput. Phys.*, 184:215–243, 2003.

[66] R. H. Dillon, L. J. Fauci, and A. L. Fogelson. Modeling biofilm processes using the immersed boundary method. *J. Comput. Phys.*, 129:57–73, 1996.

[67] M. Dumett and J. Keener. An immersed interface method for anisotropic elliptic problems on irregular domains in 2D. *Numer. Methods Partial Differential Equations*, 21:397–420, 2005.

[68] D. L. Dwoyer and F. C. Thames. Accuracy and stability of time-split finite-difference schemes. In 5th *AIAA Computational Fluid Dynamics Conference*, Palo Alto, CA, 1981.

[69] W. E and J.-G. Liu. Projection method I. Convergence and numerical boundary layers. *SIAM J. Numer. Anal.*, 32:1017–1057, 1995.

[70] W. E and J. Liu. Vorticity boundary condition and related issues for finite difference schemes. *J. Comput. Phys.*, 124:368–382, 1996.

[71] D. Enright, R. J. Fedkiw Ferziger, and I. Mitchell. A hybrid particle level set method for improved interface capturing. *J. Comput. Phys.*, 183:83–116, 2002.

[72] G. H. Gottet and E. Maitre. A level-set formulation of immersed boundary methods for fluid-structure interaction problem. *C. R. Acad. Sci. Paris*, 338:581–586, 2004.

[73] L. C. Evans. *Partial Differential Equations*. AMS, Providence, RI, 1998.

[74] R. Ewing, O. Iliev, and R. Lazarov. A modified finite volume approximation of second-order elliptic equations with discontinuous coefficients. *SIAM J. Sci. Comput.*, 23:1335–1351, 2001.

[75] R. Ewing, Z. Li, T. Lin, and Y. Lin. The immersed finite volume element method for the elliptic interface problems. *Math. Comput. Simulation*, 50:63–76, 1999.

[76] D. Eyre and A. Fogelson. IBIS: A software system for immersed boundary and interface simulations. http://www.math.utah.edu/IBIS.

[77] L. J. Fauci. Interaction of oscillating filaments: A computational study. *J. Comput. Phys.*, 86:294–313, 1990.

[78] B. A. Finlayson. *Numerical Methods for Problems with Moving Fronts*. Ravenna Park Publishing, Inc., Seattle, 1992.

[79] A. L. Fogelson. A mathematical model and numerical method for studying platelet adhesion and aggregation during blood clotting. *J. Comput. Phys.*, 56:111–134, 1984.

[80] A. L. Fogelson. Continuum models of platelet aggregation: Formulation and mechanical properties. *SIAM J. Appl. Math.*, 52: 1089–1110, 1992.

[81] A. L. Fogelson and J. P. Keener. Immersed interface methods for Neumann and related problems in two and three dimensions. *SIAM J. Sci. Comput.*, 22:1630–1654, 2000.

[82] A. L. Fogelson and C. S. Peskin. A fast numerical method for solving the three dimensional Stokes equations in the presence of suspended particles. *J. Comput. Phys.*, 79: 50–69, 1988.

[83] B. Fornberg and R. Meyer-Spasche. A finite difference procedure for a class of free boundary problems. *J. Comput. Phys.*, 102:72–77, 1992.

[84] R. M. Furzeland. A comparative study of numerical methods for moving boundary problems. *J. Inst. Math. Appl.*, 26:411–429, 1980.

[85] A. Garbey and W. Shyy. A least square extrapolation method for improving solution accuracy of PDE computations. *J. Comput. Phys.*, 186:1–23, 2003.

[86] F. Gibou, R. Fedkiw, L.-T. Cheng, and M. Kang. A second-order-accurate symmetric discretization of the Poisson equation on irregular domains. *J. Comput. Phys.*, 176:2002, 205–227.

[87] J. Glimm, J. W. Grove, X. L. Li, K.-M. Shyue, Y. Zengi, and Q. Zhang Three-dimensional front tracking. *SIAM J. Sci. Comput.*, 19:703–727, 1998.

[88] J. Glimm, J. W. Grove, X. L. Li, and D. C. Tan. Robust computational algorithms for dynamic interface tracking in three dimensions. *SIAM J. Sci. Comput.*, 21:2240–2256, 2000.

[89] J. Glimm, X. Li, Y. Liu, Z. Xu, and N. Zhao. Conservative front tracking with improved accuracy. *SIAM J. Numer. Anal.*, 41:1926–1947, 2003.

[90] J. Glimm, D. Marchesin, and O. McBryan. Subgrid resolution of fluid discontinuities II. *J. Comput. Phys.*, 37:336–354, 1980.

[91] J. Glimm and O. McBryan. A computational model for interfaces. *Adv. Appl. Math.*, 6:422–435, 1985.

[92] R. Glowinski. *Numerical Methods for Nonlinear Variational Problems*. Springer, New York, 1984.

[93] Y. Gong and Z. Li. A conforming piecewise quadratic immersed finite element method for elliptic interface problems. Preprint, 2005.

[94] M. A. Grayson. The heat equation shrinks embedded plane curves to round points. *J. Differential Geom.*, 26:285–314, 1987.

[95] A. Greenbaum, L. Greengard, and A. Mayo. On the numerical solution of the biharmonic equation in the plane. *Phys. D*, 60:216–225, 1992.

[96] L. Greengard and M. C. Kropinski. An integral equation approach to the incompressible Navier–Stokes equations in two dimensions. *SIAM J. Sci. Comput.*, 20:318–336, 1998.

[97] L. Greengard, M. C. Kropinski, and A. Mayo. Integral equation methods for Stokes flow and isotropic elasticity in the plane. *J. Comput. Phys.*, 125:403–414, 1996.

[98] L. Greengard and M. Moura. On the numerical evaluation of electrostatic fields in composite materials. *Acta Numerica*, pages 379–410, Cambridge University Press, Cambridge, UK, 1994.

[99] J. L. Guermond and J. Shen. A new class of truly consistent splitting schemes for incompressible flows. *J. Comput. Phys.*, 192:262–276, 2003.

[100] H. Han. The numerical solutions of the interface problems by infinite element methods. *Numer. Math.*, 39:39–50, 1982.

[101] J. Haslinger and P. Neittanmaaki. *Finite Element Approximation for Optimal Shape Design: Theory and Applications*. J. Wiley, Chichester, 1988.

[102] B. T. Helebrook and C. K. Law. The role of Landau-Darrieus instability in large scale flows. *Combustion and Flame*, 117:155–169, 1999.

[103] B. T. Helebrook, L. Martinelli, and C. K. Law. A numerical method for solving incompressible flow problems with a surface of discontinuity. *J. Comput. Phys.*, 148:366–396, 1999.

[104] J. D. Hoffman. *Numerical Methods for Engineers and Scientists*. McGraw–Hill, New York, 1991.

[105] S. Hou and X. Liu. A numerical method for solving variable coefficient elliptic equation with interfaces. *J. Comput. Phys.*, 202:411–445, 2005.

[106] T. Hou. Private communication, 2000.

[107] T. Hou, Z. Li, S. Osher, and H. Zhao. A hybrid method for moving interface problems with application to the Hele-Shaw flow. *J. Comput. Phys.*, 134:236–252, 1997.

[108] T. Y. Hou, J. S. Lowengrub, and M. J. Shelley. Removing the stiffness from interfacial flows with surface tension. *J. Comput. Phys.*, 114:312–338, 1994.

[109] T. Y. Hou and B. T. R. Wetton. Convergence of a finite difference scheme for the Navier–Stokes equations using vorticity boundary conditions. *SIAM J. Numer. Anal.*, 29:615–639, 1992.

[110] H. Huang and Z. Li. Convergence analysis of the immersed interface method. *IMA J. Numer. Anal.*, 19:583–608, 1999.

[111] J. Huang and J. Zou. A mortar element method for elliptic problems with discontinuous coefficients. *IMA J Numer. Anal.*, 22:549–576, 2002.

[112] J. Huang and J. Zou. Some new a priori estimates for second-order elliptic and parabolic interface problems. *J. Differential Equations*, 184:570–586, 2002.

[113] J. Hunter, Z. Li, and H. Zhao. Autophobic spreading of drops,. *J. Comput. Phys.*, 183:335–366, 2002.

[114] K. Hutter, A. Zryd, and H. Röthlisberger. On the numerical solution of Stefan problems in temperate ice. *J. Glaciology*, 36:41–48, 1990.

[115] K. Ito and K. Kunisch. BV-type regularization methods for convoluted objects with edge, flat and grey scales. *Inverse Problems*, 16:909–928, 2000.

[116] K. Ito, K. Kunisch, and Z. Li. Level-set function approach to an inverse interface problem. *Inverse Problems*, 17:1225–1242, 2001.

[117] K. Ito, Z. Li, and Y. Kyei. Higher-order, Cartesian grid based finite difference schemes for elliptic equations on irregular domains. *SIAM J. Sci. Comput.*, 27:346–367, 2005.

[118] K. Ito and Z. Li. Interface relations for Stokes equations with discontinuous viscosity and singular sources. *Appl. Math. Lett.*, 19:229–234, 2006.

[119] K. Ito and Z. Li. Solving a nonlinear problem in magneto-rheological fluids using the immersed interface method. *J. Sci. Comput.*, 19:253–266, 2003.

[120] K. Ito and J. Toivanen. Preconditioned iterative methods on sparse subspaces. *Apl. Math. Lett.*, 19:1191–1197, 2006.

[121] J. E. Dendy Jr., Black box multigrid. *J. Comput. Phys.*, 48:336–386, 1982.

[122] J. E. Dennis Jr., and R. B. Schnabel. *Numerical Methods for Unconstrained Optimization and Nonlinear Equations*. SIAM, Philadelphia, 1996.

[123] G.-S. Jiang and D. Peng. Weighted ENO schemes for Hamilton–Jacobi equations. *SIAM J. Sci. Comput.*, 21:2126–2143, 2000.

[124] G. Jiang and C. Shu. Efficient implementation of weighted ENO schemes. *J. Comput. Phys.*, 126:202–228, 1996.

[125] S. Jin and X. Wang. Robust numerical simulation of porosity evolution in chemical vapor infiltration. *J. Comput. Phys.*, 179:557–577, 2002.

[126] H. Johansen and P. Colella. A Cartesian grid embedded boundary method for Poisson equations on irregular domains. *J. Comput. Phys.*, 147:60–85, 1998.

[127] C. Johnson. *Numerical Solution of Partial Differential Equations by the Finite Element Method*. Cambridge University Press, Cambridge, UK, 1987.

[128] H. Johnston and J. Liu. Accurate, stable and efficient Navier-Stokes solvers based on explicit treatment of the pressure term. *J. Comput. Phys.*, 188:221–259, 2004.

[129] D. D. Joseph and Y. Y. Renardy. *Fundamentals of two-fluid dynamics.* Springer-Verlag, New York, 1993.

[130] B. S. Jovanović and L. G. Vulkov. On the convergence of finite difference schemes for parabolic problems with concentrated data. *Numer. Math*, 89:715–734, 2001.

[131] B. S. Jovanović and L. G. Vulkov. On the convergence of finite difference schemes for hyperbolic problems with concentrated data. *SIAM J. Numer. Anal.*, 41:516–538, 2003.

[132] D. Juric and G. Tryggvason. A front-tracking method for dentritic solidification. *J. Comput. Phys.*, 123:127–1487, 1996.

[133] K. Kafafy, T. Lin, Y. Lin, and J. Wang. Three-dimensional immersed finite element methods for electric field simulation in composite materials. *Internat. J. Numer. Methods Engrg.*, 64:940–972, 2005.

[134] J. Van Kan. A second-order accurate pressure-correction scheme for viscous incompressible flow. *SIAM J. Sci. Comput.*, 7:870–891, 1986.

[135] J. D. Kandilarov and L. G. Vulkov. Analysis of immersed interface difference schemes for reaction-diffusion problems with singular sources. *Comput. Methods Appl. Math.*, 3:253–273, 2003.

[136] M. Kang, R. Fedkiw, and X. Liu. A boundary condition capturing method for multiphase incompressible flow. *J. Sci. Comput.*, 15:323–360, 2000.

[137] M. R. Kenedy, C. Pozrikidis, and R. Skalak. Motion and deformation of liquid drops, and the rheology of dilute emulsions in shear flow. *Comput. & Fluids*, 23:251–278, 1994.

[138] J. Kevorkian. *Partial Differential Equations.* Wadsworth & Brooks/Cole, Pacific Grove, CA, 1990.

[139] J. Kim and P. Moin. Application of a fractional-step method to incompressible Navier-Stokes equations. *J. Comput. Phys.*, 59:308–323, 1985.

[140] Y. A. Kuznetsov. Matrix iterative methods in subspaces. In *Proceedings of the International Congress of Mathematicians*, Warsaw, volume 1-2, pages 1509–1521, 1984.

[141] O. A. Ladyzhenskaya. *The Mathematical Theory of Viscous Incompressible Flow.* Gordon & Breach, New York, 1963.

[142] M. Lai and C. S. Peskin. An immersed boundary method with formal second-order accuracy and reduced numerical viscosity. *J. Comput. Phys.*, 160:705–719, 2000.

[143] M. Lai and W.-C. Wang. Fast direct solvers for Poisson equation on 2D polar and spherical geometries. *Numer. Methods Partial Differential Equations*, 18:56–58, 2002.

[144] M.-C. Lai and Z. Li. A remark on jump conditions for the three-dimensional Navier-Stokes equations involving an immersed moving membrane. *Appl. Math. Lett.*, 14:149–154, 2001.

[145] M.-C. Lai, Z. Li, and X.-B. Lin. Fast solvers for 3D Poisson equations involving interfaces in a finite or the infinite domain. *J. Comput. Appl. Math.*, 191:106–125, 2006.

[146] J. S. Langer. Instabilities and pattern formation in crystal growth. *Rev. Modern Phys.*, 52:1–28, 1980.

[147] E. Langlois. *Slow Viscous Flow*. Macmillan, New York, 1964.

[148] D.-V. Le, B.-C. Khoo, and J. Peraire. An immersed interface method for the incompressible Navier-Stokes equations. DSpace at MIT. https://dspace.mit.edu/index.jsp, 2004.

[149] D.-V. Le, B.-C. Khoo, and J. Peraire. An immersed interface method for the incompressible Navier-Stokes equations in irregular domains. DSpace at MIT. https://dspace.mit.edu/index.jsp, 2005.

[150] L. Lee and R. J. LeVeque. An immersed interface method for incompressible Navier–Stokes equations. *SIAM J. Sci. Comput.*, 25:832–856, 2003.

[151] R. B. Lehoucq and D. C. Sorensen. Deflation techniques for an implicitly restarted Arnoldi iteration. *SIAM J. Matrix Anal. Appl.*, 17:789–821, 1996.

[152] R. B. Lehoucq, D.C. Sorensen, and C. Yang. *ARPACK Users' Guide: Solution of Large-Scale Eigenvalue Problems with Implicitly Restarted Arnoldi Methods*. SIAM, Philadelphia, 1998.

[153] R. J. LeVeque. Clawpack and Amrclaw—Software for high-resolution Godunov methods. In *4th Int. Conf. on Wave Propagation*, Golden, Colorado, 1998.

[154] R. J. LeVeque and Z. Li. The immersed interface method for elliptic equations with discontinuous coefficients and singular sources. *SIAM J. Numer. Anal.*, 31:1019–1044, 1994.

[155] R. J. LeVeque and Z. Li. Simulation of bubbles in creeping flow using the immersed interface method. In *Proc. Sixth International Symposium on Computational Fluid Dynamics*, Lake Tahoe, NV, pp. 688–693, 1995.

[156] R. J. LeVeque and Z. Li. Immersed interface methods for Stokes flow with elastic boundaries or surface tension. *SIAM J. Sci. Comput.*, 18:709–735, 1997.

[157] R. J. LeVeque, C. S. Peskin, and P. D. Lax. Solution of a two-dimensional cochlea model with fluid viscosity. *SIAM J. Appl. Math.*, 48:191–213, 1988.

[158] R. J. LeVeque and C. Zhang. Immersed interface methods for wave equations with discontinuous coefficients. *Wave Motion*, 25:237–263, 1997.

[159] R. J. LeVeque. *Finite Difference Methods for Ordinary and Partial Differential Equations: Steady-State and Time-Dependent Problems*, SIAM, Philadelphia, 2007.

[160] Z. Li. *The Immersed Interface Method—A Numerical Approach for Partial Differential Equations with Interfaces*. Ph.D. thesis, University of Washington, 1994.

[161] Z. Li. A note on immersed interface methods for three-dimensional elliptic equations. *Comput. Math. Appl.*, 31:9–17, 1996.

[162] Z. Li. Immersed interface method for moving interface problems. *Numer. Algorithms*, 14:269–293, 1997.

[163] Z. Li, T. Lin, Y. Lin, and R. C. Rogers, An immersed finite element space and its approximation capability, Numerical Methods for Partial Differential Equations, 20(3) 338-367,2004. DOI: 10.1002/num.10092.

[164] Z. Li. The immersed interface method using a finite element formulation. *Appl. Numer. Math.*, 27:253–267, 1998.

[165] Z. Li. IIMPACK: A collection of fortran codes for interface problems. Anonymous ftp at ftp.ncsu.edu/pub/math/zhilin/Package and http://www4.ncsu.edu/~zhilin/IIM, last updated: 2005.

[166] Z. Li and K. Ito. Maximum principle preserving schemes for interface problems with discontinuous coefficients. *SIAM J. Sci. Comput.*, 23:339–361, 2001.

[167] Z. Li, K. Ito, and M.-C. Lai. An augmented approach for Stokes equations with a discontinuous viscosity and singular forces. NCSU-CRSC Tech. Report: CRSC-TR04-23, North Carolina State Univeristy, 2004.

[168] Z. Li and M.-C. Lai. The immersed interface method for the Navier-Stokes equations with singular forces. *J. Comput. Phys.*, 171:822–842, 2001.

[169] Z. Li, T. Lin, Y. Lin, and R. C. Rogers. Error estimates of an immersed finite element method for interface problems. *Numer. Methods Partial Differential Equations*, 12:338–367, 2004.

[170] Z. Li, T. Lin, and X. Wu. New Cartesian grid methods for interface problem using finite element formulation. *Numer. Math.*, 96:61–98, 2003.

[171] Z. Li, X. Lin, M. Torres, and H. Zhao. Generalized Snell's law for weighted minimal surface. *Methods Appl. Anal.*, 10:199–214, 2003.

[172] Z. Li and A. Mayo. ADI methods for heat equations with discontinuties along an arbitrary interface. In *Proc. Symp. Appl. Math.*, W. Gautschi, editor, volume 48, pages 311–315, AMS, Providence, RI, 1993.

[173] Z. Li, D. McTigue, and J. Heine. A numerical method for diffusive transport with moving boundaries and discontinuous material properties. *Internat. J. Numer. Anal. Methods Geomech.*, 21:653–662, 1997.

[174] Z. Li and Y. Shen. Numerical method for simulation of bubbles flowing through another fluid. In Z.-C. Shi, W. M. Xue, M. Mu, and J. Zou, editors, *Advances in Scientific Computing*, pages 74–81. China Science Publisher, Beijing, 2000.

[175] Z. Li and B. Soni. Fast and accurate numerical approaches for Stefan problems and crystal growth. *Numerical Heat Transfer, B: Fundamentals*, 35:461–484, 1999.

[176] Z. Li, X. Wan, K. Ito, and S. Lubkin. An augmented pressure boundary condition for a Stokes flow with a non-slip boundary condition. *Commun. Comput. Phys.*, 1:874–885, 2006.

[177] Z. Li and C. Wang. A fast finite difference method for solving Navier-Stokes equations on irregular domains. *Commun. Math. Sci.*, 1:180–196, 2003.

[178] Z. Li, W.-C. Wang, I.-L. Chern, and M.-C. Lai. New formulations for interface problems in polar coordinates. *SIAM J. Sci. Comput.*, 25:224–245, 2003.

[179] Z. Li and X. Yang. An immersed finite element method for elasticity equations with interfaces. In *Proc. Symp. Appl. Math.,* Z.-C. Shi et. al., editors, AMS, Providence, RI, 2004.

[180] Z. Li, H. Zhao, and H. Gao. A numerical study of electro-migration voiding by evolving level set functions on a fixed cartesian grid. *J. Comput. Phys.*, 152:281–304, 1999.

[181] Z. Li and J. Zou. Theoretical and numerical analysis on a thermo-elastic system with discontinuities. *J. Comput. Appl. Math.*, 91:1–22, 1998.

[182] M. N. Linnick and H. F. Fasel. A high-order immersed interface method for simulating unsteady incompressible flows on irregular domains. *J. Comput. Phys.*, 204:157–192, 2005.

[183] X. Liu, R. Fedkiw, and M. Kang. A boundary condition capturing method for Poisson's equation on irregular domain. *J. Comput. Phys.*, 160:151–178, 2000.

[184] X. Liu and T. C. Sideris. Convergence of the ghost fluid method for elliptic equations with interfaces. *Math. Comp.*, 72:1731–1746, 2003.

[185] S. R. Lubkin and Z. Li. Force and deformation on branching rudiments: Cleaving between hypotheses. *Biomechanics and Modeling in Mechanobiology*, 1:5–16, 2002.

[186] H. Ly, M. Jolly, F. Reitich, H. T. Banks, and K. Ito. Simulations of particle dynamics in magnetorheological fluids. *J. Comput. Phys.*, 155:160–177, 1999.

[187] H. Ly, M. Jolly, F. Reitich, K. Ito, and H. T. Banks. Piecewise linear models for field-responsive fluids. *IEEE Trans. Magnetics*, 37:558–560, 2001.

[188] A. Mayo. The fast solution of Poisson's and the biharmonic equations on irregular regions. *SIAM J. Numer. Anal.*, 21:285–299, 1984.

[189] A. Mayo. A decomposition finite difference method for the fourth order accurate solution of Poisson's equation on general regions. *Int. J. High Speed Comput.*, 3:89–105, 1991.

[190] A. Mayo. The rapid evaluation of volume integrals of potential theory on general regions. *J. Comput. Phys.*, 100:236–245, 1992.

[191] A. Mayo. Rapid, accurate methods for the solution of the stokes problem in the presence of an immersed boundary. IBM Research Report, 1993.

[192] A. Mayo. High order accurate particular solutions of the biharmonic equation on general regions. *Contemp. Math.*, 323:233–243, 2003.

[193] A. Mayo. Rapid, fourth order accurate solution of the steady Navier Stokes equations on general regions. *Dyn. Contin. Discrete Impuls. Syst. Ser. B Appl. Algorithms*, 12:59–72, 2005.

[194] A. Mayo and A. Greenbaum. Fast parallel iterative solution of Poisson's and the biharmonic equations on irregular regions. *SIAM J. Sci. Statist. Comput.*, 13:101–118, 1992.

[195] A. Mayo and A. Greenbaum. Rapid parallel evaluation of integrals in potential theory on general three-dimensional regions. *J. Comput. Phys.*, 145:731–742, 1998.

[196] A. Mayo and C. S. Peskin. An implicit numerical method for fluid dynamics problems with immersed elastic boundaries. *Contemp. Math.*, 141:261–277, 1991.

[197] M. McCracken and C. Peskin. A vortex method for blood flow through heart valves. *J. Comput. Phys.*, 35:183–205, 1980.

[198] A. McKenney, L. Greengard, and Anita Mayo. A fast Poisson solver for complex geometries. *J. Comput. Phys.*, 118:348–355, 1995.

[199] D. M. McQueen and C. S. Peskin. Shared-memory parallel vector implementation of the immersed boundary method for the computation of blood flow in the beating mammalian heart. *J. Supercomputing*, 11:213–236, 1997.

[200] E. Meiburg and G. Homsy. Nonlinear unstable viscous fingers in Hele-Shaw flows. II. Numerical simulation. *Phys. Fluids.*, 31:429–439, 1988.

[201] A. R. Mitchell. *Computational Methods in Partial Differential Equations*. John Wiley and Sons, Ltd., London, 1969.

[202] K. W. Morton and D. F. Mayers. *Numerical Solution of Partial Differential Equations*. Cambridge University Press, Cambridge, UK, 1995.

[203] D. Nguyen, R. Fedkiw, and M. Kang. A boundary condition capturing method for incompressible flame discontinuities. *J. Comput. Phys.*, 176:205–227, 2002.

[204] R. H. Nochetto, M. Paolini, and C. Verdi. An adaptive finite element method for two-phase Stefan problems in two space dimensions. Part II: Implementation and Numerical Experiments *SIAM J. Sci. Statist. Comput.*, 12:1207–1244, 1991.

[205] S. Osher and R. Fedkiw. Level set methods: An overview and some recent results. *J. Comput. Phys.*, 169:272–288, 2001.

[206] S. Osher and R. Fedkiw. *Level Set Methods and Dynamic Implicit Surfaces*. Springer-Verlag, New York, 2002.

[207] S. Osher and J. A. Sethian. Fronts propagating with curvature-dependent speed: Algorithms based on Hamilton-Jacobi formulations. *J. Comput. Phys.*, 79:12–49, 1988.

[208] D. Peng, B. Merriman, S. Osher, H. Zhao, and M. Kang. A PDE-based fast local level set method. *J. Comput. Phys.*, 155:410–438, 1999.

[209] C. S. Peskin. Numerical analysis of blood flow in the heart. *J. Comput. Phys.*, 25:220–252, 1977.

[210] C. S. Peskin. Lectures on mathematical aspects of physiology. *Lectures in Appl. Math.*, 19:69–107, 1981.

[211] C. S. Peskin. The immersed boundary method. *Acta Numer.*, 11:479–517, 2002.

[212] C. S. Peskin and D. M. McQueen. Modeling prosthetic heart valves for numerical analysis of blood flow in the heart. *J. Comput. Phys.*, 37:113–132, 1980.

[213] C. S. Peskin and D. M. McQueen. A three-dimensional computational method for blood flow in the heart. I. Immersed elastic fibers in a viscous incompressible fluid. *J. Comput. Phys.*, 81:372–405, 1989.

[214] C. S. Peskin and D. M. McQueen. A three-dimensional computational method for blood flow in the heart. II. Contractile fibers. *J. Comput. Phys.*, 82:289–297, 1989.

[215] C. S. Peskin and D. M. McQueen. A general method for the computer simulation of biological systems interacting with fluids. In *Symposia of the Society for Experimental Biology*, volume 49, 1995.

[216] C. S. Peskin and B. F. Printz. Improved volume conservation in the computation of flows with immersed elastic boundaries. *J. Comput. Phys.*, 105:33–46, 1993.

[217] R. Peyret and T. D. Taylor. *Computational Methods for Fluid Flow*. Springer, New York, 1983.

[218] M. A. Pinsky. *Partial Differential Equations and Boundary-Value Problems with Applications*. Third Edition, International Series in Pure and Applied Mathematics, WCB McGraw–Hill, New York, 1998.

[219] J. Piraux and B. Lombard. A new interface method for hyperbolic problems with discontinuous coefficients: One-dimensional acoustic example. *J. Comput. Phys.*, 168:227–248, 2001.

[220] C. Pozrikidis. *Boundary Integral and Singularity Methods for Linearized Viscous Flow*. Cambridge University Press, Cambridge, UK, 1992.

[221] W. Proskurowski and O. Widlund. On the numerical solution of Helmholtz's equation by the capacitance matrix method. *Math. Comp.*, 30:433–468, 1976.

[222] E. G. Puckett, A. S. Almgren, J. B. Bell, D. L. Marcus, and W. J. Rider. A high-order projection method for tracking fluid interfaces in variable density incompressible flows. *J. Comput. Phys.*, 130:269–282, 1997.

[223] D. A. Reinelt. Interface conditions for two-phase displacement in Hele-Shaw cells. *J. Fluid Mech.*, 183:219–234, 1987.

[224] A. Roma. *A Multi-Level Self Adaptive Version of the Immersed Boundary Method.* Ph.D. thesis, New York University, 1996.

[225] A. Roma, C. S. Peskin, and M. Berger. An adaptive version of the immersed boundary method. *J. Comput. Phys.*, 153:509–534, 1999.

[226] M. E. Rose. An enthalpy scheme for Stefan problem in several dimensions. *Appl. Numer. Math.*, 12:229–238, 1993.

[227] J. W. Ruge and K. Stüben. Algebraic multigrid. In S.F. McCormick, editor, *Multigrid Methods*, volume 3 of *Frontiers in Applied Mathematics*, chapter 4, pages 73–130. SIAM, Philadelphia, 1987.

[228] V. Rutka. *A Fast Numerical Method for the Computation of Effective Elastic Moduli of 3D Fibrous Miscrostructures.* MS. Thesis, University of Kaiserslautern, 2001.

[229] V. Rutka. *Immersed Interface Methods for Elliptic Boundary Value Problems.* Ph.D. thesis, University of Kaiserslautern, 2005.

[230] V. Rutka and A. Wiegmann. Explicit jump immersed interface method for virtual material design of the effective elastic moduli of composite materials. Technical Report 73, Fraunhofer ITWM Kaiserslautern, 2005. Available online at http://www.itwm.fhg.de/en/zentral/index.

[231] Y. Saad and M. H. Schultz. GMRES: A generalized minimal residual algorithm for solving nonsymmetric linear systems. *SIAM J. Sci. Statist. Comput.*, 7:856–869, 1986.

[232] P. Sackinger, P. Schunk, and R. Rao. A Newton-Raphson pseudo-solid domain mapping technique for free and moving boundary problems: A finite element implementation. *J. Comput. Phys.*, 125:83–103, 1996.

[233] P. G. Saffman and G. I. Taylor. The penetration of a fluid into a porous medium or Hele-Shaw cell containing a more viscous liquid. *Proc. Roy. Soc. Lond. Ser. A*, 245:312–329, 1958.

[234] K. Schittkowski. QL-quadratic Programming, version 1.5, 1991. http://www.uni-bayreuth.de/departments/math/~kschittkowski/ql.htm.

[235] A. Schmidt. Computation of three dimensional dendrites with finite elements. *J. Comput. Phys.*, 125:293–312, 1996.

[236] J. Sethian. Crystal growth and dendritic solidification. *J. Comput. Phys.*, 98:231–2539, 1992.

[237] J. Sethian and A. Wiegmann. Structural boundary design via level set and immersed interface methods. *J. Comput. Phys.*, 163:489–528, 2000.

[238] J. A. Sethian. *Level Set Methods and Fast Marching methods*. 2nd edition, Cambridge University Press, Cambridge, UK,1999.

[239] J. Shen. On error estimates of the projection methods for the Navier-Stokes equations: Second-order scheme. *Math. Comp.*, 65:1039–1065, 1996.

[240] G. R. Shubin and J. B. Bell. An analysis of the grid orientation effect in numerical simulation of miscible displacement. *Comput. Methods Appl. Mech. Engrg.*, 47:47–71, 1984.

[241] J. Sokolowski and J. P. Zolesio. *Introduction to Shape Optimization*. Springer-Verlag, New York, 1991.

[242] J. M. Stockie and B. T. R. Wetton. Stability analysis for the immersed fiber problem. *SIAM J. Appl. Math.*, 55:1577–1591, 1995.

[243] J. Stockie and B. T. R. Wetton. Analysis of stiffness in the immersed boundary method and implications for time-stepping schemes. *J. Comput. Phys.*, 147:41–64, 1999.

[244] H. Stone. Dynamics of drop formulation and breakup in viscous flows. *Annu. Rev. Fluid Mech.*, 26:65–102, 1994.

[245] J. C. Strikwerda. *Finite Difference Scheme and Partial Differential Equations*. 2nd edition, SIAM, Philadelphia, 2004.

[246] D. Sulsky and J. U. Brackbill. A numerical method for suspension flow. *J. Comput. Phys.*, 96:339–368, 1991.

[247] M. Sussman and E. G. Puckett. A coupled level set and volume-of-fluid method for computing 3D and axisymmetric incompressible two-phase flows. *J. Comput. Phys.*, 162:301–337, 2000.

[248] M. Sussman. A second order coupled level set and volume-of-fluid method for computing the growth and collapse of vapor bubbles. *J. Comput. Phys.*, 187:110–136, 2003.

[249] M. Sussman, A. Almgren, J. B. Bell, P. Colella, L. H Howell, and M. L. Welcome. An adaptive level set approach for incompressible two-phase flows. *J. Comput. Phys.*, 148:81–124, 1999.

[250] M. Sussman and E. Fatemi. An efficient, interface-preserving level set redistancing algorithm and its application to interfacial incompressible fluid flow. *SIAM J. Sci. Comput.*, 20:1165–1191, 1999.

[251] M. Sussman, P. Smereka, and S. Osher. A level set approach for computing solutions to incompressible two-phase flow. *J. Comput. Phys.*, 114:146–159, 1994.

[252] Paul N. Swarztrauber. Fast Poisson solver. In G. H. Golub, editor, *Studies in Numerical Analysis*, volume 24, pages 319–370. MAA, 1984.

[253] R. Teman. *Navier-Stokes Equations*. North–Holland, New York, 1977.

[254] R. Teman. Remark on the pressure boundary condition for the projection method. *Theoret. Comput. Fluid Dynamics*, 3:181–184, 1991.

[255] A. N. Tikhonov and A. A. Samarskii. Homogeneous difference schemes. *USSR Comput. Math. Math. Phys.*, 1:5–67, 1962.

[256] A. K. Tornberg. *A Finite Element Based Level Set Method for Multiphase Flow Application*. Ph.D. thesis, NADA, KTH, Stockholm, Sweden, 1998.

[257] A.-K. Tornberg. Regularization techniques for singular source terms in differential equations. In *ECM Stockholm*, European Mathematical Society, pages 477–499, 2005.

[258] A.-K. Tornberg and B. Engquisty. Regularization techniques for numerical approximation of PDES with singularities. *J. Sci. Comput.*, 19:527–552, 2003.

[259] G. Tryggvason and H. Aref. Numerical experiments on Hele-Shaw flow with a sharp interface. *J. Fluid Mech.*, 136:1–20, 1983.

[260] G. Tryggvason, B. Bunner, A. Esmaeeli, D. Juric, N. Al-Rawahi, W. Tauber, J. Han, S. Nas, and Y.-J. Jan. A front-tracking method for the computations of multiphase flow. *J. Comput. Phys.*, 169:708–759, 2001.

[261] Y.-H. R. Tsai, L.-T. Cheng, S. Osher, and H.-K. Zhao. Fast sweeping algorithms for a class of Hamilton–Jacobi equations. *SIAM J. Numer. Anal.*, 41:673–694, 2003.

[262] C. Tu and C. S. Peskin. Stability and instability in the computation of flows with moving immersed boundaries: A comparison of three methods. *SIAM J. Sci. Statist. Comput.*, 13:1361–1376, 1992.

[263] S. O. Unverdi and G. Tryggvason. A front-tracking method for viscous, incompressible, multi-fluid flows. *J. Comput. Phys.*, 100:25–37, 1992.

[264] C. Wang and J. Liu. Analysis of finite difference schemes for unsteady Navier-Stokes equations in vorticity formulation. *Comput. Fluids*, 33:223–255, 2004.

[265] N. Wang and A. L. Fogelson. Computational methods for continuum models of platelet aggregation. *J. Comput. Phys.*, 151:649–675, 1999.

[266] X. Wang. *Level Set Model of Microstructure Evolution in the Chemical Vapor Infiltration Process*. Ph.D. thesis, Georgia Institute of Technology, 2002.

[267] N. Whitaker. Some numerical methods for the Hele-Shaw equations. *J. Comput. Phys.*, 111:81–88, 1994.

[268] A. Wiegmann. *The Explicit Jump Immersed Interface Method and Interface Problems for Differential Equations*. Ph.D. thesis, University of Washington, 1998.

[269] A. Wiegmann and K. P. Bube. The immersed interface method for nonlinear differential equations with discontinuous coefficients and singular sources. *SIAM J. Numer. Anal.*, 35:177–200, 1998.

[270] A. Wiegmann and K. P. Bube. The explicit-jump immersed interface method: Finite difference methods for PDEs with piecewise smooth solutions. *SIAM J. Numer. Anal.*, 37:827–862, 2000.

[271] L. C. Wrobel and C. A. Brebbla (editors). *Computational Modelling of Free and Moving Boundary Problems II*. Computational Mechanics Publications, Southampton, 1993.

[272] L. Xia, A. F. Bower, Z. Suo, and C. Shih. A finite element analysis of the motion and evolution of voids due to strain and electromigration induced surface diffusion. *J. Mech. Phys. Solids*, 45:1473–1493, 1997.

[273] J. Xu. Error estimates of the finite element method for the 2nd order elliptic equations with discontinuous coefficients. *J. Xiangtan University*, No. 1:1–5, 1982.

[274] J. Xu and H. Zhao. An Eulerian formulation for solving partial differential equations along a moving interface. *J. Sci. Comput.*, 19:573–594, 2003.

[275] S. Xu and Z. J. Wang. An immersed interface method for simulating the interaction of a fluid with moving boundaries. *J. Comput. Phys.*, 216:454–493, 2006.

[276] S. Xu and Z. J. Wang. Systematic derivation of jump conditions for the immersed interface method in three-dimensional flow simulation. *SIAM J. Sci. Comput.*, 27:1948–1980, 2006.

[277] N. N. Yanenko. *The Method of Fractional Steps*. Springer-Verlag, New York, 1971.

[278] X. Yang, B. Li, and Z. Li. The immersed interface method for elasticity problems with interface. *Dyn. Contin. Discrete Impuls. Syst. Ser. A Math. Anal.*, 10:783–808, 2003.

[279] Z. Yang. *A Cartesian Grid Method for Elliptic Boundary Value Problems in Irregular Regions*. Ph.D. thesis, University of Washington, 1996.

[280] S. Yusef. Variations on Arnoldi's method for computing eigenelements of large unsymmetric matrices. *Linear Algebra Appl.*, 34:269–295, 1980.

[281] C. Zhang. *Immersed Interface Method for Hyperbolic System of PDEs with Discontinuous Coefficients*. Ph.D. thesis, University of Washington, 1996.

[282] H. Zhao, T. Chan, B. Merriman, and S. Osher. A variational level set approach to multiphase motion. *J. Comput. Phys.*, 127:179–195, 1996.

[283] H. Zhao, B. Merriman, S. Osher, and L. Wang. Capturing the behavior of bubbles and drops using the variational level set approach. *J. Comput. Phys.*, 143:495–518, 1998.

[284] H. Zhao and S. Osher. Visualization, analysis and shape reconstruction of unorganized data sets. In N. Paragios et. al, editors, *Geometric Level Set Methods in Imaging, Vision and Graphics*. Springer-Verlag, New York, 2002.

[285] H. Zhao, S. Osher, B. Merriman, and M. Kang. Implicit and non-parametric shape reconstruction from unorganized points using variational level set method. *Computer Vision and Image Understanding*, 80:295–319, 2000.

[286] S. Zhao and G. W. Wei. High-order FDTD methods via derivative matching for Maxwell's equations with material interfaces. *J. Comput. Phys.*, 200:60–103, 2004.

[287] S. Zhao and G. W. Wei. Tensor product derivative matching for wave propagation in inhomogeneous media. *Microwave and Optical Technology Letters*, 43:69–77, 2004.

[288] S. Zhao and G. W. Wei. On the implementation of boundary conditions in high order methods. Michigan State University, East Lansing, MI. Preprint, 2005. http://www.math.msu.edu/~wei/pub-mib.html

[289] S. Zhao, G. W. Wei, and Y. Xiang. DSC analysis of free-edged beams by an iteratively matched boundary method. *Journal of Sound and Vibration*, 284:487–493, 2005.

[290] Y. C. Zhou, S. Zhao, M. Feig, and G. W. Wei. High order matched interface and boundary method for elliptic equations with discontinuous coefficients and singular sources. *J. Comput. Phys.*, 213:1–30, 2006.

[291] L. Zhu and C. Peskin. Simulation of a flapping flexible filament in a flowing soap film by the immersed boundary method. *J. Comput. Phys.*, 179:452–468, 2002.

Index